VOLUME SEVENTY

ADVANCES IN BOTANICAL RESEARCH

Fungi

ADVANCES IN BOTANICAL RESEARCH

Series Editors

VOLUME SEVENTY

ADVANCES IN BOTANICAL RESEARCH

Fungi

Volume Editor

FRANCIS M. MARTIN

Institut National de la Recherche Agronomique (INRA), UMR 1136 INRA/Lorraine University, Interactions Arbres/Micro-organismes, Centre de Nancy, Champenoux, France

ELSEVIER

AMSTERDAM • BOSTON • HEIDELBERG • LONDON
NEW YORK • OXFORD • PARIS • SAN DIEGO
SAN FRANCISCO • SINGAPORE • SYDNEY • TOKYO
Academic Press is an imprint of Elsevier

Academic Press is an imprint of Elsevier
32 Jamestown Road, London NW1 7BY, UK
Radarweg 29, PO Box 211, 1000 AE Amsterdam, The Netherlands
The Boulevard, Langford Lane, Kidlington, Oxford, OX5 1GB, UK
225 Wyman Street, Waltham, MA 02451, USA
525 B Street, Suite 1800, San Diego, CA 92101-4495, USA

First edition 2014

ISBN: 978-0-12-397940-7
ISSN: 0065-2296

Printed in the United States of America

CONTENTS

CONTRIBUTORS

Guus Bakkeren
Pacific Agri-Food Research Centre, Agriculture and Agri-Food Canada, Summerland, British Columbia, V0H 1Z0 Canada

Marie-Helene Balesdent
INRA Bioger, Thiverval Grignon, France

Gregory Bonito
Royal Botanic Gardens, Melbourne, South Yarra, Australia

Brian Bushnell
US Department of Energy Joint Genome Institute, Walnut Creek, California, USA

Murray P. Cox
Institute of Fundamental Sciences, Massey University, Palmerston North, New Zealand

Ronald P. de Vries
CBS-KNAW Fungal Biodiversity Centre, Utrecht, The Netherlands

Gunther Doehlemann
Department of Organismic Interactions, Max-Planck-Institute for Terrestrial Microbiology, Marburg, Germany

Sébastien Duplessis
Institut National de la Recherche Agronomique (INRA), UMR 1136 INRA/Lorraine University, Interactions Arbres/Micro-organismes, Centre de Nancy, Champenoux, France

Pierre-Yves Dupont
Institute of Fundamental Sciences, Massey University, Palmerston North, New Zealand

Mark L. Farman
Department of Plant Pathology, University of Kentucky, Lexington, Kentucky, USA

Simona Florea
Department of Plant Pathology, University of Kentucky, Lexington, Kentucky, USA

Jonathan Grandaubert
INRA-Bioger, Thiverval-Grignon, France

Igor V. Grigoriev
US Department of Energy Joint Genome Institute, Walnut Creek, California, USA

Stéphane Hacquard
Department of Plant-Microbe Interactions, Max Planck Institute for Plant Breeding Research, Cologne, Germany

Richard Hamelin
Natural Resources Canada, Canadian Forest Service, Laurentian Forestry Centre, Quebec, and Department of Forest and Conservation Sciences, The University of British Columbia, Vancouver, British Columbia, Canada

Jaqueline Hess
Section for Genetics and Evolutionary Biology, University of Oslo, Oslo, Norway

Kristiina S. Hildén
Division of Microbiology and Biotechnology, Department of Food and Environmental Sciences, University of Helsinki, Helsinki, Finland

Jerzy W. Jaromczyk
Department of Computer Science, University of Kentucky, Lexington, Kentucky, USA

Jolanta Jaromczyk
Department of Plant Pathology, University of Kentucky, Lexington, Kentucky, USA

Jörg Kämper
Institute for Applied Biosciences, Karlsruhe Institute of Technology (KIT), Karlsruhe, Germany

Annegret Kohler
Institut National de la Recherche Agronomique (INRA), UMR 1136 INRA/Lorraine University, Interactions Arbres/Micro-organismes, Centre de Nancy, Champenoux, France

Nicholas Krom
The Samuel Roberts Noble Foundation, Ardmore, Oklahoma, USA

Alan Kuo
US Department of Energy Joint Genome Institute, Walnut Creek, California, USA

Mårten Lind
Department of Forest Mycology and Plant Pathology, Swedish University of Agricultural Sciences, Uppsala, Sweden

Taina K. Lundell
Division of Microbiology and Biotechnology, Department of Food and Environmental Sciences, University of Helsinki, Helsinki, Finland

Mathilde Malbreil
Université de Toulouse, UPS, UMR5546, Laboratoire de recherche en Sciences Végétales, and CNRS, UMR5546, BP 42617, F-31326, Castanet-Tolosan Cedex, France

Francis Martin
Institut National de la Recherche Agronomique (INRA), UMR 1136 INRA/Lorraine University, Interactions Arbres/Micro-organismes, Centre de Nancy, Champenoux, France

Miia R. Mäkelä
Division of Microbiology and Biotechnology, Department of Food and Environmental Sciences, University of Helsinki, Helsinki, Finland

Neil Moore
Department of Computer Science, University of Kentucky, Lexington, Kentucky, USA

Claude Murat
Institut National de la Recherche Agronomique (INRA), UMR 1136 INRA/Lorraine University, Interactions Arbres/Micro-organismes, Centre de Nancy, Champenoux, France

Åke Olson
Department of Forest Mycology and Plant Pathology, Swedish University of Agricultural Sciences, Uppsala, Sweden

Juan Pan
Department of Plant Pathology, University of Kentucky, Lexington, Kentucky, USA

Thibaut Payen
Institut National de la Recherche Agronomique (INRA), UMR 1136 INRA/Lorraine University, Interactions Arbres/Micro-organismes, Centre de Nancy, Champenoux, France

Anne Pringle
Organismic and Evolutionary Biology, Harvard University, Cambridge, Massachusetts, USA

Christophe Roux
Université de Toulouse, UPS, UMR5546, Laboratoire de recherche en Sciences Végétales, and CNRS, UMR5546, BP 42617, F-31326, Castanet-Tolosan Cedex, France

Thierry Rouxel
INRA-Bioger, Thiverval-Grignon, France

Christopher L. Schardl
Department of Plant Pathology, University of Kentucky, Lexington, Kentucky, USA

Jan Schirawski
Microbial Genetics, Aachen Biology and Biotechnology, Institute of Applied Microbiology, RWTH Aachen University, Aachen, Germany

Jan Stenlid
Department of Forest Mycology and Plant Pathology, Swedish University of Agricultural Sciences, Uppsala, Sweden

Eva H. Stukenbrock
MPRG Fungal Biodiversity, Max-Planck-Institute for Terrestrial Microbiology, Marburg, Germany

Emilie Tisserant
Institut National de la Recherche Agronomique (INRA), UMR 1136 INRA/Lorraine University, Interactions Arbres/Micro-organismes, Centre de Nancy, Champenoux, France

Jennifer S. Webb
Department of Plant Pathology, University of Kentucky, Lexington, Kentucky, USA

Carolyn A. Young
The Samuel Roberts Noble Foundation, Ardmore, Oklahoma, USA

PREFACE

FUNGAL–PLANT INTERACTIONS: A NEVER ENDING STORY OF FRIENDS AND FOES

The fungi are a very large and highly diverse group of microorganisms. They have successfully exploited many different habitats to obtain preformed organic compounds as nutrients as they are heterotrophic for carbon. Amongst these habitats, plants are rich sources of carbon and other nutrients that are, to no one's surprise, host to diverse fungal communities both above and below the ground. Arbuscular mycorrhizal fungi are likely to have accompanied the ancestral plants that emigrated from water to land 400 hundred million years ago. Many of their descendant contemporary fungi are adapted to take advantage of the nutrient niches afforded to them by the huge diversity of plants all over the earth. As saprophytes, they break down and decompose every parts of plants through the release of a large set of efficient degradative enzymes. Solubilized compounds are then absorbed and used for growth and development. Many fungi interact with plants, providing supplies of vital nutrients, such as nitrogen and phosphorus, which increased improve plant health and promote plant growth. Others use the resources of living plants directly for their own needs causing damage, disease, and even the death of those plants. Through most of history, people have feared fungi. Despite crop protection chemicals, available disease-resistant crop varieties, and novel cultural practices, an estimated 15% of global crop production is lost to pre-harvest plant disease. A prerequisite for learning how fungi interact with plants to promote their growth or cause damage is to have their complete blueprint, specifically the genome sequence that reveals the gene repertoire and the genes expressed during the interaction. In this volume, we address some of the genomic insights gained into the biology of world's most important groups of fungi interacting with plants and update the state of knowledge of the genomics of major fungal–plant interactions.

The first two chapters (Kohler & Tisserant, 2014; Kuo, Bushnell, & Grigoriev, 2014) provide up-to-date overviews of the next-generation sequencing techniques and bioinformatic tools deployed for studying fungal genomes and transcriptome—paving the way towards the 1000 fungal genomes. Chapter 3 reviews how genomics enables a glimpse into the putative role of transposable elements in the evolution and environmental adaptation of plant–interacting fungi (Grandaubert, Balesdent, & Rouxel, 2014).

The following chapters are dedicated to case studies of the genomics of devastating plant fungal pathogens. In Chapter 4, Hacquard (2014) describes how comparative genomics of powdery mildews was used to highlight the impact of the reproductive mode and the evolutionary history of these plant parasites on the evolution of obligate biotrophy. As shown in Chapter 5, comparative analysis of the genomes of smut fungi, as well as transcriptome profiling approaches, allowed the identification of regulatory elements controlling pathogenic development as well as of secreted effector genes with important functions in plant interaction (Doehleman, Schirawski, & Kämper, 2014). In Chapter 6, Duplessis, Bakkeren, and Hamelin (2014) provide an overview of the current knowledge on rust genomics and highlight how next-generation sequencing techniques are moving this field forward, providing new avenues in the understanding of fungal biotrophy.

Symbiotic fungi are the focus of four chapters. The evolutionary history and genomes of the ectomycorrhizal truffles are deciphered by Payen, Murat, and Bonito (2014) in Chapter 7. In Chapter 8, Hess and Pringle (2014) describe the natural histories and the genomes of *Amanita* species. The long-awaited genome of the model arbuscular mycorrhizal fungus *Rhizophagus irregularis* is the focus of Malbreil, Tisserant, Martin, and Roux's (2014). In Chapter 10, Schardl et al. (2014) provide a wide-ranging coverage of members of the Clavicipitaceae in Chapter 11 and delve deeply into evolutionary scenarios relating to the origins of ergot alkaloids synthesis.

Chapter 11 (Lundell, Mäkelä, de Vries, & Hilden, 2014) and Chapter 12 (Lind, Stenlid, & Olson, 2014) are devoted to wood-decaying and litter-decomposing fungi. They describe the formidable complexity of the wood-decay-related machineries in Basidiomycota and their key role in carbon cycling in terrestrial ecosystems.

In the final chapter, Stukenbrock (2014) demonstrates how the history of species and the underlying genetics associated with the establishment of reproductive isolation can be unravelled by using large-scale comparative genomics.

I am especially indebted and grateful to these authors for the high quality of their contributions. Thanks to their effort, we have produced the most complete and up-to-date treatment of the genomics of plant–fungal interactions. This book should provide guidance for future research.

I hope that this book will serve as a primary research reference for researchers and research managers working in the expanding field of mycology, fungal genomics, as well as plant–microbe interactions. It should provide a useful resource for experienced as well as new researchers and students.

ACKNOWLEDGEMENT

F. M. M.'s research group is part of the Laboratory of Excellence ARBRE (ANR-11-LABX-0002-01).

Francis M. Martin
Institut National de la Recherche Agronomique (INRA),
UMR 1136 INRA/Lorraine University,
Interactions Arbres/Micro-organismes,
Centre de Nancy, Champenoux,
France

REFERENCES

Doehleman, G., Schirawski, J., & Kämper, J. (2014). Functional genomics of smut fungi—From genome sequencing to protein function. *Advances in Botanical Research, 70*, 143–172.

Duplessis, S., Bakkeren, G., & Hamelin, R. (2014). Advancing knowledge on biology of rust fungi through genomics. *Advances in Botanical Research, 70*, 173–209.

Grandaubert, J., Balesdent, M. H., & Rouxel, T. (2014). Evolutionary and adaptive role of transposable elements in fungal genomes. *Advances in Botanical Research, 70*, 79–107.

Hacquard, S. (2014). The genomics of powdery mildew fungi: Past achievements, present status and future prospects. *Advances in Botanical Research, 70*, 109–142.

Hess, J., & Pringle, A. (2014). The natural histories of species and their genomes: Asymbiotic and ectomycorrhizal Amanita fungi. *Advances in Botanical Research, 70*, 235–257.

Kohler, A., & Tisserant, E. (2014). Exploring the transcriptome of mycorrhizal interactions. *Advances in Botanical Research, 70*, 53–78.

Kuo, A., Bushnell, B., & Grigoriev, I. (2014). Fungal genomics: Sequencing and annotation. *Advances in Botanical Research, 70*, 1–52.

Lind, M., Stenlid, J., & Olson, A. (2014). *Heterobasidion annosum s.l.* genomics. *Advances in Botanical Research, 70*, 371–396.

Lundell, T., Mäkelä, M., de Vries, R., & Hilden, K. (2014). Genomics, life-styles and future prospects of wood-decaying and litter-decomposing Basidiomycota. *Advances in Botanical Research, 70*, 329–370.

Malbreil, M., Tisserant, E., Martin, F., & Roux, C. (2014). Genomics of arbuscular mycorrhizal fungi: Out of the shadows. *Advances in Botanical Research, 70*, 259–290.

Payen, T., Murat, C., & Bonito, G. (2014). Truffle phylogenomics: New insights into truffle evolution and truffle life cycle. *Advances in Botanical Research, 70*, 211–234.

Schardl, C. L., Young, C. A., Moore, N., Krom, N., Dupont, P. Y., Pan, J., et al. (2014). Genomes of plant-associated Clavicipitaceae. *Advances in Botanical Research, 70*, 291–327.

Stukenbrock, E. (2014). Speciation genomics of fungal plant pathogens. *Advances in Botanical Research, 70*, 397–423.

CHAPTER ONE

Fungal Genomics: Sequencing and Annotation

Alan Kuo, Brian Bushnell, Igor V. Grigoriev[1]
US Department of Energy Joint Genome Institute, Walnut Creek, California, USA
[1]Corresponding author: e-mail address: ivgrigoriev@lbl.gov

Contents

Abstract

Plants grow in close associations with fungi—pathogens, symbionts, and endophytes—and depend on these associations for success. Therefore, understanding the mechanisms of these interactions is important in order to create optimal conditions for plant growth. Genomics is a powerful tool not only to discover the potential encoded in genomes of both plants and plant-associated fungi but also to monitor dynamics of their interactions through gene expression and other genome-enabled analyses.

Advances in Botanical Research, Volume 70
ISSN 0065-2296
http://dx.doi.org/10.1016/B978-0-12-397940-7.00001-X

Next-generation sequencing technologies offer tools to explore nature's wide array of plant–fungal interactions in unprecedented depth, and the sheer volume of data produced is enormous. Several approaches to assemble, annotate, and analyse this wealth of genomics data are discussed in this chapter.

1. INTRODUCTION

Plants are dependent on the environment in which they grow as well as the interactions with local microbial and fungal community. They live in a constant arms race with plant pathogens, which can cause significant losses to agricultural crops, as with the southern corn leaf blight of the 1970s caused by the Ascomycota *Bipolaris maydis*. On the other hand, 80% of plants cannot live without mycorrhizal associations, which provide nutrients for growth. The roles of various endophytes hosted inside plant tissues are not well understood, but they are often beneficial for the success of plants. In addition, many soil fungi like *Trichoderma* species may affect plant growth indirectly by attacking other plant parasites, fungi, nematodes, or others. An organism's genome encodes the complete set of instructions defining its lifestyle and behaviour in a changing environment. Plant genomics and molecular genetics played critical roles in dramatically accelerating breeding programmes, allowing efficient targeted selection for specific traits. Independent of plant genomics, genomics of fungi is being developed with an amazing speed reaching an unprecedented scale with projects like the 1000 Fungal Genomes Project (Grigoriev et al., 2014).

Genomics is a discipline in genetics, which combines DNA sequencing and bioinformatics to assemble, annotate, and analyse genomes. The term "genomics" was introduced by Tom Roderick in 1986 (Yadav, 2007) and has overgrown its initial definition to include various sequence-based techniques such as transcriptomics or epigenomics.

In 1975, a sequencing procedure using DNA polymerase with radiolabelled nucleotides was introduced and later used to sequence the 5 kbp genome of bacteriophage φX174, the first fully sequenced DNA-based genome (Sanger, Air, et al., 1977). The refinement of this technique resulted in the Sanger sequencing method. After sequencing chromosome III of brewer's yeast *Saccharomyces cerevisiae* in 1992 (Oliver et al., 1992), its entire genome was published in 1996 by a consortium of researchers from laboratories around the world. This was the first sequenced eukaryotic genome and the beginning of fungal genomics.

The next several important milestones in fungal genomics were achieved with sequenced genomes of *Schizosaccharomyces pombe* (Wood et al., 2002), *Neurospora crassa* (Galagan et al., 2003), and *Phanerochaete chrysosporium* (Martinez et al., 2004). In 2000, the Broad Institute launched the Fungal Genome Initiative (FGI; Cuomo & Birren, 2010), a large-scale effort to sequence the genomes throughout Kingdom Fungi. Many of these were primarily for medical studies. In 2004, Génolevures published its large-scale comparative analysis of five yeast genomes (Dujon et al., 2004). In 2009, the US Department of Energy Joint Genome Institute (JGI) started its fungal programme to explore fungi in the context of energy and the environment, including many plant-associated fungi (Grigoriev et al., 2011). One of the latest projects of this programme is the 1000 Fungal Genomes Project (Grigoriev et al., 2014), which aims to comprehensively document fungal genomic diversity on the family level to inform evolutionary and ecogenomics studies.

The very first fungal genome was sequenced by a large international consortium. After that, most of the sequencing was done in large sequencing centres like the Broad Institute and the JGI, which produced dozens of terabases of sequence each year. Many smaller-scale genomics core facilities and centres have been developed at universities and research institutes. As sequencing technology continues to advance, a new generation of benchtop sequencers has become available for many academic laboratories (Quail et al., 2012). However, the data produced by these centres or benchtop machines may be overwhelming. The purpose of this chapter is to review the basic principles and tools for genome assembly, annotation, and analysis and to walk through important computational steps following data generation.

2. WHAT IS GENOME ASSEMBLY, AND HOW TO DO IT

2.1. Purpose of assembly

The single most important tool in the genomic analysis of an organism is the linear template of its genomic sequence—a literal file containing the letters A, C, G, and T in the same order as the actual DNA and in the same number of contiguous pieces as the organism has chromosomes or plasmids. With this information, genes can be annotated as to their function, promoters can be correlated with genes, loci can be correlated with phenotype, and phylogenetic trees can be generated, all relatively accurately and often even automatically. Without the genome, many of these things can still be done, but it's more akin to trekking through the jungle without a map or compass or searching for oil without seismic data. With a complete and correct genome, more advanced genomic studies are possible in ways that are more straightforward and less reliant on chance and intuition. Though modern techniques can rarely achieve a perfectly accurate genome for anything larger than a bacterium (~4 megabase pairs, or Mbp), even an imperfect genome is adequate for most purposes. Hence, creating one should be the first priority when studying a new organism.

Modern sequencing methods cannot determine the content of a single, chromosome-length DNA molecule, though methods are being developed that promise to move closer to this goal (Luan, Martyna, & Stolovitzky, 2011; Timp, Comer, & Aksimentiev, 2012). Rather, sequencing is analogous to shredding a book and then digitizing the contents of the individual paper shreds, called reads (Fig. 1.1A). Genome assembly, then, is analogous to the reconstruction of a book from its shreds. There are different algorithms for doing so, but all share the basic premise of finding shreds that overlap by a sufficient margin, assuming they were originally connected. As adjacent shreds from a single book do not overlap, this is only possible when multiple copies are shredded, with breakpoints randomly distributed between different copies. Thus, to perform genome assembly, the genome must be oversampled and randomly sheared. Once these data have been digitized (as literal sequences of A, C, G, and T), assembly can begin. The following sections will discuss different aspects of assembly: the challenges, sequencing platforms, algorithms, computational limits, use of paired reads, generation of contigs and scaffolds, and finally assessment of quality.

2.2. Challenges of assembly

Before discussing assembly strategy, it's important to discuss the problems encountered during assembly. The two largest obstacles are the presence of error in sequencing (and polymorphism in organisms, which causes the same issues and is often indistinguishable from error) and repetition in genomes. Much of bioinformatics—including assembly—would be fairly trivial in the presence of error-free data from repeat-free genomes, but algorithms that work perfectly in such cases often do not extend well into the real world. For example, a basic "greedy" assembler can start with a read and iteratively extend both ends of this read by scanning through all other reads and concatenating the ones that overlap perfectly, yielding a complete

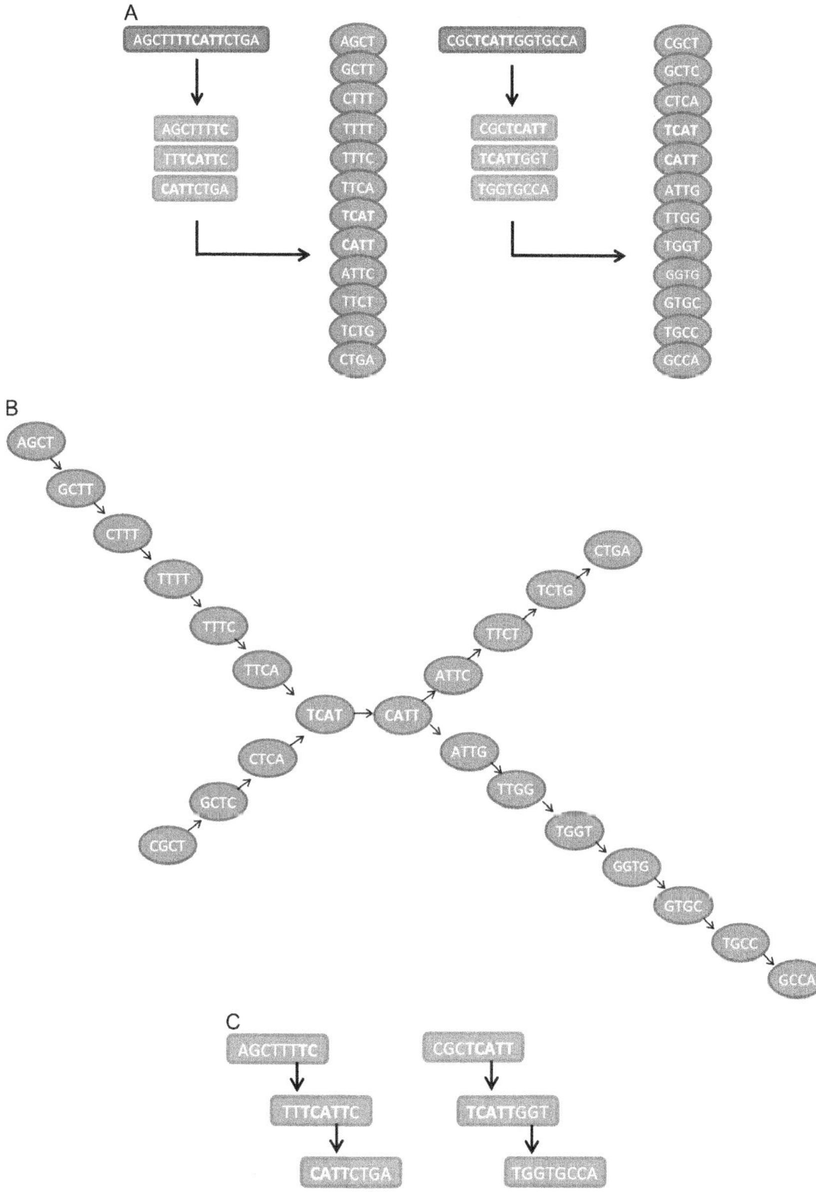

Figure 1.1 Genome assembly steps. (A) These two 15 bp genomic sequences (long rectangles) from *Escherichia coli* each generate three 8 bp reads (short rectangles). The reads can be further decomposed into a larger number of 4 bp *k*mers (ovals). The longest subsequence shared between the two large sequences is 5 bp long, shown in bold. For simplicity reverse-complement sequences are ignored. (B) de Bruijn graph made from *k*mers. Because there is a repeat sequence (TCATT) of at least length *k* (4), the graph makes a false join of two unrelated sequences at the duplicate *k*mers. Note that each *k*mer is joined only to other *k*mers that share a 3 bp suffix or prefix. (C) This read overlap graph is formed by greedily selecting the longest nonconflicting overlap until no read overlaps remain longer than some threshold (in this case, 1). Unlike the de Bruijn graph, it is able to correctly resolve the two original sequences despite the 5 bp duplicate subsequence.

(*Continued*)

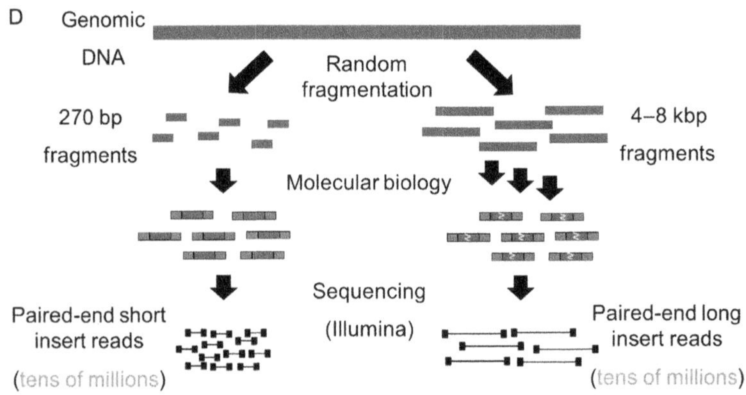

Figure 1.1—Cont'd (D) Genomic DNA is randomly sheared to a target length range. These pieces are processed with various enzymes and chemicals before being fed to the sequencing machines. Only the sequence of the left and right tips (150 bp) is discovered and reported as digital letters; so if the pieces are larger than 300 bp, the middle portion remains unknown.

scaffold when there are no more overlapping reads. This is easy to write and will always give the correct answer for error-free data on repeat-free genomes but is robust against neither errors nor repeats.

No existing or anticipated platform can produce strictly error-free reads of DNA molecules. The errors may have an unbounded number of causes, such as unwanted chemical reactions, imprecise enzyme speed, or noisy optical sensors, but their results can generally be described using the following operations:

Substitutions replace one or more bases with another: "Money" → "Monet"

Insertions add one or more bases: "Pop" → "Prop"

Deletions remove one or more bases: "Town" → "Ton"

Chimerism joins two unrelated sequences: "Be", "Low" → "Below"

Which error types dominate depends on the sequencing platform and sample preparation.

Repeats complicate assembly by creating ambiguity as to the correct origin of a read (Nagarajan & Pop, 2009). In fact, the minimal length of reads needed to yield an unambiguously correct genome assembly is proportional to the length of the longest exact repeat in that genome. Consider the sentence "A tandem repeat can occur over and over and over". With a read length of 8, it is impossible to determine how many repeats there are. To span the repeat, you need reads long enough to contain a character on either side—namely, the period at the end and the "r" from "occur"—for a length of 25. The similar sentence "A tandem repeat can occur many times" can be reconstructed unambiguously with a read length of 4, as the longest repeated subsequence has length 2. For this reason, the very long reads can be invaluable, particularly in more repetitive organisms.

2.3. Sequencing platforms

Many platforms are currently used for sequencing DNA, and their characteristics differ much more than, say, different brands of cameras or printers (Schuster, 2008). The first major technology, still considered the "gold standard", is the Sanger sequencing (Sanger, Nicklen, & Coulson, 1977). This produces long (potentially over 1000 base pairs, or bp) low-error reads and was the primary method used to complete the human genome. However, it is very slow and expensive per sequenced base pair and relies on an awkward cloning step and thus has largely been supplanted by massively parallel high-throughput sequencing, also called the second-generation or next-generation sequencing (NGS). NGS machines produce shorter reads at a higher error rate but yield thousands of times more output at a tiny fraction of the cost per base pair. These include SOLiD (http://marketing.appliedbiosystems.com/images/Product/Solid_Knowledge/flash/102207/solid.html), with paired reads typically in the 25–75 bp range and a high rate of substitution-type errors, with the errors scattered throughout the read; Illumina/Solexa (Bentley et al., 2008), with paired reads in the 36–250 bp range and an error-model also substitution-dominated but with the errors concentrated toward the tail of the read; 454 (Margulies et al., 2006), with single reads in the 500 bp range and error-model dominated by insertions and deletions (indels) typically in homopolymer areas, so a genomic sequence "CAAAAT" might be incorrectly read as "CAAAT" or "CAAAAAT"; and Ion Torrent, with similar characteristics to 454 but a shorter maximum read length of around 400 bp. Pacific Biosciences (PacBio) (Eid et al., 2009) produces a third-generation or single-molecule sequencing platform. These reads are much longer than anything else—up to 20 kilobase pairs (kbp)—but have a substantially higher error rate than the second-generation reads, with a mix of insertions, deletions, and substitutions at a rate of around 15%, in addition to genomic chimerism. PacBio reads are expensive per base—somewhere between Sanger and Illumina (http://www.genome.gov/sequencingcosts/)—but compared to the Illumina platforms, in which genomic coverage is highly biased toward against

Table 1.1 Comparison of sequencing platforms

Platform	Read length	Error rate	Error type	Paired ends	Cost per base pair
Sanger	Up to 1 kbp	Very low	Read end; various	Yes	Very high
Illumina	36–250 bp	Low	Read end substitutions	Yes	Very low
SOLiD	25–75 bp	Medium	Random substitutions	Yes	Low
454	Up to 500 bp	Low	Homopolymer indels	No	Medium
PacBio	Up to 20 kbp	High	Random subs and indels	No	Medium

GC-rich regions, PacBio coverage is virtually unbiased and thus less coverage is needed. Most platforms also exhibit some degree of platform-specific chimerism—a read may contain partly genomic DNA and partly sequence from some artificial molecule attached to the read during library construction. Reads produced by these platforms are sufficiently different (Table 1.1) that the assembly algorithms and strategy are typically dictated by the sequencing platform far more than any other factor, such as the organism itself.

2.4. Assembly strategy and algorithms

There are two primary methods of finding overlaps between DNA reads. The more straightforward is to directly compare reads to each other. This works well when there are a small number of reads, as you might get from an isolated plasmid. However, the number of read pairwise comparisons increases with the square of the number of reads. A 40 Mbp genome covered at 40 × with 150 bp reads needs 10 million reads; the number pair combinations for R reads are $(R^* (R-1))/2$, or about 50 trillion. As each pairwise alignment takes hundreds or thousands of computer cycles, this borders on computational infeasibility, and it becomes progressively less viable with larger genomes. Read overlaps can be used to either extend reads immediately (as in greedy assemblers) (Miller, Koren, & Sutton, 2010) or to build a graph in which the nodes are reads and edges indicate overlaps between reads (Fig. 1.1B). This overlap graph is used by overlap–layout–consensus (OLC) assemblers (Li et al., 2012). Note that graph in this context is not a statistical device such as a scatterplot or bar chart but a computer science graph, which describes structures in terms of nodes and edges. A simple example would be a family tree—each family member is a node, and a relationship such as father/son is an edge.

The other method (Zerbino & Birney, 2007; Zerbino, McEwen, Margulies, & Birney, 2009) is linear in time (rather than quadratic) with the input data size and thus has the potential to scale better to large data sets. It uses *k*mers, or subsequences of length *k*; for example, the word "yellow" can be decomposed into three *k*mers when $k=4$: "yell", "ello", and "llow". If two (error-free) reads overlap by at least *k* bases, they will share some *k*mers; hence, if two reads share a *k*mer, they are likely to overlap, with some probability that increases with *k*. If reads are indexed by their constituent *k*mers, it is possible to find sets of potentially overlapping reads very quickly. Alternatively, the reads may be decomposed into *k*mers and then discarded altogether. This process increases the memory requirement by a constant factor over storing the bases. The *k*mers are then used to build a de Bruijn graph (DBG, Fig. 1.1C), in which the nodes are *k*mers and the edges are overlaps between *k*mers. Because DNA has an alphabet of only four symbols, there are only eight possible edges out from a *k*mer (4 per end). For example, the four-mer ACGT can be extended to the right to create CGTA, CGTC, CGTG, or CGTT. Since there are at most eight edges per node, and each edge is shared by two nodes, the graph is bounded at B nodes and 4 B edges for an input data set of B bases, whereas an overlap graph is bounded by *R* nodes and $(R^2)/2$ edges for an input of *R* reads. Furthermore, the edges in a DBG do not have to be explicit. With an alphabet of size 4, you can find all forward edges from a node simply by testing the presence of all four possible next *k*mers. As a result, DBGs can be very concise to represent and quick to traverse.

However, errors and repeats cause greater problems in *k*mer space than in base space. The largest resolvable repeat shrinks down to proportionality with *k* rather than read length, and *k* is necessarily shorter than the read length. Further, any single error can spawn *k* false *k*mers, bloating and tangling the graph. For a given error rate *E*, the chance of a *k*mer being correct decreases exponentially with *k*—specifically, $P_{\text{correct}}=(1-E)^k$. Therefore, *k*mers are less useful in data sets with high error rates. Because of the high error rate and low number of very long reads, PacBio reads are most suitable for OLC-type assembly using an overlap graph. Illumina reads are too short for good OLC assembly and too numerous for it to be computationally feasible but, due to their low-error rate, can be used effectively to generate *k*mers for DGB assembly. Sanger reads are few and long and have a very low-error rate and thus are generally used in OLC. They could be used for DBG assembly, but devolving them into *k*mers loses information—OLC is generally a better approach when possible. If multiple different libraries are available, the best approach may be to use multiple assemblers; at the JGI, fungal assemblies are often generated using Illumina reads with a DBG assembler and improved with PacBio reads using an OLC-like algorithm.

2.5. Computational limitations on algorithm choice

In addition to computer time, assembly can take a lot of memory (Kleftogiannis, Kalnis, & Bajic, 2013). Assembly of a 4 megabase (Mbp) bacteria is often possible in the few gigabytes of RAM present on a typical laptop computer. But as genome sizes increase to fungal and larger, the amount of memory needed can quickly become the limiting factor, and while it is possible to store information on (much larger) hard disk instead, this is rarely practical as hard disk speeds are over 1000 × slower than RAM. Let's assume we have sequenced a 40 Mbp fungus to a depth of 100×—on average, there are 100 reads overlapping every genomic location, for 4 gigabase pairs (Gbp) of sequence. For simplicity, let's assume a read length of 100 bp, or 40 million reads. OLC assemblers need access to all of the read sequence; when stored as plain text, this takes 4 GB (more if metadata like base qualities are kept), which is easily accomplished. But if a full overlap graph is stored, the edges (read overlaps) quickly dwarf this number. Assume all true overlaps are stored, and an overlap can be represented in 8 bytes. If the coverage is uniform, every read will overlap 198 other reads—99 to the left and 99 to the right; let's round that to 200. So in the best case, there will be (200 overlaps per read) * (40 million reads) = 8 billion overlaps, needing 64 GB of memory. But the coverage is usually uneven (especially in the case of Illumina), particularly if the library is highly amplified. If 91% of the genome has a coverage of 10 × and 9% has a coverage of 1000 ×, the average coverage is still about 100 ×. But 90% of the reads now fall in regions of 1000 × coverage and thus have 1998 overlaps, for a total of ((0.10 * 18) + (0.90 * 1998)) * 40,000,000 = 72 billion overlaps, needing 576 GB of memory, which is extremely large.

DBG assemblers, on the other hand, do not need to store read sequence in memory, and for error-free input, the memory use is proportional to the genome size (Philippe et al., 2011), rather than the input data size. At 24 bytes per *k*mer, this would translate to (40 million unique *k*mers) * (24 B/*k*mer) 960 megabytes, which is quite small. But consider an error rate of 100%, in which all reads are random sequence. In this case, virtually all *k*mers will be unique. If $k = 50$ (the length of *k*mer used), there are about 50 *k*mers per 100 bp read; (40 million reads) * (50 *k*mers/read) * (24 B/*k*mer) = 48 GB of RAM. This number will not drop much until the error rate falls below 1 error per 50 bp (the *k*mer length) on average, or 2%. Even at a low 0.1% error rate, the majority of unique *k*mers will be contributed by sequencing error rather than the genome.

In summary, OLC assembler memory requirements are a function of the number of reads and coverage and thus are most suitable to small numbers of long reads with very even coverage (such as PacBio); DBG memory requirements are a function of total sequence and error rate and thus are more suitable for short reads with very low error rates (such as Illumina). In order to run successfully in the limited available memory, both types of assemblers may need their input data preprocessed using subsampling to reduce the input size, depth normalization to flatten the coverage distribution, or error correction to fix incorrect bases using statistics.

2.6. Paired reads

The discussion in the preceding text has considered reads abstractly as contiguous sequences of DNA. But many platforms, including Sanger, Illumina, and SOLiD, offer paired-end reads (Peng et al., 2012). In each case, the sequencing chemistry decreases in efficiency and accuracy with each progressive base, limiting the total number of consecutive bases sequenced to some number typically ranging from 25 to 250 (using the latest Illumina MiSeq chemistry) before accuracy drops too low (Nakamura et al., 2011). However, it is possible to get a fresh start by sequencing the opposite end of the same (or complementary) molecule. This yields two paired sequences, in which read 1 and read 2 start from opposite ends of the molecule and represent opposite strands. The "insert size", or length of the molecule being sequenced, is unknown, so these sequences cannot be trivially joined to form a single long sequence. Still, the insert-size range can be determined using a variety of techniques such as by examining the actual molecules in the library, via mapping to a reference genome, or by attempting to rejoin the paired reads using overlapping bases or *k*mer walking. During assembly, paired-end reads can add information in the form of additional placement constraints (typically for overlap-based assemblers), or paired reads can be merged into single longer reads when overlap, which allows a larger k to be used in *k*mer-based assemblers. Due to physical constraints, the maximum paired-end insert size on Illumina platforms is somewhere around 1000 bp. Assemblers can theoretically resolve some repeats proportional to the insert size, rather than read length, if paired-end information is used optimally.

2.7. Generation of contigs and scaffolds

The assembly process does not typically create a single, complete contig (contiguous sequence of bases) in a single step. Instead, many small contigs are grown. This is a result of several factors. First, genomic sequence may have highly variable depth (the number of reads originating from the same genomic location). Where the depth is sufficient, contigs can be made; when the depth drops too low for there to be any overlaps, contigs will necessarily be truncated. Second, assemblers may break apart contigs at repeat boundaries. Consider three distinct long sequences, referred to as A, B, and C. If the original genome contains the sequence ABACA, and A is longer than a read, the assembler may be able to tell that A is connected to B and C, but it cannot necessarily tell how many copies there are of A, B, or C, and furthermore, it can't even determine their order—it could be ABACA or ACABA. Therefore, they may be broken into contigs A, B, and C, possibly with multiple copies of A being the output. Third, it's difficult to distinguish between sequencing errors, homologous regions, and polyallelic sites; all of them manifest as reads that almost match, but not quite. Thus, additional duplicate contigs may be created when an assembler interprets a read with an error as a valid read originating from a homologous sequence. Lastly, modern assemblers are typically written to run on parallel computers, in which each individual thread works with a subset of the data. This allows much greater speed (and much more available memory, for distributed computing) but may cause greater fragmentation, as each thread generates contigs without necessarily having full global knowledge of the data. The end result is that assemblies of an organism with a single chromosome might yield 2, or 20, or 200,000 contigs, in no particular order.

Some assemblers terminate after writing the contigs. Others, such as ALLPATHS-LG (Gnerre et al., 2011), attempt to further improve the assembly using scaffolding and gap filling. This often necessitates a different library type, such as long mate pair (LMP, Fig. 1.1D). Paired-end fragment libraries are sequenced off of both ends of an intact molecule, and thus, their insert size is limited by physical constraints of the platform. LMP libraries are prepared differently—long DNA fragments of typically 4 kbp, 8 kbp, or more are circularized, with the two tips joined by a linker sequence. Then these circles are cleaved on either side of the linker, such that the resulting fragment is a few hundred bp (short enough to sequence on the platform) and the vast majority is washed away. Unfortunately, this requires much more input DNA than normal fragment libraries, has a lower yield, and produces more artefacts (such as chimeric reads with linker sequence present). Furthermore, it requires higher quality DNA, because DNA can degrade when sitting around, reducing the average fragment length until it is too short to circularize. So it is generally more difficult and more expensive. But by mapping these LMP libraries to the unordered contigs, the contigs can be associated with each other. If one read in a pair maps to contig A and the other read maps to contig B, you do not simply know that A and B are nearby; you also know that they are approximately 4 kbp apart (in the case of a 4 kbp LMP library), their order, and their relative strand orientation. This allows the construction of a scaffold, in which sequences A and B are placed approximately 4 kbp apart and the gap between them (called a "captured gap") is filled with Ns. Despite containing the same number of sequenced bases, scaffolded genomes are more valuable than unscaffolded ones in many studies, particularly those associating traits or genes with chromosomes and genomic locations.

Once scaffolding is complete, it is possible to improve the assembly even more with gap filling. Some assemblers and stand-alone programs will attempt to replace the Ns in the gaps between contigs with sequenced bases by mapping paired reads to the scaffolds. If one read in a pair is on the edge of a contig, and the other can be inferred by the insert-size distribution to fall within the gap, it may be possible to generate a consensus sequence, filling or at least reducing the gap. Though it may seem that this should have been performed during the contig-generating phase, filling scaffolded gaps is often easier because there is more information available—both the size of the gap and "anchoring" contigs on both sides of it, rather than just one side. There are also stand-alone programs such as PBJelly (English et al., 2012) that use very long reads from PacBio to fill the gaps directly, relying on the read being longer than the gap and thus being anchored on both sides. In either case, the sequence in the filled gap may be somewhat less accurate than the sequence of the original contigs, as the gap was probably originally caused due to ambiguity, low coverage, or some other problems with the local sequence.

2.8. Assembly quality evaluation

The determination of the quality of an assembly is perhaps more difficult than assembly itself. The most trivial way to rate an assembly is in terms of basic, objective statistics that can be gathered purely from the assembly itself. These include the total assembled size, number of contigs and scaffolds, percent gap in scaffolds, and contig or scaffold N50—the minimum number of pieces needed to represent 50% of the assembled genome, which is an important metric for genome contiguity. The total assembled size is not particularly useful alone, but typically, there will be other information available (such as genomes of related organisms) that will give a rough indication of the expected genome size. However, none of these statistics give any indication of the correctness of the assembly (Salzberg, 2005).

To determine correctness, mapping is typically used—the reads from which the assembly was created are mapped back to the assembly, such that they align to the site maximizing some scoring function based on read bases matching the assembly bases—in other words, the place in the genome from which that read was most likely to originate. Then additional statistics can be gathered, though unlike the prior statistics, all statistics based on mapping are subjective, affected by biases in the particular mapping software. The percent of reads successfully mapped can give an indication of genome completeness; the number of mapped bases that do not match the assembly, or the presence of indels in the reads relative to the assembly, can give an idea of assembly base-level accuracy; assembly locations to which no reads map can indicate misassemblies, such as improper joins between sequences that are not connected in the actual genome; and parts of the assembly with unusually high or low coverage can indicate improperly replicated sequence, or actual genomic repeats that were condensed into too few copies. There are some tools such as ALE (Clark, Egan, Frazier, & Wang, 2013; Ghodsi et al., 2013; Medvedev & Brudno, 2009; Vezzi, Narzisi, & Mishra, 2012) that statistically quantify the mapping statistics into an overall quality score for the assembly, although this is only a useful number as compared to other assemblies from the same set of reads, mapped with the same mapping program. Other methods of determining assembly quality through mapping are to map an assembly to the reference genome of a related species (Peterson, Wessler, & Paterson, 2002); to look for the presence of genes that are known to exist as a single copy in most organisms; and to sequence and assemble the RNA of the organism and then map these sequences (known as ESTs, or expressed sequence tags) back to the assembly, since a complete genome should contain all of its RNA. An assembly quality can be also assessed by genome annotation.

3. WHAT IS ANNOTATION, AND WHY DO IT

3.1. The purpose of annotation

The term "annotation" generally means the appending of additional information ("metadata") to an existing object, such as notes in the margin of a manuscript, labels added to a photograph, and computer-friendly mark-up for otherwise human-friendly text. In genomics, annotation in principle refers to the identification of features of interest on assembled genomic sequence, including but not limited to protein-coding genes, noncoding RNA genes, promoters and other *cis*-acting elements, and nucleosomes. In practice, the vast majority of annotation effort is devoted to finding protein-coding genes (coding sequences or CDSs and associated entities) and then characterizing them. Therefore, in this chapter, except where specifically indicated otherwise, the term "gene" refers to a protein-coding gene.

Of the many important statistics that are generated by a genome-sequencing project, perhaps the one of greatest interest and most immediate accessibility to the general scientific community and public is the gene count. Within the community of scientists directly working with the data, for many if not most of them, having a reliable and robust list of the likely genes is the starting point for subsequent research. The focus on genes, even in a field where the dominant experimental technology is the sequencing of nucleotides, follows from the central theoretical role of genes in biology. In a very practical sense, the annotation provides the tangible link between megabasepairs of DNA sequence on the one hand and biological knowledge on the other.

Fundamentally, there are two ways to annotate genes: manually (by biology experts) and automatically (by computer programs). Each has its strengths and weaknesses. Manual annotation by experts is often considered the "gold standard" for defining a gene but is inherently laborious and time-consuming and thus rarely appropriate for whole-genome annotation. During the Human Genome Project, several teams of trained specialists in different institutes and countries manually curated all predicted genes in all chromosomes. This required cross-training and development of shared standards within and between the groups. In contrast, automated annotation is limited only by computational resources in the numbers of genomes that may be processed. However, the quality of those gene predictions (and predictions is what they are) varies widely from genome to genome and even from gene to gene within a single genome, depending on the availability of the various kinds of inputs used by the automatic methods. These automatic methods are the subject for this chapter, and henceforth, "annotation", except where specifically indicated otherwise, refers to automated efforts to locate genes on a sequenced genome and characterize their structure and function.

Even in this specific context, we can make further distinctions. There are two conceptually distinct aspects of genome annotation: structural annotation and functional annotation. Structural annotation is finding genes and surmising their exon–intron structures, start and stop codons, and untranslated regions (UTRs). Functional annotation is inferring the biological and physicochemical functions, properties, activities, locations, pathways, and expression patterns of the proteins encoded by those genes. While distinct, the two kinds of annotation can also complement and interoperate with each other, as functional annotation obviously builds on the results of structural annotation, but can also be used to refine and improve structural annotation.

Both kinds of annotation can in themselves be very complex. Individual computer programs used in annotation often require dedicated informatics infrastructure and specific genomic data inputs, and also produce appropriate data outputs. The totality of a whole-genome annotation is of sufficient complexity to encourage the development of "annotation pipelines", in effect computer programs that call upon each of many individual programs involved in annotation in the correct order and across appropriate informatics architecture. With a pipeline developed and in hand, there is also a strong incentive to add new features (in the form of additional annotation programs) and to expand it at its termini (thus automating preprocessing of inputs and postprocessing of outputs).

The major input of whole-genome annotation is necessarily an existing whole-genome assembly, which may be produced by a variety of sequencing and assembly methods. The diversity of these methodologies together with the great diversity of fungal genomes leads to a wide quality range of the assemblies that an annotation pipeline may be asked to process. Additional inputs often (and ideally) include a partial or whole transcriptome of the same fungal species and proteins of the same and related species. These too may be of highly variable quality. The assessment of input quality and any resulting preprocessing is also increasingly the purview of annotation pipelines.

The major output of whole-genome annotation is a more or less complete set of protein-coding genes, their conceptual translations, and a rich set of descriptors attached to each gene/protein. In principle, these may be available as flat files, but increasingly pipelines output to databases, which themselves may undergird complex feature-rich portals. Web databases of this sort provide both a powerful means of distributing the output over the web as well as a robust platform for the conduct of downstream analyses. Tying portals of individual genome annotations together provides even greater analytic power.

The core of structural annotation is the prediction of genes. We will not here delve into the theory of gene modelling, but we will review the most common methods currently used to automatically predict genes across whole fungal genomes. The various methods use different kinds of evidences or apply different kinds of algorithms and thus may predict different gene structures at a single locus (some methods even systematically propose a selection of gene structures at a single locus). The different methods have attendant strengths and weaknesses, but as these may be complementary, annotation pipelines often deploy multiple gene predictors using different methods and then choose from or combine the multiple predictions into a single "best" gene prediction at each locus. The basis for choice is often the functional annotation of the rival gene predictions, illustrating the interconnections between structural and functional annotation.

An individual gene prediction at a specific locus, with a particular exon–intron structure and start and stop sites, necessarily has a conceptual transcription and translation. From these sequences, many different kinds of functional annotation may be applied, depending on evidence available. The objective of these procedures is usually to determine properties of the gene/transcript/protein that allow inference of its biological role and function. For example, the prediction of protein domains may allow inference of binding sites or catalytic activity, the prediction of signal peptides may allow inference of cellular localization, and alignment with RNA sequences may allow inference of conditional or developmental expression. The objective is to provide a rich, if tentative, prediction of the biological significance of the newly predicted gene.

3.2. How to annotate

The specific tasks involved in annotation may vary from genome project to project, depending on the specific research objectives and interests as well as the availability of input data. Nevertheless, most whole-genome projects involve the following steps:

Gather inputs

- Get genomic assembly.
- Get transcriptomic data (ESTs and RNA-Seq) and assemblies thereof.
- Get known repeats and TEs, cDNAs, and proteins of related organisms.
- Assess quality, especially genomic assembly.
- Find repeats and mask.

Structural annotation

- Align cDNAs, ESTs, transcriptome data, and proteins.
- Train gene predictors.
- Predict genes.

Combine, choose, and filter predictions based on RNA and protein evidence.

Functional annotation

Align with known proteins.

Predict domains, transmembrane modules, signal peptides, etc.

Apply data from conditional expression experiments.

Cluster to find gene families, homologs, and paralogs.

Name genes.

Distribute outputs

Build one's own database or website ("portal").

Deposit with GenBank and other public databases.

Link portals with each other.

Note that most of the steps are potentially programmable. This, together with the large number of steps and potentially complex interactions between steps, lends the whole process to automation as a single annotation pipeline.

An example of a currently active annotation pipeline, and the one with which the authors have the most practical experience, is the JGI Annotation Pipeline (Fig. 1.2; Grigoriev, Martinez, & Salamov, 2006). Other major annotation pipelines built to handle large numbers of genes and genomes are similarly run internally by NCBI, Ensembl, and other biological data centres (Curwen et al., 2004; Thibaud-Nissen, Souvorov, Murphy, DiCuccio, & Kitts, 2013). On the other extreme is MAKER, an open-source implementation designed with a relatively small individual laboratory in mind (Cantarel et al., 2008). The JGI and Broad FGI Annotation Pipelines (Grigoriev et al., 2006; Haas, Zeng, Pearson, Cuomo, & Wortman, 2011) are notable for being optimized for the peculiar characteristics of fungal genomes. A brief list of some popular annotation software, including pipelines, may be found in Table 1.2.

The pipeline typically operates in a computational environment that balances speed, scalability, and stability against software and hardware limitations. The JGI Annotation Pipeline uses a large compute cluster and sophisticated job scheduler to ensure fast data processing that is robust to external and internal failures (e.g. dead compute node and illegal input data). Most intermediate and output data are written to a relational database, which provides data storage that is both stable and easily retrievable by the pipeline, by other software (e.g. MycoCosm; see in the succeeding text), and by humans.

3.2.1 Assessing and preparing the assembly

The extremely well-known model organism *S. cerevisiae* was the first fungus to have its genome sequenced, assembled, and annotated. Not

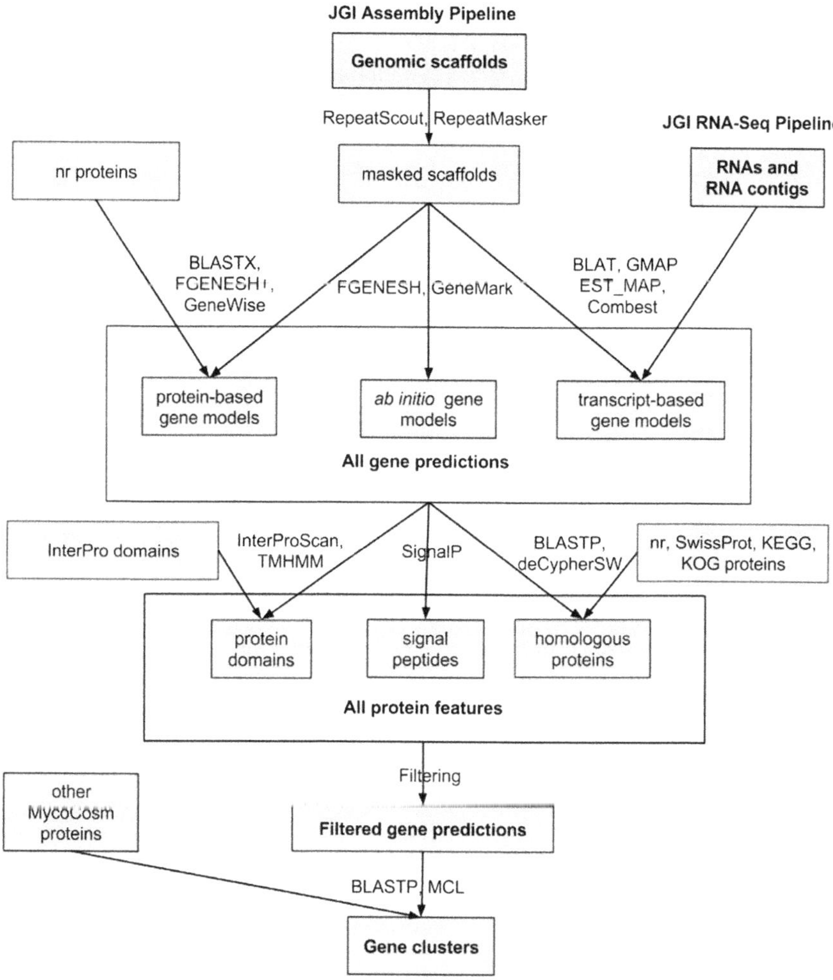

Figure 1.2 The JGI Annotation Pipeline workflow. Boxes indicate major data sets (input, intermediate, and output). Arrows indicate processing between data sets, labelled with major bioinformatic program used. Not shown are peripheral pipeline processes (e.g. portal creation and tRNA detection) that are not discussed in the text.

coincidentally, this model organism has a relatively simple genome not representative of the vast majority of fungi. Most fungal genomes sequenced to date have been significantly larger and more repetitive, with genes that are longer and more intron-rich. Many fungal clades demonstrate significantly more gene duplication or alternative splicing. Also, in most clades outside of Ascomycota, it is difficult to isolate a haploid or monokaryotic culture (in many clades, such are not even known), thus introducing complications

Table 1.2 Some popular bioinformatic programs used for automated whole-genome structural and functional annotation of fungi

Function	Bioinformatic program	Citation or URL
Annotation pipeline	JGI Annotation Pipeline	Grigoriev et al. (2006)
	Broad FGI Annotation Pipeline	Haas et al. (2011)
	MAKER	Cantarel et al. (2008)
Repeat or TE finder	RepeatScout	Price, Jones, and Pevzner (2005)
	RepeatMasker, RepeatModeler	http://www.repeatmasker.org/
	TransposonPSI	http://transposonpsi.sourceforge.net
Ab initio gene predictor	geneid	Guigó (1998)
	FGENESH	Salamov and Solovyev (2000)
	AUGUSTUS	Stanke and Waack (2003)
	SNAP	Korf (2004)
	GeneMark-ES	Ter-Hovhannisyan et al. (2008)
	GAZE	Howe, Chothia, and Durbin (2002)
Transcript-based gene predictor	EST_MAP	http://www.softberry.com
	FindEstOrf	http://www.broadinstitute.org/science
	PASA	Haas et al. (2003)
	Cufflinks	Trapnell et al. (2010)
Protein-based gene predictor	GeneWise	Birney and Durbin (2000)
	FGENESH+	Salamov and Solovyev (2000)
Gene prediction combiner	GLEAN	Elsik et al. (2007)
	EVidenceModeler	Haas et al. (2008)
Transcript or protein aligner	BLASTX, BLASTP	Altschul, Gish, Miller, Myers, and Lipman (1990)
	BLAT	Kent (2002)
	GMAP	Wu and Watanabe (2005)
	AAT	Huang, Adams, Zhou, and Kerlavage (1997)
	exonerate	Slater and Birney (2005)
	TopHat	Trapnell, Pachter, and Salzberg (2009)
	DeCypherSW	http://www.timelogic.com/

Table 1.2 Some popular bioinformatic programs used for automated whole-genome structural and functional annotation of fungi—cont'd

Function	Bioinformatic program	Citation or URL
Domain finder or predictor	HMMER	Bateman et al. (1999)
	TMHMM	Sonnhammer, von Heijne, and Krogh (1998)
	SignalP	Nielsen, Brunak, and von Heijne (1999)
	Phobius	Käll, Krogh, and Sonnhammer (2004)
	TargetP	Emanuelsson, Nielsen, Brunak, and von Heijne (2000)
Compartment predictor	WoLF PSORT	Horton et al. (2007)
Cluster generator	TRIBE-MCL	Enright, Van Dongen, and Ouzounis (2002)

from polymorphism. On the other hand, fungal genomes are often not as challenging as many very large, highly repetitive, and polyploid plant and animal genomes. In any case, just as with other eukaryotes, a fungal genome annotation project must take into account repeats, duplications, and (if diploid or dikaryotic) polymorphism.

The quality of a whole-genome annotation is limited by the quality of the input whole-genome assembly. In particular, if the assembly is so fragmented that the average contig is shorter than the average gene, then the annotation will produce large numbers of partial genes and gene fragments. The two most common biological causes of fragmentation that we discussed earlier are high repeat content and high polymorphism rate. Besides fragmentation, other common assembly artefacts that are inimical to quality annotation are misassembly (chimeric sequence), incomplete assembly (missing sequence), separated alleles (redundant sequence), and contamination (foreign sequence). Ideally, these are resolved (if technically possible) before annotation begins. However, often they are not, and annotation can ironically be a powerful tool to detect the artefacts, in effect a complex tool for quality assessment.

An assembly that is good enough to annotate must still be prepared for the actual gene predictors. The most important of these preparations is the masking of repeats, and this may easily be built into the annotation pipeline. Though repetitive genomic sequence is itself a burgeoning area of biological research, the primary goal of repeat discovery in most genome annotation projects is to remove them from consideration by downstream steps in the annotation process. By convention, this is done by "masking" the discovered repetitive sequence by conversion of the nucleotide symbol either to lowercase or to "N". In practice, two distinct kinds of logic are brought to bear in repeat discovery. One kind is to locate known repetitive elements, especially transposons and viral genes, whether or not they are actually repeated in large numbers in the genome. The other kind is to discover *de novo* all repetitive sequences, whether or not they are known transposons or viral genes. The first approach is unable to detect transposable elements (TEs) or other repetitive sequences that are heretofore unknown. The second approach risks classifying *bona fide* gene duplications and homopolymeric genes as repeats and masking them. Thus, pipelines often use both approaches in combination. For example, the JGI Annotation Pipeline uses the *de novo* repeat-finder RepeatScout to build a library of putative repeats (Price et al., 2005), adds this to an existing expert-curated libraries of fungal TEs, and supplies both to RepeatMasker for masking the genomic assembly (http://www.repeatmasker.org/). Other popular or recently used repeat-finding and TE-finding programs are listed in Table 1.2.

3.2.2 Finding the genes (structural annotation)

After preparing the assembly sequence, the typical pipeline runs multiple gene predictors. Some of these predictors may be informed by nongenomic transcript or protein sequence, in which case these are aligned to the assembly by the pipeline first. After all gene predictions are made, the pipeline then combines or chooses between them at each locus to produce a non-redundant set of best gene predictions over the whole genome.

Gene prediction methods are generally of two types: *ab initio* predictors, using only the genomic assembly as input, and evidence-based predictors, using transcript and/or protein sequences in addition to the genomic assembly. The latter may be differentiated further between those that rely on transcripts (e.g. cDNAs, ESTs, and RNA-Seq and assemblies thereof), usually from the same species or even culture as the DNA source, and those that rely on proteins, usually from closely related but often from quite distantly related species. *Ab initio* predictors model gene exon–intron structures and CDS starts and stops by deriving parameters from a training set of known genes or through a "bootstrap" process of iterative prediction. Transcript-based predictors and protein-based predictors require first the accurate alignment of RNA sequences and amino acid sequences, respectively, to the genomic DNA sequences. The predictor then uses the alignments as seed to model the gene's structure.

Whatever the prediction method used, the predictions may then be confirmed, corrected, or updated by comparison with data of transcriptomic (ESTs or RNAs), proteomic (peptides), and genomic (syntenic orthologs from very closely related already-annotated genomes) origin. Splice sites may be changed, UTRs added, and new genes constructed at loci where other gene predictors failed.

Finally, multiple predictions at a single locus can be combined into a single "best" prediction, or the "best" can be chosen to represent the locus. Either way, various criteria for "best" are applied, such as agreement with transcript and peptide data and similarity of the translated protein with known homologs. Congruence with these external data is embodied by a scoring system, and the highest scoring predictions or combinations "win".

For example, the JGI Annotation Pipeline uses BLASTX (Altschul et al., 1990) to align protein sequences to the genomic assembly; BLAT (Kent, 2002) and GMAP (Wu & Watanabe, 2005) to similarly align transcript sequences; the *ab initio* FGENESH (Salamov & Solovyev, 2000) and GeneMark-ES (Ter-Hovhannisyan, Lomsadze, Chernoff, & Borodovsky, 2008), the transcript-based EST_MAP (http://www.softberry.com/) and Combest (unpublished), and the protein-based GeneWise (Birney & Durbin, 2000) and FGENESH+ (Salamov & Solovyev, 2000) to predict genes, and a custom filtering algorithm (unpublished) to choose the best prediction at each locus. Other popular or recently used annotation software are listed in Table 1.2.

3.2.3 Characterizing the genes (functional annotation)

Programs to perform functional annotation are as diverse as the biological properties that they wish to predict. The most commonly predicted properties seek to address the most basic questions pertaining to the function of a newly discovered feature: What does the gene/protein do, where does it do it, and when does it do it?

What does the gene/protein do? Molecular binding or catalytic activity may be inferred from protein family domains. The most universally used public domain databases are Pfam (Sonnhammer, Eddy, & Durbin, 1997), NCBI's CDD (Wheeler et al., 2005), and InterPro (Apweiler et al., 2001), which includes Pfam as well as a number of other smaller databases that mostly use less rigorous but nevertheless complementary methods of compilation. Each of these specially formatted databases requires an appropriate tool for querying. For example, Pfam is an HMM database and so requires an HMM tool such as HMMER (Bateman et al., 1999) to be queried. In addition to domains, a less rigorous but simpler way of inferring function is to align the predicted protein sequence with other better-described proteins of other better-known species. Protein predictions may be queried against public protein sequence databases such as GenBank's RefSeq and nr (Benson, Lipman, & Ostell, 1993), European Bioinformatic Institute's UniProt (SwissProt; Bairoch & Boeckmann, 1991), Kyoto Encyclopedia of Genes and Genomes (KEGG; Ogata et al., 1999), clusters of euKaryotic Orthologous Groups (KOG; Tatusov et al., 2003), or against custom data sets of specific applicability to the genome. By far the most commonly used sequence alignment tool is BLASTP (Altschul et al., 1990), due to its speed and sensitivity. Often used in a complementary fashion are tools based on the Smith–Waterman algorithm (Smith & Waterman, 1981), which are more specific and reliable but slower. The last property can require specialized hardware, as is used by the JGI Annotation Pipeline's DeCypherSW implementation (http://www.timelogic.com/).

Where does it do it? Secretory pathways and cellular compartment localization may be inferred from signal peptides, transit peptides, transmembrane domains, protease cleavage sites, glycophosphatidylinositol anchors, disulphide bonds, and other features. Many of these features may be described with a sequence motif or set of motifs, and the tools deployed by pipelines seek to exploit this knowledge. For example, TMHMM predicts transmembrane domains (Sonnhammer et al., 1998), SignalP and Phobius predict a signal peptide and thus whether a protein enters the classical secretory pathway (Käll et al., 2004; Nielsen et al., 1999), TargetP predicts a transit peptide and thus whether a protein is bound for an organelle (Emanuelsson et al., 2000), and WoLF PSORT considers a list of known motifs and predicts the likeliest target cellular compartment of a protein (Horton et al., 2007). All of these predictions are exactly that, providing candidates for experimental confirmation.

When does it do it? Expression under various environmental conditions, developmental stages, or genetic backgrounds may be inferred by annotation with differential transcriptomic or proteomic data. Depending on the sequencing methodology used, the transcripts are mapped to the genome using the same alignment tools mentioned in the "transcript-base prediction" section in the preceding text, and congruence of the alignments with the genes is examined. If the sequencing methodology is quantitative, then levels of expression may be measured. Peptides from proteomics experiments may be used in a similar manner. Combined with genomic sequence, expression data may also be used to define *cis*-acting genomic motifs regulating transcription. Transcript and peptide data may also discover alternative splice variants.

The plethora of functional annotations may be used to characterize and classify the genes, to add higher-order annotations (e.g. KOGs, GO terms (Ashburner et al., 2000), EC numbers (Bairoch, 1994), metabolic and regulatory pathways, and gene families), and to provide a comprehensive basis for downstream analysis (e.g. pathway comparisons, regulatory motif discovery, phylogenomics, and population genomics). The functional annotations may also be used to inform structural annotation, both manual and automatic. For example, the JGI Annotation Pipeline explicitly uses transcript coverage, protein similarity (to proteins in nr, SwissProt, KEGG, and KOG), InterPro domains, transmembrane domain predictions, signal peptide predictions, and paralog predictions (multigene family membership) to inform the choice of "best" gene prediction, as well as to detect and exclude TEs. Functional annotations may also supply ready-made metrics as to the quality of the structural annotation. Similarly, the JGI Genome Portal displays structural and functional information together with tools for human annotators to assess and if necessary edit or discard automatic gene predictions (Grigoriev et al., 2012).

3.2.4 Using and distributing the annotation

The forms of the large amount of data generated by a genome annotation pipeline ought to be dictated by the ways the data are to be used, which may vary immensely by project, by scientific community, and by individual scientists (users) within that community. At one extreme, a user may be interested in a very specific gene's presence or absence and its structure and function, with the intent to exploit the sequence for wet lab experiments; at another extreme, a user may be interested in no specific genes, only in the properties of all of the genes in aggregate. The second type of user often wishes to apply computational tools to relatively raw data and so is satisfied with downloadable text files following standard or at least simple (mainly tab-delimited text) formats. The first type of user may be best served by a rich graphical interface with tools to browse, search, and analyse the annotation, an interface most easily implemented as a web accessible database. The JGI Annotation Pipeline automatically writes its outputs into database tables and text files and instantiates a genome portal with tools for viewing, browsing, searching, and downloading. In addition to providing multiple tools for accessing and analysing the new annotation, the new genome portal also provides the input genome assembly and transcripts and metainformation about the organism and the project.

Each new JGI Genome Portal is automatically added to MycoCosm, a "super portal" that contains all of the genome portals created by the JGI Annotation Pipeline for each fungal genome annotation, organized into a hierarchy of phylogenetic nodes (Fig. 1.3A; Grigoriev et al., 2014). Each node corresponds to a standard fungal taxon (the highest and biggest node is "Fungi") and is a gateway to a "PhyloGroup portal" (Fig. 1.3B). The latter is in effect a collective portal for the individual member genomes of the node, providing a common toolset for bulk access and comparative analysis of the member genomes. MycoCosm also supports "EcoGroup portals", whose members are not phylogenetically related but share an ecological niche, habitat, or lifestyle. Currently, the JGI Annotation Pipeline does not automatically insert each new JGI Genome Portal into an EcoGroup, but once it is inserted manually, the same analytic tools become available as in a PhyloGroup. To further enhance the power of this platform, genomes not annotated by JGI are being systematically imported into MycoCosm as genome portals (Grigoriev et al., 2014).

4. WHAT TO DO WITH THE ANNOTATION ONCE YOU'VE GOT IT

The previous discussion is from the point of view of the producer of annotation. From the point of view of the above-described consumer or user, the availability of a whole-genome annotation of reasonable quality and completeness immediately allows two distinct and complementary kinds of uses: gene-specific analysis and genome-wide analysis. Genome-wide analysis is processing of all of the newly annotated genes simultaneously and in parallel, with the purpose of either elucidating genome-wide phenomena or screening for novel subsets of genes, including those with no known function. Gene-specific analysis typically involves targeted investigation of single or small sets of genes (or their encoded transcripts or proteins) based on a usually already-known property of interest. Examples of such properties of interest to mycologists include catalytic function (e.g. carbohydrate-active enzymes or "CAZymes"), cellular compartment (e.g. extracellular secreted proteins), and genomic location (e.g. genes on a dispensable chromosome). If a field of biological study has a specific gene/transcript/protein focus, then the field may be immediately advanced by

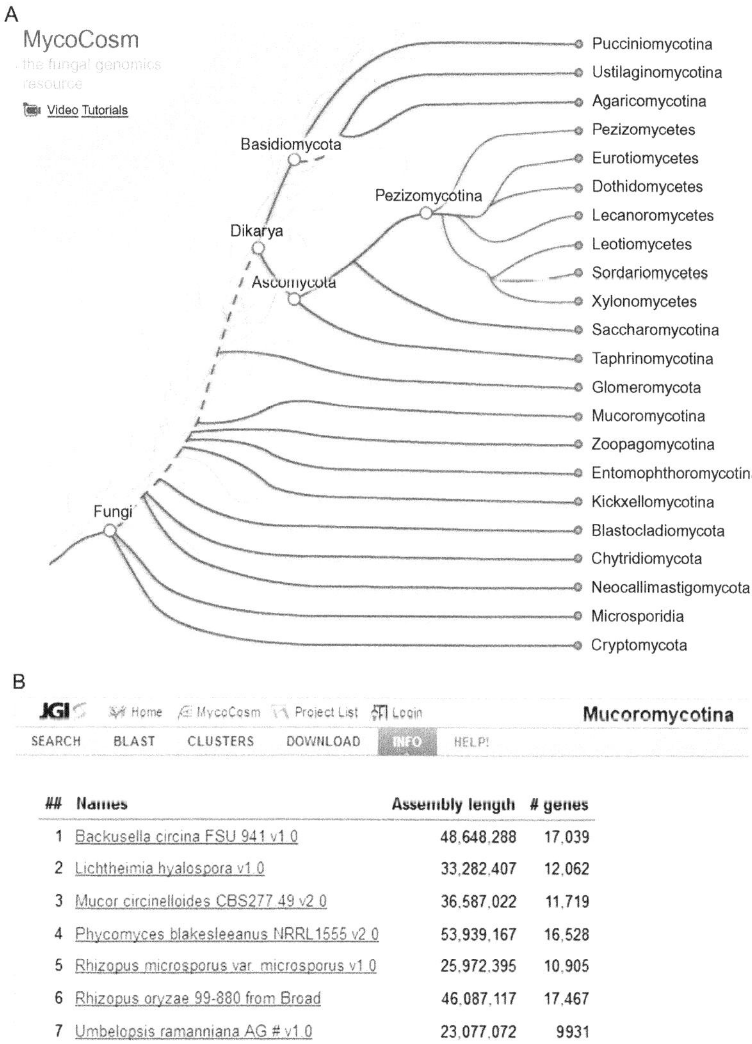

Figure 1.3 JGI MycoCosm portal. (A) The tree illustrates phylogenetic relationships between the major groups of fungi with sequenced genomes. Each node is linked to individual genome portals and a comparative PhyloGroup portal with analytic tools. (B) PhyloGroup portal has tools for comparative genomic analysis as well as links to the individual genome portals of each group member.

genomic annotation. New examples of genes relevant to the field, even if tentative in structure, may be cloned or used to build probes, for validation and for the enormous battery of experimental and computational tools well developed in molecular biology, biochemistry, genetics, cell biology, systematics, etc. In a sense, genome annotation merely provides (potentially large amounts of) additional raw material for gene-specific analysis. There are so many examples of this in the modern study of fungi that we will not try to list any.

In contrast to gene-specific analysis, genome-wide analysis is only possible with whole-genome sequencing. Furthermore, most (but not all) analysis requires or is facilitated by knowledge of the genes and thus is predicated on a whole-genome annotation. A growing number of whole-genome analytic methods are now routinely applied to genome projects and are described in the growing number of genome papers. Many of these methods originate from attempts to extend gene/transcript/protein-specific methods to a high-throughput context, while others are genuinely novel methods that could be conceived only after the advent of whole-genome sequencing.

The following are brief introductions to the most commonly used of these whole-genome analyses in fungal projects other than *S. cerevisiae* and *S. pombe*, which as model eukaryotes have served as test beds for a large number of high-throughput methods that are unlikely to be quickly or easily transferred to other nonmodel fungi. Each introduction includes a description of the method and a simple example from the often extensive literature. A comprehensive survey of the whole-genome analysis literature *per se* is not within the scope of this review. The examples are largely drawn from JGI-enabled studies.

4.1. Comparing the annotation to itself: Duplications and higher-order structure

Whole-genome annotation allows discovery of higher-order genomic structures that may be missed by analysis at the nucleotide level. This is because amino acid sequences may be more conserved than the underlying nucleotide sequences, making protein alignment tools more sensitive than their DNA alignment counterparts. A very common method is to align each protein against all of the others using BLASTP and then use the high-quality alignment similarity scores to inform a clustering tool such as TRIBE-MCL (Enright et al., 2002), which uses a Markov clustering algorithm (MCL; http://www.micans.org/mcl/). The clusters may then be interpreted as a draft set of gene families, with members of clusters as potential paralogs. Another common method is to follow BLASTP with a neighbourhood analysis tool such as DAGchainer to identify syntenic regions (Haas, Delcher, Wortman, & Salzberg, 2004). These may be segmental duplications, tandem repeats, or in the case of dikaryons even polymorphic alleles artefactually separated by the assembly process. If the organism is known to be monokaryotic, most of the genes appear to be duplicated, and the duplicates appear to be syntenic, and a whole-genome duplication (WGD) or polyploidization event in the recent evolution of the lineage may be inferred.

For example, the monokaryotic *Rhizopus delemar* was the first member of the "basal" lineage Mucoromycotina (archetypal of the former "Zygomycota") to have its genome sequenced. The Broad FGI Annotation Pipeline predicted 13,895 genes (Ma et al., 2009). All genes were aligned with each other using BLAST. For each gene, the highest scoring alignment was determined, with the other gene its "best hit". If the first gene was the best hit of the other gene, then the two genes are necessarily a 1:1 pairing and may be called a "best bidirectional blast hit", or BBB. In the *Rhizopus* study, if the members of at least three BBBs were found on the same genomic segment and preserved gene order and orientation, then the two genomic segments and their genes were treated as duplicates. This method found 648 duplicated genes on 256 duplicated genomic segments. As these constituted 12% of the genome and were spread across the entire genome, it was concluded that an ancestor of *R. delemar* experienced a WGD. In comparison with another Mucoromyotina, the JGI-sequenced *Phycomyces blakesleeanus* (http://genome.jgi.doe.gov/phycomyces/) suggested that the WGD event happened after divergence between the *Rhizopus* and *Phycomyces* lineages. A similar argument was made for the dothideomycete *Hortaea werneckii* (Lenassi et al., 2013).

4.2. Comparing the annotation with experimental data: other -omics

The functional annotation allows easy search of the gene predictions for properties of interest. Some subset genes are so routinely extracted from eukaryotic genome annotations that they have earned an -omic name: "kinome" for all proteins predicted to have a kinase domain and "secretome" for all proteins predicted to have a cleaved signal peptide. In addition, fungal biologists have exceptional interest (as shown in recent publications; see in the succeeding text) in CAZymes, mating loci, secondary metabolism genes, pathogenicity factors, and symbiosis effectors. The latter two are often enriched among "small secreted proteins" (SSPs), which are simplistically a subset of the secretome. Such analyses gain even more power if combined with data from other genomes (comparative genomics; see in the succeeding text) or with data from laboratory experiments.

Unfortunately, most gene/protein-specific laboratory methods are not (yet) high throughput for any fungus except for *S. cerevisiae* and *S. pombe*, though that may change with the maturation of more fungal model systems (e.g. *Neurospora, Aspergillus, Fusarium, Coprinopsis*, and *Schizophyllum*). Currently, the most common additional high-throughput data supplementing a fungal genome project are from transcriptome sequencing and genome "resequencing".

4.2.1 Transcriptomics

The immense and still-expanding variety of ways of interrogating RNA samples and processing the resulting transcriptomic data are described in much greater depth in the next chapter. Here, we will discuss the ways that transcriptomics informs and is informed by whole-genome annotation.

The transcript sequences derived from a whole-genome annotation may be used to design probes for microarrays that are used to interrogate RNA samples. In principle, with a high-quality annotation and RNA samples from diverse conditions, a whole transcriptome may be described. Conversely, RNA samples or cDNA libraries derived thereof may be directly sequenced (serial analysis of gene expression or SAGE, massively parallel signature sequencing or MPSS, and RNA sequencing or RNA-Seq; see the next chapter) and used to define a transcriptome independently from a genome project. As sequencing technology has become cheaper, RNA-Seq in particular has become very popular. While the reads may be assembled into putative transcripts in the absence of a genome project, RNA-Seq becomes even more powerful when conducted in tandem with genome sequencing. The reads may be directly aligned to the genomic assembly and the alignments then assembled into transcripts (i.e. transcript-based gene prediction, discussed previously) or directly incorporated into existing gene predictions.

Whatever their methodological source, if the transcriptomic data from different biological conditions are analysed separately, they may be used to identify differentially expressed genes, differentially spliced transcripts, and even antisense (presumptively regulatory) small RNAs. This has recently been used in fungi to identify genes involved in fruiting body development (see in the succeeding text), mycorrhizal symbiosis (Martin et al., 2008, 2010), xylose fermentation (Wohlbach et al., 2011), biomass degradation (Berka et al., 2011), mycoparasitism (Atanasova et al., 2013), and pathogenicity (Duplessis et al., 2011), among others.

For example, *Schizophyllum commune* is being actively developed to become a model organism for Basidiomycota, with the full panoply of standard laboratory molecular biology techniques. Complementing the wet lab efforts, this mushroom-forming agaricomycete has also had its genome and transcriptome sequenced. The JGI Annotation Pipeline predicted 13,210 genes (Ohm et al., 2010). RNA was sampled from four different stages of mushroom development, from nonfruiting monokaryons to mature dikaryotic fruiting bodies and interrogated by MPSS. Sixty percent of the genes were associated with MPSS tags, meaning they were expressed in at least one developmental stage. Interestingly, 42% of the genes were associated with an antisense transcript. Three hundred and sixty-six genes were both highly expressed (large numbers of tags) and differentially expressed (different numbers of tags between the four stages). Furthermore, of 311 transcription factors (TFs) identified from the annotation and shown to be expressed, 44% were developmentally regulated. Taking advantage of the molecular techniques now available for *S. commune*, two of the later-stage-specific TFs were deleted, and in one case, the null mutant did not fruit. Analogous studies using genome annotation and transcriptomics identified differentiation and development TFs in the fellow agaricomycetes *Coprinopsis cinerea* (Stajich et al., 2010) and *Agaricus bisporus* (Morin et al., 2012) and the ascomycete *Pyronema confluens* (Traeger et al., 2013).

4.2.2 Proteomics

Increasingly high-throughput analytic chemistry techniques for separation (chromatography or electrophoresis) and detection (mass spectrometry) of peptides have allowed systematic interrogation of protein samples. However, the resulting mass spectra are ambiguous, so a reference protein set is needed for deducing each peptide's amino acid sequence. Whole-genome annotation produces such a reference set, so in effect, genomics has enabled proteomics. The reference set database is typically searched with mass spectra by a tool such as Mascot or Sequest. Conversely, the peptide sequences so identified may be used in a manner analogous to RNA-Seq: to correct predicted gene structures, to confirm the presence of genes, and to quantify the expression of genes at the translational level. Marrying genomics and proteomics with cell biology techniques to separate cellular compartments (e.g. centrifugation) can provide even more resolution. A comprehensive review of fungal proteomics is beyond the scope of this review (but see Bouws, Wattenberg, & Zorn, 2008; Kim, Nandakumar, & Marten, 2007; Zhou, Panisko, Magnuson, Baker, & Grigoriev, 2009), so we will limit ourselves to a few examples.

For example, *Aspergillus niger* is a key platform for the agrochemical and biotechnology industries, and two different strains were duly sequenced by DSM Food Specialties (Pel et al., 2009) and JGI (Andersen et al., 2011). The JGI Annotation Pipeline predicted 11,200 genes in *A. niger* ATCC 1015. Mass spectra were used to query the reference protein database (Wright et al., 2009). Peptides were successfully mapped to 214 genes, of which 13 were inconsistent with the proteomics data and subsequently improved (Fig. 1.4). Secretion mechanisms are of special interest in a biotechnology platform, so a more targeted investigation of secreted proteins in *A. niger* has also been done (Tsang, Butler, Powlowski, Panisko, & Baker, 2009). Mass spectra were generated from cell-free culture media under six different growth conditions and then used to query the reference database. Two hundred and twenty-two proteins were confirmed by the peptides, including 74 proteins expressed under only one of the conditions. Twenty genes were corrected. Interestingly, only 20% of genes predicted to be secreted by SignalP were detected by the proteomics experiments.

The genome of the important mycorrhizal symbiont *Laccaria bicolor* was sequenced by JGI (Martin et al., 2008), with 20,614 genes predicted and used for proteomic analysis of the secretome (Vincent et al., 2012). The 224 secreted proteins thus identified included CAZymes and proteases; the genes of most were also shown to be transcribed *in vivo*, demonstrating the power of combining proteomics with transcriptomics and genomics. Since then, proteomics has been applied to more and more fungi, with mass spectra representing 2545 *Candida albicans* proteins recently deposited in the public proteomics database PeptideAtlas (Vialas et al., 2013). *C. albicans* is PeptideAtlas' first fungus other than the models *S. cerevisiae* and *S. pombe*.

4.2.3 Metabolomics

By analogy with the transcriptome and proteome, the metabolome is a comprehensive description of the small molecules of an organism. Also like the transcriptome and proteome, while in principle metabolomics data may be gathered with methodologies that do not require a sequenced genome, in practice, the use of the data is greatly enhanced by having a whole-genome annotation in hand. Unlike the other -omics, there is not (yet) a single methodology that can fully profile the small molecules of even a single sample. The most common methodologies deploy a combination of high-throughput analytic chemistry techniques for separation (e.g. gas and liquid chromatography) and detection (e.g. mass spectrometry and nuclear magnetic resonance spectroscopy). Using these, specific subcategories of

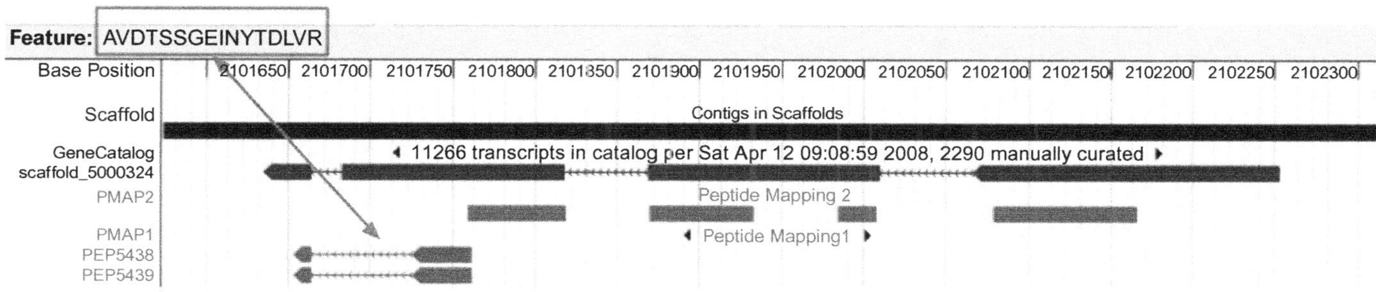

Figure 1.4 An example of an automated gene prediction refuted by a proteomics experiment. Browser tracks in the JGI *Aspergillus niger* Portal show that most peptides from proteomics experiments agree with the gene prediction ("GeneCatalog" track), but two peptides ("Peptide Mapping 1" tracks) do not agree with the 3′-intron. The Browser's "Feature" box shows the peptide's sequence (box and arrow). (See the colour plate.)

small molecules may be quantified and compared between samples. A subcategory of special interest in fungi are secondary metabolites (SMs), which are usually specific to a fungal taxon (sometimes a single strain) and are often demonstrably of adaptive significance, mediating competitive, cooperative, or antagonistic interactions with other organisms. As such, SMs are a proven source of antibiotics, toxins, and other substances of significance to humans. Unlike most fungal genes, SM genes are usually physically clustered together by metabolite. So far, the major way that metabolomics studies have been able to leverage genomics is to use a whole-genome annotation to predict potential SM genes in experimentally tractable organisms (Forseth et al., 2011; Gummer, Trengove, Oliver, & Solomon, 2013; Smith & Bluhm, 2011), knockout the genes with traditional low-throughput molecular techniques, and compare metabolomics profiles to identify missing SMs. In a move toward a more comprehensive -omics approach, metabolomics profiles of 42 known SMs in the model fungus *Aspergillus nidulans* were used to optimize the culture conditions for examining differential coexpression across SM gene clusters (Andersen et al., 2012). Even more recently, liquid chromatography/mass spectrometry profiles of alfalfa roots with and without *Rhizophagus intraradices* mycorrhizae were compared to find symbiosis-specific small molecules. Of 71 analytes thus detected, 2 affected transcription of some of 96 tested *Rhizophagus* genes (Laparre et al., 2013).

4.2.4 Epigenomics

Epigenetic modifications of DNA are heritable but do not involve changes in the primary nucleotide sequence. The best-characterized are histone modification and DNA methylation and are implicated in the regulation of transcription. Combining chromatin immunoprecipitation with microarrays (ChIP-chip) or with sequencing (ChIP-Seq) allows whole-genome detection of histone modifications, and bisulphite-conversion sequencing allows whole-genome detection of methylated cytosines. The epigenomic data may then be combined with whole-genome annotation to discover the landscape of epigenetic *cis*-regulatory targets under different modification regimes. This approach is being used to test the hypothesis that the physical clustering of SM genes may be linked to common regulation through epigenetic modifications.

For example, the genome of the industrially important sordariomycete *Trichoderma reesei* was sequenced, and 9129 genes were predicted by JGI (Martinez et al., 2008). Epigenomics studies using ChIP-Seq demonstrated that the homolog to the *A. nidulans* histone methyltransferase does regulate multiple gene clusters but not all of these are mediated by histone methylation (Karimi-Aghcheh et al., 2013; Seiboth et al., 2012). Meanwhile, methylation of repeats has been documented in diverse fungi, including Ascomycota, Basidiomycota, and Mucoromycotina (Zemach, McDaniel, Silva, & Zilberman, 2010).

4.2.5 Population genomics

Another way to exploit transcriptomic methods is to treat the genome assembly as a "reference genome", sequence transcriptomes of strains other than the reference genome, and map the data onto the reference genome to detect polymorphisms. Similarly and even more easily, the genomes of non-reference strains may be sequenced to low coverage ("resequencing") and the reads mapped. When associated with the reference genome's annotation, the polymorphisms may be used to characterize and explain known mutant and strain phenotypes, to detect pseudogenes and alleles with no visible phenotype, and to calculate nonsynonymous/synonymous substitution rates and thus identify genes under natural selection.

For example, *N. crassa* was arguably the original model fungus, the system in which was formulated the "1 gene-1 enzyme" hypothesis, the direct precursor to the "central dogma" of molecular biology. Though surpassed by *S. cerevisiae* and *S. pombe* as dominant model organisms, *N. crassa* continues to play a role in fundamental biology and was sequenced, with the Broad FGI Annotation Pipeline predicting 9907 genes (Galagan et al., 2003). Hundreds of genetic markers and associated phenotypes were discovered or created before genome sequencing. Now with the assembly and annotation of one strain on hand as a reference genome, it is possible to rapidly characterize these mutations on the nucleotide and amino acid levels. Without further crossing or cloning, 18 *N. crassa* classical mutants were resequenced, their reads mapped on the reference genome, and polymorphisms catalogued (McCluskey et al., 2011). By limiting consideration of polymorphisms to regions between known genetic markers and excluding polymorphisms in common between mutants with different phenotypes, each mutant phenotype could be strongly associated with a specific gene and allele (Fig. 1.5). Moving in the opposite direction of polymorphism to phenotype, a resequencing study of 48 wild *N. crassa* isolates discovered distinct subpopulations as well as alleles of specific genes that may be adaptive to the populations' different environmental conditions (temperature and latitude) (Ellison et al., 2011). Resequencing of multiple strains of pathogenic Ascomycota *Blumeria graminis*, *Cochliobolus heterostrophus*, and *Pyrenophora tritici-repentis*, and of serial isolates from a single patient of *Cryptococcus*

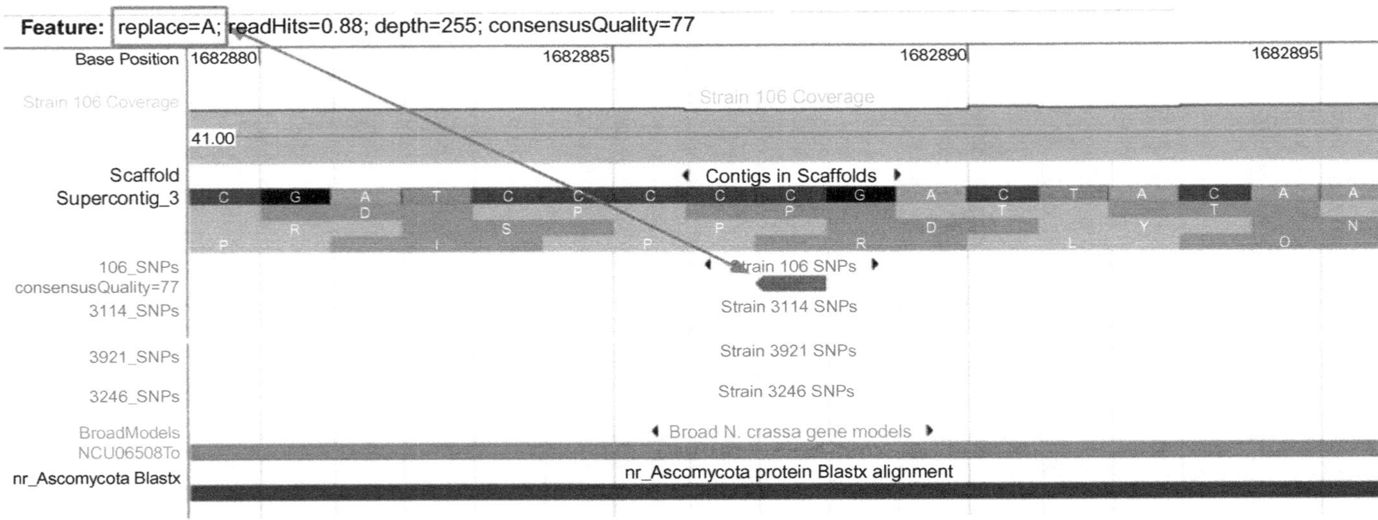

Figure 1.5 An example of a mutant's underlying mutation discovered by a resequencing experiment. Browser tracks in the JGI *Neurospora crassa* Broad reference strain Portal shows that a *compact* mutant has a single-nucleotide polymorphism ("Strain 106 SNPs" track) within a Broad gene prediction ("Broad N. crassa gene models" track). The C → T ("Contigs in Scaffolds" nucleotide track) results in a P → L mutation ("Contigs in Scaffolds" amino acid track). The browser's "Feature" box shows the resequencing read's SNP, which is A as the read is on the opposite DNA strand (box and arrow). (See the colour plate.)

neoformans, pointed to potential effector, virulence, or pathogenicity genes (Condon et al., 2013; Manning et al., 2013; Ormerod et al., 2013; Wicker et al., 2013). Similarly, resequencing of multiple isolates helped establish gene flow between pathogenic *Coccidioides* sp. (Neafsey et al., 2010).

4.3. Comparing the annotation to those of other genomes

4.3.1 Comparative genomics

Armed with a new gene sequence, a researcher would often probe DNA extracts of other strains and species for the presence of the gene. The computational analog of the procedure is now universal, with immediate results, ever since public deposition of newly sequenced genes became the norm and BLAST was implemented to query them over the web. Whole-genome annotation has taken this to an even higher level. First, it has greatly increased the numbers of new gene and protein sequences deposited in public databases. Second, and more revolutionary, it has allowed direct mutual interrogation of entire gene/protein complements between two or more species. With two genomes, BBBs may be used to define a set of putative orthologs. With more than 2 genomes, the preferred method is to apply an alignment and then a clustering tool, as described with BLASTP and MCL earlier, or some variation thereof. The clusters may then be interpreted as a draft set of gene families and thus used to distinguish putative orthologous groups and paralogs. Common sets of "core" genes may be identified, and conversely lineage-specific gene gains and losses and gene family expansions and contractions may be inferred. Lineage-specific genes may thus be identified as potentially adaptive for that lineage, even if the genes are "unknown" in that they have no other functional annotation.

For example, *Candida tenuis*, *Pichia stipitis*, and *Spathaspora passalidarum* are gut symbionts of wood-boring beetles. They differ from other saccharomycetes in their ability to ferment xylose, second only to glucose in abundance in wood. Xylose fermentation presumably plays a role in the symbiosis and certainly makes these yeasts candidates for industrial processing of wood products for energy. JGI sequenced and annotated these three genomes and compared them to five other yeasts that grow on but do not ferment xylose and six other yeasts that do not grow on xylose at all (Jeffries et al., 2007; Wohlbach et al., 2011). Clustering and definition of orthologous groups allowed identification of 3 genes unique to the xylose fermenters and 15 genes common to both the fermenters and the respirers (Fig. 1.6). More recently, comparative genomics, transcriptomics, and proteomics were combined in the wood rot *Ceriporiopsis subvermispora* to

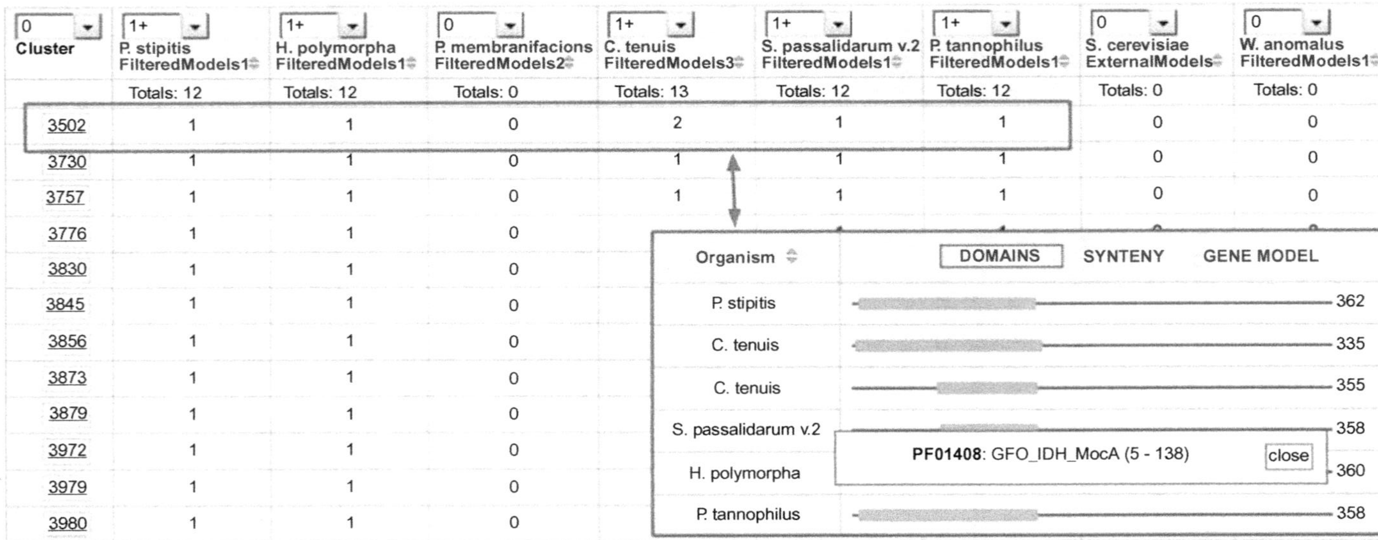

Figure 1.6 An example of potential function-specific genes discovered by clustering and filtering. Large image: Clustering run Viewer in the JGI *Spathaspora passalidarum* Portal shows gene clusters between five xylose-fermenting yeasts and three nonfermenters. The viewer's filtering tools were then used to find 12 clusters that are specific to the xylose fermenters. The largest cluster (upper box) is detailed in the inset image: Cluster Viewer shows domain structure of each member of the cluster. The Pfam domain prediction suggests that the proteins are oxidoreductases.

identify expansions of manganese peroxidases and fatty acid desaturases that explain this agaricomycetes' selective ligninolysis compared to its indiscriminate relative *P. chrysosporium* (Fernandez-Fueyo et al., 2012). Similarly, the comparison between *Cladosporium fulvum* and *Dothistroma septosporum*, two closely related dothideomycetes with radically different hosts and modes of pathogenicity, revealed many shared genes (toxins and CAZymes) that are not expressed or pseudogenized in one species but not the other, as appropriate for its host or lifestyle (de Witt et al., 2012).

4.3.2 Phylogenomics

Clustering groups genes into candidate gene families but does not describe the internal structure of those families. The internal structure most commonly of interest is the phylogenetic relationships between the members of the family, especially orthology and paralogy. Assuming that membership in the cluster indicates homology (common evolutionary origin), a gene phylogeny may be proposed by a three-step process of aligning the multiple sequences of a cluster and then computing a hierarchy or tree of relationships assuming an evolutionary model and using the aligned nucleotides or amino acids as evolving characters. If such a gene tree is then reconciled with a known species tree for the taxa involved, then gene losses may be inferred and in-paralogs and out-paralogs may be distinguished from each other. JGI has implemented such a tree-reconciliation pipeline currently separate from the JGI Annotation Pipeline. After the latter pipeline generates BLASTP/MCL clusters, the tree-reconciliation pipeline performs a multiple sequence alignment (MSA) on the proteins of each cluster using MAFFT (Katoh, Misawa, Kuma, & Miyata, 2002), builds a maximum likelihood tree using FastTree (Price, Dehal, & Arkin, 2009), and finally infers duplications and losses in the gene tree by reconciling it with a separately generated species tree (see in the succeeding text) using Notung (Chen, Durand, & Farach-Colton, 2000).

Reconstructing species phylogenies is a special case of building gene trees. Often a single gene family is used for character analysis, chosen because of its ubiquity (present in all or nearly all taxa to be included in the phylogeny), its unambiguous orthology (usually one copy per genome), and an appropriate level of sequence conservation. The gene family is then treated as in the preceding text, with the resulting gene tree lacking any losses or paralogs and thus allowing a species branching to be inferred from the gene branching. More than one gene may be used, often entailing concatenation of the genes before the MSA, in effect increasing the number of characters informing the phylogenetic reconstruction. Whole-genome annotation of large numbers of genomes allows this methodology to be taken to its logical extreme, as orders of magnitude more orthologous groups may be used. However, one drawback of the larger number of orthologous groups is usually a smaller number of taxa considered, as only a tiny fraction of known fungi have had their genomes sequenced and annotated. Naturally, this drawback is increasingly mitigated as genome projects continue to proliferate. To take advantage of large numbers of new genomes, an automated species-tree building feature was added to the JGI Annotation Pipeline. After clustering all of the proteins across selected genomes using BLASTP and MCL as described in the preceding text, the pipeline selects a subset of clusters with 1 protein from each genome as putative orthologous groups, concatenates the proteins of each genome, creates an MSA from the concatenation using MAFFT, and finally builds a maximum likelihood tree using RAxML (Stamatakis, Ludwig, & Meier, 2005). The species tree may then be used for estimating gene family sizes at each node using a tool such as CAFÉ (De Bie, Cristianini, Demuth, & Hahn, 2006) or for building species-reconciled gene trees with a tool such as Notung, as described in the preceding text.

For example, Dothideomycetes is the most speciose fungal class (so far as mycologists know). It is ecologically diverse, embracing saprobes, biotrophic pathogens, and necrotrophic pathogens, and in this last regard, it is also economically significant. JGI sequenced and annotated nine dothideomycete genomes and imported nine others into the MycoCosm platform on which to perform comparative analysis (Ohm et al., 2012). After clustering of gene families as described in the preceding text (including *A. nidulans* as outgroup), 51 gene families with 1 member from each of the 19 taxa were treated as orthologous sets and subject to MSA. The MSA was used to build a maximum likelihood tree. This tree was then used as a phylogenetic framework for interpreting the distribution and abundance of various categories of genes likely to be involved in pathogenesis, namely, SSPs, SM gene clusters, CAZymes, proteases, and lipases. Despite similar numbers of genome-wide gene counts, the mostly necrotrophic Pleosporales had more of the putatively pathogenic candidates than did the largely biotrophic Capnodiales. In the specific case of CAZymes, certain subcategories were enriched in one or the other order. A smaller phylogenomic analysis of three *Trichoderma* sp., the saprobic *T. reesei* and the mycoparasitic *T. virens* and *T. atroviride*, demonstrated that mycoparasitism is ancestral to the genus and that saprotrophy is likely a derived condition (Fig. 1.7; Kubicek et al., 2011).

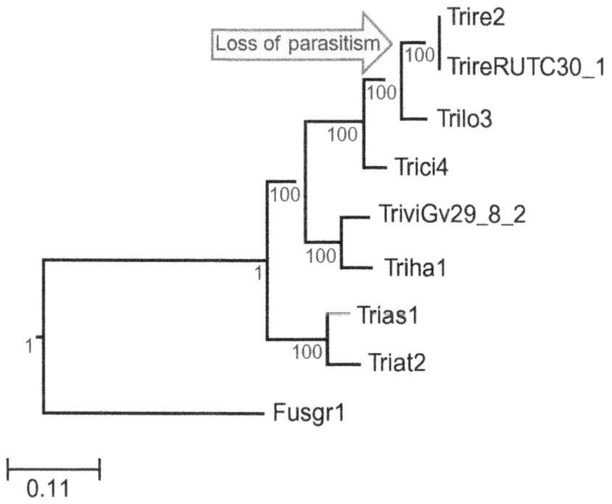

Figure 1.7 An example of a phylogenetic tree inferred from gene clusters across a PhyloGroup. Species tree Viewer suggests that the mycoparasitism lifestyle was likely ancestral to *Trichoderma* and lost by the *T. reesei* lineage of Trire2 and TrireRUTC30_1 (arrow). *T. reesei* is a saprobe, all six other *Trichoderma* sp. on the tree are mycoparasites. Trilo3 is *T. longibrachiatum*, Trici4 is *T. citrinoviride*, TriviGv29_8_2 is *T. virens*, Triha1 is *T. harzianum*, Trias1 is *T. asperellum*, Triat2 is *T. atroviride*, and Fusgr1 is *Fusarium graminearum*, an outgroup.

4.3.3 Ecogenomics

Ecogenomics does not really exist yet but could exist if given (1) a rigorous typology of ecological niches, (2) large numbers of quality genomes and genes, and (3) rigorous functional annotation of those genes, especially those that mediate interactions of the fungus with its neighbours, hosts, and environments. Categories of such genes include:

- surface and secreted proteins with an extracellular biological activity, for example, CAZymes, proteases, lipases, pathogenesis and symbiosis effectors, mating pheromones, adhesins, and hydrophobins;
- enzymes of secondary metabolism, for example, polyketide synthases, non-ribosomal peptide synthetases, terpene cylases, and dimethylallyl tryptophan synthetases, which synthesize small biologically active molecules;
- transporters of small molecules and ions of environmental significance, e.g. symbiotic exchange of nutrients, halotolerance, and drug efflux;
- response proteins, that is, signal transduction pathways from extracellular receptors through G-proteins and protein kinases to TFs.

In future, one would like to be able to use the functional annotation of an otherwise uncharacterized species to predict the ecological niche of that species.

Wood rot fungi play a crucial role in wood decay and are also candidates for breaking down cellulose as an energy source. The wood rot fungi may be classified as "white rot" and "brown rot", with the former more completely degrading all woody components, especially the highly resistant lignin. Interestingly, both ecotypes are scattered among different orders of Agaricomycetes. To explore the differences between the ecotypes, JGI sequenced and annotated nine white rot fungi and seven brown rot fungi, and compared them with each other and with other non-wood rot fungi (Floudas et al., 2012). Comparative genomics and phylogenomics reiterated the dispersal of both wood rot ecotypes (plus non-wood rot ecotypes) amongst and within different orders of Agaricomycetes. The brown rot fungi had smaller numbers of certain CAZyme families, with smaller families, than white rot fungi, and non-rot fungi even fewer. Specific classes of ligninolytic peroxidases were associated with the different ecotypes. The origination of the classes from the manganese peroxidase class was inferred, as was the presence of this class (and thus also the white rot ecotype) in the ancestral agaricomycetes. Subsequent diversification of ecotypes was associated with expansion of ligninolytic peroxidases (white rot) or contraction and loss (all others). Using fossils to calibrate a molecular clock analysis, the initial white rot expansion was estimated to occur at the end of the Carboniferous Period, intriguingly the great age of unequalled coal-producing swamp forests.

5. GENOMICS OF PLANT–MICROBE INTERACTIONS: WHAT'S NEXT?

5.1. Systems biology

Experimental data described in the preceding text allow us not only to understand the function of individual genes but also to reconstruct a global view of the processes running a cell. These processes are better understood for model organisms like baker's yeast *S. cerevisiae,* model filamentous ascomycete *N. crassa*, and model wood decay basidiomycete *S. commune*, for which extensive collections of deletion mutants along with microarrays and other functional genomics resources have been developed (Dunlap et al., 2007; Giaever et al., 2002; Winzeler et al., 1999).

Transcriptomics and proteomics of pairs of interacting organisms can shed light on these interactions during fungal plant infection (Skibbe, Doehlemann, Fernandes, & Walbot, 2010), mycoparasitism (Atanasova et al., 2013), and mycorrhizal symbiosis (Larsen et al., 2011). Similar approaches can be developed for analysis of multi-component systems, which involve several naturally interacting organisms with sequenced genomes. One such system includes a poplar tree (Tuskan et al., 2006) and associated ectomycorrhizal symbiont *L. bicolor* (Martin et al., 2008), arbuscular endomycorrhizal symbiont *Rhizophagus irregularis* (Tisserant et al., 2013), rust pathogen *Melampsora laricis-populina* (Duplessis et al., 2011), and others. Interestingly, genomes of these symbiotic and pathogenic fungi have many things in common: large genomes inflated with repeats and expanded gene families, including lineage-specific SSPs, which are also among the most expressed genes *in planta*, during infection of the plant host.

In nature, however, interactions are more complex than the pairwise interactions in model systems. The same plant defends itself from numerous types of pathogens at once using different molecular mechanisms. A fungal pathogen or a symbiont establishes contact with its host while at the same time fighting other fungi for resources or survival. Soil, for example, is inhabited by thousands of fungal species and tens of thousands of bacteria and offers a complexity that requires new genomics approaches.

5.2. Metagenomics

Metagenomics is a method to understand composition and interactions within a microbial community via genome sequencing. The very first step to understand composition of a microbial community is sequencing marker genes like 16S ribosomal RNA (rRNA) for prokaryotes, 18S rRNA for eukaryotes, or ribosomal Internal Transcribed Sequences (ITS) for fungi. The lengths of these markers may pose some challenges. While long Sanger reads were sufficient to cover the entire $\sim$1 kbp long ITS, 454 pyrosequencing and most recently 2×250 bp Illumina reads (which if overlap efficiently form a 500 bp read) from MiSeq or HiSeq 2500 only allow sequencing approximately half of ITS: ITS1 or ITS2 regions (Buée et al., 2009). Comparing these markers with collections of previously sequenced 16S, 18S, and ITS from other organisms and communities helps identify members of the community under study at different levels of resolution: species, genus, phylum, etc.

The next and more complex step is to sequence metagenomes in order to assemble genes from different organisms. While prokaryotic communities have been successfully sequenced (Kalyuzhnaya et al., 2008; Tringe & Rubin, 2005), the complex organization and gene structure of eukaryotic genomes present a stumbling block for fungal metagenomics. Assembled DNA fragments are too short for *ab initio* gene predictors, while homology-based gene predictors require a representative collection of reference genomes. Except for very simple communities containing a scant handful of species, eukaryotic metagenomics is not yet a feasible option for analysis.

Finally, metatranscriptomics—sequencing pooled RNA from an entire community—can provide a functional portrait of the community and help monitor the dynamics of changes in gene expression of the most active genes. One challenge with complex communities like that of soil, where fungal transcripts make up just a few percent of the entire microbial transcriptome, is that polyA enrichment may not yield sufficient eukaryotic mRNA for sequencing. Furthermore, only a small fraction of total eukaryotic RNA consists of mRNA. As with metagenomics, simpler communities with lower species composition make easier targets for analysis. In addition, though metatranscriptomic approaches bypass decoding the complex exon-intron structure of eukaryotic genes, an extensive collection of reference genomes is still required to assign specific transcripts to specific taxa with confidence.

5.3. The 1000 Fungal Genomes Project

Despite the growing number of fungal genomic projects, the phylogenetic diversity of sequenced fungi is still very limited, dominated by Ascomycota, and with poor representation of basal fungi among the currently available reference genomes. The 300–400 sequenced and annotated genomes at present are a small fraction of the total number of fungal species, estimated to be between 1.5 and 5 million (Blackwell, 2011; Hawksworth, 1991). This sparse sampling of the Fungal Tree of Life limits our ability to understand the functional diversity and evolution of fungi, and hampers metagenomic studies of fungal communities.

The goal of JGI's 1000 Fungal Genomes Project is to address these problems by obtaining a comprehensive set of highly diverse reference fungal genomes. This entails sequencing at least 1 genome from each of the roughly 500 known fungal families. Started in close collaborations with several culture collections and research groups providing DNA and RNA samples, the 1000 Fungal Genomes Project is open to all research groups worldwide (Grigoriev et al., 2014).

The scale of this project and difficulties with obtaining sufficient amounts of DNA and RNA for every species promotes development of new genomic technologies. High-throughput technologies for DNA and

RNA extraction are being developed. However, a relatively small fraction of fungi grows in culture. Microdissection and DNA separation and amplification may be required for obligate biotrophs. Single cell genomics may provide another approach to extract DNA from fungal spores instead of contaminated biomass. Fruiting bodies, fresh from forests, may be a source of DNA but require metagenomic-like approaches for separating the fungal genome from bacterial contaminants and cohabitants. The ability to efficiently process environmental samples will be game-changing.

6. CONCLUSION

Several decades ago genomics brought a revolution to biology, enabling scientists to ask new questions. Nowadays, new sequencing technologies supersede each other yearly, starting a revolution within genomics itself and changing the scale of genomic and biological questions that can be answered. One can sequence an entire genome or collection of genomes in order to understand function of just a small group of genes. Democratization of sequencing makes these resources available not only for large research communities assembled around large sequencing centres but also for many individual research groups. The types and quantity of sequence produced by new platforms require greater investment into computing infrastructure and bioinformatics than in sequencing itself. Data types and analytic tools will change, but the basic principles of genomic data analysis described in this chapter will continue to serve as a foundation for new analytic platforms.

ACKNOWLEDGEMENTS

The work conducted by the US Department of Energy Joint Genome Institute is supported by the Office of Science of the US Department of Energy under Contract No. DE-AC02-05CH11231.

REFERENCES

Altschul, S. F., Gish, W., Miller, W., Myers, E. W., & Lipman, D. J. (1990). Basic local alignment search tool. *Journal of Molecular Biology*, *215*, 403–410.

Andersen, M. R., Nielsen, J. B., Klitgaard, A., Petersen, L. M., Zachariasen, M., Hansen, T. J., et al. (2012). Accurate prediction of secondary metabolite gene clusters in filamentous fungi. *Proceedings of the National Academy of Sciences of the United States of America*, *110*, e99–e107.

Andersen, M. R., Salazar, M. P., Schaap, P. J., van de Vondervoort, P. J., Culley, D., Thykaer, J., et al. (2011). Comparative genomics of citric-acid-producing *Aspergillus niger* ATCC 1015 versus enzyme-producing CBS 513.88. *Genome Research*, *21*, 885–897.

Apweiler, R., Attwood, T. K., Bairoch, A., Bateman, A., Birney, E., Biswas, M., et al. (2001). The InterPro database, an integrated documentation resource for protein families, domains and functional sites. *Nucleic Acids Research*, *29*, 37–40.

Ashburner, M., Ball, C. A., Blake, J. A., Botstein, D., Butler, H., Cherry, J. M., et al. (2000). Gene ontology: Tool for the unification of biology. The Gene Ontology Consortium. *Nature Genetics*, *25*, 25–29.

Atanasova, L., Le Crom, S., Gruber, S., Coulpier, F., Seidl-Seiboth, V., Kubicek, C. P., et al. (2013). Comparative transcriptomics reveals different strategies of *Trichoderma* mycoparasitism. *BMC Genomics*, *14*, 121.

Bairoch, A. (1994). The ENZYME data bank. *Nucleic Acids Research*, *22*, 3626–3627.

Bairoch, A., & Boeckmann, B. (1991). The SWISS-PROT protein sequence data bank. *Nucleic Acids Research*, *19*(Suppl.), 2247–2249.

Bateman, A., Birney, E., Durbin, R., Eddy, S. R., Finn, R. D., & Sonnhammer, E. L. (1999). Pfam 3.1: 1313 multiple alignments and profile HMMs match the majority of proteins. *Nucleic Acids Research*, *27*, 260–262.

Benson, D., Lipman, D. J., & Ostell, J. (1993). GenBank. *Nucleic Acids Research*, *21*, 2963–2965.

Bentley, D. R., Balasubramanian, S., Swerdlow, H. P., Smith, G. P., Milton, J., Brown, C. G., et al. (2008). Accurate whole human genome sequencing using reversible terminator chemistry. *Nature*, *456*, 53–59.

Berka, R. M., Grigoriev, I. V., Otillar, R., Salamov, A., Grimwood, J., Reid, I., et al. (2011). Comparative genomic analysis of the thermophilic biomass-degrading fungi *Myceliophthora thermophila* and *Thielavia terrestris*. *Nature Biotechnology*, *29*, 922–927.

Birney, E., & Durbin, R. (2000). Using GeneWise in the *Drosophila* annotation experiment. *Genome Research*, *10*, 547–548.

Blackwell, M. (2011). The fungi: 1, 2, 3 . . . 5.1 million species? *American Journal of Botany*, *98*, 426–438.

Bouws, H., Wattenberg, A., & Zorn, H. (2008). Fungal secretomes—Nature's toolbox for white biotechnology. *Applied Microbiology and Biotechnology*, *80*, 381–388.

Buée, M., Reich, M., Murat, C., Morin, E., Nilsson, R. H., Uroz, S., et al. (2009). 454 pyrosequencing analyses of forest soils reveal an unexpectedly high fungal diversity. *New Phytologist*, *184*, 449–456.

Cantarel, B. L., Korf, I., Robb, S. M., Parra, G., Ross, E., Moore, B., et al. (2008). MAKER: An easy-to-use annotation pipeline designed for emerging model organism genomes. *Genome Research*, *18*, 188–196.

Chen, K., Durand, D., & Farach-Colton, M. (2000). NOTUNG: A program for dating gene duplications and optimizing gene family trees. *Journal of Computational Biology*, *7*, 429–447.

Clark, S. C., Egan, R., Frazier, P. I., & Wang, Z. (2013). ALE: A generic assembly likelihood evaluation framework for assessing the accuracy of genome and metagenome assemblies. *Bioinformatics*, *29*, 435–443.

Condon, B. J., Leng, Y., Wu, D., Bushley, K. E., Ohm, R. A., Otillar, R., et al. (2013). Comparative genome structure, secondary metabolite, and effector coding capacity across *Cochliobolus* pathogens. *PLoS Genetics*, *9*, e1003233.

Cuomo, C. A., & Birren, B. W. (2010). The fungal genome initiative and lessons learned from genome sequencing. *Methods in Enzymology*, *470*, 833–855.

Curwen, V., Eyras, E., Andrews, T. D., Clarke, L., Mongin, E., Searle, S. M., et al. (2004). The Ensembl automatic gene annotation system. *Genome Research*, *14*, 942–950.

De Bie, T., Cristianini, N., Demuth, J. P., & Hahn, M. W. (2006). CAFE: A computational tool for the study of gene family evolution. *Bioinformatics*, *22*, 1269–1271.

de Witt, P. J., van der Burgt, A., Ökmen, B., Stergiopoulos, I., Abd-Elsalam, K. A., Aerts, A. L., et al. (2012). The genomes of the fungal plant pathogens *Cladosporium fulvum*

and *Dothistroma septosporum* reveal adaptation to different hosts and lifestyles but also signatures of common ancestry. *PLoS Genetics*, *8*, e1003088.
Dujon, B., Sherman, D., Fischer, G., Durrens, P., Casaregola, S., Lafontaine, I., et al. (2004). Genome evolution in yeasts. *Nature*, *430*, 35–44.
Dunlap, J. C., Borkovich, K. A., Henn, M. R., Turner, G. E., Sachs, M. S., Glass, N. L., et al. (2007). Enabling a community to dissect an organism: Overview of the *Neurospora* functional genomics project. *Advances in Genetics*, *57*, 49–96.
Duplessis, S., Cuomo, C. A., Lin, Y. C., Aerts, A., Tisserant, E., Veneault-Fourrey, C., et al. (2011). Obligate biotrophy features unraveled by the genomic analysis of rust fungi. *Proceedings of the National Academy of Sciences of the United States of America*, *108*, 9166 9171.
Eid, J., Fehr, A., Gray, J., Luong, K., Lyle, J., Otto, G., et al. (2009). Real-time DNA sequencing from single polymerase molecules. *Science*, *323*, 133–138.
Ellison, C. E., Hall, C., Kowbel, D., Welch, J., Brem, R. B., Glass, N. L., et al. (2011). Population genomics and local adaptation in wild isolates of a model microbial eukaryote. *Proceedings of the National Academy of Sciences of the United States of America*, *108*, 2831–2836.
Elsik, C. G., Mackey, A. J., Reese, J. T., Milshina, N. V., Roos, D. S., & Weinstock, G. M. (2007). Creating a honey bee consensus gene set. *Genome Biology*, *8*, R13.
Emanuelsson, O., Nielsen, H., Brunak, S., & von Heijne, G. (2000). Predicting subcellular localization of proteins based on their N-terminal amino acid sequence. *Journal of Molecular Biology*, *300*, 1005–1016.
English, A. C., Richards, S., Han, Y., Wang, M., Vee, V., Qu, J., et al. (2012). Mind the gap: Upgrading genomes with Pacific biosciences RS long-read sequencing technology. *PLoS One*, 7, e47768.
Enright, A. J., Van Dongen, S., & Ouzounis, C. A. (2002). An efficient algorithm for large-scale detection of protein families. *Nucleic Acids Research*, *30*, 1575–1584.
Fernandez-Fueyo, E., Ruiz-Dueñas, F. J., Ferreira, P., Floudas, D., Hibbett, D. S., Canessa, P., et al. (2012). Comparative genomics of *Ceriporiopsis subvermispora* and *Phanerochaete chrysosporium* provide insight into selective ligninolysis. *Proceedings of the National Academy of Sciences of the United States of America*, *109*, 5458–5463.
Floudas, D., Binder, M., Riley, R., Barry, K., Blanchette, R. A., Henrissat, B., et al. (2012). The Paleozoic origin of enzymatic lignin decomposition reconstructed from 31 fungal genomes. *Science*, *336*, 1715–1719.
Forseth, R. R., Fox, E. M., Chung, D., Howlett, B. J., Keller, N. P., & Schroeder, F. C. (2011). Identification of cryptic products of the gliotoxin gene cluster using NMR-based comparative metabolomics and a model for gliotoxin biosynthesis. *Journal of the American Chemical Society*, *133*, 9678–9681.
Galagan, J. E., Calvo, S. E., Borkovich, K. A., Selker, E. U., Read, N. D., Jaffe, D., et al. (2003). The genome sequence of the filamentous fungus *Neurospora crassa*. *Nature*, *422*, 859–868.
Ghodsi, M., Hill, C., Astrovskaya, I., Lin, H., Sommer, D., Koren, S., et al. (2013). De novo likelihood-based measures for comparing genome assemblies. *BMC Research Notes*, *6*, 334.
Giaever, G., Chu, A. M., Ni, L., Connelly, C., Riles, L., Véronneau, S., et al. (2002). Functional profiling of the *Saccharomyces cerevisiae* genome. *Nature*, *418*, 387–391.
Gnerre, S., Maccallum, I., Przybylski, D., Ribeiro, F. J., Burton, J. N., Walker, B. J., et al. (2011). High-quality draft assemblies of mammalian genomes from massively parallel sequence data. *Proceedings of the National Academy of Sciences of the United States of America*, *108*, 1513–1518.
Grigoriev, I. V., Cullen, D., Goodwin, S. B., Hibbett, D., Jeffries, T. W., Kubicek, C. P., et al. (2011). Fueling the future with fungal genomics. *Mycology*, *2*, 192–209.

Grigoriev, I. V., Martinez, D. A., & Salamov, A. A. (2006). Fungal genomic annotation. In Aurora, D. K., Berka, R. M., Singh, G. B. (Vol. Eds.), *Applied mycology and biotechnology: Vol. 6. Bioinformatics* (pp. 123–142). Amsterdam: Elsevier.

Grigoriev, I. V., Nikitin, R., Haridas, S., Kuo, A., Ohm, R., Otillar, R., et al. (2014). MycoCosm portal: Gearing up for 1000 fungal genomes. *Nucleic Acids Research, 42,* D699–D704.

Grigoriev, I. V., Nordberg, H., Shabalov, I., Aerts, A., Cantor, M., Goodstein, D., et al. (2012). The genome portal of the Department of Energy Joint Genome Institute. *Nucleic Acids Research, 40,* D26–D32.

Guigó, R. (1998). Assembling genes from predicted exons in linear time with dynamic programming. *Journal of Computational Biology, 5,* 681–702.

Gummer, J. P., Trengove, R. D., Oliver, R. P., & Solomon, P. S. (2013). Dissecting the role of G-protein signalling in primary metabolism in the wheat pathogen *Stagonospora nodorum. Microbiology, 159,* 1972–1985.

Haas, B. J., Delcher, A. L., Mount, S. M., Wortman, J. R., Smith, R. K., Jr., Hannick, L. I., et al. (2003). Improving the *Arabidopsis* genome annotation using maximal transcript alignment assemblies. *Nucleic Acids Research, 31,* 5654–5666.

Haas, B. J., Delcher, A. L., Wortman, J. R., & Salzberg, S. L. (2004). DAGchainer: A tool for mining segmental genome duplications and synteny. *Bioinformatics, 20,* 3643–3646.

Haas, B. J., Salzberg, S. L., Zhu, W., Pertea, M., Allen, J. E., Orvis, J., et al. (2008). Automated eukaryotic gene structure annotation using EVidenceModeler and the Program to Assemble Spliced Alignments. *Genome Biology, 9,* R7.

Haas, B. J., Zeng, Q., Pearson, M. D., Cuomo, C. A., & Wortman, J. R. (2011). Approaches to fungal genome annotation. *Mycology, 2,* 118–141.

Hawksworth, D. L. (1991). The fungal dimension of biodiversity—Magnitude, significance, and conservation. *Mycological Research, 95,* 641–655.

Horton, P., Park, K. J., Obayashi, T., Fujita, N., Harada, H., Adams-Collier, C. J., et al. (2007). WoLF PSORT: Protein localization predictor. *Nucleic Acids Research, 35*(Web Server issue), W585–W587.

Howe, K. L., Chothia, T., & Durbin, R. (2002). GAZE: A generic framework for the integration of gene-prediction data by dynamic programming. *Genome Research, 12,* 1418–1427.

Huang, X., Adams, M. D., Zhou, H., & Kerlavage, A. R. (1997). A tool for analyzing and annotating genomic sequences. *Genomics, 46,* 37–45.

Jeffries, T. W., Grigoriev, I. V., Grimwood, J., Laplaza, J. M., Aerts, A., Salamov, A., et al. (2007). Genome sequence of the lignocellulose-bioconverting and xylose-fermenting yeast *Pichia stipitis. Nature Biotechnology, 25,* 319–326.

Käll, L., Krogh, A., & Sonnhammer, E. L. (2004). A combined transmembrane topology and signal peptide prediction method. *Journal of Molecular Biology, 338,* 1027–1036.

Kalyuzhnaya, M. G., Lapidus, A., Ivanova, N., Copeland, A. C., McHardy, A. C., Szeto, E., et al. (2008). High-resolution metagenomics targets specific functional types in complex microbial communities. *Nature Biotechnology, 26,* 1029–1034.

Karimi-Aghcheh, R., Bok, J. W., Phatale, P. A., Smith, K. M., Baker, S. E., Lichius, A., et al. (2013). Functional analyses of *Trichoderma reesei* LAE1 reveal conserved and contrasting roles of this regulator. *G3 (Bethesda, MD), 3,* 369–378.

Katoh, K., Misawa, K., Kuma, K., & Miyata, T. (2002). MAFFT: A novel method for rapid multiple sequence alignment based on fast Fourier transform. *Nucleic Acids Research, 30,* 3059–3066.

Kent, W. J. (2002). BLAT—The BLAST-like alignment tool. *Genome Research, 12,* 656–664.

Kim, Y., Nandakumar, M. P., & Marten, M. R. (2007). Proteomics of filamentous fungi. *Trends in Biotechnology, 25,* 395–400.

Kleftogiannis, D., Kalnis, P., & Bajic, V. B. (2013). Comparing memory-efficient genome assemblers on stand-alone and cloud infrastructures. *PLoS One*, *8*, e75505.

Korf, I. (2004). Gene finding in novel genomes. *BMC Bioinformatics*, *5*, 59.

Kubicek, C. P., Herrera-Estrella, A., Seidl-Seiboth, V., Martinez, D. A., Druzhinina, I. S., Thon, M., et al. (2011). Comparative genome sequence analysis underscores mycoparasitism as the ancestral life style of *Trichoderma*. *Genome Biology*, *12*, R40.

Laparre, J., Malbreil, M., Letisse, F., Portais, J. C., Roux, C., Bécard, G., et al. (2013). Combining metabolomics and gene expression analysis reveals that propionyl- and butyryl-carnitines are involved in late stages of arbuscular mycorrhizal symbiosis. *Molecular Plant*, *7*, 554–566.

Larsen, P. E., Sreedasyam, A., Trivedi, G., Podila, G. K., Cseke, L. J., & Collart, F. R. (2011). Using next generation transcriptome sequencing to predict an ectomycorrhizal metabolome. *BMC Systems Biology*, *5*, 70.

Lenassi, M., Gostinčar, C., Jackman, S., Turk, M., Sadowski, I., Nislow, C., et al. (2013). Whole genome duplication and enrichment of metal cation transporters revealed by *de novo* genome sequencing of extremely halotolerant black yeast *Hortaea werneckii*. *PLoS One*, *8*, e71328.

Li, Z., Chen, Y., Mu, D., Yuan, J., Shi, Y., Zhang, H., et al. (2012). Comparison of the two major classes of assembly algorithms: Overlap–layout–consensus and de-bruijn-graph. *Briefings in Functional Genomics*, *11*, 25–37.

Luan, B., Martyna, G., & Stolovitzky, G. (2011). Characterizing and controlling the motion of ssDNA in a solid-state nanopore. *Biophysical Journal*, *101*, 2214–2222.

Ma, L. J., Ibrahim, A. S., Skory, C., Grabherr, M. G., Burger, G., Butler, M., et al. (2009). Genomic analysis of the basal lineage fungus *Rhizopus oryzae* reveals a whole-genome duplication. *PLoS Genetics*, *5*, e1000549.

Manning, V. A., Pandelova, I., Dhillon, B., Wilhelm, L. J., Goodwin, S. B., Berlin, A. M., et al. (2013). Comparative genomics of a plant-pathogenic fungus, *Pyrenophora tritici-repentis*, reveals transduplication and the impact of repeat elements on pathogenicity and population divergence. *G3 (Bethesda, MD)*, *3*, 41–63.

Margulies, M., Egholm, M., Altman, W. E., Attiya, S., Bader, J. S., Bemben, L. A., et al. (2006). Corrigendum: Genome sequencing in microfabricated high-density picolitre reactors. *Nature*, *441*, 120.

Martin, F., Aerts, A., Ahrén, D., Brun, A., Danchin, E. G., Duchaussoy, F., et al. (2008). The genome of *Laccaria bicolor* provides insights into mycorrhizal symbiosis. *Nature*, *452*, 88–92.

Martin, F., Kohler, A., Murat, C., Balestrini, R., Coutinho, P. M., Jaillon, O., et al. (2010). Périgord black truffle genome uncovers evolutionary origins and mechanisms of symbiosis. *Nature*, *464*, 1033–1038.

Martinez, D., Berka, R. M., Henrissat, B., Saloheimo, M., Arvas, M., Baker, S. E., et al. (2008). Genome sequencing and analysis of the biomass-degrading fungus *Trichoderma reesei* (syn. *Hypocrea jecorina*). *Nature Biotechnology*, *26*, 553–560.

Martinez, D., Larrondo, L. F., Putnam, N., Gelpke, M. D., Huang, K., Chapman, J., et al. (2004). Genome sequence of the lignocellulose degrading fungus *Phanerochaete chrysosporium* strain RP78. *Nature Biotechnology*, *22*, 695–700.

McCluskey, K., Wiest, A. E., Grigoriev, I. V., Lipzen, A., Martin, J., Schackwitz, W., et al. (2011). Rediscovery by whole genome sequencing: Classical mutations and genome polymorphisms in *Neurospora crassa*. *G3 (Bethesda, MD)*, *1*, 303–316.

Medvedev, P., & Brudno, M. (2009). Maximum likelihood genome assembly. *Journal of Computational Biology*, *16*, 1101–1116.

Miller, J. R., Koren, S., & Sutton, G. (2010). Assembly algorithms for next-generation sequencing data. *Genomics*, *95*, 315–327.

Morin, E., Kohler, A., Baker, A. R., Foulongne-Oriol, M., Lombard, V., Nagy, L. G., et al. (2012). Genome sequence of the button mushroom *Agaricus bisporus* reveals mechanisms governing adaptation to a humic-rich ecological niche. *Proceedings of the National Academy of Sciences of the United States of America, 109*, 17501–17506.

Nagarajan, N., & Pop, M. (2009). Parametric complexity of sequence assembly: Theory and applications to next generation sequencing. *Journal of Computational Biology, 16*, 897–908.

Nakamura, K., Oshima, T., Morimoto, T., Ikeda, S., Yoshikawa, H., Shiwa, Y., et al. (2011). Sequence-specific error profile of Illumina sequencers. *Nucleic Acids Research, 39*, e90.

Neafsey, D. E., Barker, B. M., Sharpton, T. J., Stajich, J. E., Park, D. J., Whiston, E., et al. (2010). Population genomic sequencing of *Coccidioides* fungi reveals recent hybridization and transposon control. *Genome Research, 20*, 938–946.

Nielsen, H., Brunak, S., & von Heijne, G. (1999). Machine learning approaches for the prediction of signal peptides and other protein sorting signals. *Protein Engineering Design and Selection, 12*, 3–9.

Ogata, H., Goto, S., Sato, K., Fujibuchi, W., Bono, H., & Kanehisa, M. (1999). KEGG: Kyoto Encyclopedia of genes and genomes. *Nucleic Acids Research, 27*, 29–34.

Ohm, R. A., de Jong, J. F., Lugones, L. G., Aerts, A., Kothe, E., Stajich, J. E., et al. (2010). Genome sequence of the model mushroom *Schizophyllum commune*. *Nature Biotechnology, 28*, 957–963.

Ohm, R. A., Feau, N., Henrissat, B., Schoch, C. L., Horwitz, B. A., Barry, K. W., et al. (2012). Diverse lifestyles and strategies of plant pathogenesis encoded in the genomes of eighteen Dothideomycetes fungi. *PLoS Pathogens, 8*, e1003037.

Oliver, S. G., Van Der Aart, Q. J. M., Agostoni-Carbone, M. L., Aigle, M., Alberghina, L., Alexandraki, D., et al. (1992). The complete DNA sequence of yeast chromosome III. *Nature, 357*, 38–46.

Ormerod, K. L., Morrow, C. A., Chow, E. W., Lee, I. R., Arras, S. D., Schirra, H. J., et al. (2013). Comparative genomics of serial isolates of *Cryptococcus neoformans* reveals gene associated with carbon utilization and virulence. *G3 (Bethesda, MD), 3*, 675–686.

Pel, H. J., de Winde, J. H., Archer, D. B., Dyer, P. S., Hofmann, G., Schaap, P. J., et al. (2009). Genome sequencing and analysis of the versatile cell factory *Aspergillus niger* CBS 513.88. *Nature Biotechnology, 25*, 221–231.

Peng, Z., Zhao, Z., Nath, N., Froula, J. L., Clum, A., Zhang, T., et al. (2012). Generation of long insert pairs using a Cre-LoxP inverse PCR approach. *PLoS One, 7*, e29437.

Peterson, D. G., Wessler, S. R., & Paterson, A. H. (2002). Efficient capture of unique sequences from eukaryotic genomes. *Trends in Genetics, 18*, 547–550.

Philippe, N., Salson, M., Lecroq, T., Leonard, M., Commes, T., & Rivals, E. (2011). Querying large read collections in main memory: A versatile data structure. *BMC Bioinformatics, 12*, 242.

Price, M. N., Dehal, P. S., & Arkin, A. P. (2009). FastTree: Computing large minimum evolution trees with profiles instead of a distance matrix. *Molecular Biology and Evolution, 26*, 1641–1650.

Price, A. L., Jones, N. C., & Pevzner, P. A. (2005). *De novo* identification of repeat families in large genomes. *Bioinformatics, 21*(Suppl. 1), i351–i358.

Quail, M., Smith, M. E., Coupland, P., Otto, T. D., Harris, S. R., Connor, T. R., et al. (2012). A tale of three next generation sequencing platforms: Comparison of Ion Torrent, Pacific Biosciences and Illumina MiSeq sequencers. *BMC Genomics, 13*, 341.

Salamov, A. A., & Solovyev, V. V. (2000). *Ab initio* gene finding in Drosophila genomic DNA. *Genome Research, 10*, 516–522.

Salzberg, S. L. (2005). Beware of mis-assembled genomes. *Bioinformatics, 21*, 4320–4321.

Sanger, F., Air, G. M., Barrell, B. G., Brown, N. L., Coulson, A. R., Fiddes, C. A., et al. (1977). Nucleotide sequence of bacteriophage phi X174 DNA. *Nature, 265*, 687–695.

Sanger, F., Nicklen, S., & Coulson, A. R. (1977). DNA sequencing with chain-terminating inhibitor. *Proceedings of the National Academy of Sciences of the United States of America, 74*, 5463–5467.

Schuster, S. C. (2008). Next-generation sequencing transforms today' s biology. *Nature Methods*, *5*, 16–18.

Seiboth, B., Karimi, R. A., Phatale, P. A., Linke, R., Hartl, L., Sauer, D. G., et al. (2012). The putative protein methyltransferase LAE1 controls cellulase gene expression in *Trichoderma reesei*. *Molecular Microbiology*, *84*, 1150–1164.

Skibbe, D. S., Doehlemann, G., Fernandes, J., & Walbot, V. (2010). Maize tumors caused by *Ustilago maydis* require organ-specific genes in host and pathogen. *Science*, *328*, 89–92.

Slater, G. S., & Birney, E. (2005). Automated generation of heuristics for biological sequence comparison. *BMC Bioinformatics*, *6*, 31.

Smith, J. E., & Bluhm, B. H. (2011). Metabolic fingerprinting in *Fusarium verticillioides* to determine gene function. *Methods in Molecular Biology*, *722*, 237–247.

Smith, T. F., & Waterman, M. S. (1981). Identification of common molecular subsequences. *Journal of Molecular Biology*, *147*, 195–197.

Sonnhammer, E. L., Eddy, S. R., & Durbin, R. (1997). Pfam: A comprehensive database of protein domain families based on seed alignments. *Proteins*, *28*, 405–420.

Sonnhammer, E. L., von Heijne, G., & Krogh, A. (1998). A hidden Markov model for predicting transmembrane helices in protein sequences. *Proceedings of the International Conference on Intelligent Systems for Molecular Biology*, *6*, 175–182.

Stajich, J. E., Wilke, S. K., Ahrén, D., Au, C. H., Birren, B. W., Borodovsky, M., et al. (2010). Insights into evolution of multicellular fungi from the assembled chromosomes of the mushroom *Coprinopsis cinerea* (*Coprinus cinereus*). *Proceedings of the National Academy of Sciences of the United States of America*, *107*, 11889–11894.

Stamatakis, A., Ludwig, T., & Meier, H. (2005). RAxML-III: A fast program for maximum likelihood-based inference of large phylogenetic trees. *Bioinformatics*, *21*, 456–463.

Stanke, M., & Waack, S. (2003). Gene prediction with a hidden Markov model and a new intron submodel. *Bioinformatics*, *19*(Suppl. 2), ii215–ii225.

Tatusov, R. L., Fedorova, N. D., Jackson, J. D., Jacobs, A. R., Kiryutin, B., Koonin, E. V., et al. (2003). The COG database: An updated version includes eukaryotes. *BMC Bioinformatics*, *4*, 41.

Ter-Hovhannisyan, V., Lomsadze, A., Chernoff, Y. O., & Borodovsky, M. (2008). Gene prediction in novel fungal genomes using an *ab initio* algorithm with unsupervised training. *Genome Research*, *18*, 1979–1990.

Thibaud-Nissen, F., Souvorov, A., Murphy, T., DiCuccio, M., & Kitts, P. (2013). *Eukaryotic genome annotation pipeline. In The NCBI handbook [Internet]* (2nd ed.) Bethesda, MD: National Center for Biotechnology Information. *http://www.ncbi.nlm.nih.gov/books/NBK169439*.

Timp, W., Comer, J., & Aksimentiev, A. (2012). DNA base-calling from a nanopore using a Viterbi algorithm. *Biophysical Journal*, *102*, L37–L39.

Tisserant, E., Malbreil, M., Kuo, A., Kohler, A., Symeonidi, A., Balestrini, R., et al. (2013). The genome of an arbuscular mycorrhizal fungus provides insights into the oldest plant symbiosis. *Proceedings of the National Academy of Sciences of the United States of America*, *110*(50), 20117–20122.

Traeger, S., Altegoer, F., Freitag, M., Gabaldon, T., Kempken, F., Kumar, A., et al. (2013). The genome and development-dependent transcriptomes of *Pyronema confluens*: A window into fungal evolution. *PLoS Genetics*, *9*, e1003820.

Trapnell, C., Pachter, L., & Salzberg, S. L. (2009). TopHat: Discovering splice junctions with RNA-Seq. *Bioinformatics*, *25*, 1105–1111.

Trapnell, C., Williams, B. A., Pertea, G., Mortazavi, A., Kwan, G., van Baren, M. J., et al. (2010). Transcript assembly and quantification by RNA-Seq reveals unannotated transcripts and isoform switching during cell differentiation. *Nature Biotechnology*, *28*, 511–515.

Tringe, S. G., & Rubin, E. M. (2005). Metagenomics: DNA sequencing of environmental samples. *Nature Reviews. Genetics*, *6*(11), 805–814.

Tsang, A., Butler, G., Powlowski, J., Panisko, E. A., & Baker, S. E. (2009). Analytical and computational approaches to define the *Aspergillus niger* secretome. *Fungal Genetics and Biology*, *46*(Suppl. 1), S153–S160.

Tuskan, G. A., Difazio, S., Jansson, S., Bohlmann, J., Grigoriev, I., Hellsten, U., et al. (2006). The genome of black cottonwood, *Populus trichocarpa* (Torr. & Gray). *Science*, *313*, 1596–1604.

Vezzi, F., Narzisi, G., & Mishra, B. (2012). Feature-by-feature—Evaluating de novo sequence assembly. *PLoS One*, *7*, e31002.

Vialas, V., Sun, Z., Loureiro, Y., Penha, C. V., Carrascal, M., Abián, J., et al. (2013). A *Candida albicans* PeptideAtlas. *Journal of Proteomics*, *97*, 62–68.

Vincent, D., Kohler, A., Claverol, S., Solier, E., Joets, J., Gibon, J., et al. (2012). Secretome of the free-living mycelium from the ectomycorrhizal basidiomycete *Laccaria bicolor*. *Journal of Proteome Research*, *11*, 157–171.

Wheeler, D. L., Barrett, T., Benson, D. A., Bryant, S. H., Canese, K., Church, D. M., et al. (2005). Database resources of the National Center for Biotechnology Information. *Nucleic Acids Research*, *33*, D39–D45.

Wicker, T., Oberhaensli, S., Parlange, F., Buchmann, J. P., Shatalina, M., Roffler, S., et al. (2013). The wheat powdery mildew genome shows the unique evolution of an obligate biotroph. *Nature Genetics*, *45*, 1092–1096.

Winzeler, E. A., Shoemaker, D. D., Astromoff, A., Liang, H., Anderson, K., Andre, B., et al. (1999). Functional characterization of the *S. cerevisiae* genome by gene deletion and parallel analysis. *Science*, *285*, 901–906.

Wohlbach, D. J., Kuo, A., Sato, T. K., Potts, K. M., Salamov, A. A., Labutti, K. M., et al. (2011). Comparative genomics of xylose-fermenting fungi for enhanced biofuel production. *Proceedings of the National Academy of Sciences of the United States of America*, *108*, 13212–13217.

Wood, V., Gwilliam, R., Rajandream, M. A., Lyne, M., Lyne, R., Stewart, A., et al. (2002). The genome sequence of *Schizosaccharomyces pombe*. *Nature*, *415*, 871–880.

Wright, J. C., Sugden, D., Francis-McIntyre, S., Riba-Garcia, I., Gaskell, S. J., Grigoriev, I. V., et al. (2009). Exploiting proteomic data for genome annotation and gene model validation in *Aspergillus niger*. *BMC Genomics*, *10*, 61.

Wu, T. D., & Watanabe, C. K. (2005). GMAP: A genomic mapping and alignment program for mRNA and EST sequences. *Bioinformatics*, *21*, 1859–1875.

Yadav, S. P. (2007). The wholeness in suffix -omics, -omes, and the word om. *Journal of Biomolecular Techniques*, *18*, 277.

Zemach, A., McDaniel, I. E., Silva, P., & Zilberman, D. (2010). Genome-wide evolutionary analysis of eukaryotic DNA methylation. *Science*, *328*, 916–919.

Zerbino, D., & Birney, E. (2007). Velvet: De novo assembly using very short reads. *Journal of Virology*, *44*, 494612.

Zerbino, D. R., McEwen, G. K., Margulies, E. H., & Birney, E. (2009). Pebble and rock band: Heuristic resolution of repeats and scaffolding in the velvet short-read de novo assembler. *PLoS One*, *4*, e8407.

Zhou, K., Panisko, E. A., Magnuson, J. K., Baker, S. E., & Grigoriev, I. V. (2009). Proteomics for validation of automated gene model predictions. *Methods in Molecular Biology*, *492*, 447–452.

CHAPTER TWO

Exploring the Transcriptome of Mycorrhizal Interactions

Annegret Kohler[1], Emilie Tisserant
Institut National de la Recherche Agronomique (INRA), UMR 1136 INRA/Lorraine University, Interactions Arbres/Micro-organismes, Centre de Nancy, Champenoux, France
[1]Corresponding author: e-mail address: kohler@nancy.inra.fr

Contents

Abstract

The transcriptome is defined as the sum of all transcripts in a given cell, tissue or organ and can be assessed by different approaches. The most recent ones, microarrays and RNA sequencing, are compared in this chapter with a focus on plant–microbe interactions. The advantages and the specific problems to monitor both partners at the same time by sequencing are discussed. We show that microdissection in combination with transcriptomics is a powerful tool to study interactions. Further, we present an overview of bioinformatic approaches and available algorithms to process RNA-Seq data. Finally, we summarize the transcriptome analyses from mycorrhizal fungi in the genome era, in particular *Laccaria bicolor*, *Tuber melanosporum*, *Rhizophagus irregularis* and *Paxillus involutus*.

Advances in Botanical Research, Volume 70
ISSN 0065-2296
http://dx.doi.org/10.1016/B978-0-12-397940-7.00002-1

1. INTRODUCTION

The transcriptome is defined as the sum of all RNA molecules (rRNA, tRNA, *mRNA* and small and other noncoding RNAs) in a cell or a cell population, tissue or organ. But often, transcriptome studies are restricted to the analysis of *messenger RNA*, especially in eukaryotic organisms like fungi or plants, where poly-A RNA enrichment techniques are used prior to gene expression analyses. The term was first used for the publication of *Characterization of the Yeast Transcriptome* in 1997 (McGettigan, 2013; Velculescu et al., 1997). The measurement of transcript concentrations has become possible in the 1970s with the discovery of reverse transcriptase (Baltimore, Huang, & Stampfer, 1970), an enzyme that converts RNA into complementary DNA (*cDNA*). The techniques used were Northern blot, RT-PCR and more recently quantitative PCR (qPCR). These methods were time-consuming and the number of genes to study was very restricted. Large-scale Sanger *expressed sequence tag* (EST) sequencing projects a decade ago made it possible to study a higher number of transcripts simultaneously by using *macroarrays*. But only with the sequencing of complete genomes it became finally possible to study whole transcriptomes. To do so, sequence information from gene models was used to define specific probes for each gene, which were then spotted or directly synthesized on oligoarray supports. Hybridization of samples allowed the monitoring of all transcript concentrations at the same time and transcriptomics was born (Fig. 2.1A). The development of next-generation sequencing (NGS) methods changed again rapidly the possibilities to study gene expression: first by the increasing number of available genomes that allowed to develop whole genome expression oligoarrays for many organisms, including fungi, and second by the possibility to use NGS directly to sequence transcriptomes. The high-throughput cDNA sequencing (RNA-Seq) has rapidly emerged as the major quantitative transcriptome profiling platform (Wang, Gerstein, & Snyder, 2009, Fig. 2.1B). In addition, it can directly be used to improve genome annotation (Denoeud et al., 2008). As for genome sequencing, several NGS techniques are available, including SOLID, Roche/454, Illumina or Ion Torrent, but all are still developing and improving and new technologies are still evolving with a tremendous pace. Here, we will summarize the current state of the art of fungal transcriptome analyses, and we will discuss the particular problems of plant–microbe interaction studies. Finally, we will give some examples of whole transcriptome studies of mycorrhizal fungi.

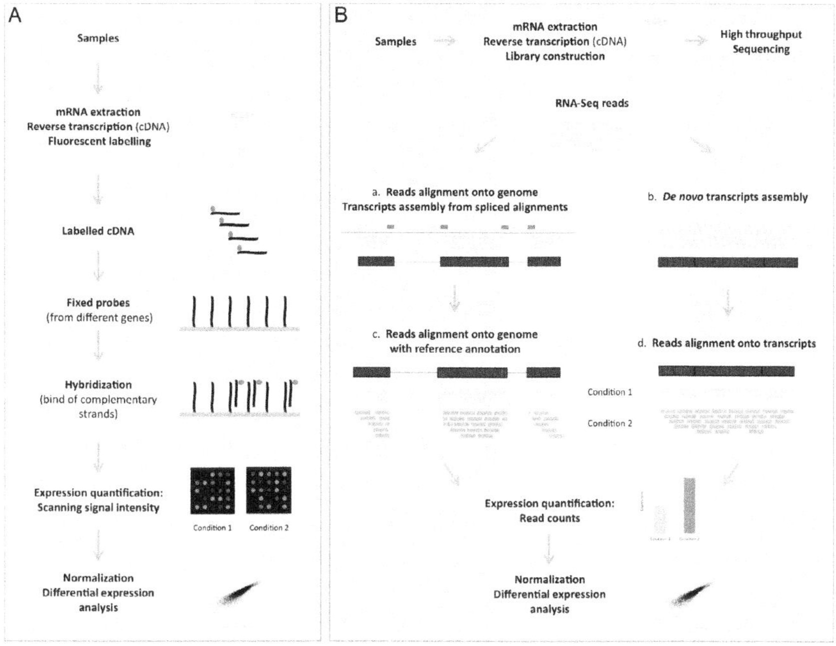

Figure 2.1 *Overview of transcriptomic approaches for exploring mycorrhizal interactions.* (A) *Hybridization-based approach.* Specific DNA sequences corresponding to a short part of a gene (probes) are synthesized on the array. These probes are used to hybridize a fluorescent-labelled cDNA (sample). Fluorescent-labelled cDNA sequences that bind to a probe generate a signal. The scanner returns signal intensities for each probe on the array, corresponding to gene expression level. (B) *RNA-Seq approach.* Based on direct sequencing of transcriptome. With this approach, it is possible to reconstruct transcriptome either (a) by aligning sequenced reads to a pre-existing genome sequence and then by reconstructing transcripts from resulting spliced alignment or (b) by directly *de novo* assembling sequenced reads into transcripts. This approach also allows quantification of gene expression (c) by mapping sequencing reads onto reference genome with reference annotation or (d) by directly mapping sequencing reads onto transcript sequences. Expression levels are then quantified by counting the number of reads mapping to a given region of a genome or a reference transcript assembly. (See the colour plate.)

2. MICROARRAYS VERSUS DEEP SEQUENCING

The development of hybridization-based approaches enabled for the first time to quantitatively survey the global transcriptome of an organism. In the last decade, whole genome expression analyses were often the starting point of major findings in biology. Transcriptome analyses were often used prior to other "omics" technologies such as proteomics or metabolomics (McGettigan, 2013).

Custom-made or commercial high-density oligoarrays are relatively high-throughput and inexpensive (about four times less expensive than sequencing a fungal sample at the moment), but they have certain limitations: first, the reliance on existing knowledge, for example, a genome reference, and, second, the need to decide a priori about the sequences to synthesize on the array. Detection of new gene models is not possible with this approach and changes in the annotation of a genome necessitate a new microarray design. Since the method is based on hybridization of fluorescently labelled samples, *cross hybridization* and background fluorescence are big issues. Further, the detection range is quite limited due to fluorescence saturation of spots. Some of the limitations can be overcome by the construction of so-called tiling microarrays (Bertone et al., 2004), where big portions or the complete genome is present on the array irrespective of any annotation, but depending on the genome size, several tiling arrays can be necessary per sample and make this type of approach quite expensive. The recent development of massively parallel sequencing methods and the drop-down of the cost per sample and of the time required for sequencing have again completely transformed the way in which quantitative transcriptomics can be done. Compared to microarrays, RNA-Seq provides increased sensitivity, *dynamic range* and better discrimination of similar transcripts (Wang et al., 2009; Westermann, Gorski, & Vogel, 2012). New genes and splice variants are easily detectable and make the full complexity of the transcriptome accessible. Transcriptome analyses are possible either by alignment to pre-existing reference genomes or by *de novo* assembly of transcripts, so no pre-existing knowledge is required anymore. Whole transcriptome analyses became for the first time possible for non-model organisms.

With hybridization-based methods, the scanner returns signal intensities for each probe on the array. In the case of RNA-Seq data, the number of *reads* mapping to any given region of a genome or a reference transcript assembly delivers a countable, almost digital signal (Wang et al., 2009; Westermann et al., 2012). In contrast to microarrays, quality control of the reads after sequencing is possible and low-quality or contaminating reads can be easily discarded from the final read set. So the generation of reliable RNA-Seq data depends very much on the proper mapping of sequencing reads to the reference. Therefore, big effort has been made in the last years to develop algorithms that allow solid mapping and quantification of millions of short reads in a relatively short time and without huge computing resources. RNA-Seq has many advantages compared to microarrays but is of course not perfect. Similar to cross-hybridization problems with microarrays, multiple or misalignments of reads to closely related genes are an issue and better quantification software are needed to reduce this source of errors. Another problem generated by RNA-Seq due to the big amount of data produced is the necessity of a good data management and sufficient storage capacity.

3. DUAL TRANSCRIPTOMICS OF PLANT–FUNGUS INTERACTION: MONITORING BOTH PARTNERS AT THE SAME TIME

The idea to monitor both partners of pathogenic or mutualistic interactions at the same time is not new. Suppression subtractive hybridization (SSH) followed by Sanger sequencing was successfully used to identify highly regulated transcripts during interactions. In mutualistic interactions, this approach was, for example, used to study the *Eucalyptus–Pisolithus* ectomycorrhizal development (Voiblet, Duplessis, Encelot, & Martin, 2001) or the *Medicago truncatula–Glomus mosseae* interaction (Brechenbacher et al., 2004). That SSH in combination with sequenced genomes is still a powerful tool for transcriptome studies has recently been shown by Vargas et al. (2012) for the *Colletotrichum*–maize interaction. Macroarrays with spotted cDNAs from mycorrhizal interactions were used for birch–*Paxillus* (Johansson et al., 2004), *Eucalyptus–Pisolithus* (Duplessis, Courty, Tagu, & Martin, 2005), or European chestnut–*Pisolithus tinctorius* (Sebastianaa et al., 2009), but these analyses were restricted to a relatively low number of genes (about 2000 cDNAs).

In principle, microarray technology also allows monitoring of gene expression in both the fungus and the host during their interaction by developing either two separate microarrays for each organism or a shared array. A shared array was, for example, used to study the maize–*Ustilago maydis* interaction and allowed the identification of organ-specific genes in host and pathogen (Skibbe, Doehlemann, Fernandes, & Walbot, 2010). A two-array strategy was used to study the interaction transcriptome of *Magnaporthe oryzae* and rice (Mosquera, Giraldo, Khang, Coughlan, & Valent, 2009) and to identify biotrophy-associated proteins. But shared arrays are technically difficult to design due to probe selection and cross-hybridization issues, and the use of one array for each partner is time-consuming and cost-intensive since each sample has to be prepared two times and the double amount of arrays has to be conducted.

Therefore, another huge advantage of sequencing compared to microarrays for plant–microbe interaction studies is the possibility to monitor the transcriptome of both partners in the same sample. Interaction studies can even be enlarged to multiple partners or be conducted without any knowledge of the microorganisms present in a system, as in the case of *metatranscriptomics* (Bailly et al., 2007).

Matsumura et al., 2003, sequenced tags >25 bp from defined positions of cDNAs by using type III restriction enzyme EcoP15I, so-called SuperSAGE, to study rice and *Magnaporthe grisea* simultaneously in infected leaves. Furthermore, they monitored gene expression in *Phytophthora* elicitor-treated *Nicotiana benthamiana* by the same method. More recently, Illumina sequencing was used for simultaneous studies of cotton–*Verticillium dahliae* (Xu et al., 2011), *M. grisea*–rice (Kawahara et al., 2012), potato tuber–*Phytophthora infestans* (Gao, Tu, Millett, & Bradeen, 2013), poplar–*Marssonina brunnea* (Zhu et al., 2013) and sorghum–*Bipolaris sorghicola* (Yazawa, Kawahigashi, Matsumoto, & Mizuno, 2013). In the case of *B. sorghicola*, the dual RNA-Seq approach was successfully combined with transcriptome *de novo* assembly. Next-generation transcriptome sequencing was first used by Larsen et al. (2011) for mycorrhizal interactions, in order to predict the ectomycorrhizal metabolome of poplar and *Laccaria bicolor.*

A problem of RNA-Seq and in particular of dual RNA-Seq can be the sequence coverage needed for saturated transcriptomes of both partners.

Often, the genome size and, related to it, the transcriptome sizes of the two partners are very different (plant genome > fungal genome). And often, the transcriptome of one partner represents only a few percentage of the RNA pool, as, for example, in the case of the early contact of pathogenic or *ectomycorrhizal fungi* with their host plants or in the case of endomycorrhizal roots, where the fungi colonize only about 7% of the root (Abba, Khouja, Martino, Archer, & Perotto, 2009). Production of enough reads for accurate quantification of low expressed transcripts can still be quite expensive for these systems as well as for large transcriptomes.

In contrast to genome sequencing, it is difficult to predict *a priori* the number of sequences needed to detect all transcripts, since the transcript concentrations will vary between tissues and treatments (Wang et al., 2009). It was estimated that more than 200 million *paired-end* reads are required to detect the full human transcriptome (Tarazona, Garcia-Alcalde, Dopazo, Ferrer, & Conesa, 2011). About 80% of the *Escherichia coli* transcriptome will be covered by sequencing of two million reads (Haas, Chin, Nusbaum, Birren, & Livny, 2012), and 80% of the yeast transcriptome will be detected (with more than four reads mapping to a gene) with four million reads (Nagalakshmi et al., 2008; Sims, Sudbery, Ilott, Heger, &

Ponting, 2014; Wang et al., 2009). For *Arabidopsis thaliana* (135 Mb, 27,000 genes), a sequencing depth of 50 million reads yields near-saturated coverage of expressed genes (Van Verk, Hickman, Pieterse, & Van Wees, 2013); for *Populus* (423 Mb, 41,335 gene models), about 50 million reads are necessary to detect 75% of the gene models (with more than four reads mapping to a gene) and about 120 million to cover 80% (unpublished data). Figure 2.2A shows an example of a filamentous fungus with 16,703 predicted genes and a genome size of 46 Mb. Sequencing of about 10 million reads was necessary to align more than four reads to 80% of the gene models (Fig. 2.2A). By sequencing 25 million reads, 88% of the genes were detected in a fungus-only sample but only 62% with a mixed fungus–plant sample that contained

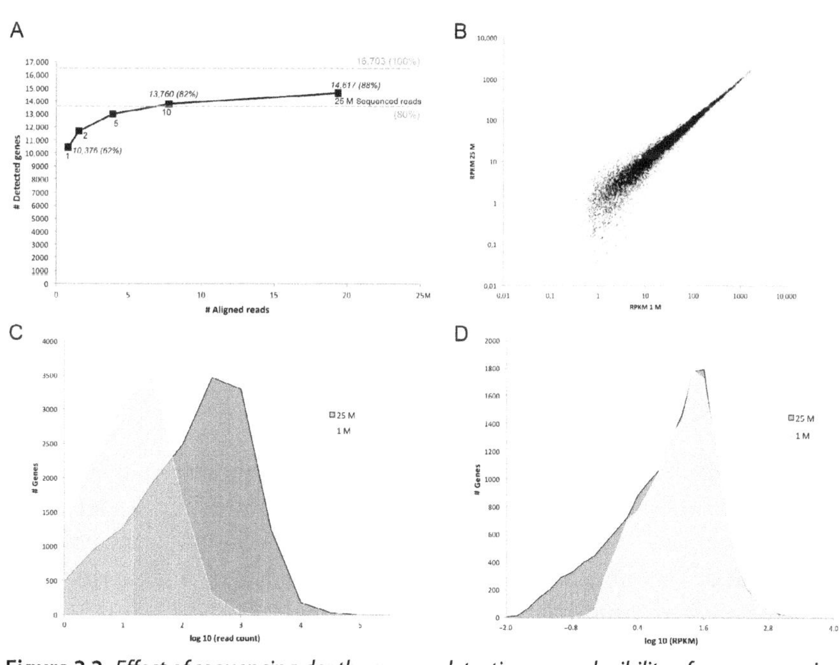

Figure 2.2 *Effect of sequencing depth on gene detection, reproducibility of gene expression values and normalization.* (A) Saturation curve for gene detection: The number of gene models detected with 1, 2, 5, 10 and 25 M sequenced reads. The number of reads aligned to the reference genome is shown for each point (about 78% of the sequenced reads). The total gene number was 16,703. (B) Comparison of RPKM values obtained with 1 M reads and 25 M reads. (C) The number of reads (log 10) detected with 1 M (mixed fungal–plant sample) reads compared to 25 M reads (fungus-only sample). (D) Distribution of expression values after normalization using RPKM method with 1 M (mixed fungal–plant sample) or 25 M reads (fungus-only sample). RPKM: reads per kilobase of transcript per million mapped reads.

about 4% fungal RNA, corresponding to about one million sequences (Fig. 2.2A). Several tools were developed to evaluate the stability of transcript detection in a given data set (Tarazona et al., 2011; Wang, Wang, & Li, 2012). Figure 2.2B shows that the normalized expression values obtained with one and 25 million reads are very variable for low expressed genes (RPKM about one). With only a few reads, it is impossible to distinguish background from expression, while with higher read coverage, the expression values become significant. But depending on the question, it is also not necessary to sequence until the lowest transcripts will be robustly detected.

Petre et al. (2012) used 454 sequencing to study early-infected poplar leaves by the rust fungus *Melampsora larici-populina* and to identify a fungal-induced sulphate transporter, but they show at the same time that both the poplar transcriptome, due to its relative big size, and the fungal transcriptome, due to the small percentage of fungal transcripts in the RNA pool, are far from being saturated. From 841,301 454 reads, 616,879 were mapped to the *Populus trichocarpa* genome and only 649 sequences to the *M. larici-populina* genome sequence (Petre et al., 2012).

The relatively high sequencing effort needed to investigate early stages of plant–fungus interactions has therefore made Illumina-HiSeq (about 180 million 100 bp reads per lane at the moment) the method of choice compared to other NGS methods, for example, Roche/454 that produces relatively low number of reads per run (about 1 million reads up to 1 kb at the moment) and higher costs per sequence.

Another problem of interaction transcriptome studies can become the correct normalization of the data, since the sequence number in infected tissues is often a multiple less than in control tissues and sometimes even impossible to normalize without artefacts. Figure 2.1C and D shows the effect of normalization on our example data set. Low expressed genes are at risk to be overestimated after normalization. In our example, RPKM normalization (Mortazavi, Williams, McCue, Schaeffer, & Wold, 2008) was used; other normalization methods exist and are discussed in Section 5, but none of these methods are capable to adjust data sets with large differences in sequence number without bias.

Further, the proper mapping of the mixed sequencing reads to the respective reference genomes is essential for correct transcript quantification. Usually, plant and fungal genomes are different enough to simply align sequences to the respective reference genomes without prefiltering, but cross alignment should always be checked. Even with only one reference genome available, the separation is possible, as shown for the *sorghum–Bipolaris sorghicola* interaction (Yazawa et al., 2013). More challenging and similar to metatranscriptome analyses with their multiple partners is the discrimination of fungal and plant transcripts if both transcriptomes have to be *de novo* assembled (see Section 5 and Fig. 2.1B(b)).

4. COMBINATION OF TRANSCRIPTOMICS AND MICRODISSECTION

Another powerful tool in plant–microbe interaction studies is the combination of *laser microdissection* (LM) with transcriptome analyses. LM can be used to isolate specific structures, tissues or even single cells from cuttings fixed on microscope slides (Gomez & Harrison, 2009). Special protocols had to be developed to find the right balance between fixing structures and keeping RNA intact and extractable (Nakazono, Qiu, Borsuk, & Schnable, 2003). In addition, the small amounts of RNA make RNA amplification necessary, often two rounds of T7-based antisense RNA (aRNA) amplification are used (Nakazono et al., 2003).

LM has been widely used not only to isolate cells from plant tissues (Gomez & Harrison, 2009) but also to monitor gene expression of pathogenic fungi *in planta* (e.g. Chandran, Inada, Hather, Kleindt, & Wildermuth, 2010; Fosu-Nyarko, Jones, & Wang, 2010; Hacquard et al., 2010; Tang, Coughlan, Crane, Beatty, & Duvick, 2006; Tremblay, Li, Sheffler, & Matthews, 2009). LM was first used in mycorrhizal studies by Balestrini, Gómez-Ariza, Lanfranco, and Bonfante (2007) to isolate cortical cells from *Glomus mosseae*-infected tomato roots. Several plant and fungal phosphate transporter transcripts associated to arbuscular structures could be identified by RT-PCR. Gomez et al. (2009) used a *Medicago* gene chip to monitor gene expression of the *arbuscular mycorrhizal* (AM) symbiosis. The combination of LM and *Tuber melanosporum* whole genome array revealed functional compartmentalization in ectomycorrhizal root tips (Hacquard et al., 2013). In particular, nitrogen and water acquisition, synthesis of secondary metabolites and detoxification mechanisms seem to be important processes in the fungal *mantle*, while transport activity was enhanced in the *Hartig net* (Hacquard et al., 2013).

So far, in mycorrhizal studies, LM has only been combined with the whole transcriptome analysis of one partner by using microarrays, but studies of the apical meristem or gametophytes of *Arabidopsis thaliana* have shown that the combination of LM with RNA-Seq is possible *in planta*. In fungi,

LM and RNA-Seq were recently applied by Teichert, Wolff, Kuck, and Nowrousian (2012) to determine gene expression pattern in developing fruiting bodies of *Sordaria macrospora*.

5. BIOINFORMATIC TOOLS FOR RNA-SEQ DATA ANALYSIS

In the following, we present an overview of bioinformatic approaches and available algorithms to process RNA-Seq data. Software development for this relative recent method is still ongoing; a comparison of free software with advantages and disadvantages was conducted by Steijger et al. (2013).

5.1. Reads mapping onto reference sequences

Alignment of reads onto reference sequences is a basic step of RNA-Seq processing, which allows further analysis, such as transcript reconstruction or expression quantification. This task, however, is a computational challenge due to the large amount of data (up to several hundreds of millions) and short read length (~36–125 bp) generated by NGS (Trapnell & Salzberg, 2009).

Several alignment programs have been developed to address this issue. They are mainly based on two different approaches: *Hash-based methods*, such as MAQ (Li, Ruan, & Durbin, 2008), and *Burrows–Wheeler transform* methods, such as BWA (Li & Durbin, 2009) and Bowtie (Langmead, Trapnell, Pop, & Salzberg, 2009). Globally, Burrows–Wheeler transform methods are significantly faster and reduce memory usage, while Hash-based approaches might achieve better sensitivity and accuracy, especially for highly polymorphic genomes.

However, these programs, called "unspliced aligners", do not allow alignments with large gaps and thus don't support mapping of *spliced reads* onto reference genome. Several methods, such as TopHat (Trapnell, Pachter, & Salzberg, 2009), MapSplice (Wang et al., 2010) and SpliceMap (Au, Jiang, Lin, Xing, & Wong, 2010) have been developed to align spliced reads onto reference genome. They first align reads to reference using an unspliced aligner. Unmapped reads are then split in shorter segments and mapped. Spliced sites are finally searched in the neighbouring genomic regions. Generally, unspliced aligners can be used for mapping reads against reference transcripts, whereas spliced aligners are suitable for alignment of reads onto reference genome.

Some genomes can contain repetitive sequences and large multigene families. In this case, some reads can be mapped to multiple locations. Removing these alignments or treating them in a random way can skew subsequent analysis. Some approaches have been proposed to address this problem, such as allocating them in proportion to the number of uniquely mapped reads (Mortazavi et al., 2008).

5.2. Transcriptome reconstruction

Transcriptome reconstruction is an important application of RNA-Seq, which makes transcript characterization, genome annotation, novel gene detection and alternative splicing discovery possible. Depending on whether the reference genome sequence is available or not, different approaches can be used to perform this task. Genome-guided approach allows reconstructing transcripts and isoforms according to the reads mapping result on the reference genome (Fig. 2.1B(a)). Especially, programs such as Cufflinks (Trapnell et al., 2010) and Scripture (Guttman et al., 2010) use spliced reads mapping to reconstruct gene structure across exons and generate genome annotation.

Another approach is *de novo* reconstruction of transcriptomes (*de novo assembly*), which uses redundancy of sequencing reads to directly assemble them into transcripts without using a reference genome (Fig. 2.1B(b)). Commonly, reads are broken into subsequences (called k-mers) and the overlaps between these k-mers are used to build *de Bruijn graphs* (Martin & Wang, 2011). Graphs are then parsed to reconstruct transcript sequences. Several programs, including Trans-ABySS (Birol et al., 2009), Trinity (Grabherr et al., 2011) and Oases (Zerbino & Birney, 2008), have been specifically suited for *de novo* transcriptome assembly. One major concern for *de novo* transcriptome reconstruction is to discriminate sequencing errors from biological variation. For genome assembly, the use of coverage information can help to solve the error problem. However, due to the large *dynamic range* of transcriptomes, this distinction is more difficult, particularly for low-abundance transcripts (Martin & Wang, 2011). Thereby, *de novo* transcriptome reconstruction is very sensitive to sequencing errors, and it is strongly advised to clean the reads before assembly. There are several software, such as FastQC (http://www.bioinformatics.babraham.ac.uk/projects/fastqc/) and FASTX-Toolkit (http://hannonlab.cshl.edu/fastx_toolkit/), that can be used for quality control and read trimming and filtering.

Globally, *de novo* approach needs more significant computing resources than genome-guided methods and requires a much higher *sequencing depth* to achieve full-length transcript or to detect transcripts expressed at low levels (Martin & Wang, 2011). However, algorithm development tends to reduce this discrepancy. As an example, Trinity (*de novo* approach) and Cufflinks (genome-guided approach) reconstruct 86% and 77%, respectively, of annotated transcripts at full-length using 50 million of paired-end reads in fission yeast (Grabherr et al., 2011). More importantly, for species that have no available reference genomes, *de novo* transcriptome reconstruction is the only way to obtain a set of transcripts. These transcripts can be used as reference sequence for RNA-Seq expression analysis and can be compared to protein database to infer functional annotation.

5.3. Gene expression quantification

A major application of RNA-Seq is the expression quantification. For RNA-Seq, the gene expression level is quantified by counting the number of reads mapped on genes or transcripts. This can be done either by using mapping result against reference genomes with a reference annotation (Fig. 2.1B(c)) or by directly mapping reads onto transcripts (Fig. 2.1B(d)). However, some genes may be alternatively spliced and may produce several isoforms during their expression, complicating the expression quantification process. Some more sophisticated tools, such as Cufflinks (Trapnell et al., 2010) and MISO (Katz, Wang, Airoldi, & Burge, 2010), are designed to estimate the expression at the isoform level by implementing a statistical model.

As the number of reads generated by a transcript is proportional to its length and to sequencing depth of sample, read counts need to be normalized to ensure that expression levels can be compared between or within samples. A common method to solve this issue is the use of the *RPKM* (reads per kilobase of transcript per million mapped reads) normalization (Mortazavi et al., 2008). However, it has been shown that this method does not completely eliminate bias (Dillies et al., 2013). Moreover, the presence of a few, highly expressed genes can significantly lower the counts for all other genes in a sample, skewing comparison to another sample having different read counts distribution. Other normalization strategies have been proposed, which attempt to correct difference in RNA composition between samples, including trimmed mean of M values (TMM) (Robinson & Oshlack, 2010) and the normalization provided in the DESeq package (Anders & Huber, 2010). Both methods are based on the assumption that the majority of genes are not differentially expressed and estimate a scaling factor from mean or median of read counts ratio across samples.

Several methods have been specifically developed to test the statistical significance of differentially expressed genes between different biological conditions for this type of count data. Most of them, such as EdgeR (Robinson, McCarthy, & Smyth, 2010), DESeq (Anders & Huber, 2010) and Cuffdiff (Trapnell et al., 2010), attempt to model biological variability by using parametric approaches based on probability distributions, such as the negative binomial distribution. Some nonparametric methods, having no assumptions about the data distribution, have been also proposed, including NOISeq (Tarazona et al., 2011) and SAMSeq (Li & Tibshirani, 2013). However, the field of differential expression analysis for RNA-Seq data is still emerging, and there is no clear consensus regarding the most effective statistical methods (Soneson & Delorenzi, 2013). Nevertheless, it clearly appears that a larger number of biological replicates greatly improve the power of differential expression detection (Robles et al., 2012). Recently, the availability of *multiplex experimental designs* allows increasing the numbers of replicates without increasing the cost of sequencing.

6. MYCORRHIZAL TRANSCRIPTOMES IN THE GENOME ERA

In 2008, the mycorrhizal genome era started with the publication of the *L. bicolor* genome draft (Martin et al., 2008; http://genome.jgi-psf.org/Lacbi2/Lacbi2.home.html), an ectomycorrhizal basidiomycete. The *L. bicolor* genome sequencing was part of a bigger project aiming to sequence fungi interacting with the model tree *Populus*, including also the AM fungus *Rhizophagus irregularis* (Tisserant et al., 2013; http://genome.jgi.doe.gov/Gloin1/Gloin1.home.html). In 2010, the genome of the first ectomycorrhiza-forming ascomycete, *T. melanosporum*, was published and allowed first insights in the evolution of ectomycorrhizal fungi (Martin et al., 2010; http://genome.jgi.doe.gov/Tubme1/Tubme1.home.html). To deepen the understanding of the origin of mycorrhizal symbiosis and to explore the molecular mechanisms behind it, a large-scale sequencing program was started by the Mycorrhizal Genomics Initiative (MGI) and the Joint Genome Institute (JGI), tending to compare more than 30 mycorrhizal genomes, including endomycorrhizal fungi-forming symbioses with orchids and *Ericaceae*, and originated from different phylogenetic clades. In the following, we will summarize the outcome of transcriptome analyses of these fungi.

6.1. Transcriptome of *L. bicolor*, a basidiomycete ectomycorrhizal fungus

L. bicolor is a member of the Hydnangiaceae, a large order of ectomycorrhizal and saprobic basidiomycetes. As mentioned earlier, *Laccaria* was chosen for sequencing not only due to its ability to form mycorrhiza with *Populus* species but also because it is a cosmopolitan and forms ectomycorrhiza with numerous trees. The ecology and physiology of *L. bicolor* were already well studied before the genome became available (Martin & Kohler, 2013). One reason was its rapid growth in pure culture and its mycorrhization under laboratory conditions. In addition, *Laccaria* is used in commercial inoculation programs in forest nurseries worldwide to enhance growth of tree seedlings (Villeneuve, LeTacon, & Bouchard, 1991). Several early symbiosis-related transcripts had been identified in a *L. bicolor*–Pinus *in vitro* system by using differential display technique (DDRT-PCR), including a secreted protein PF6.2 (Kim, Zheng, Hiremath, & Podila, 1998), an autophagocytosis protein Aut7 (Kim, Bernreuther, Thumm, & Podila, 1999) and a malate synthase (Balasubramanian, Kim, & Podila, 2002). *L. bicolor* cDNA resources were produced using Sanger EST sequencing (Peter et al., 2003) and then used for the generation of a macroarray. This macroarray was used to monitor for the first time gene expression during interaction of *L. bicolor* with mycorrhiza helper bacteria (Deveau et al., 2007). But only with the genome draft, it became possible to study the complete transcriptome of *L. bicolor* in free-living mycelium, fruiting bodies and *ectomycorrhizal root tips* (Martin et al., 2008). The expression of nearly 80% of the predicted genes was detected in one of these tissues by using a NimbleGen whole genome oligoarray (Martin et al., 2008). The most striking finding in the transcriptome of *L. bicolor* was the upregulation of the expression of several *small secreted proteins* (SSPs) in ectomycorrhizal root tips (Martin et al., 2008). Five of the 20 most highly upregulated fungal transcripts coded for these SSPs. For one of these mycorrhiza-induced SSPs (MiSSPs), MiSSP7, Plett et al. (2011) showed recently that it acts, similar to what have been shown for pathogens, as an effector. It enters into the plant cell and into the nucleus to interfere with the host's metabolism.

In the *L. bicolor* genome, a large number of predicted transporter genes were found (Martin et al., 2008), for instance, multiple genes coding for ammonium transporter. The expression of one of these ammonium transporters, AMT2.2, is highly upregulated in ectomycorrhiza, showing an increased nitrogen uptake potential in *L. bicolor* that is probably related to its symbiotic lifestyle (Lucic et al., 2008; Martin et al., 2008). Further, transcript levels of proteins involved in cell wall remodelling (e.g. CAZymes) were highly increased in differentiating tissues of *L. bicolor* as ectomycorrhizal root tips or fruiting bodies.

Larsen et al. (2010) used RNA-Seq to improve the annotation of the *Laccaria* genome. In addition, RNA-Seq data from fully formed ectomycorrhiza between *L. bicolor* and poplar roots were used to predict the ectomycorrhizal metabolome (Larsen et al., 2011). Larsen et al. generated a mycorrhizal metabolome model that predicts, for example, the synthesis of glycine, glutamate and allantoin by *L. bicolor*. These compounds or their metabolites may be then used for the benefit of aspen in exchange for the photosynthetically derived sugars fructose and glucose. More RNA-Seq analyses of the *Laccaria*–poplar interaction are ongoing, and the possibility of monitoring gene expression in both partners at the same time will certainly help to understand the fine-tuning of this model symbiosis.

6.2. Transcriptome of *T. melanosporum*, an ascomycete ectomycorrhizal fungus

The genome of *T. melanosporum*, an economically important species due to its tasty and relatively rare fruiting body, the so-called black truffle, was published in 2010 and allowed the comparison of two ectomycorrhizal genomes (Martin et al., 2010). Transcriptome analyses from free-living mycelium, truffles and ectomycorrhizal root tips were conducted by using a NimbleGen whole genome oligoarray (Martin et al., 2010) and RNA-Seq (Tisserant et al., 2011). Both ectomycorrhiza development and truffle development induce important alterations in gene expression, affecting major cellular processes, such as mating (Rubini et al., 2011), transcription (Montanini et al., 2011), metabolic pathways (Ceccaroli et al., 2011) and cell wall remodelling (Balestrini et al., 2012; Sillo et al., 2013). Gene co-expression network analysis revealed tissue-specific clusters (Hacquard et al., 2013). ECM clusters were enriched in genes coding for transport and carbohydrate metabolic processes, while truffle clusters showed enhanced expression of genes coding for sulphur compound metabolism, a pathway probably necessary to generate the typical truffle perfume, as well as in amino acid and lipid metabolism-related transcripts (Hacquard et al., 2013). Inorganic ion transport was found enriched in both ECM and fruiting bodies. An LM approach to separate mantle and Hartig net in the ectomycorrhizal root tip revealed that expression of genes coding for transport activity was enhanced in the Hartig net, while transcripts related to nitrogen and water acquisition, synthesis of secondary metabolites and detoxification mechanisms were detected in high concentrations in the outer mantle (Hacquard et al., 2013).

Surprisingly, mycorrhiza-induced SSPs, shown as a major feature for the *L. bicolor* mycorrhization, could not be identified in *T. melanosporum* (Martin et al., 2010). Together with other differences, as, for example, the upregulation of genes encoding for lipases and multicopper oxidases in *T. melanosporum* ECMs, this led to the hypothesis that the symbiosis toolboxes of ascomycetes and basidiomycetes evolved in different ways and independently (Martin et al., 2010; Plett & Martin, 2011).

6.3. Transcriptome of AM fungi

The AM symbiosis, which involves the basal fungal lineage of Glomeromycota and vast majority of plant species, is one of the oldest and most widespread plant–fungus associations. Transcriptome of the host plant during AM symbiosis has been broadly explored, including by using high-throughput expression profiling in several plant models, such as *Oryza sativa* and *Medicago truncatula* (Gaude, Bortfeld, Duensing, Lohse, & Krajinski, 2012; Gomez et al., 2009; Grunwald et al., 2009; Guether, Balestrini, Hannah, Udvardi, & Bonfante, 2009; Güimil et al., 2005; Hogekamp & Küster, 2013). However, owing to the absence of genome sequence for the Glomeromycota phylum, the regulation of gene expression in AM fungal partner was mainly investigated by targeted approaches. These studies mainly focused on genes involved in nutrient metabolism and transport (Bago et al., 2003; Balestrini et al., 2007; Balestrini & Lanfranco, 2006; Benedetto, Magurno, Bonfante, & Lanfranco, 2005; Kuznetsova et al., 2010; Seddas et al., 2009).

Recently, the development of NGS technologies made the first analysis of transcriptome at genome-wide scale of an AM fungus, *R. irregularis* DAOM 197198, possible (Tisserant et al., 2012). This was achieved by sequencing cDNA libraries from germinated spores, arbuscules, extraradical and intraradical mycelium using Sanger and 454 technologies. The resulting sequences have been *de novo* assembled into a set of 25,906 nonredundant virtual transcripts (NRVTs), providing a comprehensive repertoire of transcripts. This study has allowed characterizing metabolic pathways, thus highlighting the absence of reduction of metabolic complexity in this

AM fungus. Lack of transcripts involved in degradation of plant cell wall polysaccharides and thiamine biosynthetic pathway, which is a feature shared with ectomycorrhizal symbionts, has also been observed. These transcript sequences were also used to construct an oligoarray for investigating transcriptome regulation at different developmental stages (Tisserant et al., 2012). Several hundreds (395) of transcripts induced in *M. truncatula* intraradical mycelium in comparison to germinated spores have thus been identified.

More recently, the first version of *R. irregularis* genome was released (Tisserant et al., 2013), and deep transcriptome sequencing using Illumina technology (RNA-Seq) was used to help gene annotation, confirming and expanding previous results. The RNA-Seq approach was also employed to explore gene expression profiling upon AM symbiosis. Of the 22,647 expressed genes, 1068 (4,7%) were induced in *Medicago*-colonized roots compared to germinating spores. Most of them are lineage-specific genes, but some are involved in known function including signal transduction mechanisms, energy production and conversion, secondary metabolism, and carbohydrate, amino acid, lipid and inorganic ion transport and metabolism. In particular, numerous SSPs were highly induced during AM symbiotic interaction, as it was observed in the ectomycorrhizal fungus *L. bicolor* (Martin et al., 2008).

These works illustrate the strength of deep transcriptome sequencing that allows transcriptome characterization for species without reference genome as well as genome annotation and gene expression profiling. Such approach allows considering many further analyses and could be applied to the other members of Glomeromycota in order to improve knowledge on this important symbiosis.

6.4. Transcriptome of *Paxillus involutus*

Among the finished but so far unpublished genomes is the one of *P. involutus*, one of the most well-studied ectomycorrhizal fungi at molecular, physiological and ecological level.

Global patterns of gene regulation associated with the ectomycorrhizal development between *P. involutus* and birch were identified using cDNA microarrays (Johansson et al., 2004; Le Quéré, Wright, Söderström, Tunlid, & Johansson, 2005). The transcriptome of the *P. involutus*–birch interaction has also been investigated in a soil microcosm and revealed spatial pattern of gene expression in extrametrical mycelium and mycorrhizal root tips (Wright, Johansson, Le Quéré, Söderström, & Tunlid, 2005).

The extrametrical mycelium is supposed to proliferate extensively within the soil and to prospect for nutrient resources as *saprotrophic* fungi. In accordance with this hypothesis, genes implicated in the glutamine synthetase/glutamate synthase and urea cycles as well as for ammonium assimilation were highly expressed in rhizomorph and nutrient patch mycelium. More recently, a NimbleGen oligoarray containing 12,214 transcripts obtained by Roche/454 and Sanger DNA sequencing was used to study the protein and plant litter degradation ability, both major N-sources in soils, of *P. involutus* in more detail (Rineau et al., 2013; Shah, Rineau, Canbäck, Johansson, & Tunlid, 2013). Transcriptional analysis of this proteolytic machinery revealed the implication of a large number of extracellular endo- and exopeptidases. Those expression levels were regulated in parallel with transporters and enzymes involved in the assimilation and metabolism of the released peptides and amino acids (Shah et al., 2013). The analysis of this genome and further whole transcriptome analyses will certainly help to complete our comprehension of the saprotrophic machinery of ectomycorrhizal fungi.

7. OUTLOOK

As mentioned earlier, the sequencing of about 30 mycorrhizal genomes is finished or ongoing. The comparison of these genomes and their transcriptomes will hopefully allow us to filter out essential mycorrhizal features and to elucidate the origin of its evolution. As illustrated, new generation sequencing techniques have revolutionized transcriptome analyses, but their development is still ongoing. Analysis tools have been adapted to this new type of data sets, each with its advantages and bias. Another revolution in transcriptome analyses would therefore certainly be the possibility to sequence transcripts without fragmentation in complete length. This would solve many alignment and assembly problems and would approach transcriptome analyses closer to "reality".

ACKNOWLEDGEMENTS

Our research was financed by the "Institut National de la Recherche Agronomique" (INRA), the Région de Lorraine and the lab of excellence ARBRE (ANR-11-LABX-0002-01). The computations were performed at the INRA Nancy Ecogenomics facilities. Part of our research was sponsored by the Genomic Science Program of the US Department of Energy, Office of Science, Biological and Environmental Research (under contract DE-AC05-00OR22725). We would like to thank Francis Martin for discussions, input and critical review of this chapter.

GLOSSARY

Arbuscular mycorrhiza (AM) is characterized by the formation of unique structures, so-called arbuscules by fungi of the phylum Glomeromycota that penetrate the cortical cells of the roots of vascular plants.

Burrows–Wheeler transform (BWT) an alignment strategy, which allows storing a memory-efficient representation of the reference genome, that is very efficient for searching perfect alignment.

cDNA Complementary DNA is synthesized from mRNA using reverse transcriptase. Starting material for microarrays and RNA-Seq.

Cross hybridization refers to complementary base pairing between two DNA molecules that are not identical in sequence. Cross hybridization usually is observed between nucleic acids derived from orthologous or paralogous genes.

De Bruijn graph a directed mathematical graph that uses a subsequence of length k to represent nodes. Pairs of nodes are connected if shifting a sequence by one character creates an exact $k-1$ overlap between the two sequences.

***De novo* assembly** strategy that leverages the redundancy of read to find overlaps between the reads and assembles them into transcripts.

Dynamic range the range of expression levels over which gene expression can be accurately quantified.

Ectomycorrhizal (ECM) fungi mutualistic fungi, belonging to the Basidiomycota or Ascomycota, whose hyphae surround plant roots to form a mantle and grow within the apoplastic space of the plant root (and form the Hartig net).

Ectomycorrhizal root tip plant lateral roots colonized by ECM fungi. Morphologically, roots are surrounded by the fungal mantle and contain a fully developed Hartig net.

EST expressed sequence tag, a partially sequenced cDNA, usually generated with the Sanger sequencing method.

Hartig net the complex structure of ECM fungal hyphae that reside within the apoplastic space of host roots. Across the Hartig net, the fungus exchanges nutrients with its host tree.

Hash-based methods an alignment strategy that first builds a hash table containing the location of each subsequence of reads (seed) within the reference genome. These algorithms then extend these seeds with more sensitive methods to find the best alignment for each read.

Laser microdissection is a method to isolate specific cells/tissues of interest from a tissue/organism by coupling a laser to a microscope.

Mantle thick, multilayered sheath of fungal hyphae surrounding plant host roots.

Macroarrays usually cDNA probes spotted to a nylon or glass support. Reverse transcribed RNA samples labelled by radioactivity or by fluorescence hybridize to these probes. Radioactivity or fluorescence intensity is measured and converted into expression values.

Metatranscriptomics focus on the characterization of gene expression pattern displayed by microbial communities.

Microarray expression microarrays are oligonucleotide probes (up to 1.5 million) attached to a solid surface. These probes hybridize to fluorescence-labelled cDNAs reverse transcribed from RNA samples. Levels of hybridization are measured by quantifying fluorescence and converted into expression values.

mRNA messenger RNA, an RNA product that is transcribed from DNA, transported to a ribosome, where it is translated into a protein.

Multiplex experimental designs approach that allows the sequencing of multiple samples in a single sequencing run. By ligating short stretches of sequence (tags) to the sample sequence fragments, sequences from individual samples can be extracted after sequencing by analysis of the sequenced tag.

Paired-end refers to the two ends of the same DNA fragment. After sequencing one end, it is turned around and the other end sequenced. The two sequences are called "paired-end reads".

Read a sequence of DNA bases generated by a sequencer.

RPKM (reads per kilobase of transcript per million mapped reads): Normalization method for read count, which attempts to correct for gene length and sequencing depth bias, such as RPKM for gene i = (# of mapped reads for gene i)/((length of gene i)/1000)/((total mapped reads)/10^6).

Saprotrophic organisms that rely on nutrients absorbed from the decomposition of dead organic substrates.

Sequencing depth the average number of reads representing a given nucleotide in the reference sequence.

Small secreted protein (SSP) protein secreted by an organism that is less than 300 amino acids in length.

Spliced reads Reads spanning intron boundaries.

Tiling microarray a DNA microarray that uses a set of overlapping oligonucleotide probes representing a part or a whole genome at high resolution.

REFERENCES

Abba, S., Khouja, H. R., Martino, E., Archer, D. B., & Perotto, S. (2009). SOD1-targeted gene disruption in the ericoid mycorrhizal fungus Oidiodendron maius reduces conidiation and the capacity for mycorrhization. *Molecular Plant-Microbe Interactions*, *22*, 1412–1421.

Anders, S., & Huber, W. (2010). Differential expression analysis for sequence count data. *Genome Biology*, *11*, R106.

Au, K. F., Jiang, H., Lin, L., Xing, Y., & Wong, W. H. (2010). Detection of splice junctions from paired-end RNA-seq data by SpliceMap. *Nucleic Acids Research*, *38*, 4570–4578.

Bago, B., Pfeffer, P. E., Abubaker, J., Jun, J., Allen, J. W., Brouillette, J., et al. (2003). Carbon export from arbuscular mycorrhizal roots involves the translocation of carbohydrate as well as lipid. *Plant Physiology*, *131*, 1496–1507.

Bailly, J., Fraissinet-Tachet, L., Verner, M. C., Debaud, J.-C., Lemaire, M., Wésolowski-Louvel, M., et al. (2007). Soil eukaryotic functional diversity, a metatranscriptomic approach. *The ISME Journal*, *1*, 632–642.

Balasubramanian, S., Kim, S.-J., & Podila, G. K. (2002). Differential expression of a malate synthase gene during the preinfection stage of symbiosis in the ectomycorrhizal fungus *Laccaria bicolor*. *New Phytologist*, *154*, 517–527.

Balestrini, R., Gómez-Ariza, J., Lanfranco, L., & Bonfante, P. (2007). Laser microdissection reveals that transcripts for five plant and one fungal phosphate transporter genes are contemporaneously present in arbusculated cells. *Molecular Plant-Microbe Interactions*, *20*, 1055–1062.

Balestrini, R., & Lanfranco, L. (2006). Fungal and plant gene expression in arbuscular mycorrhizal symbiosis. *Mycorrhiza*, *16*, 509–524.

Balestrini, R., Sillo, F., Kohler, A., Schneider, G., Faccio, A., Tisserant, E., et al. (2012). Genome-wide analysis of cell wall-related genes in *Tuber melanosporum*. *Current Genetics*, *58*, 165–177.

Baltimore, D., Huang, A. S., & Stampfer, M. (1970). Ribonucleic acid synthesis of vesicular stomatitis virus, II. An RNA polymerase in the virion. *PNAS*, *66*(2), 572–576.

Benedetto, A., Magurno, F., Bonfante, P., & Lanfranco, L. (2005). Expression profiles of a phosphate transporter gene (GmosPT) from the endomycorrhizal fungus Glomus mosseae. *Mycorrhiza*, *15*, 620–627.

Bertone, P., Stolc, V., Royce, T. E., Rozowsky, J. S., Urban, A. E., Zhu, X., et al. (2004). Global identification of human transcribed sequences with genome tiling arrays. *Science*, *306*, 2242–2246.

Birol, I., Jackman, S. D., Nielsen, C. B., Qian, J. Q., Varhol, R., Stazyk, G., et al. (2009). De novo transcriptome assembly with ABySS. *Bioinformatics*, *25*, 2872–2877.

Brechenbacher, L., Weidman, S., Van Tuinen, D., Chatagnier, O., Gianinazzi, S., Franken, P., et al. (2004). Expression profiling of up-regulated plant and fungal genes in early and late stages of Medicago truncatula-Glomus mosseae interactions. *Mycorrhiza*, *14*, 253–262.

Ceccaroli, P., Buffalini, M., Saltarelli, R., Barbieri, E., Polidori, E., Ottonello, S., et al. (2011). Genomic profiling of carbohydrate metabolism in the ectomycorrhizal fungus *Tuber melanosporum*. *New Phytologist*, *189*, 751–764.

Chandran, D., Inada, N., Hather, G., Kleindt, C. K., & Wildermuth, M. C. (2010). Laser microdissection of Arabidopsis cells at the powdery mildew infection site reveals site-specific processes and regulators. *Proceedings of the National Academy of Sciences of the United States of America*, *107*, 460–465.

Denoeud, F., Aury, J.-M., Da Silva, C., Noel, B., Rogier, O., Delledonne, M., et al. (2008). Annotating genomes with massive-scale RNA sequencing. *Genome Biology*, *9*, R175.

Deveau, A., Palin, B., Delaruelle, C., Peter, M., Kohler, A., Pierrat, J. C., et al. (2007). The mycorrhiza helper Pseudomonas fluorescens BBc6R8 has a specific priming effect on the growth, morphology and gene expression of the ectomycorrhizal fungus *Laccaria bicolor* S238N. *New Phytologist*, *175*, 743–755.

Dillies, M. A., Rau, A., Aubert, J., Hennequet-Antier, C., Jeanmougin, M., Servant, N., et al. (2013). Comprehensive evaluation of normalization methods for Illumina high-throughput RNA sequencing data analysis. *Briefings in Bioinformatics*, *14*(6), 671–683.

Duplessis, S., Courty, P., Tagu, D., & Martin, F. (2005). Transcript patterns associated with ectomycorrhiza development in *Eucalyptus globulus* and *Pisolithus microcarpus*. *The New Phytologist*, *165*, 599–611.

Fosu-Nyarko, J., Jones, M. G., & Wang, Z. (2010). Application of laser microdissection to study plant-fungal pathogen interactions. *Methods in Molecular Biology*, *638*, 153–163.

Gao, L., Tu, Z. J., Millett, B. P., & Bradeen, J. M. (2013). Insights into organ-specific pathogen defense responses in plants: RNA-seq analysis of potato tuber-Phytophthora infestans interactions. *BMC Genomics*, *14*, 340.

Gaude, N., Bortfeld, S., Duensing, N., Lohse, M., & Krajinski, F. (2012). Arbuscule-containing and non-colonized cortical cells of mycorrhizal roots undergo extensive and specific reprogramming during arbuscular mycorrhizal development. *The Plant Journal*, *69*(3), 510–528.

Gomez, S. K., & Harrison, M. J. (2009). Laser microdissection and its application to analyze gene expression in arbuscular mycorrhizal symbiosis. *Pest Management Science*, *65*, 504–511.

Gomez, S. K., Javot, H., Deewatthanawong, P., Torres-Jerez, I., Tang, Y., Blancaflor, E. B., et al. (2009). Medicago truncatula and Glomus intraradices gene expression in cortical cells harboring arbuscules in the arbuscular mycorrhizal symbiosis. *BMC Plant Biology*, *22*, 10.

Grabherr, M. G., Haas, B. J., Yassour, M., Levin, J. Z., Thompson, D. A., Amit, I., et al. (2011). Full-length transcriptome assembly from RNA-Seq data without a reference genome. *Nature Biotechnology*, *29*, 644–652.

Grunwald, U., Guo, W., Fischer, K., Isayenkov, S., Ludwig-Müller, J., Hause, B., et al. (2009). Overlapping expression patterns and differential transcript levels of phosphate transporter genes in arbuscular mycorrhizal, Pi-fertilised and phytohormone-treated Medicago truncatula roots. *Planta*, *229*, 1023–1034.

Guether, M., Balestrini, R., Hannah, M. A., Udvardi, M. K., & Bonfante, P. (2009). Genome-wide reprogramming of regulatory networks, transport, cell wall and membrane biogenesis during arbuscular mycorrhizal symbiosis in Lotus japonicus. *New Phytologist*, *182*, 200–212.

Güimil, S., Chang, H. S., Zhu, T., Sesma, A., Osbourn, A., Roux, C., et al. (2005). Comparative transcriptomics of rice reveals an ancient pattern of response to microbial colonization. *Proceedings of the National Academy of Sciences of the United States of America*, *102*, 8066–8070.

Guttman, M., Garber, M., Levin, J. Z., Donaghey, J., Robinson, J., Adiconis, X., et al. (2010). Ab initio reconstruction of cell type-specific transcriptomes in mouse reveals the conserved multi-exonic structure of lincRNAs. *Nature Biotechnology*, *28*, 503–510.

Haas, B. J., Chin, M., Nusbaum, C., Birren, B. W., & Livny, J. (2012). How deep is deep enough for RNA-seq profiling of bacterial transcriptomes? *BMC Genomics*, *13*, 734.

Hacquard, S., Delaruelle, C., Legué, V., Tisserant, E., Kohler, A., Frey, P., et al. (2010). Laser capture microdissection of uredinia formed by Melampsora larici-populina revealed a transcriptional switch between biotrophy and sporulation. *Molecular Plant-Microbe Interactions*, *23*, 1275–1286.

Hacquard, S., Tisserant, E., Brun, A., Legué, V., Martin, F., & Kohler, A. (2013). Laser microdissection and microarray analysis of *Tuber melanosporum* ectomycorrhizas reveal functional heterogeneity between mantle and Hartig net compartments. *Environmental Microbiology*, *15*(6), 1853–1869.

Hogekamp, C., & Küster, H. (2013). A roadmap of cell-type specific gene expression during sequential stages of the arbuscular mycorrhiza symbiosis. *BMC Genomics*, *14*(1), 306.

Johansson, T., Le Quere, A., Ahren, D., Soderstrom, B., Erlandsson, R., Lundeberg, J., et al. (2004). Transcriptional responses of *Paxillus involutus* and *Betula pendula* during formation of ectomycorrhizal root tissue. *Molecular Plant-Microbe Interactions*, *17*, 202–215.

Katz, Y., Wang, E. T., Airoldi, E. M., & Burge, C. B. (2010). Analysis and design of RNA sequencing experiments for identifying isoform regulation. *Nature Methods*, *7*, 1009–1015.

Kawahara, Y., Oono, Y., Kanamori, H., Matsumoto, T., Itoh, T., & Minami, E. (2012). Simultaneous RNA-Seq analysis of a mixed transcriptome of rice and blast fungus interaction. *PLoS ONE*, *7*, e49423.

Kim, S. J., Bernreuther, D., Thumm, M., & Podila, G. K. (1999). LBAUT7, a novel symbiosis regulated gene from an ectomycorrhizal fungus, *Laccaria bicolor*, is functionally related to vesicular transport and autophagocytosis. *Journal of Bacteriology*, *181*, 1963–1967.

Kim, S. J., Zheng, J., Hiremath, S. T., & Podila, G. K. (1998). Cloning and characterization of a symbiosis-related gene from an ectomycorrhizal fungus *Laccaria bicolor*. *Gene*, *222*, 203–212.

Kuznetsova, E., Seddas-Dozolme, P. M. A., Arnould, C., Tollot, M., van Tuinen, D., Borisov, A., et al. (2010). Symbiosis-related pea genes modulate fungal and plant gene expression during the arbuscule stage of mycorrhiza with Glomus intraradices. *Mycorrhiza*, *20*, 427–443.

Langmead, B., Trapnell, C., Pop, M., & Salzberg, S. L. (2009). Ultrafast and memory-efficient alignment of short DNA sequences to the human genome. *Genome Biology*, *10*, R25.

Larsen, P. E., Sreedasyam, A., Trivedi, G., Podila, G. K., Cseke, L. J., & Collart, F. R. (2011). Using next generation transcriptome sequencing to predict an ectomycorrhizal metabolome. *BMC Systems Biology*, *5*, 70.

Larsen, P. E., Trivedi, G., Sreedasyam, A., Lu, V., Podila, G. K., & Collart, F. R. (2010). Using deep RNA sequencing for the structural annotation of the *Laccaria bicolor* mycorrhizal transcriptome. *PLoS ONE*, *5*(7), e9780.

Le Quéré, A., Wright, D., Söderström, B., Tunlid, A., & Johansson, T. (2005). Global patterns of gene regulation associated with the development of ectomycorrhiza between birch (*Betula pendula* Roth.) and *Paxillus involutus* (Batsch) Fr. *Molecular Plant-Microbe Interaction*, *18*(7), 659–673.

Li, H., & Durbin, R. (2009). Fast and accurate short read alignment with Burrows-Wheeler transform. *Bioinformatics*, *25*, 1754–1760.

Li, H., Ruan, J., & Durbin, R. (2008). Mapping short DNA sequencing reads and calling variants using mapping quality scores. *Genome Research*, *18*, 1851–1858.

Li, J., & Tibshirani, R. (2013). Finding consistent patterns: A nonparametric approach for identifying differential expression in RNA-Seq data. *Statistical Methods in Medical Research*, *22*(5), 519–536.

Lucic, E., Fourrey, C., Kohler, A., Martin, F., Chalot, M., & Brun-Jacob, A. (2008). A gene repertoire for nitrogen transporters in *Laccaria bicolor*. *New Phytologist*, *180*, 343–364.

Martin, F., & Kohler, A. (2013). *The mycorrhizal symbiosis genomics in the ecological genomics of fungi*. (F. Martin, Ed.), Hoboken, NJ: John Wiley & Sons, Inc.

Martin, J. A., & Wang, Z. (2011). Next-generation transcriptome assembly. *Nature Reviews Genetics*, *12*(10), 671–682.

Martin, F., Aerts, A., Ahr, D., Brun, A., Danchin, E. G. J., Duchaussoy, F., et al. (2008). The genome of *Laccaria bicolor* provides insights into mycorrhizal symbiosis. *Nature*, *452*, 88–92.

Martin, F., Kohler, A., Murat, C., Balestrini, R., Coutinho, P. M., Jaillon, O., et al. (2010). Périgord black truffle genome uncovers evolutionary origins and mechanisms of symbiosis. *Nature*, *464*, 1033–1038.

Matsumura, H., Reich, S., Ito, A., Saitoh, H., Kamoun, S., Winter, P., et al. (2003). Gene expression analysis of plant host–pathogen interactions by SuperSAGE. *PNAS*, *100*(26), 15718–15723.

McGettigan, P. A. (2013). Transcriptomics in the RNA-Seq era. *Current Opinion in Chemical Biology*, *17*, 4–11.

Montanini, B., Levati, E., Bolchi, A., Kohler, A., Morin, E., Tisserant, E., et al. (2011). Genome-wide search and functional identification of transcription factors in the mycorrhizal fungus *Tuber melanosporum*. *New Phytologist*, *189*, 736–750.

Mortazavi, A., Williams, B. A., McCue, K., Schaeffer, L., & Wold, B. (2008). Mapping and quantifying mammalian transcriptomes by RNA-seq. *Nature Methods*, *5*, 621–628.

Mosquera, G., Giraldo, M. C., Khang, C. H., Coughlan, S., & Valent, B. (2009). Interaction transcriptome analysis identifies Magnaporthe oryzae BAS1-4 as biotrophy-associated secreted proteins in rice blast disease. *Plant Cell*, *21*, 1273–1290.

Nagalakshmi, U., Wang, Z., Waern, K., Shou, C., Raha, D., Gerstein, M., et al. (2008). The transcriptional landscape of the yeast genome defined by RNA sequencing. *Science*, *320*, 1344–1349.

Nakazono, M., Qiu, F., Borsuk, L. A., & Schnable, P. S. (2003). Laser-capture microdissection, a tool for the global analysis of gene expression in specific plant cell types: Identification of genes expressed differentially in epidermal cells or vascular tissues of maize. *Plant Cell*, *15*, 583–596.

Peter, M., Courty, P.-E., Kohler, A., Delaruelle, C., Martin, D., Tagu, D., et al. (2003). Analysis of expressed sequence tags from the ectomycorrhizal basidiomycetes *Laccaria bicolor* and *Pisolithus microcarpus*. *New Phytologist*, *159*, 117–129.

Petre, B., Morin, E., Tisserant, E., Hacquard, S., Da Silva, C., Poulain, J., et al. (2012). RNA-Seq of early-infected poplar leaves by the rust pathogen melampsora larici-populina uncovers PtSultr3;5, a fungal-induced host sulfate transporter. *PLoS ONE*, 7(8), e44408.

Plett, J. M., Kemppainen, M., Kale, S. D., Kohler, A., Legué, V., Brun, A., et al. (2011). A secreted effector protein of *Laccaria bicolor* is required for symbiosis development. *Current Biology*, *21*, 1197–1203.

Plett, J. M., & Martin, F. (2011). Blurred boundaries: Lifestyle lessons from ectomycorrhizal fungal genomes. *Trends in Genetics*, *27*, 14–22.

Rineau, F., Shah, F., Smits, M. M., Persson, P., Johansson, T., Carleer, R., et al. (2013). Carbon availability triggers the decomposition of plant litter and assimilation of nitrogen by an ectomycorrhizal fungus. *ISME Journal*, 7, 2010–2022.

Robinson, M. D., McCarthy, D. J., & Smyth, G. K. (2010). edgeR: A bioconductor package for differential expression analysis of digital gene expression data. *Bioinformatics*, *26*, 139–140.

Robinson, M. D., & Oshlack, A. (2010). A scaling normalization method for differential expression analysis of RNA-seq data. *Genome Biology*, *11*, R25.

Robles, J. A., Qureshi, S. E., Stephen, S. J., Wilson, S. R., Burden, C. J., & Taylor, J. M. (2012). Efficient experimental design and analysis strategies for the detection of differential expression using RNA-sequencing. *BMC Genomics*, *13*, 484.

Rubini, A., Belfiori, B., Riccioni, C., Tisserant, E., Arcioni, S., Martin, F., et al. (2011). Isolation and characterization of MAT genes in the symbiotic ascomycete *Tuber melanosporum*. *New Phytologist*, *189*, 710–722.

Sebastianaa, M., Figueiredoa, A., Aciolia, B., Sousab, L., Pessoaa, F., Baldéa, A., et al. (2009). Identification of plant genes involved on the initial contact between ectomycorrhizal symbionts (Castanea sativa—European chestnut and *Pisolithus tinctorius*). *European Journal of Soil Biology*, *45*(3), 275–282.

Seddas, P. M., Arias, C., Arnould, C., van Tuinen, D., Godfroy, O., Benhassou, H., et al. (2009). Symbiosis-related plant genes modulate molecular responses in an arbuscular mycorrhizal fungus during early root interactions. *Molecular Plant-Microbe Interactions*, *22*, 341–351.

Shah, F., Rineau, F., Canbäck, B., Johansson, T., & Tunlid, A. (2013). The molecular components of the extracellular protein-degradation pathways of the ectomycorrhizal fungus *Paxillus involutus*. *New Phytologist*, *200*, 875–887.

Sillo, F., Gissib, C., Chignolia, D., Ragnib, E., Popolob, L., & Balestrini, R. (2013). Expression and phylogenetic analyses of the Gel/Gas proteins of *Tuber melanosporum* provide insights into the function and evolution of glucan remodeling enzymes in fungi. *Fungal Genetics and Biology*, *53*, 10–21.

Sims, D., Sudbery, I., Ilott, N. E., Heger, A., & Ponting, C. P. (2014). Sequencing depth and coverage: Key considerations in genomic analyses. *Nature Reviews. Genetics*, *15*, 121–132.

Skibbe, D. S., Doehlemann, G., Fernandes, J., & Walbot, V. (2010). Maize tumors caused by Ustilago maydis require organ-specific genes in host and pathogen. *Science*, *328*, 89–92.

Soneson, C., & Delorenzi, M. (2013). A comparison of methods for differential expression analysis of RNA-seq data. *BMC Bioinformatics*, *14*, 91.

Steijger, T., Abril, J. F., Engström, P. G., Kokocinski, F., The RGASP Consortium, Hubbard, T. J., et al. (2013). Assessment of transcript reconstruction methods for RNA-seq. *Nature Methods*, *10*, 1177–1184. http://dx.doi.org/10.1038/nmeth.2714.

Tang, W., Coughlan, S., Crane, E., Beatty, M., & Duvick, J. (2006). The application of laser microdissection to in planta gene expression profiling of the maize anthracnose stalk rot fungus Colletotrichum graminicola. *Molecular Plant-Microbe Interactions*, *19*, 1240–1250.

Tarazona, S., Garcia-Alcalde, F., Dopazo, J., Ferrer, A., & Conesa, A. (2011). Differential expression in RNA-seq: A matter of depth. *Genome Research, 21*, 2213–2223.

Teichert, I., Wolff, G., Kuck, U., & Nowrousian, M. (2012). Combining laser microdissection and RNA-seq to chart the transcriptional landscape of fungal development. *BMC Genomics, 13*, 511.

Tisserant, E., Da Silva, C., Kohler, A., Morin, E., Wincker, P., & Martin, F. (2011). Deep RNA sequencing improved the structural annotation of the *Tuber melanosporum* transcriptome. *New Phytologist, 189*, 883–891.

Tisserant, E., Kohler, A., Dozolme-Seddas, P., Balestrini, R., Benabdellah, K., Colard, A., et al. (2012). The transcriptome of the arbuscular mycorrhizal fungus Glomus intraradices (DAOM 197198) reveals functional tradeoffs in an obligate symbiont. *The New Phytologist, 193*(3), 755–769.

Tisserant, E., Malbreil, M., Kuo, A., Kohler, A., Symeonidi, A., Balestrini, R., et al. (2013). Genome of an arbuscular mycorrhizal fungus provides insight into the oldest plant symbiosis. *Proceedings of the National Academy of Sciences of the United States of America, 110*(50), 20117–20122.

Trapnell, C., Pachter, L., & Salzberg, S. L. (2009). TopHat: Discovering splice junctions with RNA-Seq. *Bioinformatics, 25*, 1105–1111.

Trapnell, C., & Salzberg, S. L. (2009). How to map billions of short reads onto genomes. *Nature Biotechnology, 27*(5), 455–457.

Trapnell, C., Williams, B. A., Pertea, G., Mortazavi, A., Kwan, G., van Baren, M. J., et al. (2010). Transcript assembly and quantification by RNA-seq reveals unannotated transcripts and isoform switching during cell differentiation. *Nature Biotechnology, 28*, 511–515.

Tremblay, A., Li, S., Sheffler, B. E., & Matthews, B. F. (2009). Laser capture microdissection and expressed sequence tag analysis of uredinia formed by Phakopsora pachyrhizi, the causal agent of Asian soybean rust. *Physiological and Molecular Plant Pathology, 73*, 163–174.

Van Verk, M. C., Hickman, R., Pieterse, C. M. J., & Van Wees, S. C. M. (2013). RNA-Seq: Revelation of the messengers. *Trends in Plant Science, 18*(4), 175–179.

Vargas, W. A., Martín, J. M., Rech, G. E., Rivera, L. P., Benito, E. P., Díaz-Mínguez, J. M., et al. (2012). Plant defense mechanisms are activated during biotrophic and necrotrophic development of Colletotricum graminicola in maize. *Plant Physiology, 158*, 1342–1358.

Velculescu, V. E., Zhang, L., Zhou, W., Vogelstein, J., Basrai, M. A., Bassett, D. E., Jr., et al. (1997). Characterization of the yeast transcriptome. *Cell, 88*(2), 243–251.

Villeneuve, N., LeTacon, F., & Bouchard, D. (1991). Survival of inoculated *Laccaria bicolor* in competition with native ectomycorrhizal fungi and effect on the growth of out-planted Douglas fir seedlings. *Plant and Soil, 135*, 97–107.

Voiblet, C., Duplessis, S., Encelot, N., & Martin, F. (2001). Identification of symbiosis-regulated genes in *Eucalyptus globulus–Pisolithus tinctorius* ectomycorrhiza by differential hybridization of arrayed cDNAs. *The Plant Journal, 25*(2), 181–191.

Wang, Z., Gerstein, M., & Snyder, M. (2009). RNA-Seq: A revolutionary tool for transcriptomics. *Nature Reviews. Genetics, 10*, 57–63.

Wang, L., Wang, S., & Li, W. (2012). RSeQC: Quality control of RNA-seq experiments. *Bioinformatics, 28*(16), 2184–2185. http://dx.doi.org/10.1093/bioinformatics/bts356.

Wang, K., Singh, D., Zeng, Z., Coleman, S. J., Huang, Y., Savich, G. L., et al. (2010). MapSplice: Accurate mapping of RNA-seq reads for splice junction discovery. *Nucleic Acids Research, 38*, e178.

Westermann, A. J., Gorski, S. A., & Vogel, J. (2012). Dual RNA-seq of pathogen and host. *Nature Reviews. Microbiology, 10*, 618–630.

Wright, D. P., Johansson, T., Le Quéré, A., Söderström, B., & Tunlid, A. (2005). Spatial patterns of gene expression in the extramatrical mycelium and mycorrhizal root tips formed by the ectomycorrhizal fungus *Paxillus involutus* in association with birch (*Betula pendula*) seedlings in soil microcosms. *New Phytologist*, *167*, 579–596.

Xu, L., Zhu, L., Tu, L., Liu, L., Yuan, D., Jin, L., et al. (2011). Lignin metabolism has a central role in the resistance of cotton to the wilt fungus Verticillium dahliae as revealed by RNA-Seq-dependent transcriptional analysis and histochemistry. *Journal of Experimental Botany*, *62*(15), 5607–5621.

Yazawa, T., Kawahigashi, H., Matsumoto, T., & Mizuno, H. (2013). Simultaneous transcriptome analysis of sorghum and bipolaris sorghicola by using RNA-seq in combination with de novo transcriptome assembly. *PLoS ONE*, *8*(4).

Zerbino, D. R., & Birney, E. (2008). Velvet: Algorithms for de novo short read assembly using de Bruijn graphs. *Genome Research*, *18*, 821–829.

Zhu, S., Dai, Y. M., Zhang, X. Y., Ye, J.-R., Wang, M.-X., & Huang, M.-R. (2013). Untangling the transcriptome from fungus-infected plant tissues. *Gene*, *519*(2), 238–244.

CHAPTER THREE

Evolutionary and Adaptive Role of Transposable Elements in Fungal Genomes

Jonathan Grandaubert, Marie-Helene Balesdent, Thierry Rouxel[1]
INRA-Bioger, Thiverval-Grignon, France
[1]Corresponding author: e-mail address: rouxel@versailles.inra.fr

Contents

Abstract

In complex eukaryotes, transposable elements (TEs), previously considered as junk DNA, are more and more acknowledged as genome shapers and as a source of gene innovation, genome plasticity and genome divergence. Fungi are simple and easy-to-manipulate eukaryotic organisms, for which ever-increasing genome information indicates that many plant-associated fungi have a tendency to have expanded genomes. This increase in genome size is mostly driven by TE expansion that eventually shapes adaptive regions of the genome. Such genome regions harbour genes involved in niche adaptation and favour accelerated evolution of these genes. The recent rise of comparative genomics in fungi now allows the use of phylogeny to date TE invasion and proliferation in genomes. This in turn provides inferences about the impact of TEs on speciation and on the rise of better-adapted species. Here, focusing on plant-associated fungi, we review our current level of knowledge on how TEs may have

Advances in Botanical Research, Volume 70
ISSN 0065-2296
http://dx.doi.org/10.1016/B978-0-12-397940-7.00003-3

contributed to speciation, the rise of two-speed genomes and the shaping of specific genome environments. Moreover, we consider the role of TEs in gene duplication and diversification that contribute to both adaptation to new hosts and adaptation to host resistance genes in gene-for-gene systems.

1. INTRODUCTION

1.1. Fungal genomes

Fungi are eukaryotic organisms that share numerous biological functions with more complex eukaryotes while having smaller and generally more compact genomes. Genes involved in these fundamental biological processes have a common origin in all eukaryotes, and because fungi typically are amenable to numerous molecular genetics manipulations in the laboratory, they provide model organisms to understand fundamental biological functions (Galagan, Henn, Ma, Cuomo, & Birren, 2005). Moreover, their genetic and biological properties (mixed reproduction regime, large population sizes, and short generation times) allow us to study the evolution of eukaryote genomes and thus to have a clearer insight into the history of life and ecosystems on Earth.

1.1.1 Fungal genomics

It began almost 20 years ago when the first eukaryote genomes to be sequenced were those of fungi, baker's yeast *Saccharomyces cerevisiae* (Goffeau et al., 1996) followed 6 years later by that of another ascomycetous yeast, *Schizosaccharomyces pombe* (Wood et al., 2002). These data allowed the first large-scale genome-wide functional and expression analyses of eukaryote genes. They also opened the path to comparative genomics, with Botstein and associates showing in 1997 that 31% of the 6000 genes of baker's yeast were shared with mammals (Botstein, Chervitz, & Cherry, 1997). However, yeasts with their simple genomes only represent a minute part of the extreme biological diversity of the kingdom Fungi. The first filamentous ascomycetes to be sequenced, namely, the model species *Neurospora crassa* and *Aspergillus nidulans* (Galagan et al., 2003, 2005), had genomes three times bigger than those of yeasts with twice as many genes. While being larger than those of yeasts, filamentous fungi still had compact genomes, a feature that was also found in the first phytopathogens to be sequenced. Ascomycetes such as *Fusarium graminearum*, *Magnaporthe oryzae* or *Stagonospora nodorum* (Cuomo et al., 2007; Dean et al., 2005; Hane et al., 2007) had genomes sized 36–40 Mb, with a similar number of genes, short intergenic regions, short introns and few transposable elements (TEs). The first basidiomycete phytopathogen sequenced, *Ustilago maydis*, even had a smaller, 20 Mb genome size (Kämper et al., 2006). Since then, with the launching of many genome initiatives in fungi, the general features of fungal genomes had to be revisited. Indeed, in many cases, fungal genomes, and more specifically those of phytopathogenic fungi, showed size expansion and variable genome architectures from one species to another, even when comparing closely related species (see the succeeding text).

1.1.2 Fungal effectors

In order to successfully colonize plant tissues, phytopathogens and mutualists must avoid or inhibit PAMP (pathogen-associated molecular patterns)-triggered immunity corresponding to recognition of conserved microbial-specific patterns such as chitin in fungi. For this purpose, phytopathogens have developed effectors and toxins (Howlett, 2006; Tyler & Rouxel, 2013; Wolpert, Dunkle, & Ciuffetti, 2002). While numerous toxins have been identified in the 'biochemistry era' decades ago, due to their effect on human health (e.g. ergot alkaloids causing ergotism produced by fungi within Clavicipitaceae or aflatoxins produced by several *Aspergillus* species), proteinaceous effectors were only recognized as major players in interactions between plants and filamentous phytopathogens with the advent of fungal and oomycete genomics (Kamoun, 2007). Toxins are considered to target central metabolic processes such as transcription and translation and/or trigger cell death, while effectors are considered to be molecules that target specific physiological processes to inhibit plant defence and enhance pathogen virulence without causing a general debilitation of cellular metabolism or cell death. Effectors are thus an important part of the armoury of weapons enabling colonization of the plant without restriction by defence responses, and they are used similarly by both phytopathogens and mutualists (Tyler & Rouxel, 2013). Toxins usually are secondary metabolites encoded by clusters of genes or small secreted proteins (proteinaceous toxins), while the vast majority of effectors are small secreted proteins (SSPs) (Martin & Kamoun, 2012; Tyler & Rouxel, 2013). Effectors represent specific adaptations to new modes of life and new hosts, over core genes and functions recycled from saprotrophic ancestors. Quite possibly, effectors evolved from molecules utilized by saprobes to suppress ecological competitors. More recently, small RNAs (Bc-sRNAs) were shown to be able to silence *Arabidopsis* and tomato genes involved in immunity during the interaction of the plant with the fungal pathogen *Botrytis cinerea* and must in the future be integrated into the general corpus of fungal effectors (Weiberg et al., 2013).

In the course of the plant–pathogen coevolutionary arms race, effectors have become specific targets for recognition by some (usually not all) plant genotypes within a species (Jones & Dangl, 2006). Such targets have been known for a long time from the 'age of formal genetics' as 'avirulence genes' and formed the basis of the gene-for-gene postulate established by Flor (1955) and are abundantly used in plant breeding strategies. The zigzag model of coevolution established by Jones and Dangl (2006) indicates that effectors have a dual role in microbe–plant interactions, both targeting plant component processes and being targeted by plant defence receptors (van't Slot & Knogge, 2002). As a result, they constitute a focal point of the dynamic evolutionary tension that exists between plants and microbes (Tyler & Rouxel, 2013).

1.2. TEs: General facts

TEs were discovered in 1950 by Barbara McClintock in maize when linking mutations to nonrandom spontaneous translocations of chromosomal fragments within the genome (McClintock, 1950). TEs were defined as mobile DNA sequences encoding all the necessary machinery required for their movement (i.e. transposition) within the genome. TEs are present in most prokaryotes and eukaryotes and may represent an important part of the genome, especially in some plant species. TEs were initially regarded as junk DNA (Ohno, 1972), or selfish/parasite DNA (Orgel & Crick, 1980), but they are increasingly considered a source of gene innovation, genome plasticity and genome divergence because of their mutational power and their contribution to the generation of new genes (TE domestication) or regulatory elements and causative agents for larger-scale chromosomal rearrangements (Biémont & Vieira, 2006; Fedoroff, 2012). Due to their small and frequently haploid genomes, short generation times and highly contrasted TE contents (even between closely related species), phytopathogenic fungi are excellent models to address the role of TEs in genome reshaping, speciation, gene innovation and adaptation to new hosts or novel host resistances.

1.2.1 Classification

TEs were initially classified according to their transposition intermediate: either an RNA intermediate leading to a copy-and-paste transposition mechanism (class I elements; retrotransposons) or a DNA intermediate

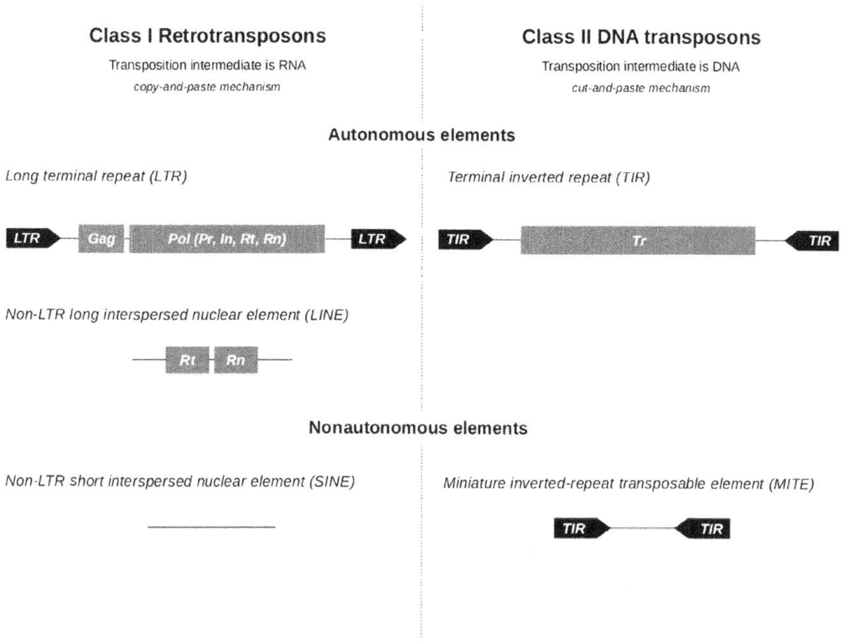

Figure 3.1 A simplified classification of transposable elements (TEs). TEs are divided into two classes according to their transposition intermediate, RNA or DNA. Within these classes, the elements can encode the enzymatic machinery needed to transpose (autonomous elements) or not (nonautonomous elements). Here are displayed the main TE classes found in fungal genomes. Pr, protease; In, integrase; Rt, reverse transcriptase; Rn, RNase H; Tr, transposase.

leading to a cut-and-paste transposition mechanism (class II elements; DNA transposons) (Finnegan, 1989) (Fig. 3.1). Later on, the discovery of new TEs that can transpose using a copy-and-paste mechanism without RNA intermediates and of nonautonomous elements led to a revised classification (Wicker et al., 2007) taking into account the mechanisms of transposition, sequence similarities and structural relationships. The new classification retains the previous class I/class II separation but defines new subclasses and TE families.

1.2.2 Retrotransposons

They are firstly replicated as an RNA molecule that is then retrotranscribed by the TE enzymatic machinery as a DNA molecule that will eventually insert itself in another genomic location. Each replication cycle thus produces a new copy of the element. Retrotransposons are usually differentiated according to the presence or absence of long terminal repeats (LTRs) at their extremities. LTR retrotransposons encode ORFs for a *gag* viral structure protein and for a polyprotein (*pol*) including a protease, a reverse transcriptase, a ribonuclease and an integrase. The organization of functional modules discriminates the different families of LTR retrotransposons (Gypsy and Copia) (Wicker et al., 2007). The organization of LTR retrotransposons resembles that of retroviruses that are suggested to originate from LTR retrotransposons via the acquisition of an envelope protein (*env*) (Biémont & Vieira, 2006; Wicker et al., 2007). Retrotransposons lacking LTR include LINEs (long interspersed nuclear elements) and SINEs (short interspersed nuclear elements). These latter elements are short non-autonomous TEs, without coding sequences, that depend on other elements from their own family or other families for replication and transposition.

1.2.3 DNA transposons

They use a DNA intermediate to transpose. They are separated into two subclasses depending on the number of DNA strands broken in the course of the transposition process. Elements belonging to subclass 1 encode a transposase causing a double-strand break during excision. They encompass repeated inverted sequences at their extremities (TIRs: terminal inverted repeats). Subclass 1 also encompasses nonautonomous elements such as MITEs (miniature inverted-repeat transposable elements). The second subclass contains helitrons and polintrons (Mavericks) with a specific 'rolling-circle' transposition mechanism (Kapitonov & Jurka, 2007; Pritham, 2009).

1.2.4 TEs in eukaryote genomes

In eukaryotes, the genome size is largely independent of the organism complexity or its gene number. The C-value (i.e. the DNA content per haploid genome) varies widely even when comparing closely related organisms. The most striking differences can be found in flowering plants whose genome size varies between 64 Mb (*Genlisea* spp.; Greilhuber et al., 2006) and 127,000 Mb (*Fritillaria* spp.; Ambrozová et al., 2011). Smaller-scale differences can be found for simpler eukaryotes such as filamentous fungi, with genome sizes ranging between 22 and 160 Mb in the two ascomycete species *Baudoinia compniacensis* and *Golovinomyces orontii*, respectively (Ohm et al., 2012; Spanu et al., 2010). In the basidiomycete fungi, *U. maydis*, and some *Hemileia vastatrix* isolates, genome size varies between 20 and 610 Mb (Kämper et al., 2006; http://bioinformatics.cenicafe.org/index.php/wiki/Genome_Size-Chromosome_Number). Among stramenopiles, genome size varies between 27 Mb in diatoms and 240 Mb in species in the *Phytophthora infestans* clade (Raffaele & Kamoun, 2012). While exceptions exist, in these two latter cases, genome expansion is usually found in phytopathogens rather than their free-living relatives (Raffaele & Kamoun, 2012). In prokaryotes, genome size expansion is mostly due to gene content expansion, often linked with events of horizontal gene transfer (Frost, Leplae, Summers, & Toussaint, 2005). A similar trend partly accounts for genome expansion in plants, where polyploidization is common (Bennetzen, Ma, & Devos, 2005), or in basidiomycete fungi causing rust diseases in which genome expansion is partly due to expansion of lineage-specific gene families (Duplessis et al., 2011; Zheng et al., 2013). However, in eukaryotes, including plants, size differences between related species are mainly due to noncoding DNA, including TEs and other repeated sequences (Biémont & Vieira, 2006). Consequently, the relationship of TEs to genome size is dramatic in plants, with 10–20% of TEs for small-genome species such as *Brachypodium distachyon* and up to 85% TEs in large-genome species such as maize or barley. This is also observed among nonmammal vertebrates, with ca. 9% TE in birds such as chicken and up to 77% TE in frogs. In mammals, in contrast, there is a limited genome size variation between species, with genomes typically 3300 Mb in size and TEs representing usually half of the genome (Gregory, 2013).

1.2.5 The main representative TE classes in genomes

They may differ depending on the organism considered. In general, partly due to their size and mechanism of transposition, retrotransposons are often overrepresented in eukaryote genomes. For example, they are the only type found in TE-poor genomes of yeasts such as *S. cerevisiae* and *S. pombe* and the amoeba *Entamoeba dispar* (Pritham, 2009). They dominate in the genomes of *Homo sapiens* and mouse, where they represent 90% of the TEs. In contrast, with rare exceptions such as *Trichomonas vaginalis* or *Entamoeba invadens*, DNA transposons only make a minor contribution to the pool of TEs. Except for the *Fusarium oxysporum* f.sp. *lycopersici* genome (see the succeeding text), they only contribute to a very minor extent to genome size variation between related species (Pritham, 2009). Where particular classes of TE predominate, they also tend to be represented by only a few families that show massive expansion. For example, *L1* LINE and *Alu* SINE dominate in the human genome and are only represented by a few subfamilies, while LTR retrotransposons are represented by a dozen different families with only a limited number of copies (Lander et al., 2001). Similarly, in the genome of the filamentous fungus *Leptosphaeria maculans* 'brassicae' in which LTR retrotransposons largely predominate, the TE expansion is attributable to only four different families of LTR retrotransposons (Rouxel et al., 2011).

1.2.6 Selfish DNA?

As mentioned in the preceding text, TEs have long been regarded as genome parasites but are now increasingly viewed as genome shapers that are at the heart of eukaryote evolution (Fedoroff, 2012). Both the sudden bursts of transposition and the prolonged cohabitation between TEs and their host genome have enabled multiple interactions with structural consequences (e.g. chromosomal rearrangements and remodelling of heterochromatin) and/or functional consequences (e.g. TE domestication and gene regulation). TEs were initially identified because of the chromosomal rearrangements they induce in the maize genome. TEs can thus cause ectopic recombinations between copies, leading to deletions, translocations or chromosomal inversions. In mammals, for example, Zhao and Bourque (2009) had shown that breakpoints bordering chromosomal inversions were enriched in LINE *L1*. In *Drosophila buzzatii*, Delprat, Negre, Puig, and Ruiz (2009) showed that the *Galileo* TE was associated with two chromosomal inversions responsible for ectopic recombination. Such recombinations can take place even if TEs are inactivated, suggesting that TEs remain genome shapers independently of their ability to transpose (Oliver & Greene, 2012). TEs also shape the chromatin landscape along chromosomes and are abundant in regions of constitutive heterochromatin such as telomeres and centromeres (Hua-Van, Le Rouzic, Boutin, Filee, & Capy, 2011). Following epigenetic methylation, TEs are also involved in the formation of facultative heterochromatin (Lewis et al., 2008). Functional modifications due to TEs include insertions into coding sequences leading to gene inactivation, altered gene expression or modification of gene structure (e.g. premature stop codon and alternative splicing). *cis*-Regulatory regions in TEs can directly modify gene expression when inserted into promoters. For example, it has been suggested that modified expression of a red pigment due to a retrotransposon insertion was the origin of white grape cultivars in *Vitis viticola* (Kobayashi, Goto-Yamamoto, & Hirochika, 2004). In the human genome, 4% of the genes and ca. 25% of the promoters contain sequences that may derive from TEs (Nekrutenko & Li, 2001; van de Lagemaat, Landry, Mager, & Medstrand, 2003).

Major genome rearrangements or more minor effects due to TEs indicate that these mobile elements could participate in speciation events.

However, this is a widely debated point, and it is always hard to determine whether TE bursts in a genome are the cause or the consequence of speciation (Biémont, 2010; Rebollo, Horard, Hubert, & Vieira, 2010).

1.3. TEs: How genomes control proliferation

1.3.1 How can genomes control TE proliferation?

In *Drosophila melanogaster*, the number of TE insertions is much higher than the number of excisions (Maside, Bartolome, Assimacopoulos, & Charlesworth, 2001). In the absence of control mechanisms, one would predict an overaccumulation of TEs in the genome with the only counterselection being due to trade-offs between the selective advantage(s) due to accumulation of TEs and the cost of maintaining these large genomes and/or natural selection counterselecting deleterious insertions (Biémont et al., 1997). However, most organisms use efficient control mechanisms to limit the spread of TEs in their genomes. The most common are epigenetic mechanisms that do not irreversibly alter the sequence of the TEs, for example, repressive protein complexes, histone methylation, RNA interference and RNA-directed DNA methylation (for a review, see Fablet & Vieira, 2011; Fedoroff, 2012; Rebollo et al., 2010). These epigenetic silencing phenomena can be divided into two categories, transcriptional gene silencing (TGS), when no transcript of the targeted gene is produced, and posttranscriptional gene silencing (PTGS), when transcripts are produced but specifically degraded before translation can occur. Phylogenetic surveys of proteins involved in TGS or PTGS have shown that they are present in all lineages of eukaryotes (Bird, 2007; Jaenisch & Bird, 2003), and thus, the ancestor of the eukaryotes was likely endowed with some sort of primitive homology-based genome defence mechanism. However, such control mechanisms are not mandatory for survival, since organisms lacking TGS and/or PTGS machineries can be found. These efficient silencings are heritable and, in most cases, reversible. TEs thus remain a potential reservoir of variability and can be reactivated under certain environmental stress conditions (Hua-Van et al., 2011). Such bursts of transposition have been described in numerous species, suggesting that epigenetic control has been temporarily lost (Rebollo et al., 2010).

1.3.2 In fungi

Although absent in *S. cerevisiae* (Drinnenberg et al., 2009), PTGS has been extensively studied in the fission yeast *S. pombe* (Moazed et al., 2006) and in

N. crassa, where the 'quelling' process is active during the vegetative phase (Cogoni et al., 1996), whereas the meiotic silencing of unpaired DNA process (Aramayo & Metzenberg, 1996) is specifically active during sexual reproduction. Two TGS mechanisms are known in fungi: methylation induced premeiotically (MIP) and repeat-induced point (RIP) mutation. MIP has been described in the ascomycete *Ascobolus immersus* (Goyon, Rossignol, & Faugeron, 1996) and in the basidiomycete *Coprinopsis cinerea* (Irelan & Selker, 1996). MIP is an epigenetic mechanism that methylates but does not irreversibly mutate duplicated sequences. It is conserved between ascomycetes and basidiomycetes and is therefore suggested to be an ancestral mechanism from which evolved a more drastic one, RIP (Galagan & Selker, 2004). In fungi, RIP is the ultimate inactivation mechanism of repeated elements, first described in the model ascomycete *N. crassa* (Galagan et al., 2003). RIP is a premeiotic event that induces C:G to T:A mutations on the two copies of a duplicated element. The process occurs during the transient dikaryon stage preceding karyogamy and meiosis in ascomycete fungi and in *N. crassa* is dependent on the *rid* (*RIP-defective*) gene, whose structure is similar to that of a cytosine methylase (Freitag, Williams, Kothe, & Selker, 2002). RIP thus irreversibly inactivates TEs and allows sequence divergence between copies, with possible consequences for recombination (Galagan & Selker, 2004). This however depends on the efficiency of the RIP process that can introduce multiple mutations in a single sexual cycle, as shown in *N. crassa* in where up to 30% of the C:G pairs in duplications can be mutated (Galagan & Selker, 2004). In contrast, RIP efficiency can be much lower as found in *Podospora anserina*, a species related to *N. crassa* but showing much lighter RIP with 1% of bases mutated over 9 kb of sequence (Graïa et al., 2001). In *N. crassa* and other fungi, the mechanism is so efficient that no intact and functional copies of TEs are present in the genome (Galagan et al., 2003; Rouxel et al., 2011). Indeed, RIP was suggested in *N. crassa* to have prevented gene duplication, gene family expansion and eventually gene innovation (Galagan & Selker, 2004). RIP has been experimentally demonstrated to be currently active in a series of ascomycetes such as *P. anserina*, *L. maculans* and *Nectria haematococca* (Coleman et al., 2009; Graïa et al., 2001; Idnurm & Howlett, 2003), and evidence of RIP has been reported in most ascomycete species, suggesting RIP is an ancestral mechanism common to (at least) ascomycete fungi (Clutterbuck, 2011). In a few ascomycetes, however, occasional patterns of secondary losses are observed. For example, in all analysed powdery mildew fungi, extreme specialization towards biotrophy is accompanied by loss of many genes including those involved in the RIP machinery (Spanu, 2012; Spanu et al., 2010), while in *Claviceps purpurea*, the *rid* gene is absent (Schardl et al., 2013). A similar trend is found in the mutualist ascomycete *Tuber melanosporum*, where it is unclear whether the RIP machinery was lost along with numerous other genes unnecessary for the symbiotic lifestyle or if RIP was acquired after the early divergence of Pezizales from other members of the Pezizomycotina (Martin et al., 2010). In fungi, an alternative way to by-pass RIP and allow TE expansion in genomes is to have an essentially asexual lifestyle, since RIP requires meiosis to be functional. This, for example, may be the case for *F. oxysporum* f.sp. *lycopersici*, where no trace of RIP is found in its genome, but the elements of the RIP machinery seem to be present (Ma et al., 2010). In addition, related sexual species, *F. graminearum* and *Fusarium verticillioides*, show evidence of RIP in their LTR retrotransposons (Ma et al., 2010).

1.4. TEs in fungal genomes

Initial sequencing of filamentous fungal genomes suggested that these were compact with very few TEs and a high gene density. With additional fungal genome sequencing, the 'compact genome' paradigm no longer holds true; many phytopathogenic and mutualist fungal species show 'invaded' genomes, with a massive presence of active or inactive TEs (De Wit et al., 2012; Duplessis et al., 2011; Gan et al., 2013; Martin et al., 2008, 2010; Rouxel et al., 2011; Schardl et al., 2013; Spanu et al., 2010; Wicker et al., 2013; Zheng et al., 2013).

1.4.1 Main families found in fungal genomes

As in other organisms, retrotransposons, and very often LTR retrotransposons of the Gypsy family, usually dominate (e.g. De Wit et al., 2012; Martin et al., 2008, 2010; Ohm et al., 2012; Rouxel et al., 2011). In TE-poor genomes, however, DNA transposons can make an important contribution to the overall pool of TEs (e.g. 31.6% of the TEs in the genome of *Leptosphaeria biglobosa* 'brassicae' or 36.4% in the genome of *P. tritici-repentis*) (Grandaubert, 2013; Manning et al., 2013). A contrasting situation is found in *F. oxysporum* f.sp. *lycopersici* in which the 60 Mb genome contains supplementary chromosomes encompassing 74% of the TEs of the genome, and these TEs mostly are DNA transposons (65% of the manually annotated TEs of the genome) of the Tc1–mariner and hAT families (Ma et al., 2010). In contrast, LTR retrotransposons are evenly distributed in the core and dispensable genome and are the main TEs in related species such as *F. graminearum* and *F. verticillioides* (Ma et al., 2010).

In *Blumeria graminis* f.sp. *hordei*, in which 64% of the 120 Mb genome is made up of TEs, non-LTR retrotransposons such as LINEs dominate over LTR retrotransposons (Spanu et al., 2010). In the genomes of some rust species, which typically contain 50% repeats, this proportion can vary with the *Melampsora larici-populina* genome being enriched for TIR DNA transposons and DIRS retrotransposons, while LTR retrotransposons dominate in *Puccinia striiformis* f.sp. *tritici* and *Puccinia graminis* f.sp. *tritici* (Duplessis et al., 2011; Zheng et al., 2013).

1.4.2 Repertoire of TEs and dating the invasion of fungal genomes

To date, very few published studies have attempted to comprehensively characterize TE families in fungal genomes or to retrace the history of genome invasion by TEs. Ma et al. (2010) showed that the drastic genome size expansion of *F. oxysporum* f.sp. *lycopersici* compared with the related *F. graminearum* and *F. verticillioides* was due to the addition of new chromosomes. The TE amplification was lineage-specific and suggested that a recent lineage-specific expansion of TEs was associated with the generation of the novel dispensable chromosomes. One other recent study provided an extensive classification and dating of TE expansion in the *L. maculans–L. biglobosa* species complex. There, TEs were manually reannotated and their occurrence was investigated in fungal genomes available in the MycoCosm database (Grandaubert, 2013). The *L. maculans–L. biglobosa* species complex encompasses two main species of crucifer pathogens, *L. maculans* and *L. biglobosa* that diverged ca. 22 mya (Fig. 3.2). Additional clades, *L. maculans* 'brassicae' (Lmb), *L. maculans* 'lepidii' (Lml), *L. biglobosa* 'brassicae' (Lbb), *L. biglobosa* 'thlaspii' (Lbt) and *L. biglobosa* 'canadensis' (Lbc), were recognized as distinct species following phylogenetic analyses. In cases of species in terminal branches of the phylogeny, separation dates were 5.1 mya between Lmb and Lml and 3.6 mya between Lbb and Lbc (Grandaubert, 2013). Repeat annotation identified 121 sequences including 24 families of retrotransposons and 21 DNA transposons. All of these were exclusively found in dothideomycete genomes, and all but two of them were confined to species belonging to the order Pleosporales and were therefore suggested to have invaded the fungal genomes 90 mya or earlier. Patterns of gains and losses could be identified at each radiation point, and 66% of the families were found to be specific of the *L. maculans–L. biglobosa* species complex, with all possible levels of specificity (genus, *L. maculans* or *L. biglobosa*, species) (Grandaubert, 2013). In this model, only one species, Lmb, shows a massive invasion of its genome by TEs, with just four families of LTR retrotransposons accounting for one

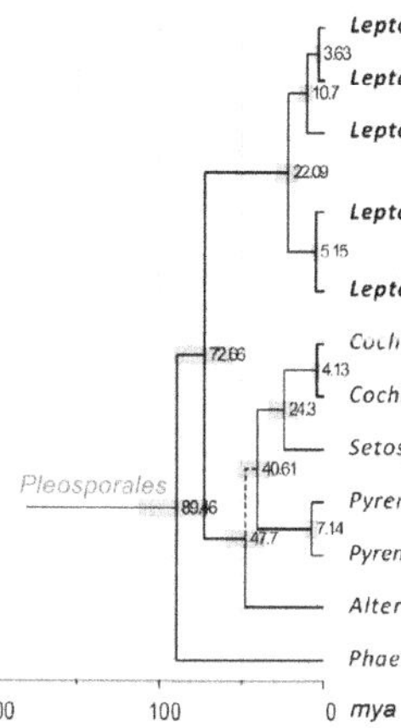

	RLG_Olly	RLG_Polly	RLC_Pholy
	Total size within genomes (kb)		
Leptosphaeria biglobosa **'brassicae'**	166	–	3
Leptosphaeria biglobosa **'canadensis'**	130	1	5
Leptosphaeria biglobosa **'thlaspii'**	174	73	3
Leptosphaeria maculans **'brassicae'**	**3080**	**2930**	**3200**
Leptosphaeria maculans **'lepidii'**	43	2	11
Cochliobolus heterostrophus	135	11	–
Cochliobolus sativus	280	18	45
Setosphaeria turcica	162	123	12
Pyrenophora tritici-repentis	38	29	22
Pyrenophora teres	12	12	6
Alternaria brassicicola	8	49	2
Phaeosphaeria nodorum	11	45	106

Figure 3.2 Presence of three LTR retrotransposon families, *RLG_Olly*, *RLG_Polly* and *RLC_Pholy*, within the Pleosporales phylogeny. These three families are found in almost all sequenced Pleosporales genomes, meaning that they were present in the phylogeny for at least 90 million years. They represent less than 200 kb in most of the genomes, except that of *Leptosphaeria maculans* 'brassicae' (Lmb). In this latter species, they account for approximately 3 Mb each. As Lmb is the only species in which massive expansion of these three TE families is observed, it is hypothesized that this expansion took place during or after the speciation event between Lmb and *L. maculans* 'lepidii,' 5 million years ago. RLG, long terminal repeat (LTR) retrotransposon from the Gypsy superfamily; RLC, LTR retrotransposon from the Copia superfamily (Wicker et al., 2007). The figures at the nodes indicate the mean node ages in million years, while the green bars indicate the error associated with the estimate (C.L. Schoch, pers. comm.).

quarter of the genome, while the closely related species, Lml, only contains 2.7% TEs. Twenty-two families were found to be specific to Lmb, suggesting that they invaded the genome 'recently' at the time of, or after, the separation between Lmb and Lml 5.1 mya (Grandaubert, 2013; Rouxel et al., 2011). In addition, three of the most expanded families of LTR retrotransposons in Lmb were present in phylogenies for at least 90 million years but are only represented by a small fraction of the genomes of other Pleosporales species or other members of the species complex (Fig. 3.2). Alignment-based dating of these TE copies in Lmb genome confirmed a burst of transposition at the time of speciation ca. 5 mya, in conjunction with genome invasion by new families (Rouxel et al., 2011).

Interestingly, Lmb is to date the only Pleosporales species in which such a TE expansion has been observed, whereas it is more common in another important order of Dothideomycetes, the Capnodiales (De Wit et al., 2012; Ohm et al., 2012). Moreover, among the four *Colletotrichum* species sequenced to date, only one, *C. orbiculare*, shows a massive TE invasion (Gan et al., 2013; O'Connell et al., 2012). These provide the only examples currently available where one terminal branch species shows genome expansion due to TE expansion while all closely related species are TE-poor. A similar TE invasion was observed in *Cladosporium fulvum* compared with *Dothistroma septosporum* (De Wit et al., 2012), but separation dates between these two species greatly predate the separation between *L. maculans* and *L. biglobosa* or between the different *Colletotrichum* species. In other cases in which comparative genomic approaches considered TE content, genome invasion by TEs is common to a series of related species, as exemplified by all related powdery mildew species, rusts species and three related endophyte or phytopathogenic Clavicipitaceae (Duplessis et al., 2011; Schardl et al., 2013; Spanu et al., 2010; Zheng et al., 2013).

1.4.3 Incidence of TEs on genome architecture

Following the initial description in *L. maculans* of a compartmentalized genome divided into GC-equilibrated, gene-rich isochores and large AT-rich blocks mainly made up of mosaics of intermingled TEs degenerated by RIP (Rouxel et al., 2011), more ascomycete species have been reported in which RIP is still active and that show a similar genome architecture. Examples include most of the endophytic Clavicipitaceae (Schardl et al., 2013), *Mycosphaerella fijiensis* (Ohm et al., 2012), *C. fulvum* (De Wit et al., 2012) or *C. orbiculare* (Gan et al., 2012). In *B. graminis* f.sp. *hordei*, and other powdery mildew fungi, in which TEs are not inactivated by RIP and remain active, TEs are evenly distributed throughout the genomes with no evidence of clustering. This lack of large blocks of TEs however may be due to the extreme fragmentation of the assembly generated, amounting to ca. 7000 scaffolds (Spanu et al., 2010). In addition, even in those species with compact genomes, some specific compartments of the genomes are enriched in TEs. These TE-rich compartments may either lie within the principal chromosomes (e.g. near telomeres and centromeres), generally resulting in smaller-sized AT-rich blocks, or represent the bulk of small conditionally dispensable chromosomes (CDCs) (Balesdent et al., 2013; Coleman et al., 2009; Ma et al., 2010). In species in which RIP is active, these dispensable chromosomes are generally also AT-rich (e.g. Balesdent et al., 2013). These compartmentalized genomes can be seen as two-speed genomes in which a core genome contributes to essential functions and an adaptive genome, corresponding either to portions of the core chromosomes or to distinct CDCs, is dedicated to niche functions such as pathogenicity. Interestingly, a similar convergent evolution towards plastic, two-speed genomes is found in phytopathogenic oomycetes such as *P. infestans* (Raffaele & Kamoun, 2012).

2. TEs AND SPECIATION IN FUNGI

2.1. The mesosynteny postulate

In contrast to what is found in other eukaryotes, previous analyses in the fungi have indicated that macrosynteny, that is, the conservation of large syntenic blocks, is rare between species. Instead, comparisons across species show no synteny or only microsyntenic regions encompassing usually five or fewer genes, typically corresponding to functionally relevant gene clusters such as secondary metabolite gene clusters. Mesosynteny, that is, the conservation within chromosomes of gene content but not order or orientation, was recently postulated to be a mode of chromosomal evolution specific to fungi, mostly found in filamentous ascomycetes and more extensively studied in fungi belonging to the class Dothideomycetes (Grandaubert, 2013; Hane et al., 2011; Ohm et al., 2012). Comparative analyses of 18 dothideomycete species and a simulation-based approach indicated that, starting from one or a few chromosome inversions, serial random inversions within the chromosome can lead with time to intense reshuffling of gene order within homologous chromosomes from one species to a distantly related one (Ohm et al., 2012). For example, both the divergence time estimates (ca. 22 mya) and the intensity of intrachromosomal inversions were comparable between *L. maculans* and *L. biglobosa* or between *Cochliobolus heterostrophus* and *Setosphaeria turcica* (Grandaubert, 2013; Ohm et al., 2012). Comparing more closely related species within the class Dothideomycetes, 30 inversions were observed in a single scaffold between *Cochliobolus sativus* and *C. heterostrophus* (Ohm et al., 2012), which diverged 4.3 mya (Grandaubert, 2013), while even fewer inversions were observed between *L. maculans* 'brassicae' and *L. maculans* 'lepidii' that diverged ca. 5.1 mya (Grandaubert, 2013). In this latter case, while large-scale translocations were not seen, 30 intrachromosomal sequence inversions sized 1.3–355 kb were identified. These inversions were scattered along all the chromosomes and encompassed one to more than a hundred genes. Focusing on short divergence times, comparative genome analyses have thus identified the first steps towards mesosynteny in perfect accordance with the modeling-based hypothesis proposed by Ohm et al. (2012).

2.2. TEs at the origin of mesosynteny?

In the Ohm et al. (2012) paper, inversion breakpoints in dothideomycete genomes were suggested to be associated with simple sequence repeats (SSR). However, a more comprehensive study proposed a divergent interpretation. Grandaubert (2013) compared genomes of species within the *L. maculans–L. biglobosa* species complex and dated the timing of transposition responsible for genome invasion and bursts of transposition. The authors observed that in *L. maculans* 'brassicae', 70% of the intra-chromosomal inversions were bordered by TEs specific to the *L. maculans* clade, which were suggested to have invaded the genome (or to have undergone a massive transposition burst) at the time of, or after, separation between the two related species *L. maculans* 'brassicae' and *L. maculans* 'lepidii' ca. 5.1 mya. These data may indicate a major role for TEs rather than SSRs in genome reshaping and reshuffling at the chromosomal level and thus generating mesosynteny in chromosomes of the class Dothideomycetes. These data are consistent with documented incidence of TEs in chromosomal inversions in more complex eukaryotes such as *Drosophila* (Delprat et al., 2009) or mammals (Zhao & Bourque, 2009). This first step towards mesosynteny clearly ascribes a role to species-specific TEs in this evolutionary mechanism that will eventually generate nonhomologous chromosome sections and isolate part of the genome from meiotic recombination.

3. TEs AND ADAPTATION TO HOST PLANT AND PLANT DISEASE RESISTANCE

The effector repertoires of plant-associated microbes comprise hundreds of candidate effector genes (SSPs and secondary metabolite genes; typically over 500 in ascomycetes). While basidiomycete rusts and ascomycete powdery mildews show expanded effector families (which is also the case for oomycetes), in other fungal species, SSP-encoding genes typically show extreme sequence diversity, with few recognizable paralogs and few or no homologues in public databases. However, both effector genes and secondary metabolite clusters often show aberrant phylogenetic distributions that are usually interpreted as a hallmark of lateral gene transfer (LGT). This suggests that the repertoire of effector genes continually evolves firstly to allow adaptation to new hosts and then to retain or improve their effectiveness in modulating plant defences while at the same time minimizing detection of the effectors by the plant. In parallel, plants are driven to retain or improve their detection of pathogens and to minimize the impact of effectors on their physiology. As a result of this coevolutionary arms race, microbial genes encoding effectors, and plant genes encoding components that interact with effectors, often display accelerated evolutionary change. By shaping specific genome environments and being themselves involved in gene duplication and diversification, TEs are the major players that contribute to both adaptation to a new host and adaptation to resistance genes in a host species on which the fungus is already adapted.

3.1. TEs associated with effector genes in fungi

Effector genes and secondary metabolite clusters are often associated with TEs in fungal genomes, suggesting that there is an evolutionary link between TEs and the birth and death, or expansion and diversification, of effectors in fungal species. As a consequence, TE-rich regions are enriched in species- and even isolate-specific genes. For example, in *L. maculans*, 25% of the genes in AT-rich blocks are species-specific, and this ratio rises to 41% when only considering genes encoding putative effectors (Grandaubert, 2013). Up to seven out of 17 avirulence genes are located in TE-rich telomere-proximal regions of the compact genome of *M. oryzae* (Farman, 2007; Orbach, Farrall, Sweigard, Chumley, & Valent, 2000; Valent & Khang, 2010); all currently known avirulence effectors of *L. maculans* are located within AT-rich blocks made up of degenerated TEs, and AT-rich isochores contain 20% of the putative effector genes compared to 4% in the rest of the genome (Balesdent et al., 2013; Fudal et al., 2007; Gout et al., 2006; Parlange et al., 2009; Rouxel et al., 2011); in the same species, two secondary metabolite gene clusters only found in Lmb, of which one is postulated to encode a host-specific toxin, are also embedded in a TE-rich genome environment (Grandaubert, 2013); in *B. graminis*, effectors are embedded in TE-rich regions of the genome and are closely associated with a few specific TE families (Spanu et al., 2010); in *F. oxysporum*, supernumerary CDCs making up the bulk of DNA transposons in the genome are enriched in strain-specific effectors accounting for the host range of each 'forma specialis' (Ma et al., 2010); in the *Clavicipitaceae* pathogens or endophytes, secondary metabolite cluster peripheries are positioned near to transposon-derived, AT-rich repeat blocks (Schardl et al., 2013). In contrast with these examples, no specific genome location is found or postulated for effectors in the TE-sparse genomes of *F. graminearum*, *P. nodorum* or

Mycosphaerella graminicola (Cuomo et al., 2007; Goodwin et al., 2011; Hane et al., 2007). While close association of effector genes with TEs and/or with TE-rich regions may tend to be the rule in ascomycetes, this does not seem to be true for TE-rich basidiomycete genomes; these genomes show only rare association of effector genes with TEs (Duplessis et al., 2011; Hacquard et al., 2011; Martin et al., 2008). In these fungi, lineage-specific expansion of numerous gene families is observed. Effector genes are among these expanded families, both in phytopathogens such as the rust-causing *Melampsora* species and in mutualists like *Laccaria bicolor* (Martin et al., 2008), and some families can contain more than 100 paralogs (Duplessis et al., 2011; Hacquard et al., 2011). However, no evidence has been detected so far for TE involvement in whole-genome duplication or large-scale dispersed segmental duplications. Interestingly, it is in small-sized basidiomycete genomes, such as the 26.1 Mb *Ustilago hordei*, 19.7 Mb *U. maydis* and 18.2 Mb *Sporisorium reilianum* genomes, that comparative genome analyses have revealed high synteny and association between TEs and effector clusters (Laurie et al., 2012).

3.2. TEs and the generation/diversification of novel pathogenicity genes

In all these species in which TEs are closely associated with effector genes, it is postulated and sometimes shown that TEs have favoured duplications and/or translocations, thereby allowing allelic diversification, neofunctionalization and eventually the birth of multigene families. TEs were also instrumental in promoting LGT between species sharing the same ecological niche and even across kingdom boundaries (Richards et al., 2011). They eventually contributed to the generation of new pathogenicity determinants.

Diversification of effector genes in some fungal species is postulated to be associated with TE-driven gene duplication and generation of multigene families. *B. graminis* is one of the few ascomycete species in which multigene families of effectors have been confidently predicted (Godfrey et al., 2010). One family of effectors in *B. graminis* is associated with a LINE family (Sacristan et al., 2009). In this species, retroduplications appear to have occurred frequently, and massive proliferation of retrotransposons associated with loss of RIP has probably contributed to the extensive losses, expansions and reshuffling of effector genes (Raffaele & Kamoun, 2012). In *M. oryzae*, *AVR-Pita* is closely associated with a retrotransposon, and variants of *AVR-Pita* are present at different chromosomal locations or on different chromosomes including dispensable chromosomes, depending on the isolate (Chuma et al., 2011). The authors postulated that other avirulence genes of *M. oryzae* may have the same behaviour and be also subjected to multiple translocations in the genome. In other ascomycete species, effector genes usually lack paralogs and unequivocal orthologs in related species, which is a signature of rapid diversification. However, in *L. maculans*, comparative genomics studies indicated that a few avirulence genes located close to TEs were translocated to other genome locations in Lmb compared with other species of the species complex in which orthologs can be identified (Grandaubert, 2013). This suggests that, during or after the time of speciation, effector genes were subjected to TE-mediated translocations in the Lmb genome, often resulting in their isolation within large AT-rich isochores. In many species and notably all Dothideomycetes and Clavicipitaceae investigated (except *C. purpurea*) in which TEs are systematically RIPed, the genes located close to TEs or embedded within large arrays of RIPed TEs also show signatures of RIP (Ohm et al., 2012; Rouxel et al., 2011; Schardl et al., 2013). Such genes are mostly genes encoding candidate effectors (SSPs and secondary metabolites). In *L. maculans*, for example, the characteristics of the effector genes strongly suggest that they were subjected to RIP as a result of their genome location (Rouxel et al., 2011). The consequent rapid sequence diversification contributes either to a loss of function due to deleterious effect of RIP or to an accelerated pace of mutation, eventually resulting in neofunctionalization and generation of novel effectors with novel specificities (Rouxel et al., 2011).

Although frequently postulated, only in a few cases were TEs demonstrated to have favoured LGT. The most evident example of that is the *SnToxA* gene of *P. nodorum* encoding a host selective proteinaceous toxin. The gene is associated with TEs, and this may have facilitated lateral transfer of this toxin gene from *P. nodorum* to *P. tritici-repentis*. In this case, the transfer of the new virulence determinant did not change the host range of *P. tritici-repentis* but resulted in a novel more severe disease of wheat that emerged in the 1950s (Friesen et al., 2006). In *F. oxysporum* species, adaptation to a new host species strongly depends upon interstrain transfer of TE-rich 'B' chromosomes (Ma et al., 2010). Similarly, *Alternaria* species can broaden their host range by lateral transfer of a single chromosome enriched in TEs and carrying a cluster of genes encoding host-specific toxins that enabled them to become pathogenic on new hosts such as apple, Japanese pear, strawberry and tomato (Mehrabi et al., 2011).

3.3. TEs and adaptation to the host in gene-for-gene systems

Fungi are the most important pathogens of cultivated plants, and whenever possible, plant genetic resistance to diseases is used as an environment-friendly strategy to control the diseases. One common approach allowing effective resistance is the incorporation of a major resistance (*R*) gene into a plant genotype facing the 'avirulent' part of the fungal population. Unfortunately, fungi show an incredible plasticity, allowing rapid emergence of novel, better-adapted strains. 'Breakdown' of resistance genes in a few years only is a common trait of fungal phytopathogens leading to a boom-and-bust cycle of commercial success of a variety with novel resistance followed by rapid change in the avirulence allele frequency, eventually leading to a complete loss of efficacy of the new resistance source (McDonald & Linde, 2002). As outlined in the preceding text, avirulence genes encode effectors, and more and more research indicates that the pathogen evolutionary potential should not be uniquely considered from a population genetics point of view, but rather must take into account the evolvability of avirulence genes as a function of their genome environment, especially location within the plastic genome compartment, and more generally their close association with TEs. In other words, aside from a role in evolution towards pathogens better adapted to their hosts or new hosts, TEs are also suggested to allow an 'immediate response' of the pathogen to the resistance gene selection. For example, the telomere-proximal *AVR-Pita* can be spontaneously lost in culture either by the complete loss of sequence or by the insertion of TEs (Orbach et al., 2000; Zhou, Jia, Singh, Correll, & Lee, 2007). In this species, a paralog of the TE-associated *AVR-Pita* (termed *AVR-Pita3*) is found at an invariant genome location, while *AVR-Pita* and other variants may be fully absent from the genome or present at different chromosomal locations or on different chromosomes, depending on the isolate (Chuma et al., 2011). In this species, both patterns of complete loss of the gene and multiple translocations were ascribed to the close association of *AVR-Pita* with a retrotransposon (Chuma et al., 2011). The adaptive advantage of these multiple translocations was considered by these authors to maintain a pool of isolates harbouring the avirulence gene when selection is exerted so that it can be easily disseminated in populations when the selection is no longer present.

In the literature, diverse molecular events resulting in breakdown of resistance gene have been described in the few fungal phytopathogens for which avirulence effectors have been identified (essentially *M. oryzae*, *C. fulvum*, *L. maculans*, *F. oxysporum*, *Melampsora lini* and *Rhynchosporium secalis*).

In most, this results in drastic modification of the coding sequence of the corresponding protein, including complete or partial deletion, which is a common feature to all known avirulence genes (Balesdent et al., 2013; Daverdin et al., 2012; Farman, 2007; Fudal et al., 2009; Gout et al., 2007; Rep, Meijer, Houterman, van der Does, & Cornelissen, 2004; Schürch, Linde, Knogge, Jackson, & McDonald, 2004; Zhou et al., 2007), except those that play a central role in fungal pathogenicity and fitness (Parlange et al., 2009; Stergiopoulos, De Kock, Lindhout, & De Wit, 2007). For example, in *C. fulvum*, most avirulence genes are dispensable except for *Avr4*: the corresponding protein binds to chitin and protects the fungus against the deleterious effects of plant chitinases. In this respect, evasion of *Cf-4*-mediated recognition is mainly achieved by single-point mutations in the *Avr4* gene leading to unstable Avr4 proteins (Joosten, Vogelsang, Cozijnsen, Verberne, & De Wit, 1997). Another reported mechanism generating polymorphism is point mutation leading to amino acid substitutions (Daverdin et al., 2012; Ellis, Dodds, & Lawrence, 2007; Parlange et al., 2009; Schürch et al., 2004; Stergiopoulos et al., 2007). For example, *AvrLm4*, known to be important for *L. maculans* fitness, escapes *Rlm4* recognition by a single-amino-acid change that is unlikely to alter the effector function of the protein (Parlange et al., 2009). Point mutations may also lead to nonfunctional or unstable proteins, for example, when mutating cysteine residues (Daverdin et al., 2012; Stergiopoulos et al., 2007), or result in the production of truncated proteins (Daverdin et al., 2012; Fudal et al., 2009; Kang, Lebrun, Farrall, & Valent, 2001; Stergiopoulos et al., 2007). These mutation events are generally rare and are mostly (except in the cases of RIP 'leakage') independent of the TE environment of the effector genes. In contrast, direct effects of TEs (or the genome environment rich in TEs) on effector genes can be observed, with recorded examples of 'insertion' of a transposon (Daverdin et al., 2012; Fudal, Böhnert, Tharreau, & Lebrun, 2005; Kang et al., 2001; Zhou et al., 2007) leading to gene inactivation. Such TE integrations within coding sequences or regulatory regions of effector genes can be observed even in the cases of severely RIPed TEs and then can be due to heterologous recombination between nonhomologous copies or TEs. In addition to a direct effect on gene structure, TE insertion can also result in alteration of the expression level of the gene, as observed in promoters of *AVR-Pita* in *M. oryzae* (Zhou et al., 2007). One last important way to rapidly escape resistance gene recognition is through massive RIP mutations that inactivate the gene and are directly linked either with a genome environment favouring RIP 'leakage' from surrounding TEs or to the canonical effect of RIP on genes that were duplicated due to their TE environment, as observed for *AvrLm4-7* of *L. maculans* (Daverdin et al., 2012; Fudal et al., 2009). In summary, most of the resistance breakdowns observed are due to drastic modifications either of the avirulence gene or of the resulting protein. Most of these events are directly linked with the TE-rich environment of avirulence genes with consequences to gene instability or increased mutational potential due to RIP. Of special relevance to understand resistance gene breakdown in the case of an avirulence gene located in a TE-rich environment, Daverdin et al. (2012) investigated the sequence of mutational events occurring in field populations of *L. maculans* subjected to three consecutive years of selection by the *Rlm7* resistance gene, to which the isolate had not been previously confronted. The corresponding avirulence gene *AvrLm4-7* is located within a large array of RIP-degenerated TEs (Parlange et al., 2009). Within the frame of the experiment, Daverdin et al. (2012) firstly observed a rapid shift in the pathogen populations, which were 100% avirulent at the beginning of the experiment and 40% virulent 3 years later at the end of the experiment. Analysis of the molecular events responsible for loss of the avirulent phenotype uncovered evidence for all the mechanisms previously reported in the literature for other fungal phytopathogens, plus a couple of new mechanisms, although this was investigated in a very small, 0.25 ha field. RIP mutations were prevalent at the start of the experiment, while partial or complete deletions of the gene predominated at the end of the experiment (Daverdin et al., 2012). This suggests that an immediate response to the selection involved RIP inactivation, due to the TE-rich genome environment promoting gene duplication and/or RIP 'leakage', and the essentially sexual reproduction regime of the fungus. At later stages, the inactivated gene and the TE-rich genome environment, in which no other gene is present, were deleted, for example, via TE-mediated intrachromosomal rearrangements or during meiosis following TE-mediated mispairing between homologous chromosomes. Eventually, this led to one dominant event of gene loss, as was observed previously for another avirulence gene of *L. maculans*, *AvrLm1* (Gout et al., 2007). It should be noted that although *L. maculans* is essentially a sexual species, essentially, asexual species such as *Verticillium dahliae* can also show accelerated rates of evolution for effectors located in a TE-rich genome environment (De Jonge et al., 2013).

4. CONCLUDING REMARKS

In fungi, genome invasion by TEs is postulated to have contributed to the success of phytopathogens, mutualists and endophytes, and as stated by Raffaele and Kamoun (2012), phytopathogens often have expanded genomes compared with their free-living relatives. In at least one system, TEs were shown to have contributed to intrachromosomal inversions, likely resulting in speciation as already described for more complex eukaryotes. The hosting of effector genes in TE-rich genome environment contributed to sequence diversification and probably adaptation to new hosts or better adaptation to the existing host. It currently contributes to a highly dynamic plasticity and diversification of these genes. In sexual species, RIP mutation increases the speed with which the detrimental genes can be mutated to extinction, rendering migration of low importance compared to the ability to generate variants locally in a single sexual cycle. The hosting of effector genes in such dynamic genome environments has a dual advantage for phytopathogens: immediate adaptation to a new source of resistance (provided the effector gene is dispensable) consistent with the speed of birth and dissemination of new virulent populations and long-term enhanced ability to duplicate and diversify effector genes to facilitate adaptation to the host or adaptation to new hosts. Recent advances suggested that the association of effector genes with TEs also has an impact on epigenetic, heterochromatin-based regulation of their expression. This, along with evolutionary consequences of this finding, provides a new and promising field for research.

ACKNOWLEDGEMENTS

J. G. was funded by the French agency Agence Nationale de la Recherche (ANR), contract ANR-09-GENM-028 (FungIsochores). The authors thank Richard O'Connell (INRA-Bioger) for the critical revision of the manuscript.

REFERENCES

Ambrozová, K., Mandakova, T., Bures, P., Neumann, P., Leitch, I. J., & Koblizkova, A. (2011). Diverse retrotransposon families and an AT-rich satellite DNA revealed in giant genomes of *Fritillaria lilies*. *Annals of Botany*, *107*, 255–268.

Aramayo, R., & Metzenberg, R. L. (1996). Meiotic transvection in fungi. *Cell*, *86*, 103–113.

Balesdent, M. H., Fudal, I., Ollivier, B., Bally, P., Grandaubert, J., Eber, F., et al. (2013). The dispensable chromosome of *Leptosphaeria maculans* shelters an effector gene conferring avirulence towards *Brassica rapa*. *The New Phytologist*, *198*, 887–898.

Bennetzen, J. L., Ma, J., & Devos, K. M. (2005). Mechanisms of recent genome size variation in flowering plants. *Current Opinion in Genetics & Development*, *15*, 621–627.

Biémont, C. (2010). A brief history of the status of transposable elements: From junk DNA to major players in evolution. *Genetics*, *186*, 1085–1093.

Biémont, C., & Vieira, C. (2006). Junk DNA as an evolutionary force. *Nature*, *443*, 521–524.

Biémont, C., Vieira, C., Hoogland, C., Cizeron, G., Loevenbruck, C., Arnault, C., et al. (1997). Maintenance of transposable copy element number in natural populations of *Drosophila melanogaster* and *D. simulans*. *Genetica*, *100*, 161–166.

Bird, A. (2007). Perceptions of epigenetics. *Nature*, *447*, 396–398.

Botstein, D., Chervitz, S. A., & Cherry, J. M. (1997). Yeast as a model organism. *Science*, *277*, 1259–1260.

Chuma, I., Isobe, C., Hotta, Y., Ibaragi, K., Futamata, N., & Kusaba, M. (2011). Multiple translocation of the *AVR-Pita* effector gene among chromosomes of the rice blast fungus *Magnaporthe oryzae* and related species. *PLoS Pathogens*, *7*, e1002147.

Clutterbuck, A. J. (2011). Genomic evidence of repeat-induced point mutation (RIP) in filamentous ascomycetes. *Fungal Genetics and Biology*, *48*, 306–326.

Cogoni, C., Irelan, J. T., Schumacher, M., Schmidhauser, T. J., Selker, E. U., & Macino, G. (1996). Transgene silencing of the *al-1* gene in vegetative cells of *Neurospora* is mediated by a cytoplasmic effector and does not depend on DNA-DNA interactions or DNA methylation. *The EMBO Journal*, *15*, 3153–3163.

Coleman, J. J., Rounsley, S. D., Rodriguez-Carres, M., Kuo, A., Wasmann, C. C., Grimwood, J., et al. (2009). The genome of *Nectria haematococca*: Contribution of supernumerary chromosomes to gene expansion. *PLoS Genetics*, *5*, e1000618.

Cuomo, C. A., Guldener, U., Xu, J. R., Trail, F., Turgeon, B. G., Di Pietro, A., et al. (2007). The *Fusarium graminearum* genome reveals a link between localized polymorphism and pathogen specialization. *Science*, *317*, 1400–1402.

Daverdin, G., Rouxel, T., Gout, L., Aubertot, J. N., Fudal, I., Meyer, M., et al. (2012). Genome structure and reproductive behaviour influence the evolutionary potential of a fungal phytopathogen. *PLoS Pathogens*, *8*, e1003020.

Dean, R. A., Talbot, N. J., Ebbole, D. J., Farman, M. L., Mitchell, T. K., Orbach, M. J., et al. (2005). The genome sequence of the rice blast fungus *Magnaporthe grisea*. *Nature*, *434*, 980–986.

De Jonge, R., Bolton, M. D., Kombrink, A., van den Berg, G. C., Yadeta, K. A., & Thomma, B. P. (2013). Extensive chromosomal reshuffling drives evolution of virulence in an asexual pathogen. *Genome Research*, *23*, 1271–1282.

Delprat, A., Negre, B., Puig, M., & Ruiz, A. (2009). The transposon *Galileo* generates natural chromosomal inversions in *Drosophila* by ectopic recombination. *PLoS One*, *4*, e7883.

De Wit, P. J., van der Burgt, A., Ökmen, B., Stergiopoulos, I., Abd-Elsalam, K. A., Aerts, A. L., et al. (2012). The genomes of the fungal plant pathogens *Cladosporium fulvum* and *Dothistroma septosporum* reveal adaptation to different hosts and lifestyles but also signatures of common ancestry. *PLoS Genetics*, *8*, e1003088.

Drinnenberg, I. A., Weinberg, D. E., Xie, K. T., Mower, J. P., Wolfe, K. H., Fink, G. R., et al. (2009). RNAi in budding yeast. *Science*, *326*, 544–550.

Duplessis, S., Cuomo, C. A., Lin, Y. C., Aerts, A., Tisserant, E., Veneault-Fourrey, C., et al. (2011). Obligate biotrophy features unraveled by the genomic analysis of rust fungi. *Proceedings of the National Academy of Sciences of the United States of America*, *108*, 9166–9171.

Ellis, J. G., Dodds, P. N., & Lawrence, G. J. (2007). The role of secreted proteins in diseases of plants caused by rust, powdery mildew and smut fungi. *Current Opinion in Microbiology*, *10*, 326–331.

Fablet, M., & Vieira, C. (2011). Evolvability, epigenetics and transposable elements. *Biomolecular Concepts*, *2*, 333–341.

Farman, M. L. (2007). Telomeres in the rice blast fungus *Magnaporthe oryzae*: The world of the end as we know it. *FEMS Microbiology Letters*, *273*, 125–132.

Fedoroff, N. V. (2012). Transposable elements, epigenetics, and genome evolution. *Science*, *338*, 758–767.

Finnegan, D. J. (1989). Eukaryotic transposable elements and genome evolution. *Trends in Genetics*, *5*, 103–107.

Flor, A. H. (1955). Host-parasite interactions in flax rust-its genetics and other implications. *Phytopathology*, *45*, 680–685.

Freitag, M., Williams, R. L., Kothe, G. O., & Selker, E. U. (2002). A cytosine methyltransferase homologue is essential for repeat-induced point mutation in *Neurospora crassa*. *Proceedings of the National Academy of Sciences of the United States of America*, *99*, 8802–8807.

Friesen, T. L., Stukenbrock, E. H., Liu, Z., Meinhardt, S., Ling, H., Faris, J. D., et al. (2006). Emergence of a new disease as a result of interspecific virulence gene transfer. *Nature Genetics*, *38*, 953–956.

Frost, L. S., Leplae, R., Summers, A. O., & Toussaint, A. (2005). Mobile genetic elements: The agents of open source evolution. *Nature Reviews. Microbiology*, *3*, 722–732.

Fudal, I., Böhnert, H. U., Tharreau, D., & Lebrun, M. H. (2005). Transposition of MINE, a composite retrotransposon, in the avirulence gene *ACE1* of the rice blast fungus *Magnaporthe grisea*. *Fungal Genetics and Biology*, *42*, 761–772.

Fudal, I., Ross, S., Brun, H., Besnard, A. L., Ermel, M., Kuhn, M. L., et al. (2009). Repeat-induced point mutation (RIP) as an alternative mechanism of evolution toward virulence in *Leptosphaeria maculans*. *Molecular Plant-Microbe Interactions*, *22*, 932–941.

Fudal, I., Ross, S., Gout, L., Blaise, F., Kuhn, M. L., Eckert, M. R., et al. (2007). Heterochromatin-like regions as ecological niches for avirulence genes in the *Leptosphaeria maculans* genome: Map-based cloning of AvrLm6. *Molecular Plant-Microbe Interactions*, *20*, 459–470.

Galagan, J. E., Calvo, S. E., Borkovich, K. A., Selker, E. U., Read, N. D., Jaffe, D., et al. (2003). The genome sequence of the filamentous fungus *Neurospora crassa*. *Nature*, *422*, 859–868.

Galagan, J. E., Calvo, S. E., Cuomo, C., Ma, L. J., Wortman, J. R., Batzoglou, S., et al. (2005). Sequencing of *Aspergillus nidulans* and comparative analysis with *A. fumigatus* and *A. oryzae*. *Nature*, *438*, 1105–1115.

Galagan, J. E., Henn, M. R., Ma, L. J., Cuomo, C. A., & Birren, B. (2005). Genomics of the fungal kingdom: Insights into eukaryotic biology. *Genome Research*, *15*, 1620–1631.

Galagan, J. E., & Selker, E. U. (2004). RIP: The evolutionary cost of genome defense. *Trends in Genetics*, *20*, 417–423.

Gan, P., Ikeda, K., Irieda, H., Narusaka, M., O'Connell, R. J., Narusaka, Y., et al. (2013). Comparative genomic and transcriptomic analyses reveal the hemibiotrophic stage shift of *Colletotrichum* fungi. *The New Phytologist*, *197*, 1236–1249.

Godfrey, D., Böhlenius, H., Pedersen, C., Zhang, Z., Emmersen, J., & Thordal-Christensen, H. (2010). Powdery mildew fungal effector candidates share N-terminal Y/F/WxC-motif. *BMC Genomics*, *11*, 317.

Goffeau, A., Barrell, B. G., Bussey, H., Davis, R. W., Dujon, B., Feldmann, H., et al. (1996). Life with 6000 genes. *Science*, *274*, 563–567.

Goodwin, S. B., M'Barek, S. B., Dhillon, B., Wittenberg, A. H., Crane, C. F., Hane, J. K., et al. (2011). Finished genome of the fungal wheat pathogen *Mycosphaerella graminicola* reveals dispensome structure, chromosome plasticity, and stealth pathogenesis. *PLoS Genetics*, 7, e1002070.

Gout, L., Fudal, I., Kuhn, M. L., Blaise, F., Eckert, M., Cattolico, L., et al. (2006). Lost in the middle of nowhere: The *AvrLm1* avirulence gene of the Dothideomycete *Leptosphaeria maculans*. *Molecular Microbiology*, *60*, 67–80.

Gout, L., Kuhn, M. L., Vincenot, L., Bernard-Samain, S., Cattolico, L., Barbetti, M., et al. (2007). Genome structure impacts molecular evolution at the *AvrLm1* avirulence locus of the plant pathogen *Leptosphaeria maculans*. *Environmental Microbiology*, *9*, 2978–2992.

Goyon, C., Rossignol, J. L., & Faugeron, G. (1996). Native DNA repeats and methylation in *Ascobolus*. *Nucleic Acids Research*, *24*, 3348–3356.

Graïa, F., Lespinet, O., Rimbault, B., Dequard-Chablat, M., Coppin, E., & Picard, M. (2001). Genome quality control: RIP (repeat-induced point mutation) comes to *Podospora*. *Molecular Microbiology*, *40*, 586–595.

Grandaubert, J. (2013). *Génomique comparative et évolutive au sein du complexe d'espèces Leptosphaeria maculans-Leptosphaeria biglobosa*. Ph.D. Thesis. France: Paris XI-Orsay University, 363 pp.

Gregory, T.R. (2013). Animal Genome Size Database. http://www.genomesize.com.

Greilhuber, J., Borsch, T., Muller, K., Worberg, A., Porembski, S., & Barthlott, W. (2006). Smallest angiosperm genomes found in *Lentibulariaceae*, with chromosomes of bacterial size. *Plant Biology, 8*, 770–777.

Hacquard, S., Petre, B., Frey, P., Hecker, A., Rouhier, N., & Duplessis, S. (2011). The poplar-poplar rust interaction: Insights from genomics and transcriptomics. *Journal of Pathogens, 2011*, 716041.

Hane, J. K., Lowe, R. G. T., Solomon, P. S., Tan, K. C., Schoch, C. L., Spatafora, J. W., et al. (2007). Dothideomycete–plant interactions illuminated by genome sequencing and EST analysis of the wheat pathogen *Stagonospora nodorum*. *The Plant Cell, 19*, 3347–3368.

Hane, J. K., Rouxel, T., Howlett, B. J., Kema, G. H., Goodwin, S. B., & Oliver, R. P. (2011). A novel mode of chromosomal evolution peculiar to filamentous Ascomycete fungi. *Genome Biology, 12*, R45.

Howlett, B. J. (2006). Secondary metabolite toxins and nutrition of plant pathogenic fungi. *Current Opinion in Plant Biology, 9*, 371–375.

Hua-Van, A., Le Rouzic, A., Boutin, T. S., Filee, J., & Capy, P. (2011). The struggle for life of the genome's selfish architects. *Biology Direct, 6*, 19.

Idnurm, A., & Howlett, B. J. (2003). Analysis of loss of pathogenicity mutants reveals that repeat-induced point mutations can occur in the Dothideomycete *Leptosphaeria maculans*. *Fungal Genetics and Biology, 39*, 31–37.

Irelan, J. T., & Selker, E. U. (1996). Gene silencing in filamentous fungi: RIP, MIP and quelling. *Journal of Genetics, 75*, 313–324.

Jaenisch, R., & Bird, A. (2003). Epigenetic regulation of gene expression: How the genome integrates intrinsic and environmental signals. *Nature Genetics, 33*(Suppl.), 245–254.

Jones, J. D., & Dangl, J. L. (2006). The plant immune system. *Nature, 444*, 323–329.

Joosten, M. H., Vogelsang, R., Cozijnsen, T. J., Verberne, M. C., & De Wit, P. J. (1997). The biotrophic fungus *Cladosporium fulvum* circumvents Cf-4-mediated resistance by producing unstable AVR4 elicitors. *The Plant Cell, 9*, 367–379.

Kamoun, S. (2007). Groovy times: Filamentous pathogen effectors revealed. *Current Opinion in Plant Biology, 10*, 358–365.

Kämper, J., Kahmann, R., Bolker, M., Ma, L. J., Brefort, T., Saville, B. J., et al. (2006). Insights from the genome of the biotrophic fungal plant pathogen *Ustilago maydis*. *Nature, 444*, 97–101.

Kang, S., Lebrun, M. H., Farrall, L., & Valent, B. (2001). Gain of virulence caused by insertion of a Pot3 transposon in a *Magnaporthe grisea* avirulence gene. *Molecular Plant-Microbe Interactions, 14*, 671–674.

Kapitonov, V. V., & Jurka, J. (2007). Helitrons on a roll: Eukaryotic rolling-circle transposons. *Trends in Genetics, 23*, 521–529.

Kobayashi, S., Goto-Yamamoto, N., & Hirochika, H. (2004). Retrotransposon-induced mutations in grape skin color. *Science, 304*, 982.

Lander, E. S., Linton, L. M., Birren, B., Nusbaum, C., Zody, M. C., Baldwin, J., et al. (2001). Initial sequencing and analysis of the human genome. *Nature, 409*, 860–921.

Laurie, J. D., Ali, S., Linning, R., Mannhaupt, G., Wong, P., Güldener, U., et al. (2012). Genome comparison of barley and maize smut fungi reveals targeted loss of RNA silencing components and species-specific presence of transposable elements. *The Plant Cell, 24*, 1733–1745.

Lewis, Z. A., Honda, S., Khlafallah, T. K., Jeffress, J. K., Freitag, M., Mohn, F., et al. (2008). Relics of repeat-induced point mutation direct heterochromatin formation in *Neurospora crassa*. *Genome Research, 19*, 427–437.

Ma, L. J., van der Does, H. C., Borkovich, K. A., Coleman, J. J., Daboussi, M. J., Di Pietro, A., et al. (2010). Comparative genomics reveals mobile pathogenicity chromosomes in *Fusarium*. *Nature, 464*, 367–373.

Manning, V. A., Pandelova, I., Dhillon, B., Wilhelm, L. J., Goodwin, S. B., Berlin, A. M., et al. (2013). Comparative genomics of a plant-pathogenic fungus, *Pyrenophora tritici-repentis*, reveals transduplication and the impact of repeat elements on pathogenicity and population divergence. *G3 (Bethesda)*, *3*, 41–63.

Martin, F., Aerts, A., Ahren, D., Brun, A., Danchin, E. G., Duchaussoy, F., et al. (2008). The genome of *Laccaria bicolor* provides insights into mycorrhizal symbiosis. *Nature*, *452*, 88–92.

Martin, F., & Kamoun, S. (2012). In *Effectors in plant-microbe interactions* (p. 426). Chichester, UK: Wiley-Blackwell (Press).

Martin, F., Kohler, A., Murat, C., Balestrini, R., Coutinho, P. M., Jaillon, O., et al. (2010). Perigord black truffle genome uncovers evolutionary origins and mechanisms of symbiosis. *Nature*, *464*, 1033–1038.

Maside, X., Bartolome, C., Assimacopoulos, S., & Charlesworth, B. (2001). Rates of movement and distribution of transposable elements in *Drosophila melanogaster*: In situ hybridization vs. Southern blotting data. *Genetical Research*, *78*, 121–136.

McClintock, B. (1950). The origin and behavior of mutable loci in maize. *Proceedings of the National Academy of Sciences of the United States of America*, *36*, 344–355.

McDonald, B. A., & Linde, C. (2002). Pathogen population genetics, evolutionary potential, and durable resistance. *Annual Review of Phytopathology*, *40*, 349–379.

Mehrabi, R., Bahkali, A. H., Abd-Elsalam, K. A., Moslem, M., Ben M'barek, S., Gohari, A. M., et al. (2011). Horizontal gene and chromosome transfer in plant pathogenic fungi affecting host range. *FEMS Microbiology Reviews*, *35*, 542–554.

Moazed, D., Bühler, M., Buker, S. M., Colmenares, S. U., Gerace, E. L., et al. (2006). Studies on the mechanism of RNAi-dependent heterochromatin assembly. *Cold Spring Harbor Symposia on Quantitative Biology*, *71*, 461–471.

Nekrutenko, A., & Li, W. H. (2001). Transposable elements are found in a large number of human protein-coding genes. *Trends in Genetics*, *17*, 619–621.

O'Connell, R. J., Thon, M. R., Hacquard, S., Amyotte, S. G., Kleemann, J., Torres, M. F., et al. (2012). Lifestyle transitions in plant pathogenic *Colletotrichum* fungi deciphered by genome and transcriptome analyses. *Nature Genetics*, *44*, 1060–1065.

Ohm, R. A., Feau, N., Henrissat, B., Schoch, C. L., Horwitz, B. A., Barry, K. W., et al. (2012). Diverse lifestyles and strategies of plant pathogenesis encoded in the genomes of eighteen Dothideomycetes fungi. *PLoS Pathogens*, *8*, e1003037.

Ohno, S. (1972). So much "junk" DNA in our genome. In H. H. Smith (Ed.), *Proceedings of the 23rd Brookhaven symposium on biology, session "evolution of genetic systems"* (pp. 366–370). New York: Gordon & Breach.

Oliver, K. R., & Greene, W. K. (2012). Transposable elements and viruses as factors in adaptation and evolution: An expansion and strengthening of the TE-thrust hypothesis. *Ecology and Evolution*, *2*, 2912–2933.

Orbach, M. J., Farrall, L., Sweigard, J. A., Chumley, F. G., & Valent, B. (2000). A telomeric avirulence gene determines efficacy for the rice blast resistance gene *Pi-ta*. *The Plant Cell*, *12*, 2019–2032.

Orgel, L. E., & Crick, F. H. C. (1980). Selfish DNA: The ultimate parasite. *Nature*, *284*, 604–607.

Parlange, F., Daverdin, G., Fudal, I., Kuhn, M. L., Balesdent, M. H., Blaise, F., et al. (2009). *Leptosphaeria maculans* avirulence gene *AvrLm4-7* confers a dual recognition specificity by the *Rlm4* and *Rlm7* resistance genes of oilseed rape, and circumvents *Rlm4*-mediated recognition through a single amino acid change. *Molecular Microbiology*, *71*, 851–863.

Pritham, E. J. (2009). Transposable elements and factors influencing their success in eukaryotes. *Journal of Heredity*, *100*, 648–655.

Raffaele, S., & Kamoun, S. (2012). Genome evolution in filamentous plant pathogens: Why bigger can be better. *Nature Reviews. Microbiology*, *10*, 417–430.

Rebollo, R., Horard, B., Hubert, B., & Vieira, C. (2010). Jumping genes and epigenetics: Towards new species. *Gene*, *454*, 1–7.

Rep, M., Meijer, M., Houterman, P. M., van der Does, H. C., & Cornelissen, B. J. (2004). A small, cysteine-rich protein secreted by *Fusarium oxysporum* during colonization of xylem vessels is required for I-3-mediated resistance in tomato. *Molecular Microbiology*, *53*, 1373–1383.

Richards, T. A., Soanes, D. M., Jones, M. D., Vasieva, O., Leonard, G., Paszkiewicz, K., et al. (2011). Horizontal gene transfer facilitated the evolution of plant parasitic mechanisms in the oomycetes. *Proceedings of the National Academy of Sciences of the United States of America*, *108*, 15258–15263.

Rouxel, T., Grandaubert, J., Hane, J. K., Hoede, C., van de Wouw, A. P., Couloux, A., et al. (2011). Effector diversification within compartments of the *Leptosphaeria maculans* genome affected by Repeat-Induced Point Mutations. *Nature Communications*, *2*, 202.

Sacristan, S., Vigouroux, M., Pedersen, C., Skamnioti, P., Thordal-Christensen, H., Micali, C., et al. (2009). Coevolution between a family of parasite virulence effectors and a class of LINE-1 retrotransposons. *PLoS One*, *4*, e7463.

Schardl, C. L., Young, C. A., Hesse, U., Amyotte, S. G., Andreeva, K., Calie, P. J., et al. (2013). Plant-symbiotic fungi as chemical engineers: Multi-genome analysis of the *Clavicipitaceae* reveals dynamics of alkaloid loci. *PLoS Genetics*, *9*, e1003323.

Schürch, S., Linde, C. C., Knogge, W., Jackson, L. F., & McDonald, B. A. (2004). Molecular population genetic analysis differentiates two virulence mechanisms of the fungal avirulence gene NIP1. *Molecular Plant-Microbe Interactions*, *17*, 1114–1125.

Spanu, P. D. (2012). The genomics of obligate (and nonobligate) biotrophs. *Annual Review of Phytopathology*, *50*, 91–109.

Spanu, P. D., Abbott, J. C., Amselem, J., Burgis, T. A., Soanes, D. M., Stuber, K., et al. (2010). Genome expansion and gene loss in powdery mildew fungi reveal tradeoffs in extreme parasitism. *Science*, *330*, 1543–1546.

Stergiopoulos, I., De Kock, M. J., Lindhout, P., & De Wit, P. J. (2007). Allelic variation in the effector genes of the tomato pathogen *Cladosporium fulvum* reveals different modes of adaptive evolution. *Molecular Plant-Microbe Interactions*, *20*, 1271–1283.

Tyler, B. M., & Rouxel, T. (2013). Effectors of fungi and oomycetes: Their virulence and avirulence functions and translocation from pathogen to host. In G. Sessa (Ed.), *Molecular plant immunity* (pp. 123–167). Ames, USA; Chichester, UK & Oxford, UK: John Wiley & Sons, Inc.

Valent, B., & Khang, C. H. (2010). Recent advances in rice blast effector research. *Current Opinion in Plant Biology*, *13*, 434–441.

van de Lagemaat, L. N., Landry, J. R., Mager, D. L., & Medstrand, P. (2003). Transposable elements in mammals promote regulatory variation and diversification of genes with specialized functions. *Trends in Genetics*, *19*, 530–536.

van't Slot, K. A. E., & Knogge, W. (2002). A dual role for microbial pathogen-derived effector proteins in plant disease and resistance. *Critical Reviews in Plant Sciences*, *21*, 229–271.

Weiberg, A., Wang, M., Lin, F. M., Zhao, H., Zhang, Z., Kaloshian, I., et al. (2013). Fungal small RNAs suppress plant immunity by hijacking host RNA interference pathways. *Science*, *342*, 118–123.

Wicker, T., Oberhaensli, S., Parlange, F., Buchmann, J. P., Shatalina, M., Roffler, S., et al. (2013). The wheat powdery mildew genome shows the unique evolution of an obligate biotroph. *Nature Genetics*, *45*, 1092–1096.

Wicker, T., Sabot, F., Hua-Van, A., Bennetzen, J. L., Capy, P., Chalhoub, B., et al. (2007). A unified classification system for eukaryotic transposable elements. *Nature Reviews. Genetics*, *8*, 973–982.

Wolpert, T. J., Dunkle, L. D., & Ciuffetti, L. M. (2002). Host-selective toxins and avirulence determinants: What's in a name? *Annual Review of Phytopathology*, *40*, 251–285.

Wood, V., Gwilliam, R., Rajandream, M. A., Lyne, M., Lyne, R., Stewart, A., et al. (2002). The genome sequence of *Schizosaccharomyces pombe*. *Nature*, *415*, 871–880.
Zhao, H., & Bourque, G. (2009). Recovering genome rearrangements in the mammalian phylogeny. *Genome Research*, *19*, 934–942.
Zheng, A., Lin, R., Zhang, D., Qin, P., Xu, L., Ai, P., et al. (2013). The evolution and pathogenic mechanisms of the rice sheath blight pathogen. *Nature Communications*, *4*, 1424.
Zhou, E., Jia, Y., Singh, P., Correll, J. C., & Lee, F. N. (2007). Instability of the *Magnaporthe oryzae* avirulence gene *AVR-Pita* alters virulence. *Fungal Genetics and Biology*, *44*, 1024–1034.

CHAPTER FOUR

The Genomics of Powdery Mildew Fungi: Past Achievements, Present Status and Future Prospects

Stéphane Hacquard[1]
Department of Plant-Microbe Interactions, Max Planck Institute for Plant Breeding Research, Cologne, Germany
[1]Corresponding author: e-mail address: hacquard@mpipz.mpg.de

Contents

Advances in Botanical Research, Volume 70
ISSN 0065-2296
http://dx.doi.org/10.1016/B978-0-12-397940-7.00004-5

Abstract

Powdery mildew fungi (Ascomycota phylum) are obligate biotrophic plant pathogens that can only grow and reproduce on living host cells. They infect a wide range of plants, including many crops and the diseases they cause are common, easily recognizable and widespread. Although functional investigations in these genetically intractable organisms have been hampered by their obligate biotrophic nature, recent advances in genomics and transcriptomics have contributed tremendously to our understanding of powdery mildew biology. Comparative genomics was a powerful tool to pinpoint what distinguishes powdery mildew fungi from other filamentous plant pathogens and helped us to better understand how obligate biotrophy evolved. Comparative genome analyses among isolates in both the wheat and the barley powdery mildew lineages revealed isolate-specific mosaic genome structures of evolutionary young and old haplogroups. In addition to providing hints into the evolutionary origin of powdery mildew fungi, the observed mosaic genome structure also reflects the reproductive mode of these pathogens and explains how the large standing genetic variation is generated in powdery mildew populations. In this chapter, I discuss how the revolution in genomics has contributed and will contribute in the future to better understand the obligate biotrophic lifestyle, the virulence arsenal, the reproductive mode and the evolutionary history of powdery mildew fungi.

1. GENERAL INTRODUCTION

During the last decade, innovations in genomic technologies have revolutionized the research in the field of plant–microbes interactions. The publication of the complete genome sequence of the first plant pathogenic fungus in 2005 (Dean et al., 2005) paved the way for the exponential increase of fungal and oomycete genome sequencing projects (Grigoriev et al., 2014; Pais et al., 2013). To date, dozens of genomes of filamentous plant pathogens have been sequenced, and ambitious projects have emerged such as the 1000 Fungal Genomes Project that aims to fill in the gaps in the fungal tree of life by sequencing at least two reference genomes from every known fungal family (http://1000.fungalgenomes.org/home/; Grigoriev et al., 2014). Next-generation genome sequencing also had a tremendous impact on the study of noncultivable and genetically intractable organisms like powdery mildew fungi. In this chapter, I describe how genome sequencing of powdery mildew fungi has changed the way we conduct research and has contributed to better understand their evolution, their reproduction and their biology. I further discuss the future challenges in powdery mildew genomics and in powdery mildew research in general.

2. BIOLOGY OF POWDERY MILDEW INFECTION

2.1. Introduction

Powdery mildew fungi are widespread plant pathogens that can infect more than 10,000 plant species including major cereals such as wheat and barley, vegetable crops such as tomato and cucurbits and ornamental species like roses (Glawe, 2008). The diseases they cause are also common on fruits and are characterized by easily recognizable patches of white to greyish, talcum powderlike growth. They have a significant impact on plant growth and yield quality. For instance, a reduction of up to 20% in grain yield has been observed in wheat fields in which susceptible cultivars severely infected by *Blumeria graminis* were grown (Conner, Kuzyk, & Su, 2003). Most powdery mildew species are host-specific or only able to infect a narrow host range, suggesting that their corresponding genomes encode distinct "toolboxes" of pathogenesis-associated genes (Schulze-Lefert & Panstruga, 2011). Importantly, all powdery mildews are obligate biotrophs, meaning that they cannot be cultivated outside their hosts. Hence, they are entirely dependent on water and nutrients supply from living host cells for their growth and reproduction (Panstruga, 2003).

2.2. Phylogeny

Powdery mildew fungi belong to the Erysiphales order of the Ascomycota phylum. The Erysiphales belong to the Leotiomycetes class (Wang et al., 2006), in which many fungal pathogens causing serious plant disease are found, including many necrotrophic fungal pathogens that have very contrasted host range and infection strategies compared with powdery mildew fungi (Amselem et al., 2011). To date, sixteen genera containing ~900 species have been described in the Erysiphales order (Braun & Cook, 2012). While 12 genera (including *Blumeria*, *Erysiphe* and *Golovinomyces*) are ectoparasites that produce vegetative mycelium and conidiospores epiphytically on the host surface, four genera are endoparasites that produce internal mycelia (Takamatsu, 2013). Only powdery mildew fungi belonging to the *Leveillula* genus produce true endophytic hyphae from which conidiospores arise and emerge through stomata (Takamatsu, 2013). Molecular phylogenetic analyses, based on the amplification and the sequencing of the ribosomal DNA and the internal transcribed spacer regions, indicate that powdery mildew fungi form a monophyletic group (Mori, Sato, &

Takamatsu, 2000; Wang et al., 2006). Therefore, the obligate biotrophic lifestyle of these pathogens may has arisen only once in their ancestry and has been further retained over evolutionary time.

2.3. Lifecycle and infection strategy

Infection by powdery mildew pathogens is initiated when an airborne ascospore (sexual spore) lands on a susceptible host plant. For most powdery mildew fungi, the ascospore germinates and differentiates a hypha that elongates to produce a swollen appressorium from which a penetration peg emerged. The penetration peg punctures the host surface using a combination of mechanical force and enzymatic degradation (Howard, 1997; Pryce-Jones, Carver, & Gurr, 1999) and differentiates a haustorium that remains separated from the host cytoplasm by the host plasma membrane (Micali, Neumann, Grunewald, Panstruga, & O'Connell, 2011; Szabo & Bushnell, 2001). The haustorium is a highly specialized structure that plays a central role in establishing and maintaining the intimate relationship with the host (Panstruga, 2003). In addition to its role in host nutrient uptake, the haustorium is also a platform for the secretion of small effector molecules that manipulate the host cell, thereby facilitating fungal colonization (Panstruga & Dodds, 2009). Once a successful interaction is established, secondary haustoria are formed and vegetative hyphae are produced epiphytically. Only few days after infection, conidiophores arise from the vegetative mycelium and produce large amounts of conidia (asexual spores) that are disseminated by winds to reinfect susceptible hosts (Glawe, 2008). Notably, germination of *B. graminis* conidia on ryegrass is triggered by C26 aldehydes, indicating that powdery mildew pathogens need chemical cues in epicuticular wax for germination (Ringelmann, Riedel, Riederer, & Hildebrandt, 2009). The polycyclic asexual development of powdery mildew fungi and their rapid generation time (time from spore germination to spore production) lead to epidemics that gradually build up over spring and summer seasons. The obligate biotrophic lifestyle of powdery mildew fungi implies that they must be able to survive on their hosts throughout seasons. In Europe, the continuity of autumn- and spring-sown crops provides a "green bridge" for the fungus, allowing an unbroken asexual cycle across seasons when the weather conditions are favourable (Wolfe & McDermott, 1994). Sexual reproduction can also occurs at the end of the growing season of the host plant, when compatible isolates mate. Cleistothecia, thought to enable overwintering of the pathogen, are then produced and release ascospores that will infect adequate susceptible host plants at the beginning of the next growing season.

3. GENOMIC INSIGHTS INTO THE OBLIGATE BIOTROPHIC LIFESTYLE

3.1. Introduction

Since decades, the obligate biotrophic lifestyle of powdery mildew fungi has been a bottleneck for functional and molecular genetic investigations. Despite many efforts (Chaure, Gurr, & Spanu, 2000; Christiansen, Knudsen, & Giese, 1995; Spanu & Panstruga, 2012), attempts to establish a reliable protocol for stable transformation of powdery mildew fungi have been hampered by the difficulty to cultivate them *in vitro*. Thus, the study of these pathogens remains challenging and many aspects of their biology have not been fully elucidated yet. So far, nine genomes of powdery mildew fungi have been sequenced (Table 4.1). Within the genus *Blumeria*, three isolates (DH14, K1 and A6) corresponding to the barley powdery mildew *Blumeria graminis* f. sp. *hordei* (*Bgh*) and four isolates (96224, 94202, JIW2 and 70) corresponding to the wheat powdery mildew *B. graminis* f. sp. *tritici* (*Bgt*) were genome-sequenced (Hacquard et al., 2013; Spanu et al., 2010; Wicker et al., 2013). Draft genomes are also available for *Erysiphe pisi* that specifically infects pea and for *Golovinomyces orontii*, a broad host range powdery mildew fungus that infects *Arabidopsis thaliana* (Spanu et al., 2010). All isolates were collected in Europe except the *Bgt* isolate 70 that was collected in Israel (Table 4.1).

3.2. Genome-size expansion

Genome sequencing of all powdery mildew fungi described in the preceding text invariably revealed increases in genome size compared with almost all nonobligate biotrophic ascomycetes sequenced so far (Fig. 4.1). Their genome sizes have been estimated at 120, 121 and 127 Mb for *Bgh* isolates DH14, A6 and K1, respectively, 151 Mb for *E. pisi*, 160 Mb for *G. orontii* and 180 Mb for the *Bgt* reference genome (Table 4.1). The size of their genomes is 3–4 × larger than most ascomycete genomes [e.g. *Mycosphaerella graminicola*, 39 Mb (Goodwin et al., 2011); *Magnaporthe oryzae*, 40 Mb (Dean et al., 2005); *Sclerotinia sclerotiorum*, 38 Mb (Amselem et al., 2011)], with the exception of *Tuber melanosporum* (125 Mb, Martin et al., 2010), a biotrophic ectomycorrhizal fungus that enters into symbiosis with the fine roots of deciduous trees and

Table 4.1 Genome characteristics of powdery mildew fungi

	***B. graminis* f. sp. *hordei* (*Bgh*[a])**			***B. graminis* f. sp. *tritici* (*Bgt*[a])**				***E. pisi*[a]**	***G. orontii*[a]**
Genus	*Blumeria*	*Blumeria*	*Blumeria*	*Blumeria*	*Blumeria*	*Blumeria*	*Blumeria*	*Erysiphe*	*Golovinomyces*
Isolate	DH14 (RG[b])	A6	K1	96224 (RG[b])	94202	JIW2	70	–	–
Collection site	England	Sweden	Germany	Switzerland	Switzerland	England	Israel	England	Germany
Host	Barley	Barley	Barley	Wheat	Wheat	Wheat	Wheat	Pea	*Arabidopsis*
Host range	Narrow	Narrow	Narrow	Narrow	Narrow	Narrow	Narrow	Narrow	Broad
Sequencing strategy	Sanger–SOLiD–454 pyrosequencing	454 pyrosequencing–Illumina	454 pyrosequencing–Illumina	454 pyrosequencing	Illumina	Illumina	Illumina	454 pyrosequencing	454 pyrosequencing
Coverage (*X*)	140	37	84	13	70	24	52	8.4	8.9
Assembly size (Mb)	88	60	65	82	72	65	77	41	65
Estimated genome size (Mb)	120	121	127	180	–	–	–	151	160
N50 length[c] (Kb)	18	2.6	3.9	48.7	–	–	–	2.2	1.2
CEGMA genes[d] (%)	97.5–99.60	90.7–94.8	98.8–99.2	95.5–98.3	–	–	–	82.6–91.5	48.7–71.7

Number of genes	5854	–	–	6540	–	–	–	–	–
Repeat content (%)	64	–	–	90	–	–	–	–	–
SNPs versus RG[e]	–	183,149	168,281	–	175,093	182,904	233,997	–	–
Reference	Spanu et al. (2010)	Hacquard et al. (2013)	Hacquard et al. (2013)	Wicker et al. (2013)	Wicker et al. (2013)	Wicker et al. (2013)	Wicker et al. (2013)	Spanu et al. (2010)	Spanu et al. (2010)

[a]*Bgh*, *Blumeria graminis* f. sp. *hordei*; *Bgt*, *Blumeria graminis* f. sp. *tritici*; *E. pisi*; *Erysiphe pisi*; *G. orontii*; *Golovinomyces orontii*.
[b]RG: Reference genome.
[c]N50 length is the length of the shortest contig/scaffold such that the sum of contigs/scaffolds of equal length or longer is at least 50% of the total length of all contigs/scaffolds.
[d]CEGMA: Core Eukaryotic Genes Mapping Approach. Percentages indicate the proportion of genes that are fully–partially covered.
[e]SNP: single-nucleotide polymorphism. Numbers of high-confidence SNPs are 116,687, 115,967 and 161,117 for *Bgt* isolates 94202, JIW2 and 70, respectively.

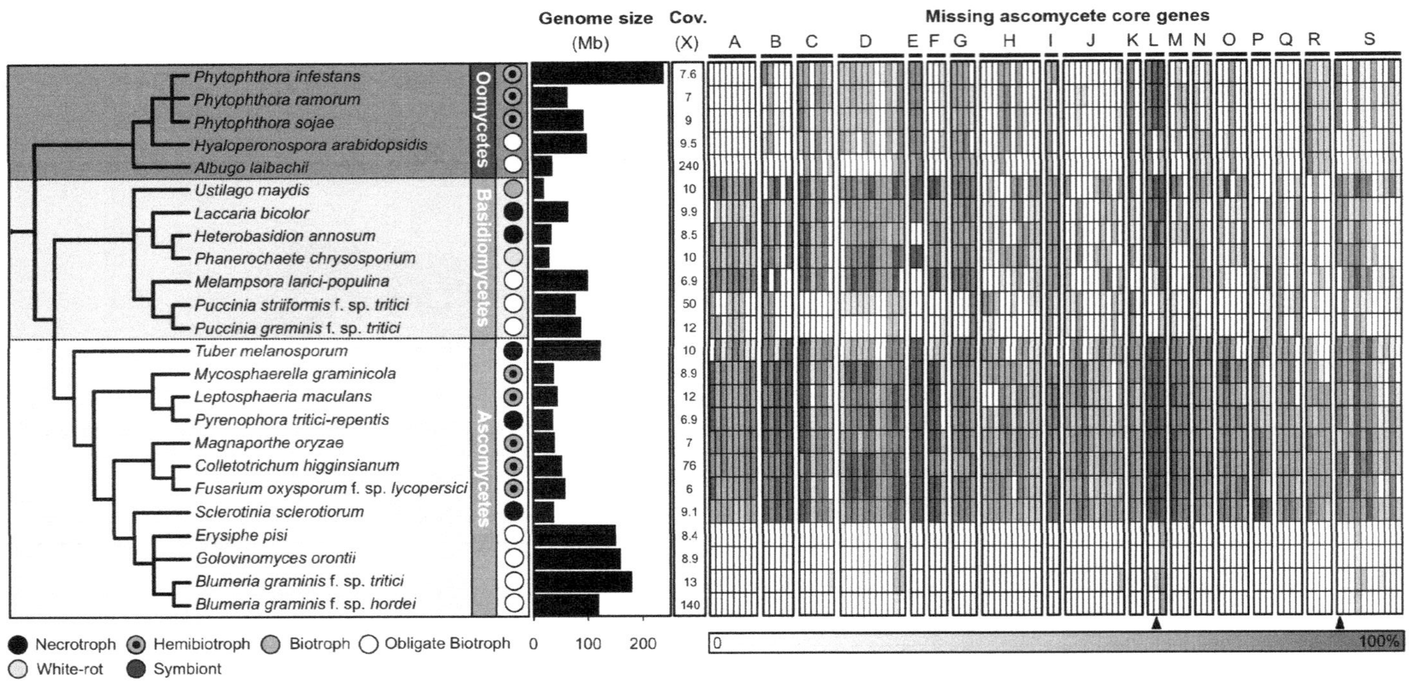

Figure 4.1 *Presence/absence and conservation profiles of 99 missing ascomycete core genes (MACGs) among three independent obligate lineages.* The phylogenetic tree has been generated using Interactive Tree Of Life (iTOL; Letunic & Bork, 2011) using NCBI taxonomy identifiers of 24 plant-interacting fungi and oomycetes for which genome sequences are available. These include 9 obligate biotrophs (white circles) belonging to the oomycete (*Hyaloperonospora arabidopsidis* and *Albugo laibachii*), the basidiomycete (*Melampsora larici-populina*, *Puccinia striiformis* f. sp. *tritici and Puccinia graminis* f. sp. *tritici*) and the ascomycete (*Erysiphe pisi*, *Golovinomyces orontii*, *Blumeria graminis* f. sp. *hordei and Blumeria graminis* f. sp. *tritici*) lineages. For each organism, the lifestyle, the genome size and the sequencing coverage depth are indicated on the right side of the tree. The 99 MAGCs were

(Continued)

Rhizophagus irregularis, an arbuscular mycorrhizal fungus having an obligate biotrophic lifestyle (>140 Mb; Tisserant et al., 2013; Lin et al., 2014; Fig. 4.1). Importantly, genome-size expansion has also been reported for obligate biotrophic rust fungi (Basidiomycota phylum) with estimated genome size ranging from 79 to 101 Mb (Cantu et al., 2011; Duplessis, Cuomo, et al., 2011; Fig. 4.1). For oomycetes, filamentous plant pathogens that are unrelated to fungi and diverged before the split of fungi from plant and animals (Rossman & Palm, 2006), significant differences in genome size have been observed among obligate biotrophs (Kemen & Jones, 2012; Raffaele & Kamoun, 2012). Indeed, the genome size of the downy mildew *Hyaloperonospora arabidopsidis* is comparable to those of rust fungi (100 Mb; Baxter et al., 2010), whereas the genome size of the obligate biotrophic pathogen *Albugo laibachii* that causes white rust on *Arabidopsis thaliana* is much smaller (37 Mb; Kemen et al., 2011). This indicates that an increase of genome size does not necessarily reflect the obligate biotrophic lifestyle (Fig. 4.1). Consistent with this, filamentous plant pathogens having the biggest genomes (>200 Mb) belong to the clade containing *Phytophthora infestans*, a hemibiotrophic oomycete pathogen responsible for the Irish potato famine (Cooke et al., 2012; Haas et al., 2009, Fig. 4.1).

3.3. Transposable elements proliferation

Genome-size inflation of powdery mildew pathogens can be explained by the proliferation of repetitive DNA and transposable elements (TEs),

Figure 4.1—Cont'd previously described and correspond to genes that were absent in the mildews but present in baker's yeast (*Saccharomyces cerevisiae*) and the phytopathogens *Colletotrichum higginsianum*, *Magnaporthe oryzae* and *Sclerotinia sclerotiorum* (Spanu et al., 2010). Conservation profiles [from 0% (white) to 100% (dark green)] of the 99 MAGCs were determined according to tBLASTn e-values. A, Thiamine metabolism/transport; B, allantoine metabolism/transport; C, methionine metabolism and (siro-)heme biosynthesis; D, alcohol metabolism/fermentation; E, glutamate metabolism; F, uracil metabolism/transport; G, glutathione metabolism; H, detoxification/stress response; I, arabinono-1,4-lactone biosynthesis; J, proteins of unknown function; K, chaperones; L, nitrate metabolism; M, proteases/peptidases; N, aromatic amino acid metabolism; O, channels/transporters; P, repeat-induced point (RIP) mutation; Q, mating type/cell cycle/budding; R, ER quality control; S, others. Gene identification numbers and functional annotation are described in Spanu (2012). Arrows indicate genes that are absent in all 9 genomes of obligate biotrophs but present in all other plant-interacting fungi and oomycetes. (See the colour plate.)

accounting for 64% and 90% of *Bgh* and *Bgt* genomes, respectively (Spanu et al., 2010; Wicker et al., 2013). Consistent with this overdose of repeats, the assembled fractions of powdery mildew genomes remain relatively fragmented and between 30% (*Bgh*) and 70% (*G. orontii*) of their sequences could not be assembled (Table 4.1). The repeat contents of *Bgh* and *Bgt* genomes are much higher than those reported in the genomes of *T. melanosporum* (58% of TEs; Martin et al., 2010), *Puccinia graminis* f. sp. *tritici* (45% of TEs; Duplessis, Cuomo, et al., 2011), *S. sclerotiorum* (8% of TEs; Amselem et al., 2011) and *M. oryzae* (10% of TEs; Dean et al., 2005). This inflation of TEs can be explain by the loss of genes required for repeat-induced point (RIP) mutations that has been observed in all powdery mildew fungi (Spanu et al., 2010; Fig. 4.1). RIP is a genome defence mechanism specific to fungi that hypermutate repetitive DNA and is suggested to prevent the accumulation of TEs (Selker, 2002). The lack of RIP may provide a potential advantage for pathogenic fungi as it facilitates genome rearrangement and duplication events caused by TE activity, thereby accelerating pathogen adaptation (Oliver & Greene, 2009). Paradoxically, RIP has also been suggested as an alternative mechanism of evolution in the ascomycete *Leptosphaeria maculans*, promoting rapid sequence diversification of effectors within AT-rich blocks of the genome (Fudal et al., 2009; Rouxel et al., 2011). Recently, it has been hypothesized that an increase in genetic variability driven by TEs activity may have conferred an advantage for pathogens such as powdery mildew fungi (Spanu, 2012). This is consistent with the fact that both barley and wheat powdery mildew fungi propagate primarily asexually (see in the succeeding text; Hacquard et al., 2013; Wicker et al., 2013) by successive polycyclic generations on their hosts. This suggests that TE proliferation in powdery mildew fungi might play a crucial role in generating extensive genetic polymorphism that contributes, together with the genetic diversity acquired by rare outbreeding events, to the rapid adaptation of the fungus to its changing environment. Thus, pathogenicity in powdery mildew pathogens may have been impacted by the activity of these elements as recently demonstrated by the resequencing of three isolates of *Pyrenophora tritici-repentis*, a necrotrophic fungus responsible for tan spot disease of wheat (Manning et al., 2013).

3.4. Gene family contraction

In powdery mildew fungi, genome-size inflation is not associated with an increase of gene repertoires since *Bgh* and *Bgt* contain, respectively, 5854 and 6540 predicted gene models (Spanu et al., 2010; Wicker et al., 2013), which are among the lowest gene sets predicted in filamentous plant pathogens (Raffaele & Kamoun, 2012). These numbers are likely accurate because CEGMA evaluation [Core Eukaryotic Genes Mapping Approach (Parra, Bradnam, & Korf, 2007)] indicated that 95.56% and 97.5% of the 248 core eukaryotic orthologous groups were full length in the wheat and barley powdery mildew reference genomes, respectively (Table 4.1). The most extreme contractions of gene families observed in powdery mildew genomes correspond to those encoding secondary metabolites and carbohydrate-active enzymes (Spanu et al., 2010). While the genomes of the relatively close species *S. sclerotiorum* and *Colletotrichum higginsianum* encode a large array of carbohydrate-active enzymes acting on plant cell walls (>100) (Amselem et al., 2011; O'Connell et al., 2012), the barley powdery mildew genome encodes only two cellulose-degrading enzymes, four hemicellulose-degrading enzymes and one pectin-degrading enzyme (Spanu et al., 2010). Importantly, many glycoside hydrolase families that are missing in both the wheat and the barley powdery mildew pathogens are also missing in the obligate biotrophic rust fungi *P. graminis* f. sp. *tritici* and *Melampsora larici-populina* (Duplessis, Cuomo, et al., 2011; Wicker et al., 2013). Similarly, a reduced set of genes encoding glycoside hydrolases have also been reported for the obligate biotrophic oomycetes *H. arabidopsidis* and *A. laibachii* (Baxter et al., 2010; Kemen et al., 2011; Zerillo et al., 2013), indicating convergence in three independent obligate lineages (McDowell, 2011). This likely reflects the absolute necessity for obligate biotrophs to remain hidden inside the host cell and to minimize cell wall-associated damage that can trigger host immunity. Consistent with their obligate biotrophic lifestyle, the set of genes encoding key secondary metabolism enzymes involved in the biosynthesis of phytotoxic compounds is also dramatically reduced in powdery mildew fungi (two key secondary metabolism enzymes in *Bgh*) compared with nonobligate biotrophic ascomycete fungi (20–103 key secondary metabolism enzymes) (O'Connell et al., 2012; Spanu et al., 2010).

3.5. Missing genes and pathways

Since decades, attempts to cultivate obligate biotrophic pathogens such as powdery mildew fungi, rust fungi or downy mildews on synthetic media were unsuccessful. The recent genome sequencing of several obligate biotrophs belonging to taxonomically independent lineages (ascomycetes, basidiomycetes and oomycetes) revealed substantial gene losses that may explain this recalcitrance towards artificial nutrient media (Baxter et al., 2010; Duplessis, Cuomo, et al., 2011; Kemen et al., 2011; Spanu et al., 2010). The careful inspection of *Bgh*, *E. pisi* and *G. orontii* genomic sequences identified a set of 99 missing ascomycete core genes (MACGs) that are absent in powdery mildew fungi and present in most autotrophic nonobligate biotrophic ascomycetes including *Saccharomyces cerevisiae*, *C. higginsianum*, *M. oryzae* and *S. sclerotiurum* (Spanu et al., 2010; Fig. 4.1). These missing genes encode various proteins or enzymes involved, for example, in the thiamine, the glutamate, the (siro)-heme, the methionine, the alcohol, the sulphate or the nitrate metabolism pathways. Importantly, some of these genes/pathways are also missing in other obligate biotrophic pathogens (Spanu, 2012; Wicker et al., 2013) and may have been lost because the corresponding metabolites can be obtained by the pathogen from the host cell during infection (Fig. 4.1). However, reanalysis of the sequence conservation profiles of the 99 MACGs among several genomes of obligate and nonobligate biotrophs indicates that evolution towards obligate biotrophy is not driven by a common and simple genomic adaptation. Although many genes that are lost in powdery mildew fungi were also identified as missing in the rust fungi *Puccinia striiformis* and *P. graminis* and in the oomycete pathogens *A. Laibachii* and *H. arabidopsidis*, numerous genes were still present in the genome of the poplar rust fungus *M. larici-populina* (Fig. 4.1). Reinspection of the genes that are invariably missing in all nine obligate biotrophic pathogens but present in all other plant-interacting organisms identifies only two genes encoding the nitrite reductase and the *S. cerevisiae* JLP1 enzyme, a Fe(II)-dependent sulfonate/alpha-ketoglutarate dioxygenase involved in sulfonate catabolism (arrows, Fig. 4.1; Crawford & Arst, 1993; Hogan, Auchtung, & Hausinger, 1999). This indicates that all obligate biotrophic pathogens sequenced so far cannot assimilate inorganic nitrogen and are also not able to use aliphatic sulfonate such as taurine, cysteate and isethionate as alternative sulphur sources. However, it is very unlikely that the loss of these two genes itself can lead to the obligate biotrophic lifestyle, which may rather be attributed to overlapping and lineage-specific gene losses. Consistent with this, the genome of the obligate symbiotic fungus *R. irregularis* encodes the nitrite reductase, illustrating that deficiency in inorganic nitrogen assimilation is not a common feature shared by symbiotic and pathogenic obligate biotrophs (Tisserant et al., 2013).

4. COMPARATIVE GENOMICS OF POWDERY MILDEW ISOLATES: INSIGHTS INTO THEIR REPRODUCTIVE MODE AND THEIR EVOLUTIONARY ORIGIN

4.1. Introduction

Comparative genomics provides a detailed view of the structural and the functional genomic relationships among different organisms. In fungi, comparative genomics among phylogenetically distant organisms (belonging to different phyla, classes or orders) has been extensively used and represents a powerful approach to identify specific genomic signatures reflecting the biology/lifestyle of a given organism (see in the preceding text). In this case, comparative analyses include comparisons of genome sizes, repeat contents, gene repertoires or gene families (Spanu, 2012). Comparative genomics among more closely related species (belonging to the same order, family and genus) identifies genetic variations that may reflect differences in host specificity, infection strategy or reproductive mode (Amselem et al., 2011; Duplessis, Cuomo, et al., 2011; O'Connell et al., 2012; Schirawski et al., 2010; Spanu et al., 2010). Finally, comparisons of different fungal isolates/strains belonging to a given species can reveal local genomic variations including single-nucleotide polymorphisms (SNPs), large deletions, duplications, chromosomal rearrangements and the presence of variable chromosome blocks that may contribute to the rapid adaptation of the pathogen to its changing environment (de Jonge et al., 2012, 2013; Hacquard et al., 2013; Manning et al., 2013; Stukenbrock, Christiansen, Hansen, Dutheil, & Schierup, 2012; Wicker et al., 2013; Xue et al., 2012).

4.2. Mosaic genome structure

Recently, comparative genome analyses of three barley powdery mildew isolates and of four wheat powdery mildew isolates revealed striking similarities between these two *formae speciales*, which diverged ~6.3 million years ago (Hacquard et al., 2013; Otto & Reid, 2013; Wicker et al., 2013). SNPs analysis revealed an overall frequency of ~1 SNP/kb for both *Bgh* and *Bgt* isolates compared with their respective reference genomes, as well as a particular genomic SNPs organization (Table 4.2; Hacquard et al., 2013; Wicker et al., 2013). Indeed, several large genomic segments are fully conserved among isolates and show low SNP frequency, whereas other regions accumulate high levels of isolate-specific SNPs (Hacquard et al., 2013; Wicker

Table 4.2 Proportion, SNP frequency and divergence time estimates of haplogroup segments among powdery mildew isolates

		***B. graminis* f. sp. *hordei* (*Bgh*[a])**		***B. graminis* f. sp. *tritici* (*Bgt*[a])**		
		A6	**K1**	**94202**	**JIW2**	**70**
H_{young}	Contribution versus RG (%)[b]	14	25	26	25	7
	SNPs frequency versus RG (SNPs/kb)[c]	0.06 ± 0.07	0.05 ± 0.06	0.11	0.11	0.22
	Divergence date versus RG (years)[d]	5700 ± 1200	4600 ± 1000	5407 ± 3241	5708 ± 3087	8690 ± 3054
H_{old}	Contribution versus RG (%)[b]	86	75	74	75	93
	SNPs frequency versus RG (SNPs/kb)[c]	1.68 ± 1.03	1.74 ± 1.11	1.20	1.11	1.31
	Divergence date versus RG (years)[d]	$137{,}500 \pm 20{,}600$	$141{,}700 \pm 21{,}200$	$60{,}439 \pm 13{,}374$	$55{,}423 \pm 12{,}156$	$63{,}157 \pm 13{,}222$

[a] *Bgh*, *Blumeria graminis* f. sp. *hordei*; *Bgt*, *Blumeria graminis* f. sp. *tritici*.
[b] Contribution of young (H_{young}) and old (H_{old}) haplogroups to the genome versus the reference genome (RG). For *Bgh* isolates, the RG is *Bgh* isolate DH14, and for *Bgt* isolates, the RG is *Bgt* isolate 96224.
[c] SNP: single-nucleotide polymorphism. Values indicate average SNP density for the haplogroup. Standard deviation is also indicated.
[d] Divergence time estimates were evaluated in these haplogroups using a mutation rate of $1.3 \times 10^{-8} \pm 2.29 \times 10^{-9}$ per site per year, which correspond to the mutation rate of intergenic regions in grasses (Ma & Bennetzen, 2004). Standard deviation is also indicated.

et al., 2013). The low SNP density blocks (0.05–0.06 SNPs/kb for *Bgh* isolates and 0.11–0.22 SNPs/kb for *Bgt* isolates) account for 14–25% of the genomes of *Bgh* isolates and for 7–26% of the genomes of *Bgt* isolates (Table 4.2). The high SNP density blocks, which accumulate 10–30 × more SNPs (1.68–1.74 SNPs/kb for *Bgh* isolates and 1.11–1.31 SNPs/kb for *Bgt* isolates), represent at least 70% of the wheat and the barley powdery mildew genomes (Table 4.2; Hacquard et al., 2013; Wicker et al., 2013). Since each isolate has its own SNP signature, authors have suggested that powdery mildew genomes are complex mosaics of different haplogroups (genetic population groups that share a common ancestor). This implies that powdery mildew genomes possess a large standing genetic variation in virulence polymorphism that may contribute to the enormous pool of genetic variation observed in mildew populations across Europe (Wolfe & McDermott, 1994).

4.3. Importance of clonal propagation

The peculiar genome structure showing large haplogroup segments (88–150 kb in the wheat powdery mildew isolates, Wicker et al., 2013) of different SNP densities indicates that mildew isolates are descended from relatively few sexual recombination events (Hacquard et al., 2013; Wicker et al., 2013). Indeed, if sexual reproduction would have been the main reproductive mode of the wheat and the barley powdery mildew pathogens, SNPs distribution would have been more homogenous along the genomes and such large haplogroup segments would not be distinguishable anymore. For the wheat powdery mildew pathogen, simulations indicated that the observed mosaic structure can be explained either by a small number of sexual generations (~100) in large populations (>1000 individuals) or by inbreeding of a very small population over a very long period of time (Wicker et al., 2013). Thus, authors hypothesized that the wheat and the barley powdery mildew pathogens primarily reproduce in a clonal manner, although near clonal reproduction that may include some inbreeding cannot be excluded (Hacquard et al., 2013; Wicker et al., 2013). The strong selection towards clonal reproduction may provide an advantage for pathogens by maintaining the virulence arsenal and the ideal genetic prerequisites necessary for successful colonization of the host (Bougnoux et al., 2008; Heitman, 2006). It is worth noting that sexual recombination between different mildew varieties has nonetheless occurred several times during the host–parasite coevolution, contributing to the observed mosaic genome structure. This not only generates genetic diversity required in the evolutionary arms race with the host but also facilitates the emergence of new mildew varieties that can infect new sorts of wheat or barley.

4.4. Evolutionary origin

Based on the number of SNPs detected in the high and low SNPs density blocks, two distinct groups corresponding to more divergent (H_{old}) and less divergent (H_{young}) haplogroups were distinguished (Hacquard et al., 2013; Wicker et al., 2013). Due to the lack of reliable common ancestor estimates via fossil data of fungi and the lack of spontaneous mutation rates in powdery mildews, divergence time estimates were evaluated in these haplogroups using a mutation rate of $1.3 \times 10^{-8} \pm 2.29 \times 10^{-9}$ per site per year, which correspond to the mutation rate of intergenic regions in grasses (Ma & Bennetzen, 2004; Oberhaensli et al., 2011). Using this calculation method, authors evaluated that H_{young} haplogroups of *Bgh* and *Bgt* isolates diverged very recently from their respective reference genomes (<8600 years ago for all European isolates and 5600–11,700 years ago for the *Bgt* isolate collected in Israel). In contrast, H_{old} haplogroups of *Bgt* and *Bgh* isolates diverged 43,000–76,000 and 117,000–163,000 years ago from their reference genomes, respectively (Table 4.2). Importantly, the H_{old} haplogroups accumulated mutation well before the onset of agriculture and the domestication of modern wheat and barley (10,000 years ago) and their divergence time estimates coincides with the last ice age (150,000–10,000 years ago). During this period, wild cereals were confined to the Fertile Crescent, a fertile arc-shaped area that extends from the eastern shore of the Mediterranean Sea to the Persian Gulf. The isolate-specific signatures of H_{old} haplogroups detected among *Bgt* (or *Bgh*) isolates may represent relics of different *Bgt* (or *Bgh*) lineages that have diverged in geographically distant area by coevolving with ancestral wheat (or barley) populations (Wicker et al., 2013). Based on the fact that H_{young} haplogroups of the Israeli *Bgt* isolate were fewer and diverged less recently compared with the European *Bgt* isolates (see in the preceding text), authors speculated that northbound agricultural migration (~10,000 years ago) may have restricted genetic exchange between these two *Bgt* lineages (Wicker et al., 2013).

5. POWDERY MILDEW EFFECTOR RESEARCH IN THE GENOMIC AREA

5.1. Introduction

In order to manipulate host defences and enable parasitic colonization, prokaryotic and eukaryotic plant pathogens deliver suites of effector proteins into different host-cell compartments (Stergiopoulos & de Wit, 2009). Although effectors primarily function as virulence factors, some of them can be recognized by intracellular plant receptors, leading to effector-triggered immunity (Jones & Dangl, 2006). The recognition of such effectors, dubbed avirulence (Avr) proteins, triggered plant immune responses that are often associated to a rapid cellular suicide of attacked plant cells (hypersensitive response). To date, two effector-like genes (AVR_{k1} and AVR_{a10}, *EKA* gene family) have been isolated from *Bgh* using a map-based cloning approach (Ridout et al., 2006). These powdery mildew effectors are atypical because they lack secretion signals, they coevolved with a family of LINE-1 retrotransposons and more than 1350 copies of *EKA* genes were detected in the genome of the barley powdery mildew pathogen (Sacristán et al., 2009; Spanu et al., 2010). So far, the avirulence function of these effectors has been validated using transient expression in MLK1 and MLA10 barley varieties and using host-induced gene silencing (Nowara et al., 2010; Ridout et al., 2006). In addition to these atypical effectors, another group of genes encoding candidate-secreted effector proteins has recently received growing attention. Recent advances in genomics have accelerated the identification of such effectors because it became possible to predict the whole complement of secreted proteins of a given organism. In powdery mildew fungi, candidate-secreted effector proteins (CSEPs) have been defined as proteins having a signal peptide but lacking a BLASTp hit outside the mildews in the nr database (Spanu et al., 2010).

5.2. Prediction and variation of CSEP repertoires

Originally, 248 CSEPs were predicted in the genome of the *Bgh* isolate DH14 (Spanu et al., 2010). However, the automated prediction of such genes encoding unknown proteins is error-prone and led in most cases to an underestimation of the predicted set of candidate effectors (Hacquard et al., 2012). Complementary strategies, based on iterative tBLASTn search using previously identified CSEPs as query sequences and *de novo* transcriptome assembly, have considerably improved the detection of additional candidate effector genes in the genome of the barley powdery mildew pathogen, exemplifying the importance of using complementary strategies for the prediction and the annotation of candidate effector genes (Pedersen et al., 2012). The current set of high-confidence *CSEPs* encoded in the genome of *Bgh* isolate DH14 is 491, all of them having been manually inspected (Fig. 4.2). Recently, isolate-specific variation in *CSEP* repertoires has been investigated among three barley powdery mildew isolates (DH14,

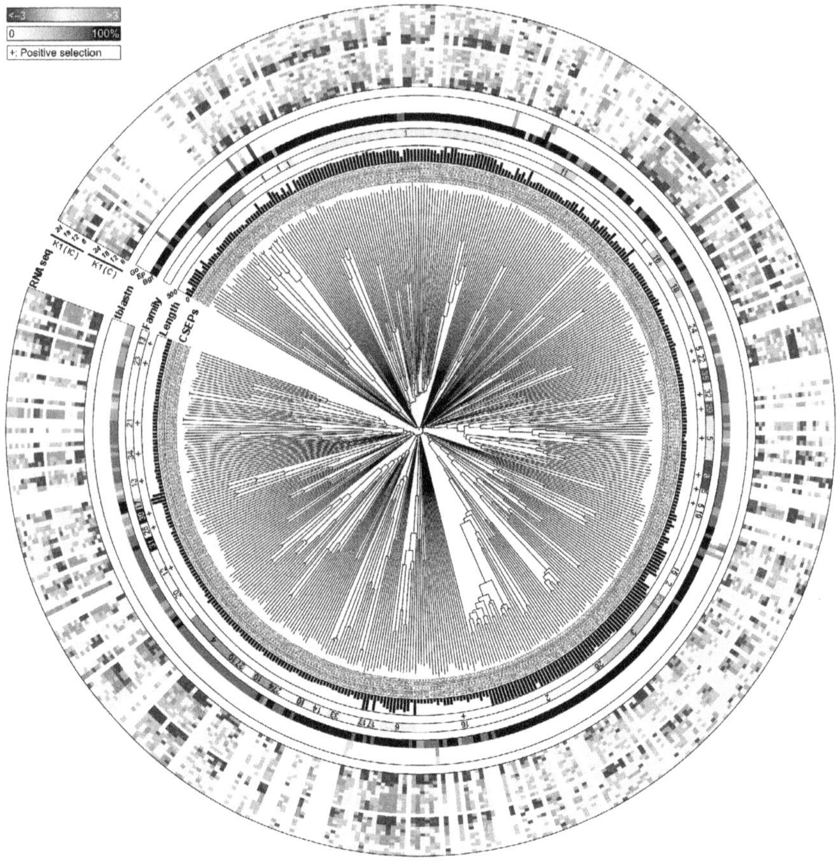

Figure 4.2 *Expression, conservation and evolution of the barley powdery mildew CSEP repertoire.* The circular plot was constructed using Interactive Tree of Life (iTOL; Letunic & Bork, 2011). At the centre of the circular plot is a dendrogram of the 491 powdery mildew *CSEPs* predicted in the reference genome of the isolate DH14 (Pedersen et al., 2012) The dendrogram was constructed based on a multiple alignment of the CSEP amino acid sequences using clustal W2, and the resulting alignment was then used to generate a neighbour-joining tree. The first ring next to the dendrogram corresponds to the CSEP length (number of amino acids) as previously reported (Pedersen et al., 2012). The second ring is colour-coded according to family affiliation so that all CSEPs belonging to the same family (according to MCL analysis) have the same colour. Only CSEP families that contain at least three members are highlighted in colour and their corresponding family identification numbers are indicated. CSEP gene families identified under diversifying selection (positive selection) are highlighted with a + (Pedersen et al., 2012). The third ring corresponds to the conservation profiles of the barley powdery mildew CSEPs in other mildew species as a function of tBLASTn e-values. Conservation profiles [from 0% (white) to 100% (black)] were deduced from the tBLASTn e-values. *Go, Golovinomyces orontii; Ep, Erysiphe pisi; Bgt, Blumeria graminis* f. sp. *tritici*. The outermost ring shows barley powdery mildew *CSEP* expression profiling (isolate K1) during

(Continued)

A6 and K1) (Hacquard et al., 2013). This revealed that a large proportion of CSEP genes present in the reference isolate DH14 have an ortholog in the isolate A6 (92%) or K1 (90%) and that only few *CSEPs* are completely lacking in one or both isolates (<1%). In the wheat powdery mildew reference genome, 437 *CSEPs* were identified plus 165 additional candidate effectors (CEPs) that have no signal peptide and no yeast homologues, leading to a total of 602 putative effector genes (Wicker et al., 2013). Importantly, the overall gene content is very similar among *Bgt* isolates and variations have been almost exclusively detected for candidate effector genes, representing 13 of the 16 missing genes (Wicker et al., 2013). Careful inspection of these regions in the wheat powdery mildew genome indicates that the observed gene losses may result from the repair of double-strand breaks (Wicker et al., 2013). CSEPs family comparison between the wheat and the barley powdery mildew pathogens indicates that most large families show differential expansion in either *Bgt* or *Bgh* (Hacquard et al., 2013; Wicker et al., 2013). Importantly, tBLASTn search using the barley powdery mildew CSEPs as queries indicates that almost all are not encoded in the genomes of the closely related powdery mildew species *G. orontii* and *E. pisi* (Pedersen et al., 2012; Spanu et al., 2010). However, most remain present, although sometimes weakly conserved, in the genome of the wheat powdery mildew pathogen (Fig. 4.2; Wicker et al., 2013). This indicates that powdery mildew pathogens belonging to different genera secrete drastically different effector arsenals to manipulate their hosts. This also likely reflects adaptation and cospeciation with their respective host plants (Spanu et al., 2010).

5.3. Evolution of CSEPs

Consistent with the model of a coevolutionary arms race between actors of the plant immune system and effectors of the pathogen, nucleotide sequences coding for candidate-secreted effectors often exhibit accelerated evolutionary rates (Dodds et al., 2006; Win et al., 2007). In powdery mildew

Figure 4.2—Cont'd colonization of partially immunocompromised *Arabidopsis* (pps: pen2, pad4 and sag101 background) expressing (IC: incompatible interaction) or not (C: compatible interaction) the cognate barley MLA1 immune receptor (Hacquard et al., 2013). 6 h postinoculation (hpi), conidiospore germination; 12 hpi, penetration of leaf epidermal cell walls; 18 hpi, formation of haustorial initials; 24 hpi, mature haustoria. Overrepresented (light red to dark red) and underrepresented transcripts (light blue to dark blue) are shown as log2-fold changes relative to the mean expression measured across all samples. (See the colour plate.)

fungi, ratio calculation of nonsynonymous versus synonymous substitution rates between *Bgt* and *Bgh* gene pairs or among *Bgh* isolates revealed elevated evolutionary rate for *CSEPs* compared with the other gene categories (Hacquard et al., 2013; Wicker et al., 2013). In the barley powdery mildew pathogen, effectors were grouped into 72 families containing between 2 and 59 members (Pedersen et al., 2012). The presence of large multigenic families indicates frequent gene duplication events during the evolution of the barley powdery mildew pathogen that can be driven by the activity of TEs. Selective pressure acting on these paralogous gene families has also been investigated, and strong evidence for diversifying (positive) selection was detected for families encoding mainly low-molecular-weight effector candidates, whereas purifying selection was observed for the largest families that encode high-molecular-weight proteins (Fig. 4.2; Pedersen et al., 2012). Notably, almost all members of the largest families ($\geq$3 members) showing evidence of diversifying selection cluster together in two separated clades in the phylogenetic tree (Fig. 4.2), indicating that they may arose from two ancestral CSEP families. Interestingly, members of gene families showing evidence of diversifying selection and encoding shorter proteins are preferentially expressed in haustoria, whereas members of the largest families are expressed more constitutively during powdery mildew infection (Fig. 4.2; Pedersen et al., 2012). Consistent with this, examination of polymorphism among *CSEP* orthologs in *Bgh* isolates revealed that haustorially expressed *CSEPs* accumulate higher rate of nonsynonymous substitutions compared with *CSEPs* that are expressed during other infection stages (Hacquard et al., 2013). Taken together, these results indicate that numerous haustorially expressed *CSEPs* become engaged in a coevolutionary arms race with the innate immune system of the host.

5.4. Structural features of CSEPs

The *CSEPs* identified in the barley powdery mildew pathogen encode mainly small secreted proteins of less than 300 amino acids (347/491). The frequency of cysteine residues is significantly higher in CSEPs than the frequency observed in non-CSEP proteins (Pedersen et al., 2012). Indeed, no less than 90% of *Bgh* CSEPs contain at least two cysteins that are potentially involved in the formation of disulfide bonds. In contrast with apoplastic pathogens, it has been hypothesized that the presence of disulphide bonds in secreted proteins of obligate biotrophs may preferentially serve as a structural or functional role rather than a protective role against apoplastic proteases (Hacquard et al., 2012). Importantly, this overrepresentation of cysteine residues is mainly due to the fact that 63% of *Bgh* CSEPs harbour a YxC motif within the first 25 amino acids of their mature protein sequences (Godfrey et al., 2010; Pedersen et al., 2012). This motif, initially detected in candidate effectors of the barley powdery mildew fungus *Bgh* (Godfrey et al., 2010), has also been identified in candidate effectors of rust fungi (Cantu et al., 2011; Godfrey et al., 2010; Hacquard et al., 2012; Saunders et al., 2012) suggesting functional and/or structural similarities among candidate effectors of haustoria-producing pathogenic fungi. At first, it was hypothesized that this motif could be involved in the translocation of effector proteins into the host cell, similarly to the RxLR motif identified in the N-terminal region of oomycete effectors (Whisson et al., 2007). Alternatively, the YxC motif might also play a crucial role for effector protein folding. Structural annotation using protein fold recognition methods indicated that only few powdery mildew CSEPs share structural similarities with other proteins for which secondary structures are known (Pedersen et al., 2012). However, no less than 72 showed sequence similarities with ribonucleases, indicating those likely evolved from an ancestral fungal gene encoding a secreted ribonuclease. However, the absence of active site residues seems to indicate that the ribonuclease activity has been lost in these CSEPs (Pedersen et al., 2012).

5.5. Functional analysis of CSEPs

Although detailed cataloguing of powdery mildew effector repertoires has been extensively reported, the biological function of individual CSEPs and their role during host infection remain poorly described (Spanu, 2012). So far, host-induced gene silencing (HIGS) has been mainly employed to characterize powdery mildew CSEP genes (Nowara et al., 2010; Pliego et al., 2013; Zhang et al., 2012). In this method, an RNA interference construct targeting a specific powdery mildew gene is transiently expressed in single host epidermal cells. The resulting double-stranded or antisense RNA molecules are exchanged from the host plant into the fungus, where target transcripts are degraded (Nowara et al., 2010). Using this strategy, 50 candidate effectors of the barley powdery mildew pathogen were recently screened and silencing of 8 *CSEPs* resulted in significant decreased pathogen development (Pliego et al., 2013). Notably, two of these genes belong to a large family encoding nonfunctional secreted ribonuclease that may interfere with the plant immune system through direct binding with host RNA molecules (Pedersen et al., 2012; Pliego et al., 2013). Consistent with this, cobombardment of barley epidermal cells with both the RNAi construct and the B-Peru/C1-expression plasmid pBC17 (Schweizer, Pokorny, Schulze-Lefert, & Dudler, 2000) revealed significant decreased number of anthocyanin-producing cells for the gene encoding a ribonuclease-like effector, underlining a possible role in suppression of plant programmed cell death (Pliego et al., 2013). Similarly, HIGS experiments also indicated that the YxC motif-containing effector CSEP0055 plays a role in host defence suppression (Zhang et al., 2012). Interestingly, this *CSEP* is expressed after 24 h, and the deduced protein physically interacts with the barley pathogenesis-related protein PR17c as demonstrated by yeast two-hybrid screen and bimolecular fluorescence complementation analyses. Notably, the apoplastic PR17c localizes to the *Bgh*-induced papilla (Christensen et al., 2002; Zhang et al., 2012) and silencing of this gene results in an almost 100% increase in *Bgh* penetration. Taken together, these data suggest that CSEP0055 sustains the fungus at sites of secondary penetration, where PR17c appears to be able to accumulate (Zhang et al., 2012). Recently, the functional characterization of five additional candidate effectors from the barley powdery mildew pathogen has been achieved by combining gene expression analysis, stably expressed transgenic *Arabidopsis* and yeast two-hybrid experiments coupled with bimolecular fluorescence complementation analyses (Schmidt et al., 2014). This multifaceted approach revealed a virulence-promoting activity for one of these candidates and suggests that the mildew conserved effector BEC5 might interfere with host vesicle trafficking (Schmidt et al., 2014).

6. TRANSCRIPTOMICS OF POWDERY MILDEW FUNGI

6.1. Introduction

Transcriptomic analyses reveal how the genetic programme of a specific organism is controlled and regulated in space and in time. This is particularly relevant during plant–pathogen interactions where fine-tuned transcriptional reprogramming occurs in both the host plant and the parasite, thereby controlling the outcome of the interaction (Wise, Moscou, Bogdanove, & Whitham, 2007). Most of the large-scale transcriptomic data obtained so far during the interaction between powdery mildew fungi and their respective host plants focused on the identification of host defence responses upon pathogen challenge (Caldo, Nettleton, & Wise, 2004; Chandran, Inada, Hather, Kleindt, & Wildermuth, 2010; Chandran et al., 2009; Gjetting et al., 2007; Maekawa et al., 2012; Moscou, Lauter, Caldo, Nettleton, & Wise, 2011). Although transcriptome analyses of early powdery mildew colonization events and in haustoria have been reported, our knowledge of the fungal transcriptome throughout the powdery mildew lifecycle remains incomplete.

6.2. Haustorial transcriptome

In powdery mildew fungi, transcriptomic studies have mainly focused on the identification of haustorially expressed genes using haustoria enrichment methods such as haustoria purification or host epidermal stripping (Godfrey et al., 2010; Micali et al., 2011; Spanu et al., 2010; Weßling et al., 2012). These studies underpin the high proportion and the high transcript levels of *CSEPs* in haustoria and the enrichment of transcripts encoding proteins involved in the detoxification of reactive oxygen species and the primary metabolism (Godfrey et al., 2010; Weßling et al., 2012). Consistent with these findings, proteomic studies also indicate that haustorial structures are enriched with small secreted proteins of unknown functions, stress-related proteins and proteins involved in carbohydrate metabolism (Bindschedler et al., 2009; Bindschedler, McGuffin, Burgis, Spanu, & Cramer, 2011; Godfrey, Zhang, Saalbach, & Thordal-Christensen, 2009).

6.3. Transcript profiling during host infection

Another interesting aspect of transcriptomic studies is the possibility to follow the dynamics of transcript activation during fungal development on host tissues (Both, Csukai, Stumpf, & Spanu, 2005; Both, Eckert, et al., 2005; Duplessis, Hacquard, et al., 2011; Hacquard et al., 2013; O'Connell et al., 2012). Powdery mildew gene expression profiling has been reported during *Bgh* infection on barley leaves using a custom microarray (Both, Csukai, Stumpf, & Spanu, 2005; Both, Eckert, et al., 2005). This revealed coordinate shifts in the expression patterns of genes encoding key enzymes of the primary metabolism (Both, Csukai, et al., 2005) and a cluster of genes encoding potential virulence determinants (Both, Eckert, et al., 2005). More recently, a dual RNA-seq approach has been carried out to monitor both fungal and host transcript activation during early stages of powdery mildew infection during both the compatible and the incompatible interactions using transgenic *Arabidopsis* as a host plant (Fig. 4.3; Hacquard et al., 2013; Maekawa et al., 2012). Although wild-type *Arabidopsis* is a strict

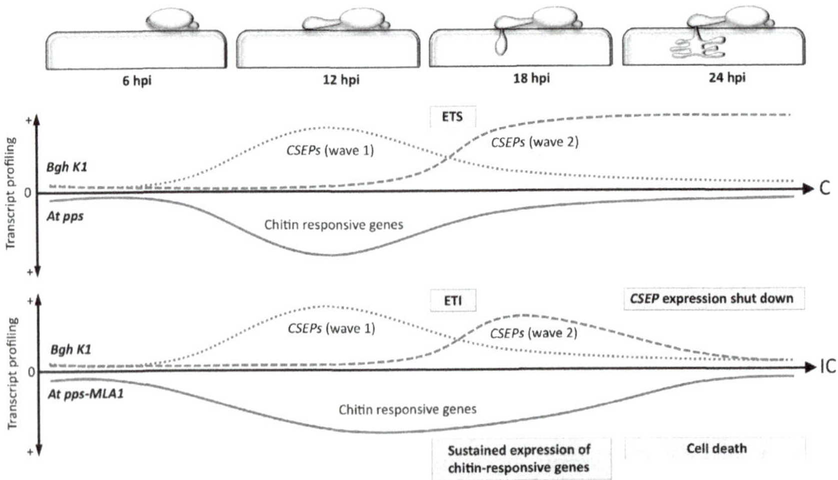

Figure 4.3 *Side by side comparison of Arabidopsis and Blumeria graminis f.* sp. *hordei gene expression profiles during compatible and MLA1-mediated incompatible interactions.* Transcriptional profiles of *B. graminis* f. sp. *hordei CSEPs* and *Arabidopsis* chitin-responsive genes are depicted based on RNA-seq data previously described (Hacquard et al., 2013; Maekawa et al., 2012). Partially immunocompromised *Arabidopsis* (*pps*: *pen2, pad4* and *sag101* background) expressing (*At pps*-MLA1) or not (*At pps*) the barley MLA1 immune receptor were infected with the *Bgh* isolate K1. C, compatible interaction; IC: incompatible interaction; 6 h postinoculation (hpi), conidiospore germination; 12 hpi, penetration of leaf epidermal cell walls; 18 hpi, formation of haustorial initials; 24 hpi, mature haustoria; ETS, effector-triggered susceptibility; ETI, effector-triggered immunity.

nonhost for the grass powdery mildew pathogen, simultaneous inactivation of the immunity components PEN2, PAD4 and SAG101 (triple mutant pps: *pen2*, *pad4* and *sag101*) renders *Arabidopsis* a host for *Bgh*, enabling completion of the asexual life cycle by conidiospore formation (Lipka et al., 2005). Furthermore, the barley MLA1 immune receptor was recently shown to be fully functional in this triple mutant background where it mediates the specific recognition of the *Bgh* isolate K1 (expressing the cognate avirulence effector AVR_{a1}), leading to the activation of race-specific immune responses including host-cell death (Maekawa et al., 2012). RNA-seq profiling on this pathosystem reveals that transcripts encoding powdery mildew effectors were barely detected in germinating conidiospores. However, distinct sets of transcripts accumulate in two successive waves during early colonization events, which coincide with the penetration of the host epidermis and the formation of haustoria, respectively (Fig. 4.3; Hacquard et al., 2013). Such dynamic transcriptional reprogramming of candidate effector genes has been also reported in rust fungi (Duplessis, Hacquard, et al., 2011), *Colletotrichum* fungi (Kleemann et al., 2012; O'Connell et al., 2012) and oomycetes (Win et al., 2012), indicating that filamentous pathogens sequentially deliver distinct suites of effector proteins during infection for efficient colonization of their host plants. The second wave of effector genes was specifically down-regulated at 24 hpi during the incompatible interaction. This shift in gene expression occurs right after the induction of host defence genes (18 hpi), among which numerous chitin-responsive genes were found (Fig. 4.3; Maekawa et al., 2012). This indicates that MLA-triggered immunity might directly or indirectly lead to the transcriptional shutdown of haustorially expressed *CSEPs* (Fig. 4.3; Hacquard et al., 2013). Transcripts encoding heat shock proteins and their corresponding proteins also seem to accumulate at very high levels in haustoria during both the compatible and the incompatible interactions (Bindschedler et al., 2009; Hacquard et al., 2013). These may play a role in the maintenance of haustorial membrane integrity and might be produced in response to stressful conditions such as those found in the host-cell environment. In contrast with the extensive transcriptional reprogramming of transport activity observed in rust haustoria (Hahn & Mendgen, 1997), only few transcripts encoding sugar and amino acid transporters were specifically expressed or significantly induced in powdery mildew haustoria (Hacquard et al., 2013; Weßling et al., 2012). This indicates that these structures may function primarily as organs for delivering small effector molecules. Notably, the transcriptional programme of *Bgh* is essentially conserved during early colonization stages on *Arabidopsis* and barley, two hosts that diverged ~200 million years ago. This suggests that during powdery mildew pathogenesis, the fungal gene expression programme is essentially robust against exposure to vastly divergent host-encoded molecules in monocots and dicots, including cell-wall constituents (Hacquard et al., 2013).

7. FUTURE CHALLENGES IN POWDERY MILDEW RESEARCH

Advances in sequencing technologies have led to the rapid decrease of sequencing costs. Thus, large-scale genome sequencing of powdery mildew fungal pathogens is now conceivable and will represent the next grand challenge in powdery mildew genome research. Among the 16 genera that cause powdery mildew (totalizing ~900 species), only the *Erysiphe*, the *Golovinomyces* and the *Blumeria* genera are represented among powdery mildew fungi that have been genome-sequenced. Selection of additional species belonging to other genera is needed to better understand the evolution, the ancestral features and the phenotypic characters of these fungi. The phenotypic variations include the morphology and the appendages, the mycelium (ectoparasite or endoparasite), the number of asci per cleistothecium and the conidiogenesis process (Takamatsu, 2013). The *de novo* genome sequencing of the endoparasitic powdery mildew fungus *Leveillula taurica (pepper pathogen, Leveillula* genus) has been initiated and will represent a unique opportunity to unravel specific features reflecting its exclusive endoparasitic mode (Kunoh, Kohno, Tashiro, & Ishizaki, 1979; Spanu, 2012).

One main question that remains unresolved and that has been enigmatic for decades is what makes an obligate biotrophic fungus obligate (Spanu, 2012). Although comparative genome analyses uncovered some convergent gene losses in three independent obligate lineages (basidiomycetes, ascomycetes and oomycetes), reanalysis of sequence conservation profiles of the MACGs indicates that evolution towards obligate biotrophy is not driven by a simple common genomic adaptation but rather by overlapping and lineage-specific gene losses. The specific loss of two genes encoding the nitrite reductase and the Fe(II)-dependent sulfonate/alpha-ketoglutarate dioxygenase in nine obligate biotrophic pathogens belonging to three independent lineages is striking and may represent a core adaptation to the living host cell. However, it is very unlikely that reintegration of these cDNA into the genomes of obligate pathogens renders them nonobligate because the corresponding regulatory networks and downstream signalling pathways might not be functional anymore. Such deficiencies in nitrogen and sulphur assimilation might be compensated by the fact that obligate biotrophs might have evolved alternative strategies to acquire these nutrients from the host plant. Consistent with this, it has been recently hypothesized based on transcriptomic data that host sulphate and nitrate transport activity might be induced upon infection with rust fungi and powdery mildew fungi, respectively, indicating a plausible link between obligate biotrophy and host transport manipulation (Petre et al., 2012; Pike et al., 2014).

Comparative genomics of powdery mildew isolates revealed isolate-specific mosaic genome structures that can be generated by rare outbreeding and frequent clonal reproduction (Hacquard et al., 2013; Wicker et al., 2013). However, authors did not exclude inbreeding by frequent sexual reproduction within a local population. To clarify this point, sequencing the genomes of dozens, if not hundreds, of geographically carefully selected isolates would give much more detailed haplogroups and their dependence on geographic distance and frequency of sexual versus asexual reproduction. Indeed, randomized but geographically stratified collections of *Bgh* or *Bgt* isolates are needed to estimate frequencies of geographically bounded sexual reproduction and clonal reproduction. Furthermore, a more detailed comparative study of the wheat and the barley powdery mildew genome sequences may uncover novel insights reflecting host specificity and cospeciation. Since these two *formae speciales* diverged 3–9 million years after the divergence of their hosts (Chalupska et al., 2008; Wicker et al., 2013), this indicates that the same pathogen may have been able to infect both wheat and barley hosts during a transition period (Oberhaensli et al., 2011).

Although there is increasing experimental evidence indicating that powdery mildew CSEPs play a role in fungal virulence (Pliego et al., 2013; Schmidt et al., 2014; Zhang et al., 2012), it remains nevertheless unknown whether some of them are recognized by cognate host resistance proteins, leading to effector-triggered immunity. In addition to the classical map-based cloning approach, Avr effectors can also be identified using EMS mutagenesis of an avirulent isolate followed by selection of EMS-derived mutants that can grow on host plants expressing the cognate resistance protein (Ridout et al., 2006). With the development of cutting-edge and high-throughput sequencing technologies, it is now possible to sequence hundreds of powdery mildew genomes. Thus, a population genomic approach can be initiated in order to understand the evolution of virulence loci in the wheat or the barley powdery mildew pathogens. This implies first to determine the virulence/avirulence pattern of each selected isolate using near-isogenic barley or wheat lines expressing different resistance specificities. Then, association-based analyses can be used to link the observed genetic variations with the virulence/avirulence phenotype. Using this strategy, it becomes possible to identify genomic segments subjected to selective sweep that may reflect major selection events such as the overcoming of a major resistance gene. So far, such genome-wide association studies have not been widely used to study virulence determinants in plant pathogens but represent a powerful approach to identify avirulence genes, as exemplified by the identification of novel Avr genes in the rice blast fungus *M. oryzae* and in the soil-borne vascular wilt fungus *Verticillium dahliae* (de Jonge et al., 2012; Yoshida et al., 2009).

In nature, the model plant *A. thaliana* can be colonized by at least four powdery mildew fungi, including *G. orontii*. This represents a unique opportunity to dissect the molecular basis of both host immune responses and powdery mildew pathogenesis (Micali, Göllner, Humphry, Consonni, & Panstruga, 2008). Furthermore, considerable efforts have been undertaken during the last 10 years in order to develop a novel tractable pathosystem using the barley powdery mildew pathogen and the nonhost plant model *A. thaliana*. Using this pathosystem, it has been possible to reconstruct MLA-triggered immunity in dicotyledonous *Arabidopsis*. This indicates that both *Bgh* effector delivery to plant cell and MLA resistance signalling must be conserved between monocots and dicots (Maekawa et al., 2012). Consistent with a functional conservation of the barley MLA1 receptor in *Arabidopsis*, auto-active MLA10 variants are also capable to trigger cell death in the dicotyledonous plant species *Nicotiana benthamiana* (Maekawa et al., 2011). The use of *Arabidopsis* as a host plant and the fact that host immune responses are conserved across plant lineages can tremendously accelerate powdery mildew effector research (Maekawa et al., 2012). For example, it becomes possible to use transgenic *N. benthamiana* or *Arabidopsis* lines expressing different immune receptors to validate the avirulence activity of some candidate AVR effectors and to understand downstream plant defence signalling pathways. It has been suggested that powdery mildew Avr effectors might directly interact with the LRR domain of the cognate immune receptors (Seeholzer et al., 2010; Yahiaoui, Brunner, & Keller, 2006). Alternatively, it is also tempting to speculate that immune receptors could indirectly recognize powdery mildew effectors through association with a host protein that serves as effector target, often called guardee (van der Hoorn & Kamoun, 2008). Within the next few years, there is no doubt that large-scale genome sequencing of powdery mildew species will contribute to answer these questions and will help us to better understand host–powdery mildew coevolution and cospeciation.

ACKNOWLEDGEMENTS

I would like to warmly thank Paul Schulze-Lefert for the constructive discussions, Takaki Maekawa for the valuable comments regarding this manuscript and Francis Martin for giving me the opportunity to write this book chapter.

REFERENCES

Amselem, J., Cuomo, C. A., van Kan, J. A., Viaud, M., Benito, E. P., Couloux, A., et al. (2011). Genomic analysis of the necrotrophic fungal pathogens *Sclerotinia sclerotiorum* and *Botrytis cinerea*. *PLoS Genetics*, 7, e1002230.

Baxter, L., Tripathy, S., Ishaque, N., Boot, N., Cabral, A., Kemen, E., et al. (2010). Signatures of adaptation to obligate biotrophy in the *Hyaloperonospora arabidopsidis* genome. *Science, 330,* 1549–1551.

Bindschedler, L. V., Burgis, T. A., Mills, D. J., Ho, J. T., Cramer, R., & Spanu, P. D. (2009). *In planta* proteomics and proteogenomics of the biotrophic barley fungal pathogen *Blumeria graminis* f. sp. *hordei*. *Molecular & Cellular Proteomics, 8,* 2368–2381.

Bindschedler, L. V., McGuffin, L. J., Burgis, T. A., Spanu, P. D., & Cramer, R. (2011). Proteogenomics and *in silico* structural and functional annotation of the barley powdery mildew *Blumeria graminis* f. sp. *hordei*. *Methods, 54,* 432–441.

Both, M., Csukai, M., Stumpf, M. P., & Spanu, P. D. (2005). Gene expression profiles of *Blumeria graminis* indicate dynamic changes to primary metabolism during development of an obligate biotrophic pathogen. *The Plant Cell, 17,* 2107–2122.

Both, M., Eckert, S. E., Csukai, M., Mueller, E., Dimopoulos, G., & Spanu, P. D. (2005). Transcript profiles of *Blumeria graminis* development during infection reveal a cluster of genes that are potential virulence determinants. *Molecular Plant-Microbe Interactions, 18,* 125–133.

Bougnoux, M. E., Pujol, C., Diogo, D., Bouchier, C., Soll, D. R., & d'Enfert, C. (2008). Mating is rare within as well as between clades of the human pathogen *Candida albicans*. *Fungal Genetics and Biology, 45,* 221–231.

Braun, U., & Cook, R. T. A. (2012). Taxonomic manual of the Erysiphales (powdery mildews). *CBS Biodiversity Series, 11,* 1–707.

Caldo, R. A., Nettleton, D., & Wise, R. P. (2004). Interaction-dependent gene expression in Mla-specified response to barley powdery mildew. *The Plant Cell, 16,* 2514–2528.

Cantu, D., Govindarajulu, M., Kozik, A., Wang, M., Chen, X., Kojima, K. K., et al. (2011). Next generation sequencing provides rapid access to the genome of *Puccinia striiformis* f. sp. *tritici*, the causal agent of wheat stripe rust. *PLoS One, 6,* e24230.

Chalupska, D., Lee, H. Y., Faris, J. D., Evrard, A., Chalhoub, B., Haselkorn, R., et al. (2008). Acc homoeoloci and the evolution of wheat genomes. *Proceedings of the National Academy of Sciences of the United States of America, 105,* 9691–9696.

Chandran, D., Inada, N., Hather, G., Kleindt, C. K., & Wildermuth, M. C. (2010). Laser microdissection of Arabidopsis cells at the powdery mildew infection site reveals site-specific processes and regulators. *Proceedings of the National Academy of Sciences of the United States of America, 107,* 460–465.

Chandran, D., Tai, Y. C., Hather, G., Dewdney, J., Denoux, C., Burgess, D. G., et al. (2009). Temporal global expression data reveal known and novel salicylate-impacted processes and regulators mediating powdery mildew growth and reproduction on Arabidopsis. *Plant Physiology, 149,* 1435–1451.

Chaure, P., Gurr, S. J., & Spanu, P. (2000). Stable transformation of *Erysiphe graminis*, an obligate biotrophic pathogen of barley. *Nature Biotechnology, 18,* 205–207.

Christensen, A. B., Cho, B. H., Næsby, M., Gregersen, P. L., Brandt, J., Madriz-Ordeñana, K., et al. (2002). The molecular characterization of two barley proteins establishes the novel PR-17 family of pathogenesis-related proteins. *Molecular Plant Pathology, 3,* 135–144.

Christiansen, S. K., Knudsen, S., & Giese, H. (1995). Biolistic transformation of the obligate plant pathogenic fungus, *Erysiphe graminis* f. sp. *hordei*. *Current Genetics, 29,* 100–102.

Conner, R. L., Kuzyk, A. D., & Su, H. (2003). Impact of powdery mildew on the yield of soft white spring wheat cultivars. *Canadian Journal of Plant Science, 83,* 725–728.

Cooke, D. E., Cano, L. M., Raffaele, S., Bain, R. A., Cooke, L. R., Etherington, G. J., et al. (2012). Genome analyses of an aggressive and invasive lineage of the Irish potato famine pathogen. *PLoS Pathogens, 8,* e1002940.

Crawford, N. M., & Arst, H. N., Jr. (1993). The molecular genetics of nitrate assimilation in fungi and plants. *Annual Review of Genetics, 27,* 115–146.

Dean, R. A., Talbot, N. J., Ebbole, D. J., Farman, M. L., Mitchell, T. K., Orbach, M. J., et al. (2005). The genome sequence of the rice blast fungus *Magnaporthe grisea*. *Nature*, *434*, 980–986.

de Jonge, R., Bolton, M. D., Kombrink, A., van den Berg, G. C., Yadeta, K. A., & Thomma, B. P. (2013). Extensive chromosomal reshuffling drives evolution of virulence in an asexual pathogen. *Genome Research*, *23*, 1271–1282.

de Jonge, R., van Esse, H. P., Maruthachalam, K., Bolton, M. D., Santhanam, P., Saber, M. K., et al. (2012). Tomato immune receptor Ve1 recognizes effector of multiple fungal pathogens uncovered by genome and RNA sequencing. *Proceedings of the National Academy of Sciences of the United States of America*, *109*, 5110–5115.

Dodds, P. N., Lawrence, G. J., Catanzariti, A. M., Teh, T., Wang, C. I. A., Ayliffe, M. A., et al. (2006). Direct protein interaction underlies gene-for-gene specificity and coevolution of the flax resistance genes and flax rust avirulence genes. *Proceedings of the National Academy of Sciences of the United States of America*, *103*, 8888–8893.

Duplessis, S., Cuomo, C. A., Lin, Y. C., Aerts, A., Tisserant, E., Veneault-Fourrey, C., et al. (2011). Obligate biotrophy features unraveled by the genomic analysis of rust fungi. *Proceedings of the National Academy of Sciences of the United States of America*, *108*, 9166–9171.

Duplessis, S., Hacquard, S., Delaruelle, C., Tisserant, E., Frey, P., Martin, F., et al. (2011). *Melampsora larici-populina* transcript profiling during germination and timecourse infection of poplar leaves reveals dynamic expression patterns associated with virulence and biotrophy. *Molecular Plant-Microbe Interactions*, *24*, 808–818.

Fudal, I., Ross, S., Brun, H., Besnard, A. L., Ermel, M., Kuhn, M. L., et al. (2009). Repeat-induced point mutation (RIP) as an alternative mechanism of evolution toward virulence in *Leptosphaeria maculans*. *Molecular Plant-Microbe Interactions*, *22*, 932–941.

Gjetting, T., Hagedorn, P. H., Schweizer, P., Thordal-Christensen, H., Carver, T. L., & Lyngkjaer, M. F. (2007). Single-cell transcript profiling of barley attacked by the powdery mildew fungus. *Molecular Plant-Microbe Interactions*, *20*, 235–246.

Glawe, D. A. (2008). The powdery mildews: A review of the world's most familiar (yet poorly known) plant pathogens. *Annual Review of Phytopathology*, *46*, 27–51.

Godfrey, D., Böhlenius, H., Pedersen, C., Zhang, Z., Emmersen, J., & Thordal-Christensen, H. (2010). Powdery mildew fungal effector candidates share N-terminal Y/F/WxC-motif. *BMC Genomics*, *11*, 317.

Godfrey, D., Zhang, Z., Saalbach, G., & Thordal-Christensen, H. (2009). A proteomics study of barley powdery mildew haustoria. *Proteomics*, *9*, 3222–3232.

Goodwin, S. B., M'barek, S. B., Dhillon, B., Wittenberg, A. H., Crane, C. F., Hane, J. K., et al. (2011). Finished genome of the fungal wheat pathogen *Mycosphaerella graminicola* reveals dispensome structure, chromosome plasticity, and stealth pathogenesis. *PLoS Genetics*, *7*, e1002070.

Grigoriev, I. V., Nikitin, R., Haridas, S., Kuo, A., Ohm, R., Otillar, R., et al. (2014). MycoCosm portal: Gearing up for 1000 fungal genomes. *Nucleic Acids Research*, *42*, D699–D704.

Haas, B. J., Kamoun, S., Zody, M. C., Jiang, R. H., Handsaker, R. E., Cano, L. M., et al. (2009). Genome sequence and analysis of the Irish potato famine pathogen *Phytophthora infestans*. *Nature*, *461*, 393–498.

Hacquard, S., Joly, D. L., Lin, Y. C., Tisserant, E., Feau, N., Delaruelle, C., et al. (2012). A comprehensive analysis of genes encoding small secreted proteins identifies candidate effectors in *Melampsora larici-populina* (poplar leaf rust). *Molecular Plant-Microbe Interactions*, *25*, 279–293.

Hacquard, S., Kracher, B., Maekawa, T., Vernaldi, S., Schulze-Lefert, P., Loren, Ver, et al. (2013). Mosaic genome structure of the barley powdery mildew pathogen and conservation of transcriptional programs in divergent hosts. *Proceedings of the National Academy of Sciences of the United States of America*, *110*, E2219–E2228.

Hahn, M., & Mendgen, K. (1997). Characterization of *in planta*-induced rust genes isolated from a haustorium-specific cDNA library. *Molecular Plant-Microbe Interactions, 10*, 427–437.

Heitman, J. (2006). Sexual reproduction and the evolution of microbial pathogens. *Current Biology, 16*, R711–R725.

Hogan, D. A., Auchtung, T. A., & Hausinger, R. P. (1999). Cloning and characterization of a sulfonate/alpha-ketoglutarate dioxygenase from *Saccharomyces cerevisiae*. *Journal of Bacteriology, 181*, 5876–5879.

Howard, R. J. (1997). Breaching the outer barriers: Cuticle and cell wall penetration. In G. C. Carroll & P. Tudzynski (Eds.), *The mycota: Vol. V* (pp. 43–60). Heidelberg: Springer-Verlag.

Jones, J. D., & Dangl, J. L. (2006). The plant immune system. *Nature, 444*, 323–329.

Kemen, E., Gardiner, A., Schultz-Larsen, T., Kemen, A. C., Balmuth, A. L., Robert-Seilaniantz, A., et al. (2011). Gene gain and loss during evolution of obligate parasitism in the white rust pathogen of *Arabidopsis thaliana*. *PLoS Biology, 9*, e1001094.

Kemen, E., & Jones, J. D. (2012). Obligate biotroph parasitism: Can we link genomes to lifestyles? *Trends in Plant Science, 17*, 448–457.

Kleemann, J., Rincon-Rivera, L. J., Takahara, H., Neumann, U., Loren, Ver, van Themaat, E., et al. (2012). Sequential delivery of host-induced virulence effectors by appressoria and intracellular hyphae of the phytopathogen *Colletotrichum higginsianum*. *PLoS Pathogens, 8*, e1002643.

Kunoh, H., Kohno, M., Tashiro, S., & Ishizaki, H. (1979). Studies of the powdery mildew fungus, *Leveillula taurica*, on green pepper. 2. Light and electron microscopic observation of the infection process. *Canadian Journal of Botany, 57*, 2501–2508.

Letunic, I., & Bork, P. (2011). Interactive Tree Of Life (iTOL): An online tool for phylogenetic tree display and annotation. *Bioinformatics, 23*, 127–128.

Lin, K., Limpens, E., Zhang, Z., Ivanov, S., Saunders, D. G., Mu, D., et al. (2014). Single nucleus genome sequencing reveals high similarity among nuclei of an endomycorrhizal fungus. *PLoS Genetics, 10*, e1004078.

Lipka, V., Dittgen, J., Bednarek, P., Bhat, R., Wiermer, M., Stein, M., et al. (2005). Pre- and postinvasion defenses both contribute to nonhost resistance in Arabidopsis. *Science, 310*, 1180–1183.

Ma, J., & Bennetzen, J. L. (2004). Rapid recent growth and divergence of rice nuclear genomes. *Proceedings of the National Academy of Sciences of the United States of America, 101*, 12404–12410.

Maekawa, T., Cheng, W., Spiridon, L. N., Töller, A., Lukasik, E., Saijo, Y., et al. (2011). Coiled-coil domain-dependent homodimerization of intracellular barley immune receptors defines a minimal functional module for triggering cell death. *Cell Host & Microbe, 9*, 187–199.

Maekawa, T., Kracher, B., Vernaldi, S., Loren, Ver, van Themaat, E., & Schulze-Lefert, P. (2012). Conservation of NLR-triggered immunity across plant lineages. *Proceedings of the National Academy of Sciences of the United States of America, 109*, 20119–20123.

Manning, V. A., Pandelova, I., Dhillon, B., Wilhelm, L. J., Goodwin, S. B., Berlin, A. M., et al. (2013). Comparative genomics of a plant-pathogenic fungus, Pyrenophora tritici-repentis, reveals transduplication and the impact of repeat elements on pathogenicity and population divergence. *G3 (Bethesda), 3*, 41–63.

Martin, F., Kohler, A., Murat, C., Balestrini, R., Coutinho, P. M., Jaillon, O., et al. (2010). Périgord black truffle genome uncovers evolutionary origins and mechanisms of symbiosis. *Nature, 464*, 1033–1038.

McDowell, J. M. (2011). Genomes of obligate plant pathogens reveal adaptations for obligate parasitism. *Proceedings of the National Academy of Sciences of the United States of America, 108*, 8921–8922.

Micali, C., Göllner, K., Humphry, M., Consonni, C., & Panstruga, R. (2008). The Powdery mildew disease of Arabidopsis: A paradigm for the interaction between plants and biotrophic fungi. *The Arabidopsis Book*, *6*, e0115.

Micali, C. O., Neumann, U., Grunewald, D., Panstruga, R., & O'Connell, R. (2011). Biogenesis of a specialized plant-fungal interface during host cell internalization of *Golovinomyces orontii* haustoria. *Cellular Microbiology*, *13*, 210–226.

Mori, Y., Sato, Y., & Takamatsu, S. (2000). Evolutionary analysis of the powdery mildew fungi using nucleotide sequences of the nuclear ribosomal DNA. *Mycologia*, *92*, 74–93.

Moscou, M. J., Lauter, N., Caldo, R. A., Nettleton, D., & Wise, R. P. (2011). Quantitative and temporal definition of the Mla transcriptional regulon during barley-powdery mildew interactions. *Molecular Plant-Microbe Interactions*, *24*, 694–705.

Nowara, D., Gay, A., Lacomme, C., Shaw, J., Ridout, C., Douchkov, D., et al. (2010). HIGS: Host-induced gene silencing in the obligate biotrophic fungal pathogen *Blumeria graminis*. *The Plant Cell*, *22*, 3130–3141.

Oberhaensli, S., Parlange, F., Buchmann, J. P., Jenny, F. H., Abbott, J. C., Burgis, T. A., et al. (2011). Comparative sequence analysis of wheat and barley powdery mildew fungi reveals gene colinearity, dates divergence and indicates host-pathogen co-evolution. *Fungal Genetics and Biology*, *48*, 327–334.

O'Connell, R. J., Thon, M. R., Hacquard, S., Amyotte, S. G., Kleemann, J., Torres, M. F., et al. (2012). Lifestyle transitions in plant pathogenic *Colletotrichum* fungi deciphered by genome and transcriptome analyses. *Nature Genetics*, *44*, 1060–1065.

Oliver, K. R., & Greene, W. K. (2009). Transposable elements: powerful facilitators of evolution. *Bioessays*, *31*, 703–714.

Otto, T. D., & Reid, A. J. (2013). Cereal killers. *Nature Reviews. Microbiology*, *11*, 744.

Pais, M., Win, J., Yoshida, K., Etherington, G. J., Cano, L. M., Raffaele, S., et al. (2013). From pathogen genomes to host plant processes: The power of plant parasitic oomycetes. *Genome Biology*, *14*, 211.

Panstruga, R. (2003). Establishing compatibility between plants and obligate biotrophic pathogens. *Current Opinion in Plant Biology*, *6*, 320–326.

Panstruga, R., & Dodds, P. N. (2009). Terrific protein traffic: The mystery of effector protein delivery by filamentous plant pathogens. *Science*, *324*, 748–750.

Parra, G., Bradnam, K., & Korf, I. (2007). CEGMA: A pipeline to accurately annotate core genes in eukaryotic genomes. *Bioinformatics*, *23*, 1061–1067.

Pedersen, C., Loren, Ver, van Themaat, E., McGuffin, L. J., Abbott, J. C., Burgis, T. A., et al. (2012). Structure and evolution of barley powdery mildew effector candidates. *BMC Genomics*, *13*, 694.

Petre, B., Morin, E., Tisserant, E., Hacquard, S., Da Silva, C., Poulain, J., et al. (2012). RNA-Seq of early-infected poplar leaves by the rust pathogen *Melampsora larici-populina* uncovers PtSultr3;5, a fungal-induced host sulfate transporter. *PLoS One*, *7*, e44408.

Pike, S., Gao, F., Kim, M. J., Kim, S. H., Schachtman, D. P., & Gassmann, W. (2014). Members of the NPF3 transporter subfamily encode pathogen-inducible nitrate/nitrite transporters in grapevine and Arabidopsis. *Plant and Cell Physiology*, *55*, 162–170.

Pliego, C., Nowara, D., Bonciani, G., Gheorghe, D. M., Xu, R., Surana, P., et al. (2013). Host-induced gene silencing in barley powdery mildew reveals a class of ribonuclease-like effectors. *Molecular Plant-Microbe Interactions*, *26*, 633–642.

Pryce-Jones, E., Carver, T., & Gurr, S. J. (1999). The roles of cellulase enzymes and mechanical force in host penetration by *Erysiphe graminis* f. sp. *hordei*. *Physiological and Molecular Plant Pathology*, *55*, 175–182.

Raffaele, S., & Kamoun, S. (2012). Genome evolution in filamentous plant pathogens: Why bigger can be better. *Nature Reviews. Microbiology*, *10*, 417–430.

Ridout, C. J., Skamnioti, P., Porritt, O., Sacristan, S., Jones, J. D., & Brown, J. K. (2006). Multiple avirulence paralogues in cereal powdery mildew fungi may contribute to parasite fitness and defeat of plant resistance. *The Plant Cell, 18*, 2402–2414.

Ringelmann, A., Riedel, M., Riederer, M., & Hildebrandt, U. (2009). Two sides of a leaf blade: *Blumeria graminis* needs chemical cues in cuticular waxes of *Lolium perenne* for germination and differentiation. *Planta, 230*, 95–105.

Rossman, A. Y., & Palm, M. E. (2006). Why are *Phytophthora* and other Oomycota not true fungi? *Outlooks on Pest Management, 17*, 217–219.

Rouxel, T., Grandaubert, J., Hane, J. K., Hoede, C., van de Wouw, A. P., Couloux, A., et al. (2011). Effector diversification within compartments of the *Leptosphaeria maculans* genome affected by Repeat-Induced Point mutations. *Nature Communications, 2*, 202.

Sacristán, S., Vigouroux, M., Pedersen, C., Skamnioti, P., Thordal-Christensen, H., Micali, C., et al. (2009). Coevolution between a family of parasite virulence effectors and a class of LINE-1 retrotransposons. *PLoS One, 4*, e7463.

Saunders, D. G., Win, J., Cano, L. M., Szabo, L. J., Kamoun, S., & Raffaele, S. (2012). Using hierarchical clustering of secreted protein families to classify and rank candidate effectors of rust fungi. *PLoS One, 7*, e29847.

Schirawski, J., Mannhaupt, G., Münch, K., Brefort, T., Schipper, K., Doehlemann, G., et al. (2010). Pathogenicity determinants in smut fungi revealed by genome comparison. *Science, 330*, 1546–1548.

Schmidt, S. M., Kuhn, H., Micali, C., Liller, C., Kwaaitaal, M., & Panstruga, R. (2014). Interaction of a *Blumeria graminis* f. sp. hordei effector candidate with a barley ARF-GAP suggests host vesicle trafficking is a fungal pathogenicity target. *Molecular Plant Pathology*. http://dx.doi.org/10.1111/mpp.12110.

Schulze-Lefert, P., & Panstruga, R. (2011). A molecular evolutionary concept connecting nonhost resistance, pathogen host range, and pathogen speciation. *Trends in Plant Science, 16*, 117–125.

Schweizer, P., Pokorny, J., Schulze-Lefert, P., & Dudler, R. (2000). Technical advance. Double-stranded RNA interferes with gene function at the single-cell level in cereals. *The Plant Journal, 24*, 895–903.

Seeholzer, S., Tsuchimatsu, T., Jordan, T., Bieri, S., Pajonk, S., Yang, W., et al. (2010). Diversity at the Mla powdery mildew resistance locus from cultivated barley reveals sites of positive selection. *Molecular Plant-Microbe Interactions, 23*, 497–509.

Selker, E. U. (2002). Repeat-induced gene silencing in fungi. *Advances in Genetics, 46*, 439–450.

Spanu, P. D. (2012). The genomics of obligate (and nonobligate) biotrophs. *Annual Review of Phytopathology, 50*, 91–109.

Spanu, P. D., Abbott, J. C., Amselem, J., Burgis, T. A., Soanes, D. M., Stüber, K., et al. (2010). Genome expansion and gene loss in powdery mildew fungi reveal tradeoffs in extreme parasitism. *Science, 330*, 1543–1546.

Spanu, P. D., & Panstruga, R. (2012). Powdery mildew genomes in the crosshairs. 2nd International Powdery Mildew Workshop and 3rd New Phytologist Workshop, in Zürich, Switzerland, February 2012. *New Phytologist, 195*, 20–22.

Stergiopoulos, I., & de Wit, P. J. (2009). Fungal effector proteins. *Annual Review of Phytopathology, 47*, 233–263.

Stukenbrock, E. H., Christiansen, F. B., Hansen, T. T., Dutheil, J. Y., & Schierup, M. H. (2012). Fusion of two divergent fungal individuals led to the recent emergence of a unique widespread pathogen species. *Proceedings of the National Academy of Sciences of the United States of America, 109*, 10954–10959.

Szabo, L. J., & Bushnell, W. R. (2001). Hidden robbers: The role of fungal haustoria in parasitism of plants. *Proceedings of the National Academy of Sciences of the United States of America, 98*, 7654–7655.

Takamatsu, S. (2013). Molecular phylogeny reveals phenotypic evolution of powdery mildews (Erysiphales, Ascomycota). *Journal of General Plant Pathology, 79*, 218–226.
Tisserant, E., Malbreil, M., Kuo, A., Kohler, A., Symeonidi, A., Balestrini, R., et al. (2013). Genome of an arbuscular mycorrhizal fungus provides insight into the oldest plant symbiosis. *Proceedings of the National Academy of Sciences of the United States of America, 110*, 20117–20122.
van der Hoorn, R. A., & Kamoun, S. (2008). From Guard to Decoy: A new model for perception of plant pathogen effectors. *The Plant Cell, 20*, 2009–2017.
Wang, Z., Binder, M., Schoch, C. L., Johnston, P. R., Spatafora, J. W., & Hibbett, D. S. (2006). Evolution of helotialean fungi (Leotiomycetes, Pezizomycotina): A nuclear rDNA phylogeny. *Molecular Phylogenetics and Evolution, 41*, 295–312.
Weßling, R., Schmidt, S. M., Micali, C. O., Knaust, F., Reinhardt, R., Neumann, U., et al. (2012). Transcriptome analysis of enriched *Golovinomyces orontii* haustoria by deep 454 pyrosequencing. *Fungal Genetics and Biology, 49*, 470–482.
Whisson, S. C., Boevink, P. C., Moleleki, L., Avrova, A. O., Morales, J. G., Gilroy, E. M., et al. (2007). A translocation signal for delivery of oomycete effector proteins into host plant cells. *Nature, 450*, 115–118.
Wicker, T., Oberhaensli, S., Parlange, F., Buchmann, J. P., Shatalina, M., Roffler, S., et al. (2013). The wheat powdery mildew genome shows the unique evolution of an obligate biotroph. *Nature Genetics, 45*, 1092–1096.
Win, J., Chaparro-Garcia, A., Belhaj, K., Saunders, D. G., Yoshida, K., Dong, S., et al. (2012). Effector biology of plant-associated organisms: Concepts and perspectives. *Cold Spring Harbor Symposia on Quantitative Biology*, 77, 235–247.
Win, J., Morgan, W., Bos, J., Krasileva, K. V., Cano, L. M., Chaparro- Garcia, A., et al. (2007). Adaptive evolution has targeted the C-terminal domain of the RXLR effectors of plant pathogenic oomycetes. *The Plant Cell, 19*, 2349–2369.
Wise, R. P., Moscou, M. J., Bogdanove, A. J., & Whitham, S. A. (2007). Transcript profiling in host-pathogen interactions. *Annual Review of Phytopathology, 45*, 329–369.
Wolfe, M. S., & McDermott, J. M. (1994). Population genetics of plant pathogen interactions: The example of the Erysiphe Graminis-Hordeum Vulgare pathosystem. *Annual Review of Phytopathology, 32*, 89–113.
Xue, M., Yang, J., Li, Z., Hu, S., Yao, N., Dean, R. A., et al. (2012). Comparative analysis of the genomes of two field isolates of the rice blast fungus *Magnaporthe oryzae*. *PLoS Genetics, 8*, e1002869.
Yahiaoui, N., Brunner, S., & Keller, B. (2006). Rapid generation of new powdery mildew resistance genes after wheat domestication. *The Plant Journal*, 47, 85–98.
Yoshida, K., Saitoh, H., Fujisawa, S., Kanzaki, H., Matsumura, H., Yoshida, K., et al. (2009). Association genetics reveals three novel avirulence genes from the rice blast fungal pathogen *Magnaporthe oryzae*. *The Plant Cell, 21*, 1573–1591.
Zerillo, M. M., Adhikari, B. N., Hamilton, J. P., Buell, C. R., Lévesque, C. A., & Tisserat, N. (2013). Carbohydrate-active enzymes in pythium and their role in plant cell wall and storage polysaccharide degradation. *PLoS One, 8*, e72572.
Zhang, W. J., Pedersen, C., Kwaaitaal, M., Gregersen, P. L., Mørch, S. M., Hanisch, S., et al. (2012). Interaction of barley powdery mildew effector candidate CSEP0055 with the defence protein PR17c. *Molecular Plant Pathology, 13*, 1110–1119.

CHAPTER FIVE

Functional Genomics of Smut Fungi: From Genome Sequencing to Protein Function

Gunther Doehlemann[*,1], **Jan Schirawski**[†], **Jörg Kämper**[‡]
[*]Department of Organismic Interactions, Max-Planck-Institute for Terrestrial Microbiology, Marburg, Germany
[†]Microbial Genetics, Aachen Biology and Biotechnology, Institute of Applied Microbiology, RWTH Aachen University, Aachen, Germany
[‡]Institute for Applied Biosciences, Karlsruhe Institute of Technology (KIT), Karlsruhe, Germany
[1]Corresponding author: e-mail address: doehlemann@mpi-marburg.mpg.de

Contents

Abstract

In contrast to most biotrophic plant pathogens, smut fungi belonging to the family of Ustilaginaceae are facultative parasites that have a yeastlike, saprophytic phase contrasting the filamentous, parasitic growth during plant infection. This dimorphic lifestyle allows easy culturing, reverse genetic modifications and cell biological analyses in the saprophytic stage; impact on virulence or pathogenicity can then been monitored easily in infection assays. Such excellent preconditions made the maize smut fungus *Ustilago maydis* the most advanced model organism for fungal biotrophs. A recent key step in *Ustilago* research was the high-quality sequencing and manual annotation of its genome, which identified a wealth of novel virulence genes that are arranged in physical gene clusters. In next steps, the genomes of the maize head smut *Sporisorium*

Advances in Botanical Research, Volume 70
ISSN 0065-2296
http://dx.doi.org/10.1016/B978-0-12-397940-7.00005-7

reilianum f. sp. *zeae* and the barley covered smut *Ustilago hordei* have been sequenced. These closely related pathogens exhibit significant differences regarding genome organization and gene equipment, likely reflecting different infection styles compared to *U. maydis*, whose unique feature is to cause tumours locally to sites of infection in vegetative tissue. Comparative analysis of these genomes as well as transcriptome profiling approaches allowed the identification of regulatory elements controlling pathogenic development as well as of secreted effector genes with important functions in plant interaction. In this chapter, we summarize the most important findings gained by genome analyses of smut fungi and define important questions that need to be addressed in the future.

1. INTRODUCTION

Smut fungi comprise more than 1200 species of widely spread biotrophic plant pathogens. They infect both monocot and dicot hosts, and among their host plants are various economically important crops such as maize, wheat and barley. Most smut fungi have a very restricted host range and are specialized on members of the *Poaceae* (grasses). Examples are *Ustilago maydis* that causes maize smut disease and *Sporisorium reilianum* of which two *formae speciales* exist that infect either maize (*S. reilianum* f. sp. *zeae*) or sorghum (*S. reilianum* f. sp. *reilianum*) (Fig. 5.1). Other relevant monocot smuts are *Ustilago hordei* that causes covered smut on barley (Fig. 5.1B) and *Sporisorium scitamineum* and *Ustilago tritici*, which parasitize sugar cane and wheat plants, respectively. Presumably, the present-day smut taxa diverged before domestication of their host plants in modern agriculture, through coevolution within natural populations of the ancestors of current crop plants (Munkacsi, Stoxen, & May, 2007).

Smut fungi display a dimorphic lifestyle and can grow as yeast or hyphae. In order to infect their host plants, two haploid yeast cells need to fuse to form an infectious dikaryotic hypha (Fig. 5.1A). For further proliferation, the dikaryon must penetrate the host plant surface, usually by formation of an appressorium. While traversing the living plant cells, the penetrating hyphae are surrounded by the plant plasma membrane, forming a tight interaction zone between the hyphal cell wall and the plant membrane. Hyphae proliferate within the plant until the two nuclei of the dikaryotic cells fuse and diploid teliospores are generated that represent the resting structures of the fungi (Fig. 5.1A). While germinating, meiosis occurs, and a septated probasidium with four haploid compartments is formed. From these

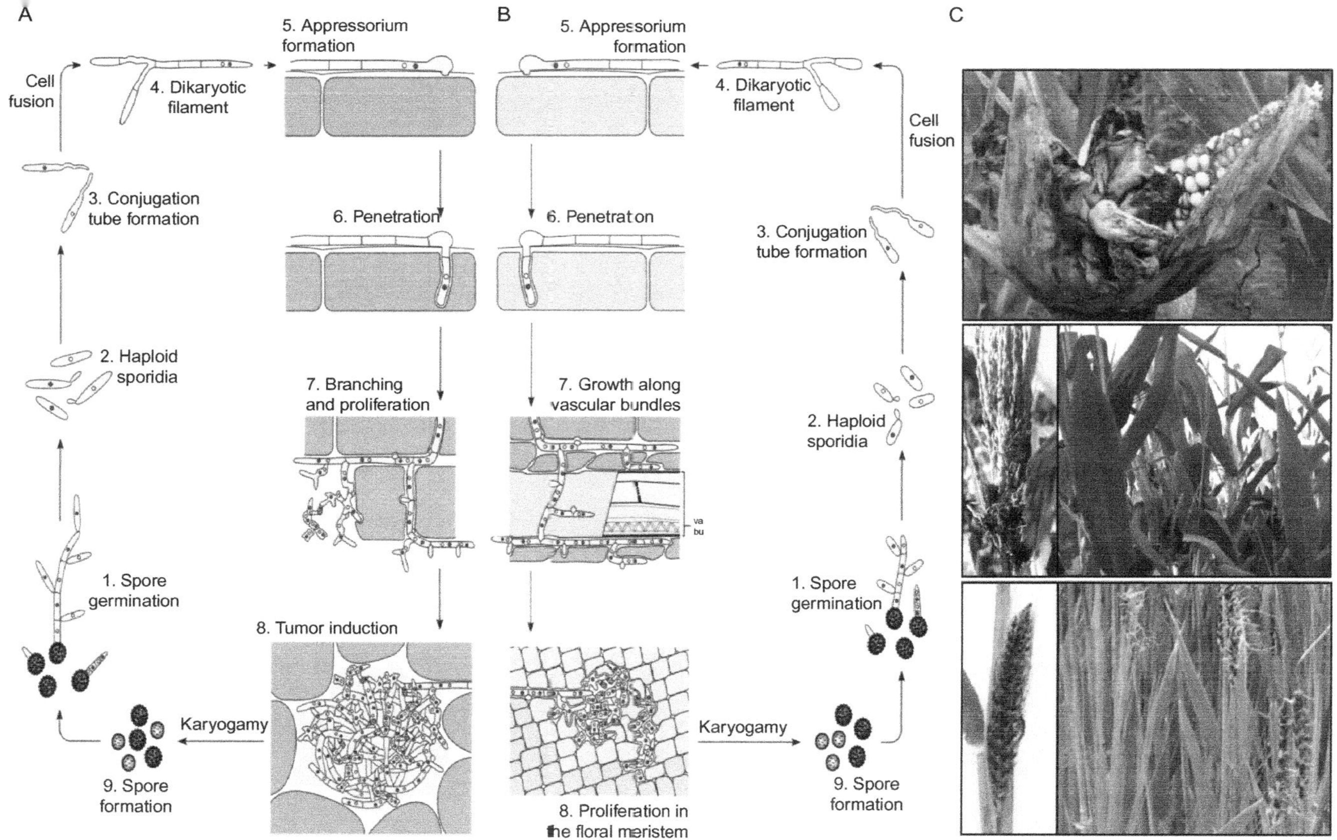

Figure 5.1 Life cycles and disease symptoms of smut fungi. (A) Life cycle of *U. maydis* compared to other Ustilaginomycetes such as *U. hordei* and *S. reilianum*. (B) *U. maydis* strongly proliferates in the host tissue close to the site of infection and causes local formation of tumors. By contrast, *U. hordei* and *S. reilianum* infections are followed by growth along the vascular bundles towards the floral meristem, where hyphal proliferation and finally spore formation take place upon development of inflorescences (C) Disease symptoms formed by smut fungi. Upper image: *U. maydis* tumors in an infected maize ear. Middle image: *S. reilianum* symptoms with dark spores in maize tassels. Lower image: *U. hordei* infection of barley plants. Dark fungal spores replaced seeds in the infected inflorescence. *(A) and (B) were modified from Hof (2013), PhD thesis. Foto taken by Tilman Schirawski.* (See the colour plate.)

probasidiae, the haploid sporidia are formed by successive budding events (reviewed in Brefort et al., 2009; Christensen, 1963; Fig. 5.1A).

Development within the host plants is quite distinct in the three smuts *U. maydis, U. hordei and S. reilianum*. Shortly after plant penetration, *U. maydis* elicits large tumours near the site of penetration that can occur on all aerial organs of the corn plant, for example, on leaves, stems or male and female flowers. Induced tumours result from an extended period of plant cell division, chromosome endoreduplication and cell expansion elicited by the fungus (Banuett & Herskowitz, 1996; Callow & Ling, 1973). In contrast, in both *S. reilianum* and *U. hordei*, plant infection occurs at the seedling stage and leads to systemically infected plants that sporulate exclusively in male or female inflorescences (Laurie et al., 2012; Schirawski et al., 2010). For *S. reilianum* f. sp. *zeae*, colonization of maize inflorescence meristems leads to tillering and the outgrowth of subapically appearing inflorescences (Ghareeb, Becker, Iven, Feussner, & Schirawski, 2011).

Smut fungi are true biotrophs, since the completion of their sexual life cycle is dependent on a living host plant. However, sporidia live as saprotrophs, and in their saprophytic stage, smut fungi can be manipulated by forward and reverse genetic methods, which makes them excellent models not only for fungal pathobiology but also for replication and cell biology (reviewed in Brefort et al., 2009; Holloman, Schirawski, & Holliday, 2008; Steinberg & Perez-Martin, 2008; Vollmeister et al., 2012).

After availability of the complete genome sequences of *U. maydis* (Kämper et al., 2006), *S. reilianum* (Schirawski et al., 2010) and *U. hordei* (Laurie et al., 2012), comparative and functional genomics can now be used to understand the different lifestyles of the three related fungi on a molecular level. Open-ended questions include: Why does *U. maydis*, but not *S. reilianum* or *U. hordei*, form tumours and spores in vegetative plant tissues? Has *U. maydis* acquired the capacity to form tumours on leaves and stems, or did the fungus lose the ability to restrict infection to floral organs, which, in the background of biotrophy, might be an evolutionary advantage? How is the extended endophytic phase between plant colonization and disease outbreak maintained in the life cycles of *S. reilianum* and *U. hordei*? How are the different stages of the life cycle, saprophytic, endophytic and pathogenic growth controlled? In this review, we will summarize the conclusions gained from analysing the genomic sequences that revealed astounding insights into the biology of these model organisms of biotrophy. In addition, by highlighting the genomic identification of individual proteins, whose involvement in specific biological processes has subsequently been experimentally validated, we will show that genome sequencing has greatly advanced our molecular understanding of fungal behaviour.

2. THE *U. MAYDIS* GENOME

The genome of *U. maydis* was the first one of the three smut genomes to be sequenced so far. Genome analysis was carried out on the haploid *U. maydis* strain 521 using two independent approaches. The initial sequencing project was started as a collaboration between the group of Regine Kahmann at the Max Planck Institute for Terrestrial Microbiology (Marburg, Germany) and Bayer Crop Science (Monheim, Germany) and was carried out at Lion Biosciences (Heidelberg, Germany). In a one-of-a-kind map-based cloning approach, 258 overlapping BAC clones were assembled into 28 large contigs for the 23 *U. maydis* chromosomes via a hybridization-based technique (Meksem et al., 2005). Subsequently, for each of the BAC clones, an ordered set of overlapping plasmid subclones was generated, again via a mapping by hybridization approach (Scholler, Benes, Voss, Ansorge, & Hoheisel, 1996; Voss et al., 1995). The resulting set of more than 45,000 overlapping plasmid clones was systematically sequenced from both ends. The laborious mapping approach led at the end to 24 physical contigs, spanning the 23 *U. maydis* chromosomes (Kämper et al., 2006). Disadvantage of the first approach was the low sequence redundancy of 2.92, which resulted in frequent sequencing errors. This constraint was overcome by a second, whole genome shotgun sequencing approach (Adams et al., 2000) under management of the Broad Institute (Cambridge, the United States). In total, these data provided an average of 10-fold sequence coverage and 63-fold physical coverage of the genome. The final assembly combining data from both approaches comprises 19.8 Mb of the estimated genome size of 20.5 Mb, with more than 99% of the sequence represented in the 24 largest scaffolds. While 22 chromosomes are covered by single large scaffolds, only chromosome 4 is covered by two scaffolds separated by the non-assembled rDNA repeats. The final assembly contains only 251 sequence gaps. For all but one end on chromosomes 5 and 7 each, the subtelomeric regions were identified, indicating almost complete sequence coverage. Based on the high quality of the draft sequence generated by the Broad Institute, automated gene predictions and annotations were performed, leading to 6522 gene models. The complete gene call set has been manually processed and individual gene models corrected at MIPS (Munich Information Center for Protein Sequences), using information from various EST libraries. In 2006, 6902 valid gene models were listed (Kämper et al., 2006). The gene models were further refined during the last years due to new evidence from the related genomes of *S. reilianum* (Schirawski et al., 2010) and *U. hordei* (Laurie et al., 2012), ESTs (Ho, Cahill, & Saville, 2007) and RNA-sequencing data (Kellner, Heimel, Obhof, Finkernagel, & Kämper, 2014), a process that is still ongoing. To date, more than 2500 genes were structurally reannotated, including gene fusions and gene splittings. The new and old gene models were annotated using the MIPS PEDANT suite (Walter et al., 2009). Today, the genome of *U. maydis* is one of the best-annotated and most complete genomes of fungi. The MUMDB (MIPS *Ustilago maydis* database; http://mips.helmholtz-muenchen.de/genre/proj/ustilago) currently lists 6784 manually curated models for protein-encoding genes. The database enables convenient genome searches on keywords and domains and provides outlinks to a graphic genome browser, related species data, etc. Current and future related genomes are processed and presented in PEDANT and GBrowse for comparative analysis (http://pedant.helmholtz-muenchen.de/genomes.jsp?category=fungal).

The 20.5 Mb genome of *U. maydis* is very compact and the number of genes (6784) is much lower compared to other plant pathogenic fungi such as *Magnaporthe grisea* (12,841), *Stagonospora nodorum* (16,591), *Fusarium graminearum* (11,640), *Puccinia graminis* (17,773) or *Melampsora larici-populina* (16,399). At present, the only known examples for a plant pathogenic fungus with lower gene content compared to *U. maydis* are the obligate powdery mildew fungi *Blumeria graminis* f.s. *graminis* (5854) and *B. graminis* f.s. *tritici* (6540) and the closely related smut *S. reilianum* (6648) (Schirawski et al., 2010; Spanu et al., 2010; Wicker et al., 2013). The small number of genes encoded by *U. maydis* is partly reflected in the absence of significant expansions of gene families. But why is the *U. maydis* genome that compact? First, there is no evidence for large-scale genome duplications that shaped the genome of, for instance, *Saccharomyces cerevisiae*. Secondly, the genome is nearly devoid of transposable elements (TEs) that contribute to large portions of the DNA in most other organisms: Only 1.1% of the genome is derived from transposon-related sequences. 70% of the *U. maydis* genes contain no intron, 17.3% contain only one, and the average intron size is only 146 bp. In total, the average number of introns per gene is 0.46, compared to 5.3, 4.7 and 2.6 introns per gene, respectively, for the related basidiomycetes *Cryptococcus neoformans*, *Coprinopsis cinerea* and *Phanerochaete chrysosporium* (Loftus et al., 2005; Broad Institute statistics *C. cinerea*: http://www.broadinstitute.org/annotation/genome/coprinus_cinereus/GeneStatsSummary.html).

Thus, it appears likely that the genome has been shaped by a massive, lineage-specific intron loss, as it has been proposed for the yeasts *S. cerevisiae* and *Schizosaccharomyces pombe* (Nielsen, Friedman, Birren, Burge, & Galagan, 2004; Rogozin, Wolf, Sorokin, Mirkin, & Koonin, 2003; Roy & Gilbert, 2005a,2005b).

2.1. Adaptation to a biotrophic lifestyle

What can we learn from the genome about the biotrophic lifestyle of *U. maydis*? The genome lacks several of the pathogenicity signatures that are associated with pathogens with a necrotrophic stage: *M. oryzae*, *F. graminearum* and *Cochliobolus heterostrophus* harbour up to 25 polyketide synthases involved in the production of secondary metabolites such as antibiotics or toxins, while *U. maydis* possesses only three (Kämper et al., 2006). Another adaption to the biotrophic lifestyle may be seen in the reduction of enzymes required for the degradation of plant cell walls, as polysaccharide hydrolases, polysaccharide lyases and pectin esterases. The *U. maydis* genome encodes only 33 hydrolytic enzymes acting on plant cell wall polysaccharides, compared to 138 and 103 for *M. grisea* and *F. graminearum*, both hemibiotrophic fungi that use these enzymes to degrade the plant tissue during their necrotrophic stage (Kämper et al., 2006). For *U. maydis*, the paucity of cell wall-degrading enzymes is advantageous, as damage to the host should be minimized for a biotrophic organism. Cell wall fragments released by lytic enzymes can act as DAMPs (damage-associated molecular patterns) that function as signals for triggering plant defence responses, which needs to be avoided by *U. maydis* (Doehlemann & Hemetsberger, 2013).

One of the major questions is why *U. maydis* is dependent on living host tissues to complete its life cycle. In other obligate biotrophic pathogens, the requirement of living host tissue can be explained by the loss of specific biosynthetic pathways that are compensated for by the host. While obligate biotrophs such as the barley powdery mildew fungus *Blumeria graminis* f. sp. *graminis* lack genes for anaerobic fermentation, biosynthesis of glycerol from glycolytic intermediates, and inorganic nitrogen (nitrate) assimilation (Spanu et al., 2010), *U. maydis* wild-type strains do not show any auxotrophies. In addition, the *U. maydis* genome is fully equipped with metabolic pathways, reflecting the biphasic lifestyle of this fungus that is not only a highly adapted biotroph but also a competitive saprotroph.

2.2. Regulation of pathogenic development

The key to understanding the biotrophic lifestyle may involve knowledge of the specific regulation of gene expression: The skill of the craftsman depends on the timely and spatial use of given tools, just as much as on the set of tools themselves. Indeed, the use of transcriptomics to monitor the gene expression profile during *in planta* growth by means of custom-designed Affymetrix microarrays has identified clustered genes for secreted effectors as "plant-induced" (Kämper et al., 2006). Microarrays were also instrumental to unravel the signalling cascades during the switch from saprophytic to biotrophic growth. The onset of the biotrophic phase is controlled by the *b* mating-type locus, which encodes two different homeodomain transcription factors, termed bE and bW (Fig. 5.2). These two proteins form a heterodimeric complex when they are derived from different alleles of the *b*-locus, for example, bE1 and bW2. It has been conclusively shown by the generation of haploid strains harbouring

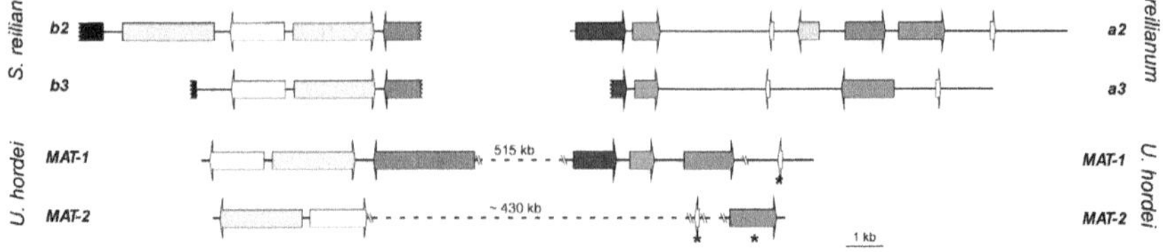

Figure 5.2 Comparison of mating-type loci of *U. maydis*, *S. reilianum* and *U. hordei*. Each *b*-locus contains the transcription factor subunit genes *bE* (yellow) and *bW* (green). Each *a*-locus contains the pheromone receptor gene (*PRA*, red) and one or two pheromone genes (*MFA*, yellow). Flanking genes are conserved and homologues are indicated by the same colour with names appearing at the top. tn is a transposon-derived sequence and star indicates that the orientation of the gene is not known. (See the colour plate.)

the two proteins in the combination bE1 and bW2 that the heterodimer is sufficient to induce pathogenic development (reviewed in Bakkeren, Kamper, & Schirawski, 2008).

By means of microarrays, a set of 350 genes have been identified that alter their expression in response to the induction of the bE/bW heterodimer in axenic culture (Heimel et al., 2010). These *b*-regulated genes are involved in cell wall remodelling, encode potential secreted effectors, but also function in cell cycle control: bE/bW induces a G2 cell cycle arrest that is released only after plant penetration. Thus, on the one hand, bE/bW induces the genes that facilitate *in planta* growth, while on the other hand, it forces the fungus to penetrate and colonize the plant by inducing a cell cycle block that is only released after penetration, which leads to the initiation of "obligate biotrophy".

Most of the genes that are differentially expressed after bE/bW induction are not regulated directly by the bE/bW heterodimer, but via a cascade of additional transcription factors, which have all been identified by means of transcript profiling based on microarray experiments. The transcriptional network was uncovered as follows: In strains lacking particular transcription factors, the *bE1/bW2* genes were induced by means of an inducible promoter, and gene expression was monitored. This approach showed whether the particular transcription factor is required for expression of b-regulated genes. To elucidate whether the transcription factor is also sufficient for the regulation of b-regulated genes, it was expressed ectopically under the control of an inducible promoter. Using this strategy, the zinc finger transcription factor Rbf1 (regulator of b-dependent filamentation 1) was identified as a master regulator that is required for the expression of most bE/bW-regulated genes (Heimel, Scherer, Vranes, et al., 2010). Rbf1 regulates an additional set of transcription factors, including the homeodomain transcription factor Hdp2 (homeodomain protein 2; Pothiratana, M. Vranes & J. Kämper, unpublished data) and the zinc finger transcription factor Biz1 (b-induced zinc finger 1; Flor-Parra, Vranes, Kämper, & Perez-Martin, 2006), which are all independently required for pathogenic development. Biz1 is required for the formation of appressoria; thus, strains deleted for *biz1* are blocked at a very early stage of plant infection (Flor-Parra et al., 2006). Comparative microarray analysis of *biz1* mutant strains to wild-type strains during growth on the plant surface during the stage prior to penetration showed that the regulatory cascade feeding into *biz1* is required for the induction of a large fraction of genes induced at the penetration stage (M. Vranes & J. Kämper, unpublished data).

The majority of these genes encode putative effectors, that is, secreted proteins with so far unknown functions. A systematic deletion analysis of *biz1*-regulated genes has identified two genes, *pst1* and *pst2*, whose deletion leads to a phenotype identical to that of the *biz1* deletion strains. Pst1 and Pst2 both encode secreted proteins that are transferred from the fungus to the plant nucleus (M. Vranes & J. Kämper, unpublished data). In summary, systematic expression profiling experiments not only led to the identification and the interconnection of the key transcription factors required for pathogenic development but also led to the identification of downstream target genes with key functions in plant infection. The advantage of such complex regulatory pathways is the dynamic integration of different signals to allow a temporal and spatial control of developmental decisions, such as the adaptation to particular environmental conditions (as the availability of nutrients) or the adaptation to specific plant tissues.

3. EFFECTOR CLUSTERS OF *U. MAYDIS*

The most thrilling finding in the *U. maydis* genome was that about 180 of its 544 secreted proteins are arranged in gene clusters that contain between 3 and up to 26 genes. 12 of the gene clusters encode putative effectors that belong to small gene families, suggesting functional redundancy (Kämper et al., 2006). Most proteins encoded in these clusters share no similarity to other known proteins and appear to be "*Ustilago*-specific". Eight of the 12 clusters contain groups of two to five related genes in tandem arrays, indicating that they have arisen by duplication. Moreover, most of the genes in these clusters are expressed only during plant infection, which suggested a concerted function for the respective proteins during pathogenic development. Indeed, the deletion of the individual clusters led in five cases to an altered disease phenotype. In four of these deletion strains, disease symptoms were alleviated, while in one, they were even increased demonstrating that secreted proteins are (1) required for fungal proliferation inside the plant host and (2) instrumental to attenuate fungal proliferation in the host to provide a functional interface between the parasite and living host tissues (Kämper et al., 2006).

Intensive studies have been done on cluster 19A, which is the largest *U. maydis* gene cluster containing 26 genes, of which 24 encode predicted effectors (Fig. 5.3). The deletion of the entire cluster results in an almost complete loss of *U. maydis*-induced tumour formation. However,

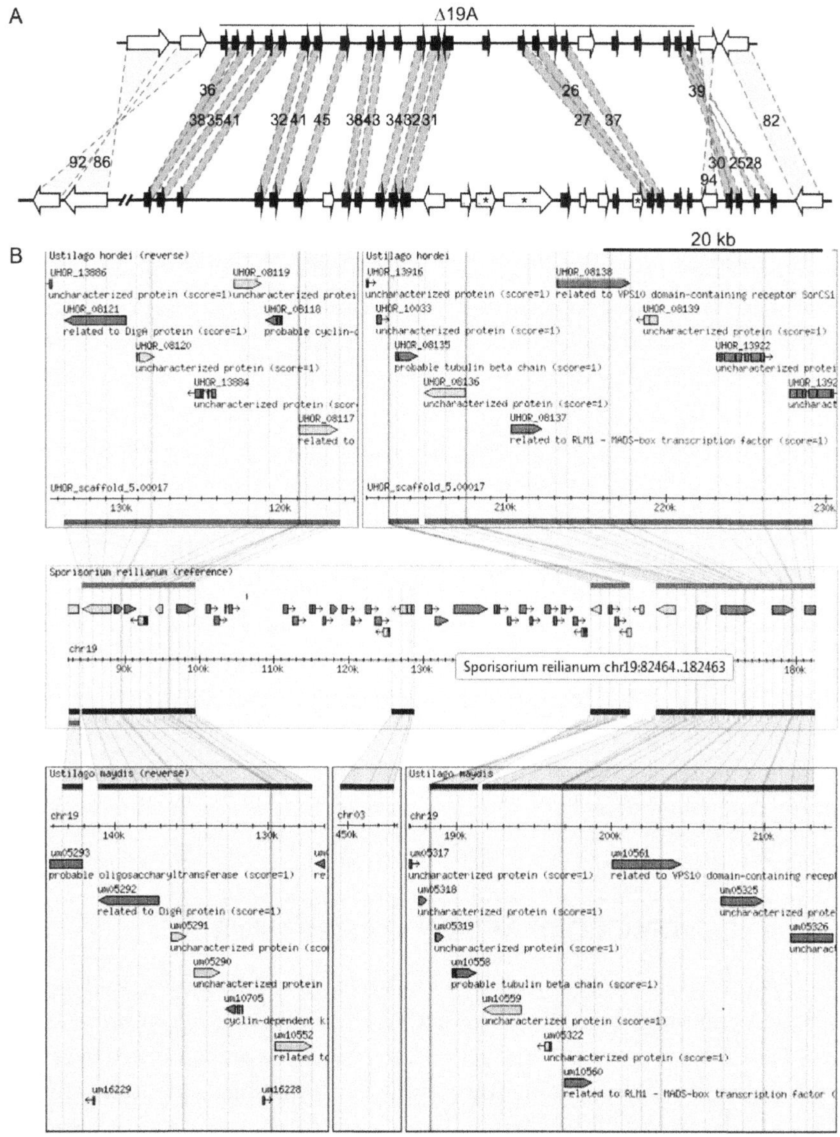

Figure 5.3 Comparison of the largest divergence region. (A) Synteny comparison of the genes of cluster 19A of *U. maydis* (top) to those of cluster 19A of *S. reilianum* (bottom). Weakly conserved genes encoding proteins with at least 20% amino acid identity are connected by blue bars. The cluster is flanked by highly conserved genes connected by yellow bars. Genes predicted to encode secreted proteins are shaded in dark blue. (B) Synteny comparison of the divergence region and flanking sequences of *S. reilianum* (middle) to the homologous sequences of *U. hordei* (top) and *U. maydis* (bottom). Highly conserved genes are in the flanking sequences. *Figure (B) is derived from the synteny viewer on the MUMDB homepage (http://mips.helmholtz-muenchen.de/genre/proj/ustilago/).* (See the colour plate.)

cluster 19A deletion strains are still able to proliferate in the plant and manage to differentiate teliospores when colonizing floral tissues, which actually results in an endophyte-like behaviour of these mutants (Brefort et al., 2014). A series of deletion mutant analyses allowed to nail down the virulence defect of the cluster mutant to a gene effector family with five members (*tin1-1* to *tin1-5*) and four individual effector genes (*tin2*, *tin3*, *tin4* and *tin5*), which contribute to tumour formation in an additive manner (Brefort et al., 2014). The most significant contribution to virulence comes from *tin3*, whose deletion results in strains strongly attenuated in tumour formation, while mutants lacking *tin2* lack the ability to induce anthocyanin formation, which is a typical hallmark of *U. maydis* wild-type infection. The virulence phenotypes of individual cluster 19A subdeletion strains are phenotypically highly similar, for example, the deletion of either *tin1*, *tin4* or *tin5* causes only marginal reduction of tumour formation. However, microarray-based transcriptome analyses of maize leaves infected by the different mutants revealed significant differences in plant responses. Such "molecular phenotypes" of the individual effector mutants are indicative of distinct molecular functions of the deleted genes, which are then reflected by biological host processes reacting specifically to the individual mutants (Brefort et al., 2014). These findings demonstrate the power of transcriptome profiling to dissect subtle effector functions, which are not visible on the macroscopic level.

4. GENOME COMPARISON OF SMUT FUNGI

Although a great heap of knowledge has been gained by sequencing and analysing the genome *U. maydis*, a large number of questions remain to be solved. The genome did not yet reveal why *U. maydis* has the capacity to form tumours with spores on any green part of the maize plant or why its virulence is restricted to maize. To answer these and other questions, genome comparison of related fungi with different characteristics is useful. The sister species *Sporisorium reilianum* f. sp. *zeae* is ideally suited for a direct comparison, since it is closely related to *U. maydis* and causes disease on the same host but has different infection characteristics. In contrast to *U. maydis* that colonizes the plant tissue locally and leads to tumour induction near the site of plant penetration, *S. reilianum* colonizes young plants, persists for a long time without causing symptoms and produces spores only in inflorescence tissues. In addition, colonization of inflorescence meristems induces outgrowth of subapically appearing inflorescences (Ghareeb et al., 2011).

4.1. Genome comparison to *S. reilianum*

Sequencing of the genome of *S. reilianum* was funded by the Max Planck Society with a grant to Regine Kahmann at a time when the next-generation sequencing technologies became first available. The fungus became one of the first eukaryotic organisms to be completely *de novo* sequenced using the Roche/454 pyrosequencing approach on a GS20 machine. The assembly was improved by adding information of a paired-end sequencing library on a GS FLX machine, which led to an overall coverage of 29 ×, and an optical map, which revealed the presence of 23 different chromosomes. Comparison of the proteins encoded in the *S. reilianum* and *U. maydis* genomes revealed an average amino acid sequence identity of 76% (Schirawski et al., 2010). Interestingly, the genes on the 23 chromosomes of *S. reilianum* lined up almost perfectly with those on the 23 chromosomes of *U. maydis*. Only one major chromosomal rearrangement was observed, exchanging the ends of chromosomes 5 and 20: A 0.78 Mb part of the genome containing 281 protein-encoding genes located on chromosome 5 in *U. maydis* was found to locate to chromosome 20 in *S. reilianum*, while a 0.13 Mb part with 58 protein-encoding genes located on chromosome 20 in *U. maydis* resides on chromosome 5 in *S. reilianum*. In addition to this single major rearrangement, only 38 additional genes show different chromosomal location in the two organisms, underlining their suitability for a comparative approach (Schirawski et al., 2010). The low degree of genome shuffling was on the one hand surprising, since *U. maydis* and *S. reilianum* diverged some 20–30 million years ago, but on the other hand, this was very useful for comparative genome analysis. Since almost all *U. maydis* genes had homologues at syntenic positions in *S. reilianum*, the two genomes could be neatly aligned and compared on a gene-by-gene basis.

This gene-by-gene analysis revealed homologous chromosomal regions where the encoded proteins showed an amino acid identity much below the average identity difference. Since an increased amino acid identity difference is the result of an increase in the number of maintained nucleotide changes in the genomic DNA, these areas represent genomic regions with an evolutionary pressure on diversification. An evolutionary pressure on diversification might result from the different lifestyles of the two pathogens where different forces act on factors involved in symptom specificity or host colonization strategy.

In total, 43 "divergence regions" were identified comprising 3–30 contiguous genes with below-average conservation of coding potential. Most of the genes within the divergence regions lack any functional annotation, and most of those are predicted to encode secreted proteins (Schirawski et al., 2010). This strengthens the assumption of evolutionary forces acting on proteins involved in interaction with the plant and leads to the hypothesis that the genes encoded in the divergence regions contribute to host colonization. To prove this hypothesis, cluster regions were deleted. For the largest divergence regions, *U. maydis* cluster deletion strains were already available: Among the identified divergence gene clusters were 7 of the 12 effector gene clusters previously found in *U. maydis* (Kämper et al., 2006) as encoding secreted proteins upregulated during plant infection. Notably, these seven divergence regions included the five previously identified clusters whose deletion affected virulence in seedling inoculations, while four of the *U. maydis* effector gene clusters whose deletion did not result in an altered virulence of *U. maydis* harboured genes with high conservation between the two organisms (Schirawski et al., 2010). This result shows that genome comparison of closely related fungi with different lifestyles leads to a more precise prediction of genes involved in plant colonization than relying alone on the search for proteins carrying secretion signal sequences.

Six additional divergence regions were tested for presence of virulence effector genes by deletion in *U. maydis* (Schirawski et al., 2010). Of these, four divergence region deletions resulted in an altered virulence phenotype. It is noteworthy that one of these virulence regions contains only genes encoding proteins lacking a recognizable secretion signal.

The fact that diversified effector genes have a function in virulence suggests a multiplicity in plant proteins that interact with the encoded fungal proteins. A temporally and spatially changing composition of host proteins that the pathogens need to interfere with may drive the evolution of different sets of effectors in closely related pathogens with different infection strategies on the same host plant. A first functional proof for this assumption comes from the analysis of the largest divergence region (cluster 19A; see Fig. 5.3A) and its effect on symptom formation of *S. reilianum* (H. Ghareeb, F. Drechsler, C. Löfke, T. Teichmann, & J. Schirawski, unpublished data). It was possible to generate *S. reilianum* deletion strains lacking the complete 30-gene-containing region.

Interestingly, the deletion mutants were still able to produce spores in inflorescences. However, they lacked a typical symptom induced by wild-type strains that interfere with apical dominance of infected inflorescences and induce inflorescence tillering. A subdeletion analysis identified a protein (named suppressor of apical dominance 1 (Sad1)) that is required for induction of subapically appearing inflorescences by *S. reilianum* (H. Ghareeb, F. Drechsler, C. Löfke, T. Teichmann, & J. Schirawski, unpublished data). N- or C-terminally tagged version of the Sad1 protein proved to be nonfunctional in *S. reilianum*, which precluded localization studies. However, yeast two-hybrid analysis identified a large repertoire of intracellular plant proteins as potential interaction partners suggesting a possible role within plant cells. Heterologous expression of the Sad1 protein in *Arabidopsis thaliana* led to an increase in secondary inflorescence branches. This supports an intracellular role of Sad1 *in planta* in the control of inflorescence branching that is conserved between monocots and dicots (H. Ghareeb, F. Drechsler, C. Löfke, T. Teichmann, & J. Schirawski, unpublished data).

4.2. Genome comparison to *U. hordei*

The *U. hordei* genome was sequenced using a combination of Titanium GS FLX Roche/454 pyrosequencing of single- and paired-end libraries, BAC-end sequencing and optical mapping. The genome turned out to be the most difficult-to-assemble genome of the three sequenced smuts because of an unusually high content of repetitive sequences that together amount to more than one megabase of sequence data. Interestingly, only a relatively small number of families of repetitive elements were found, suggesting a recent expansion of a few related elements newly introduced in the *U. hordei* lineage after separation from an ancestor common to *U. maydis* and *S. reilianum* (Laurie et al., 2012).

The *U. hordei* genome shows signs of active repeat-induced point mutation (RIP), a mechanism to reduce the chances of deleterious genome rearrangements due to homologous recombination between identical repeat elements. The lack of this mechanism in combination with a highly effective homologous recombination machinery in *U. maydis* probably resulted in the clean deletion of particular genes from the genome of *U. maydis*. These particular genes are necessary for the regulation of heterochromatin formation and include the RNA-dependent RNA polymerase and dicer and argonaute proteins functioning in RNAi (Laurie et al., 2012). Transcriptional silencing using RNAi has been shown to be functional in *U. hordei* and lacking in *U. maydis* (Laurie et al., 2008). The RNA silencing components present in the genome of *U. hordei* are conserved in *S. reilianum* (Laurie et al., 2012; Schirawski et al., 2010). It is unknown whether RNA silencing works in *S. reilianum*. However, it has been shown that the dicer protein does not contribute to symptom formation of the fungus (Schirawski et al., 2010). Therefore, the lack of RNAi does not explain the different behaviour of *U. maydis* on proliferation and symptom formation relative to the other two smuts.

The mating-type loci of the three smuts contain conserved genes (Fig. 5.2). The two *a* loci of *U. maydis* and *U. hordei* contain one pheromone receptor and one pheromone gene each. *S. reilianum* has three different *a* loci that each contain an additional pheromone gene for recognition by the third mating partner. The *b*-locus genes are also conserved: Each *b* allele of *U. maydis*, *S. reilianum* and *U. hordei* encodes both homeodomain proteins bW and bE. For *U. maydis*, 19 *b* alleles have been described (J. Kämper, unpublished data), and for *S. reilianum* six alleles are known (J. Schirawski unpublished data; Schirawski et al., 2005), while for *U. hordei*, only two *b* alleles are known, and one *b* allele is so tightly linked with one of the *a* alleles that they never separate (Bakkeren & Kronstad, 1994) (Fig. 5.2). Linking of the *a* and *b* mating-type loci in *U. hordei* likely resulted from a major genome rearrangement that had a profound impact on the biology of *U. hordei*. During evolution of *U. hordei*, an exchange of chromosomal arms of chromosomes 1 and 20 of the progenitor (Laurie et al., 2012) seems to have taken place, which resulted in the linkage of the *a* and *b* mating-type loci on the same chromosome. Physical linkage of mating-type loci imposed a bipolar instead of a tetrapolar mating behaviour on *U. hordei*, as a first step towards the evolution of a sex chromosome (Bakkeren et al., 2008).

Using the sequence information from the mating-type loci, solopathogenic *U. hordei* strains have been constructed that form filaments without prior mating (D. Schwammbach & G. Doehlemann, unpublished data). Introducing compatible *a* and *b* mating-type loci into a haploid wild-type strain was achieved using the FLP recombinase system that has been established for *U. maydis* (Khrunyk, Munch, Schipper, Lupas, & Kahmann, 2010). In *U. hordei* (Fig. 5.4) as well as in *S. reilianum* and in contrast to *U. maydis*, the sole presence of compatible *b* loci (*U. hordei* strain DS199; see Fig. 5.4) allowed induction of filamentous growth in the absence of a mating partner and to a similar extent in the presence of both compatible *a* and *b* mating-type genes (*U. hordei* strain DS200; see Fig. 5.4). In plant infection

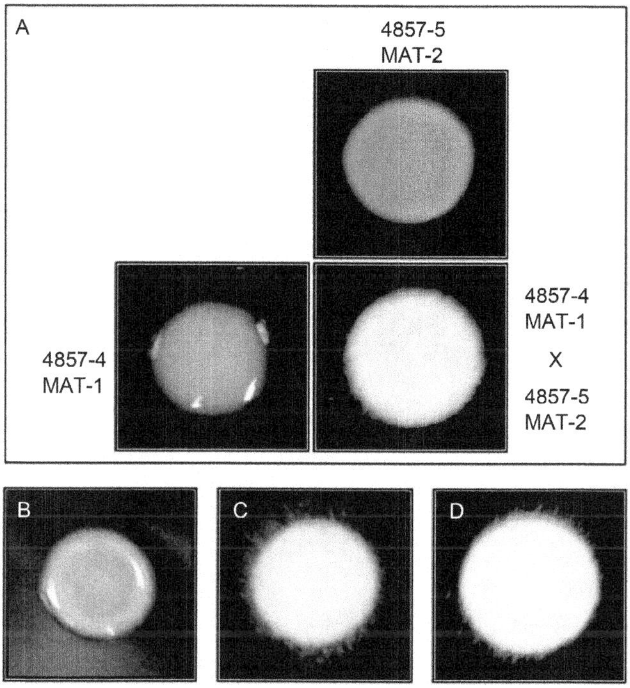

Figure 5.4 Mating of *U. hordei* on charcoal agar plates. (A) Two compatible wild-type strains 4857-4 (mating-type MAT-1, *a1b1*) and 4857-5 (mating-type MAT-2, *a2b2*) form filaments on charcoal plates only upon mating. (B) Strain DS198 (genotype: *a1mfa2b1*) that was derived from 4857-4 MAT-1. (C) Strain DS199 (a1bW1bE2). (D) DS200 (a1mfa2bW1bE2). Compatible *b*-genes (DS199 and DS200) but not compatible *a* (DS198) allow filament formation on charcoal medium without mating.

experiments, both *a*-compatible and *a*- and *b*-compatible strains of both *U. hordei* and *S. reilianum* showed appressorium formation and epidermal penetration to similar levels as mating-compatible wild-type strains (Schirawski, unpublished data; Schwammbach & Doehlemann, unpublished data).

Genome rearrangements due to TE activity also affected evolution of effector loci in *U. hordei* impacting fungal biology. Possibly, TE activity helped in increasing the speed of evolution of effector proteins that are under augmented evolutionary pressures resulting from effector exposure to host proteins. One very interesting example is a family of effector genes related to the *mig1* gene of *U. maydis*. Mig1 carries characteristics of a secreted, apoplastic fungal avirulence protein that is extremely highly expressed during biotrophic growth (Basse, Stumpferl, & Kahmann, 2000). The deletion of the three *mig1*-related genes from the genome of *U. maydis* resulted in hypervirulent strains (Schirawski et al., 2010), congruent with a function in avirulence. *U. hordei* contains 19 *mig1*-related genes (Laurie et al., 2012), and in contrast to the three copies in *U. maydis* and the eight *mig1*-related genes in *S. reilianum* that are located in a single locus in the two fungi (Schirawski et al., 2010), the 19 *mig1*-related genes are spread all throughout the genome in *U. hordei*. It is currently unknown whether they are identical to any of the six genetically identified avirulence genes of *U. hordei*.

5. ORGAN SPECIFICITY IN THE *U. MAYDIS*: MAIZE INTERACTION

Smut fungi of monocotyledonous host show a symptom-free, endophytic colonization style and cause symptoms of disease only in the inflorescences of their respective hosts. This is in clear contrast to *U. maydis*, which induces tumours locally at sites of infection within a few days on all aerial parts of the maize plant. Of note, it should be mentioned here that *U. maydis* is not capable to infect adult organs of the plant. It depends on inoculation of primordial, for example, juvenile, leaves within the leaf whirl of the plant or the developing female inflorescence that can be infected via the silks.

At present, one can only speculate on the genetic basis of this special feature of *U. maydis*, which makes it independent from flowering of the host plant and allows to complete its pathogenic cycle within only 2 weeks. However, the transformation of different host tissues to tumours is a remarkable ability, which requires to modulate organ development in fundamentally different environments. A microarray-based, parallel transcriptome analysis of maize and *U. maydis* during colonization of seedling leaves, adult leaves and tassels revealed a surprisingly high transcriptional plasticity of both interaction partners (Skibbe, Doehlemann, Fernandes, & Walbot, 2010). During early biotrophic development 3 days after infection, about 35% of the fungal transcriptome is expressed in an organ-specific manner. When looking at secretome genes, this holds true for even 45% of the *U. maydis* transcripts (Skibbe et al., 2010). This highly variable expression pattern of fungal genes inside the host suggests that *U. maydis* senses the individual tissues it is colonizing. It modulates its own metabolism to the respective environment and deploys well-adapted sets of effectors to manipulate either vegetative or floral maize primordia to form tumours. Functional proof for this assumption came from the effector-cluster mutants that had been identified previously during *U. maydis* genome annotation (Kämper et al.,

2006). Five of the 12 cluster mutants showed different virulence depending on the organ infected (Skibbe et al., 2010). Particularly, the cluster 19A mutant, which forms no tumours in seedling leaves, was found to induce such symptoms in maize tassels, although symptoms did not reach wild-type levels. Strikingly, the recently identified main effector genes contained in cluster 19A, *tin2* (um05302), *tin3* (um10556) and *tin4* (um05318) (Brefort et al., 2014), show a significantly organ-specific expression and are highly induced in seedling leaves compared to tassel infection (Skibbe et al., 2010). This expression pattern is perfectly in line with the observed mutant phenotype and suggests that the tumour-inducing function of *U. maydis* cluster 19A is mainly required in vegetative tissue.

Based on the organ-specific regulation of *U. maydis* effector genes, it was hypothesized that individual effectors could own tissue-specific virulence functions, that is, contribute to virulence in an organ-specific manner. Based on differential gene expression between infected maize organs, 21 effector candidate genes were selected to generate *U. maydis* deletion mutants. Intriguingly, 12 of these mutants showed a significant reduction in tumour formation when tested in leaf and tassel infections. In 11 cases, these virulence defects were organ-specific either for leaves or tassel (Schilling et al., 2014). All these 12 organ-specific effector mutants were fully virulent in either floral or vegetative tissue and did do not show growth defects in axenic culture (Schilling et al., 2014). Such highly specific mutant phenotypes, which exclusively involve a failure in tumour induction in a specific host tissue, suggest that organ-specific effectors do not act as general virulence factors. Most likely, they do not act as suppressors of innate immunity, but rather modulate specific cellular processes, which are required to accommodate and induce tumours in a certain plant tissue. In an evolutionary context, one might speculate that these kinds of fungal effectors are a consequence of long coevolution with the host leading to gradual increase of pathogen fitness and reflect an exceptionally high level of adaptation of *U. maydis* to its host plant.

6. FUNCTIONAL EFFECTOR BIOLOGY

Sequencing and comparative analyses of smut genomes as well as transcriptional profiling during host infection facilitated the identification of effector genes, which contribute to different aspects of fungal virulence. While functional characterization of individual genes from 12 effector clusters that originally were identified in frame of the *U. maydis* genome analysis is still in progress, three effectors that were identified in independent approaches have been studied in detail. The *U. maydis* effector that was described first to have an essential function for virulence is Pep1 (protein essential for penetration 1) (Doehlemann et al., 2009). The *pep1* gene was identified in a candidate-gene approach, based on its specific transcriptional induction during plant infection. Consequently, the deletion of *pep1* does not have any detectable effects on saprophytic growth, stress resistance and general fitness of *U. maydis*. Similarly, the early pathogenic development, that is, perception of plant surface signals leading to filamentation and formation of appressoria, does not require Pep1. Immediately upon penetration of maize epidermal cells, *pep1* deletion mutants are blocked in their biotrophic development, and at the same time, strong immune responses of the plant get activated (Doehlemann et al., 2009). Cytological analysis showed that, in the absence of Pep1, an oxidative burst is induced, followed by the formation of plant cell wall fortifications, that is, papilla (Hemetsberger, Herrberger, Zechmann, Hillmer, & Doehlemann, 2012). This typical PTI (PAMP-triggered immunity) response of the maize plant to the *pep1* deletion mutant is explained by the function of Pep1, which is to inhibit type III peroxidases. These enzymes are required for the generation of the extracellular oxidative burst (O'Brien et al., 2012). In maize leaves infected with the *pep1* deletion mutant, transcripts for the oxidative burst peroxidase POX12 are strongly induced. Direct interaction of Pep1 with POX12 was shown *in planta*, and transient shutdown of POX12 using VIGS (virus-induced gene silencing) could restore epidermal penetration of the *pep1* deletion mutant (Hemetsberger et al., 2012). This defines Pep1 as an essential core effector of *Ustilago maydis*, which serves as a suppressor of PTI. The fundamental role of Pep1 for suppression of host immunity is also reflected by the unusual sequence conservation of this effector in other smut species such as *Ustilago hordei* (Doehlemann et al., 2009; C. Hemetsberger & G. Doehlemann, unpublished data). An intriguing question for upcoming studies will be whether unrelated biotrophs such as rusts or mildew fungi and mycorrhizal symbionts also deploy Pep1-like effectors to suppress peroxidase activity or whether alternative mechanisms evolved to suppress this first line of plant defence.

A small gene cluster of four genes (the *pit* (protein involved in tumours) genes), of which three genes encode novel proteins that are strictly upregulated during plant infection, was identified on chromosome three of *U. maydis*. The deletion of the whole cluster does not impair the ability of *U. maydis* to penetrate the host tissue, but results in a complete loss of tumour formation (Doehlemann, Reissmann, Assmann, Fleckenstein, & Kahmann, 2011). Subdeletions of the individual cluster genes identified two genes, *pit1* and *pit2*, being both responsible for the virulence defect of the cluster mutant. Strikingly, *pit1* and *pit2* do not show additive effects, that is, they function independently from each other. The phenotypes caused by the deletion of either *pit1* or *pit2* could not be distinguished, which is even true for the transcriptional responses of maize plants towards the two mutants (Doehlemann et al., 2011). The *pit1* gene encodes not an effector but a membrane protein, which is shuttled via endosomes to the tips of biotrophic hyphae, and subsequently gets recycled to the vacuole. The molecular function of Pit1 has not been solved so far. Pit2 is a small secreted, which localizes to the biotrophic interface and the plant apoplast (Doehlemann et al., 2011). It inhibits and directly interacts with apoplastic, papain-like maize cysteine proteases CP1, CP2 and XCP2 (Mueller, Ziemann, Treitschke, Assmann, & Doehlemann, 2013). These proteases had been found in an independent study as important players in maize defence signalling associated with salicylic acid (SA) (van der Linde, Hemetsberger et al., 2012; van der Linde, Mueller et al., 2012). In line with this, infections with *pit2* deletion strains led to an elevated apoplastic cysteine protease activity and the induction of SA-associated defence responses, including plant cell death. The finding of Pit2 as an inhibitor of host cysteine proteases with an essential function for virulence is a striking parallel to other pathosystems. Papain-like cysteine proteases were found in other plants to be involved in defence to microbial pathogens; they are the targets of different effectors, as the tomato proteases C14 (inhibited by the *Phytophthora infestans* cystatins EPIC1/2B, (Tian et al., 2007)) and RCR3 and PIP2 (inhibited by the *Cladosporium fulvum* effector Avr2 (Shabab et al., 2008)). Recently, also the root nematode *Globodera rostochiensis* was shown to secrete a protein, the allergen-like effector VAP1, which inhibits RCR3 (Lozano-Torres et al., 2012), which further substantiates the important role of protease inhibitors as virulence factors in plant interactions.

A completely different mode of action was found for the *U. maydis* effector Cmu1. Unlike "typical" effectors, which often are novel proteins without any known domains, Cmu1 is a functional chorismate mutase (Djamei et al., 2011). During plant infection, the enzyme is secreted from biotrophic hyphae and translocates into the plant cytoplasm. Chorismate is a precursor of SA, which is synthesized via the shikimate pathway in the chloroplasts. In the cytosol of plant cells, Cmu1 acts cooperatively with plant chorismate mutases to feed into different phenylpropanoid compounds and therefore to reduce the available chorismate pool to build SA (Djamei & Kahmann, 2012; Djamei et al., 2011). This function in rechanneling the host metabolism to dampen SA-induced defences presents a novel strategy of a biotroph to subvert the host immune system.

The examples of functionally characterized effectors given in this chapter show that the combination of (comparative) genomics with transcriptome analysis is a successful strategy to gain fundamental novel insights in the molecular mechanisms of fungal biotrophy. On the other hand, the little number of proteins that have been successfully studied so far also shows a strong need for new tools and more efficient experimental pipelines to get from genome sequence of to protein function.

7. ADAPTATION TO THE HOST ENVIRONMENT

The biotrophic lifestyle of *U. maydis* not only ensures a sustainable supply of nutrients but also requires a high degree of specialization and adaptation to its host maize. The uptake of nutrients, carbohydrates and nitrogen sources needs to be adapted to the host plant not only to ensure a sustainable supply for the fungus but also to certify that the host metabolism is altered in a way that assures survival of the fungus. The specific function of hexose transporters during pathogenic development was addressed in *U. maydis* via a systematic reverse genetic approach. Potential hexose transporters with similarities to the *S. cerevisiae* hexose transporter family were identified by BLAST analysis, and each of the 19 genes was individually deleted. Three of the genes were induced *in planta*, which implied a role during biotrophic development (R. Wahl & J. Kämper, unpublished data). Only one of the genes showed specific expression only after plant penetration. Indeed, strains deleted for the latter gene showed a significant reduction in virulence. Further analysis revealed that the encoded transporter, Srt1, was the founding member of a novel class of fungal sucrose transporters (Wahl, Wippel, Goos, Kämper, & Sauer, 2010). It allows the fungus the direct utilization of sucrose without extracellular hydrolysis into monosaccharides, the carbon form more commonly taken up by pathogenic fungi. Srt1 is highly specific for sucrose, and its affinity largely exceeds that of plant sucrose transporters. The high affinity provides advantages for the carbon acquisition by the pathogen. The direct uptake of sucrose quite likely offers a mechanism to prevent induction of plant defence responses known to occur upon apoplastic sucrose hydrolysis by fungal invertases (Wahl et al., 2010).

The second transporter gene with impact on pathogenic development of *U. maydis* was named *hxt1* due to the function of the encoded protein as hexose transporter. Hxt1 is most similar to a high-affinity glucose/arabinose transporter of *Amanita muscaria* that is upregulated in ectomycorrhiza (Nehls, Wiese, Guttenberger, & Hampp, 1998) and to a high-affinity glucose/fructose transporter of *Uromyces viciae-fabae* specifically expressed in haustoria (Voegele, Struck, Hahn, & Mendgen, 2001). When compared to *S. cerevisiae*, Hxt1 is most similar to the glucose sensors Snf3 and Rgt2 (Ozcan & Johnston, 1999); however, it does not exhibit the C-terminal extension implicated in signal transduction in yeast. In contrast to *srt1*, *hxt1* is highly and constitutively expressed. Hxt1 was characterized as H^+/glucose symporter with high affinity for glucose and lower affinities for fructose, xylose and galactose. A sensor function of Hxt1 for hexoses during pathogenic development is supported by the isolation of different *hxt1* mutant alleles that, upon ectopic expression, lead, in contrast to the deletion, to a complete loss of pathogenic development. Microarray analysis during the initial stages of development indicates an alteration of glucose-dependent gene expression in the fungal cells, which may interfere with the nutritional adaptation of the fungus at the switch from saprophytic to biotrophic growth (D. Schuler, R. Wahl & J. Kämper, unpublished data).

8. SUMMARY AND FUTURE PERSPECTIVES

Doubtless, the excellent quality of the genome sequence and its manually curated annotation built the most important groundwork for the present molecular studies of *U. maydis* biology. The *U. maydis* genome also served as a blueprint for the genome projects of *S. reilianum* and *U. hordei*. Together, these three genomes are an excellent resource for functional genetics. The combination of these tools with transcriptome analyses allowed identifying several new virulence factors that have been studied on the functional level (Table 5.1). At present, this lines these smut fungi among the best-studied eukaryotic plant pathogens. Nevertheless, key questions such as what defines the special ability of *U. maydis* to locally induce tumours and what defines host (and organ) specificity of smut still remain unanswered.

The most straightforward strategy to gain more hints towards these questions is the massive use of genome sequencing, which is getting available at rapidly decreasing costs. Genome sequencing of smut fungi should be

Table 5.1 Smut genes that were functionally characterized based on genomic/transcriptomic analyses

Name	Encoded protein	Function	Conservation Um–Sr–Uh	Citation
Pep1	Secreted effector	Peroxidase inhibitor, essential for *U. maydis* penetration	Highly conserved	Doehlemann et al. (2009) and Hemetsberger et al. (2012)
Pit1	Membrane protein	Unknown function, required for tumour formation in *U. maydis*	Conserved	Doehlemann et al. (2011)
Pit2	Secreted effector	Protease inhibitor, required for tumour formation *in U. maydis*	Low conservation but conserved functional domain (PID14)	Doehlemann et al. (2011) and Mueller et al. (2013)
Tin2	Secreted effector	Unknown, required for full virulence	Low conservation (divergence cluster 19-1)	Kämper et al. (2006), Schirawski et al. (2010), and Brefort et al. (2014)
Tin3	Secreted effector	Unknown, required for full virulence	Low conservation (divergence cluster 19-1)	Kämper et al. (2006), Schirawski et al. (2010), and Brefort et al. (2014)
Tin4	Secreted effector	Unknown, required for full virulence	Low conservation (divergence cluster 19-1)	Kämper et al. (2006), Schirawski et al. (2010), and Brefort et al. (2014)
Tin5	Secreted effector	Unknown, required for full virulence	Low conservation (divergence cluster 19-1)	Kämper et al. (2006), Schirawski et al. (2010), and Brefort et al. (2014)
Pst1	Secreted effector	Unknown, required for pathogenicity	Conserved	M. Vranes and J. Kämper (unpublished data)

Table 5.1 Smut genes that were functionally characterized based on genomic/transcriptomic analyses—cont'd

Name	Encoded protein	Function	Conservation Um–Sr–Uh	Citation
Pst2	Secreted effector	Unknown, required for pathogenicity	Conserved	M. Vranes and J. Kämper (unpublished data)
Srt1	Sucrose transporter	Transport of sucrose, required for full virulence	Highly conserved	Wahl et al. (2010)
Hxt1	Hexose transporter	High affinity glucose transporter, required for full virulence	Highly conserved	D. Schuler, R. Wahl, and J. Kämper (unpublished data)
Sad1	Secreted effector	Required for suppression of apical dominance	Low conservation (divergence cluster 19-1)	H. Ghareeb, F. Drechsler, C. Löfke, T. Teichmann, and J. Schirawski (unpublished data)
Rbf1	Zinc finger transcription factor	Apathogenic; required for expression of pathogenicity genes	Highly conserved	Heimel, Scherer, Vranes, et al. (2010)
Hdp2	Homeodomain transcription factor	Apathogenic; required for expression of pathogenicity genes	Highly conserved	M. Vranes and J. Kämper (unpublished data)
Biz1	Zinc finger transcription factor	Apathogenic; required for expression of pathogenicity genes	Highly conserved	Flor-Parra et al. (2006) and M. Vranes and J. Kämper (unpublished data)
Clp1	Regulator	Required for cell cycle release at begin of biotrophic stage	Highly conserved	Heimel, Scherer, Schuler, and Kämper (2010)

Um, *Ustilago maydis*; Sr, *Sporisorium reilianum*; Uh, *Ustilago hordei*.

directed in two directions. First, genomes of more unrelated species should be sequenced to cover smuts with a broad variety of different host. In this regard, the genomes of *Melanopsichium pennsylvanicum* (M. Thines, personal communication) and *Exobasidium vaccinii* (US Joint Genome Institute project, headed by J. Spatafora and F. Martin, in collaboration with G. Doehlemann) are currently underway. *M. pennsilvanicum* parasitizes *Polygonum* and therefore will be the first dicot-specific member of the Ustilaginomycetes to be sequenced. *E. vaccinii* causes smut of blueberries and belongs to the Exobasidiomycetes, which is a sister class to the Ustilaginomycetes, within the suborder of Ustilaginomycota. These two examples, however, should be only starting points for further sequencing approaches to gain a comprehensive coverage of smut genomic information. Besides evolutionary relationships and hints for mechanisms of speciation and host adaptation, such data will be very helpful to identify conserved virulence factors, that is, core effectors, which interfere with plant innate immunity and are expected to be conserved among pathogens. In addition to broad sequencing of new species, comparative genomics will also be applied in resequencing of *U. maydis* wild-type strains to discover coevolution during maize domestication, genetic variability of transcontinental isolates and evolutionary pressure on virulence-related effector genes. So far, there is surprisingly little known on positive selection of *Ustilago* effectors, although this is a striking issue in the light that so far no *AVR–R* gene interaction has been identified in the *U. maydis*–maize interaction.

An intriguing new perspective comes from recent genome sequencing of *Pseudozyma flocculosa*, a strictly anamorphic, nonpathogenic biocontrol agent that is closely related to *U. maydis* (Lefebvre et al., 2013). *P. flocculosa* shares most of its genes with the three plant pathogenic smut fungi, including mating and meiosis loci. Strikingly, this holds not true for the effector gene clusters that were identified in *U. maydis*. For example, none of the cluster 19A genes is present in *P. flocculosa* (Lefebvre et al., 2013). This information can serve as a precious source for studying evolution of pathogenicity in smut fungi. Isolating related epiphytic species from plants, which are hosts of smut fungi, would be a very interesting tool to study the origin of fungal biotrophy.

The increasing amount of sequence data will be instrumental for further generation of specific hypotheses. Nevertheless, further technical improvements are necessary to put forward the functional analysis of virulence genes. This implies large-scale knockout approaches to generate deletion strain libraries as resources open to the research community. Due to the high frequency of homologous integration, *U. maydis* is ideally suited for such a deletion project. Multiple deletions of gene families are already aided by the FLP recombinase system but could be significantly improved by marker-less transformation techniques.

ACKNOWLEDGEMENTS

Our research is funded by Max Planck Society, the Karlsruhe Institute of Technology, the RWTH Aachen University and the German Research Foundation (DFG). *U. hordei* strains DS198, DS199 and DS200 were generated by Daniela Schwammbach.

REFERENCES

Adams, M. D., Celniker, S. E., Holt, R. A., Evans, C. A., Gocayne, J. D., Amanatides, P. G., et al. (2000). The genome sequence of *Drosophila melanogaster*. *Science*, *287*, 2185–2195.

Bakkeren, G., Kamper, J., & Schirawski, J. (2008). Sex in smut fungi: Structure, function and evolution of mating-type complexes. *Fungal Genetics and Biology*, *45*, S15–S21.

Bakkeren, G., & Kronstad, J. W. (1994). Linkage of mating-type loci distinguishes bipolar from tetrapolar mating in basidiomycetous smut fungi. *Proceedings of the National Academy of Sciences of the United States of America*, *91*, 7085–7089.

Banuett, F., & Herskowitz, I. (1996). Discrete developmental stages during teliospore formation in the corn smut fungus, *Ustilago maydis*. *Development*, *122*, 2965–2976.

Basse, C. W., Stumpferl, S., & Kahmann, R. (2000). Characterization of a *Ustilago maydis* gene specifically induced during the biotrophic phase: Evidence for negative as well as positive regulation. *Molecular and Cellular Biology*, *20*, 329–339.

Brefort, T., Doehlemann, G., Mendoza-Mendoza, A., Reissmann, S., Djamei, A., & Kahmann, R. (2009). *Ustilago maydis* as a pathogen. *Annual Review of Phytopathology*, *47*, 423–445.

Brefort, T., Tanaka, S., Neidig, N., Doehlemann, G., Vincon, V., & Kahmann, R. (2014). Characterization of the largest effector gene cluster of Ustilago maydis. *PLoS Pathogens*, (in press).

Callow, J., & Ling, I. (1973). Histology of neoplasms and chlorotic lesions in maize seedlings following the injection of sporidia of *Ustilago maydis* (DC) Corda. *Physiological Plant Pathology*, *3*, 489–494.

Christensen, J. J. (1963). Corn smut induced by Ustilago maydis. *American Phytopathological Society Monograph*, *2*, 1–141.

Djamei, A., & Kahmann, R. (2012). *Ustilago maydis*: Dissecting the molecular interface between pathogen and plant. *PLoS Pathogens*, *8*, e1002955.

Djamei, A., Schipper, K., Rabe, F., Ghosh, A., Vincon, V., Kahnt, J., et al. (2011). Metabolic priming by a secreted fungal effector. *Nature*, *478*, 395–398.

Doehlemann, G., & Hemetsberger, C. (2013). Apoplastic immunity and its suppression by filamentous plant pathogens. *The New Phytologist*, *198*, 1001–1016.

Doehlemann, G., Reissmann, S., Assmann, D., Fleckenstein, M., & Kahmann, R. (2011). Two linked genes encoding a secreted effector and a membrane protein are essential for *Ustilago maydis*-induced tumour formation. *Molecular Microbiology*, *81*, 751–766.

Doehlemann, G., van der Linde, K., Assmann, D., Schwammbach, D., Hof, A., Mohanty, A., et al. (2009). Pep1, a secreted effector protein of *Ustilago maydis*, is required for successful invasion of plant cells. *PLoS Pathogens*, *5*, e1000290.

Flor-Parra, I., Vranes, M., Kämper, J., & Perez-Martin, J. (2006). Biz1, a zinc finger protein required for plant invasion by *Ustilago maydis*, regulates the levels of a mitotic cyclin. *Plant Cell, 18*, 2369–2387.

Ghareeb, H., Becker, A., Iven, T., Feussner, I., & Schirawski, J. (2011). Sporisorium reilianum infection changes inflorescence and branching architectures of maize. *Plant Physiology, 156*, 2037–2052.

Heimel, K., Scherer, M., Schuler, D., & Kämper, J. (2010). The *Ustilago maydis* Clp1 protein orchestrates pheromone and *b*-dependent signaling pathways to coordinate the cell cycle and pathogenic development. *Plant Cell, 22*, 2908–2922.

Heimel, K., Scherer, M., Vranes, M., Wahl, R., Pothiratana, C., Schuler, D., et al. (2010). The transcription factor Rbf1 is the master regulator for *b*-mating type controlled pathogenic development in *Ustilago maydis*. *PLoS Pathogens, 6*, e1001035.

Hemetsberger, C., Herrberger, C., Zechmann, B., Hillmer, M., & Doehlemann, G. (2012). The *Ustilago maydis* effector Pep1 suppresses plant immunity by inhibition of host peroxidase activity. *PLoS Pathogens, 8*, e1002684.

Ho, E. C., Cahill, M. J., & Saville, B. J. (2007). Gene discovery and transcript analyses in the corn smut pathogen *Ustilago maydis*: Expressed sequence tag and genome sequence comparison. *BMC Genomics, 8*, 334.

Hof, A. (2013). *Die Rolle von programmiertem Zelltod in der epidermalen Resistenz von Gerste.* PhD Thesis, Philipps-Universität Marburg.

Holloman, W. K., Schirawski, J., & Holliday, R. (2008). The homologous recombination system of *Ustilago maydis*. *Fungal Genetics and Biology, 45*(Suppl. 1), S31–S39.

Kämper, J., Kahmann, R., Bölker, M., Ma, L. J., Brefort, T., Saville, B. J., et al. (2006). Insights from the genome of the biotrophic fungal plant pathogen *Ustilago maydis*. *Nature, 444*, 97–101.

Kellner, N., Heimel, K., Obhof, T., Finkernagel, F., & Kämper, J. (2014). The SPF27 homologue Num1 connects splicing and Kinesin 1-dependent cytoplasmic trafficking in *Ustilago maydis*. *PLoS Genetics, 10*(1), e1004046.

Khrunyk, Y., Munch, K., Schipper, K., Lupas, A. N., & Kahmann, R. (2010). The use of FLP-mediated recombination for the functional analysis of an effector gene family in the biotrophic smut fungus *Ustilago maydis*. *The New Phytologist, 187*, 957–968.

Laurie, J. D., Linning, R., & Bakkeren, G. (2008). Hallmarks of RNA silencing are found in the smut fungus Ustilago hordei but not in its close relative Ustilago maydis. *Current Genetics, 53*(1), 49–58.

Laurie, J. D., Ali, S., Linning, R., Mannhaupt, G., Wong, P., Guldener, U., et al. (2012). Genome comparison of barley and maize smut fungi reveals targeted loss of RNA silencing components and species-specific presence of transposable elements. *Plant Cell, 24*, 1733–1745.

Lefebvre, F., Joly, D. L., Labbe, C., Teichmann, B., Linning, R., Belzile, F., et al. (2013). The transition from a phytopathogenic smut ancestor to an anamorphic biocontrol agent deciphered by comparative whole-genome analysis. *Plant Cell, 25*, 1946–1959.

Loftus, B. J., Fung, E., Roncaglia, P., Rowley, D., Amedeo, P., Bruno, D., et al. (2005). The genome of the basidiomycetous yeast and human pathogen *Cryptococcus neoformans*. *Science, 307*, 1321–1324.

Lozano-Torres, J. L., Wilbers, R. H., Gawronski, P., Boshoven, J. C., Finkers-Tomczak, A., Cordewener, J. H., et al. (2012). Dual disease resistance mediated by the immune receptor Cf-2 in tomato requires a common virulence target of a fungus and a nematode. *Proceedings of the National Academy of Sciences of the United States of America, 109*, 10119–10124.

Meksem, K., Shultz, J., Tebbji, F., Jamai, A., Henrich, J., Kranz, H., et al. (2005). A bacterial artificial chromosome based physical map of the *Ustilago maydis* genome. *Genome, 48*, 207–216.

Mueller, A. N., Ziemann, S., Treitschke, S., Assmann, D., & Doehlemann, G. (2013). Compatibility in the Ustilago maydis-maize interaction requires inhibition of host cysteine proteases by the fungal effector Pit2. *PLoS Pathogens*, *9*, e1003177.

Munkacsi, A. B., Stoxen, S., & May, G. (2007). Domestication of maize, sorghum, and sugarcane did not drive the divergence of their smut pathogens. *Evolution: International Journal of Organic Evolution*, *61*, 388–403.

Nehls, U., Wiese, J., Guttenberger, M., & Hampp, R. (1998). Carbon allocation in ectomycorrhizas: Identification and expression analysis of an *Amanita muscaria* monosaccharide transporter. *Molecular Plant-Microbe Interactions*, *11*, 167–176.

Nielsen, C. B., Friedman, B., Birren, B., Burge, C. B., & Galagan, J. E. (2004). Patterns of intron gain and loss in fungi. *PLoS Biology*, *2*, e422.

O'Brien, J. A., Daudi, A., Finch, P., Butt, V. S., Whitelegge, J. P., Souda, P., et al. (2012). A peroxidase-dependent apoplastic oxidative burst in cultured Arabidopsis cells functions in MAMP-elicited defense. *Plant Physiology*, *158*, 2013–2027.

Ozcan, S., & Johnston, M. (1999). Function and regulation of yeast hexose transporters. *Microbiology and Molecular Biology Reviews*, *63*, 554–569.

Rogozin, I. B., Wolf, Y. I., Sorokin, A. V., Mirkin, B. G., & Koonin, E. V. (2003). Remarkable interkingdom conservation of intron positions and massive, lineage-specific intron loss and gain in eukaryotic evolution. *Current Biology*, *13*, 1512–1517.

Roy, S. W., & Gilbert, W. (2005a). The pattern of intron loss. *Proceedings of the National Academy of Sciences of the United States of America*, *102*, 713–718.

Roy, S. W., & Gilbert, W. (2005b). Rates of intron loss and gain: Implications for early eukaryotic evolution. *Proceedings of the National Academy of Sciences of the United States of America*, *102*, 5773–5778.

Schilling, L., Matei, A., Redkar, A., Walbot, V., Doehlemann, G. (2014). Induction of plant tumors by Ustilago maydis is shaped by organ-specific effectors.

Schirawski, J., Heinze, B., Wagenknecht, M., & Kahmann, R. (2005). Mating type loci of Sporisorium reilianum: Novel pattern with three a and multiple b specificities. *Eukaryotic Cell*, *4*(8), 1317–1327.

Schirawski, J., Mannhaupt, G., Munch, K., Brefort, T., Schipper, K., Doehlemann, G., et al. (2010). Pathogenicity determinants in smut fungi revealed by genome comparison. *Science*, *330*, 1546–1548.

Scholler, P., Benes, V., Voss, H., Ansorge, W., & Hoheisel, J. D. (1996). Sequencing 39,350 bp of *Saccharomyces cerevisiae* chromosome XII utilizing ordered shotgun libraries. *DNA Sequence*, *6*, 257–262.

Shabab, M., Shindo, T., Gu, C., Kaschani, F., Pansuriya, T., Chintha, R., et al. (2008). Fungal effector protein Avr2 targets diversifying defense-related cys proteases of tomato. *The Plant Cell*, *20*, 1169–1183.

Skibbe, D. S., Doehlemann, G., Fernandes, J., & Walbot, V. (2010). Maize tumors caused by *Ustilago maydis* require organ-specific genes in host and pathogen. *Science*, *328*, 89–92.

Spanu, P. D., Abbott, J. C., Amselem, J., Burgis, T. A., Soanes, D. M., Stuber, K., et al. (2010). Genome expansion and gene loss in powdery mildew fungi reveal tradeoffs in extreme parasitism. *Science*, *330*, 1543–1546.

Steinberg, G., & Perez-Martin, J. (2008). *Ustilago maydis*, a new fungal model system for cell biology. *Trends in Cell Biology*, *18*, 61–67.

Tian, M., Win, J., Song, J., van der Hoorn, R., van der Knaap, E., & Kamoun, S. (2007). A *Phytophthora infestans* cystatin-like protein targets a novel tomato papain-like apoplastic protease. *Plant Physiology*, *143*, 364–377.

van der Linde, K., Hemetsberger, C., Kastner, C., Kaschani, F., van der Hoorn, R. A., Kumlehn, J., et al. (2012). A maize cystatin suppresses host immunity by inhibiting apoplastic cysteine proteases. *Plant Cell*, *24*, 1285–1300.

van der Linde, K., Mueller, A., Hemetsberger, C., Kaschani, F., Van der Hoorn, R. A., & Doehlemann, G. (2012). The maize cystatin CC9 interacts with apoplastic cysteine proteases. *Plant Signaling & Behavior*, 7, 1397–1401.

Voegele, R. T., Struck, C., Hahn, M., & Mendgen, K. (2001). The role of haustoria in sugar supply during infection of broad bean by the rust fungus *Uromyces fabae*. *Proceedings of the National Academy of Sciences of the United States of America*, *98*, 8133–8138.

Vollmeister, E., Schipper, K., Baumann, S., Haag, C., Pohlmann, T., Stock, J., et al. (2012). Fungal development of the plant pathogen *Ustilago maydis*. *FEMS Microbiology Reviews*, *36*, 59–77.

Voss, H., Schwager, C., Wiemann, S., Zimmermann, J., Stegemann, J., Erfle, H., et al. (1995). Efficient low redundancy large-scale DNA sequencing at EMBL. *Journal of Biotechnology*, *41*, 121–129.

Wahl, R., Wippel, K., Goos, S., Kämper, J., & Sauer, N. (2010). A novel high-affinity sucrose transporter is required for virulence of the plant pathogen *Ustilago maydis*. *PLoS Biology*, *8*, e1000303.

Walter, M. C., Rattei, T., Arnold, R., Guldener, U., Munsterkotter, M., Nenova, K., et al. (2009). PEDANT covers all complete RefSeq genomes. *Nucleic Acids Research*, *37*, D408–D411.

Wicker, T., Oberhaensli, S., Parlange, F., Buchmann, J. P., Shatalina, M., Roffler, S., et al. (2013). The wheat powdery mildew genome shows the unique evolution of an obligate biotroph. *Nature Genetics*, *45*, 1092–1096.

CHAPTER SIX

Advancing Knowledge on Biology of Rust Fungi Through Genomics

Sébastien Duplessis[*,1], **Guus Bakkeren**[†], **Richard Hamelin**[‡,§]

[*]Institut National de la Recherche Agronomique (INRA), UMR 1136 INRA/Lorraine University, Interactions Arbres/Micro-organismes, Centre de Nancy, Champenoux, France
[†]Pacific Agri-Food Research Centre, Agriculture and Agri-Food Canada, Summerland, British Columbia, V0H 1Z0 Canada
[‡]Natural Resources Canada, Canadian Forest Service, Laurentian Forestry Centre, Quebec, Canada
[§]Department of Forest and Conservation Sciences, The University of British Columbia, Vancouver, British Columbia, Canada
[1]Corresponding author: e-mail address: duplessi@nancy.inra.fr

Contents

Abstract

Pucciniales are an important group of fungal plant pathogens that cause rust diseases in a diverse group of hosts including ecologically and economically important crops and trees. Rust fungi have intriguing and complex life cycles and are obligate biotrophs. Because of their biological features, these fungi are very difficult to study under laboratory conditions. The recent advances in genomics and transcriptomics have opened great perspectives for making progress in the study of this group of fungi and more particularly to dissect the genetic determinants underlying the host infection process. In this chapter, we provide an overview of the current knowledge on rust genomics and

Advances in Botanical Research, Volume 70
ISSN 0065-2296
http://dx.doi.org/10.1016/B978-0-12-397940-7.00006-9

we particularly highlight how next-generation sequencing technologies are moving this field forward, providing new avenues in the understanding of fungal biotrophy.

1. INTRODUCTION

Rust fungi (order Pucciniales) are an important and diverse group of plant pathogens that can affect a diverse group of hosts and cause a remarkable range of symptoms such as defoliation, cankers, and witch brooms (Cummins & Hiratsuka, 2003). Rust fungi cause diseases that are responsible for some of the most severe economic losses of trees and crops, including pines, poplars, eucalyptus, wheat, coffee, and soybean. In addition to causing economic losses, rusts are responsible for impacting ecosystems.

Rusts possess unique biological features and ecology. They are strict biotrophic fungi that require a living host to complete their life cycle. Rust fungi display a diversity of life cycles. Heteroecious macrocyclic rusts possess one of the most intriguing life cycles in the fungal kingdom. They must alternate between telial and aecial hosts and produce five different spore types in order to complete their life cycle (Fig. 6.1). The symptoms caused by heteroecious rusts on telial and aecial hosts are usually different and can occur on completely unrelated hosts. For example, *Cronartium ribicola* causes branch and stem cankers on the pine aecial host but only leaf infections on the telial *Ribes* spp. hosts. *Puccinia triticina*, the wheat leaf rust fungus, produces uredospores and the survival structures, the teliospores, on wheat, whereas the sexual stage occurs on a completely unrelated plant, meadow rue (*Thalictrum speciosissimum*). Other rusts have reduced life cycles. Autoecious rusts can complete their life cycle on a single host, while demicyclic and microcyclic rusts have reduced number of spore stages.

In spite of their economic importance, there are unresolved questions about rust fungi biology, epidemiology, and host–pathogen interactions that could provide critical knowledge and impact disease management. The need for host alternation in heteroecious rusts is still not clearly understood. What are the evolutionary advantages or compromises of host alternation? Do heteroecious rusts possess alternate sets of effector genes that allow them to interact with different hosts or are the same effectors acting in different hosts? Do autoecious rusts have a reduced gene set compared with their heteroecious counterparts? Another interesting and unresolved set of questions relates to host specificity. Most rusts have a relatively narrow host range

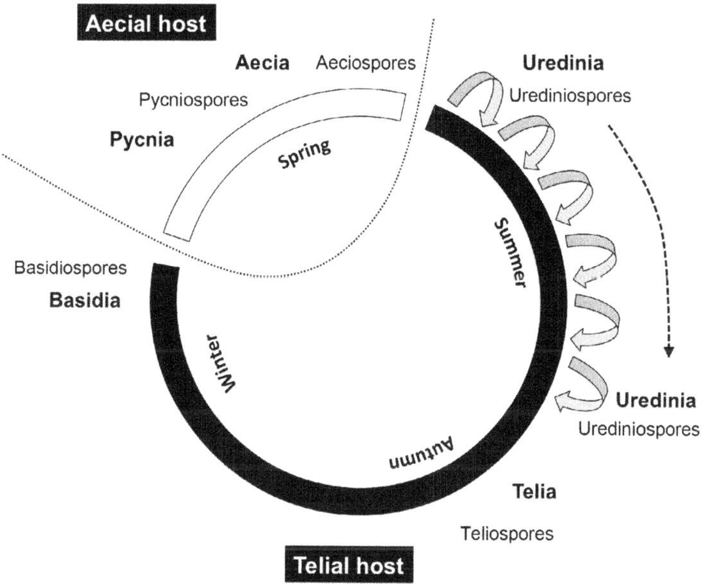

Figure 6.1 Schematic representation of the typical biological cycle of heteroecious and macrocyclic rust fungi. In spring, basidiospores infect the aecial host. Mycelia formed by the rust fungus produce pycnia, and pycniospores are released in nectar droplets in which fertilization can occur between spores and receptive hypha of compatible mating types. Spores are transmitted between nectars by insects, rain, or wind. Once dikaryons are formed, aecia are established and aeciospores are released. This leads to the telial host infection in spring. Infection of the telial host occurs throughout summer by repeated infection cycles producing uredinia, a structure that releases huge amounts of uredospores. This stage often leads to strong rust epidemics on the telial host over the growing season. At the end of summer/early autumn, black structures called telia are formed in the senescent tissues of the host. Teliospores differentiate during autumn and winter in these overwintering structures. After karyogamy and meiosis, teliospores germinate and form basidia that release basidiospores for infection of the aecial host, completing the cycle. Only the basidiospores are haploids; spores are dikaryotic at all other stages. Autoecious rust fungi proceed through their life cycle on a single host and microcyclic rust fungi only produce some of the spore types, for example, uredospores and teliospores.

on either the aecial or telial hosts. What are the determinants of this host specificity? What allows some rusts to have narrow host specificity on the aecial host but a broad host range on the telial host, or vice versa? What gives rust fungi the ability to rapidly adapt following the deployment of resistant host genotypes? At the core of the set of important questions is what drives the host–pathogen interactions. The gene-for-gene hypothesis states that each resistance gene in a host plant corresponds to an avirulent gene counterpart in the pathogen and their interaction triggers an immune response.

The experimental demonstration of this hypothesis was proven for the first time in the flax rust (Flor, 1959). These questions can be addressed using a combination of novel genomics approaches that can be used to dissect these interactions and identify genetic determinants in both the host and the pathogen.

There have been exciting new developments in the past decade in our understanding of genetics and genomics of rust fungi that is informing us on their biology, host–pathogen interactions, evolution, and epidemiology. This chapter aims at describing the latest developments in rust genomics and to take a look at future developments.

2. RUST FUNGI IN THE GENOMICS ERA

2.1. Genomics of plant-interacting fungi

Most fungal genomes exhibit a small size of 30–60 Mb compared with other eukaryotes (Raffaele & Kamoun, 2012). Fungi with smaller and compact genomes have been reported, such as Pucciniomycotina *Mixia osmundae* or members of the Ustilaginomycotina *Sporisorium reilianum*, *Ustilago maydis*, or *Ustilago hordei* (Kämper et al., 2006; Laurie et al., 2012; Schirawski et al., 2010; Toome et al., 2014). But a few are beyond this range, such as in the ascomycete *Tuber melanosporum*, the gourmet-favorite black truffle, *Blumeria* spp. causing powdery mildew diseases on plants, and the rust fungi (Duplessis, Cuomo, et al., 2011; Martin et al., 2010; Spanu et al., 2010; Wicker et al., 2013). Recent genomics reviews examined the reasons behind this wide variation in genome sizes in plant-interacting fungi and fungal-like organisms. Among the factors considered are content in repetitive elements including transposable elements (TE) and gene losses shared between different phyla of obligate biotrophic pathogens (Kemen & Jones, 2012; Martin, 2014; Raffaele & Kamoun, 2012; Spanu, 2012). The genomes of fungi interacting with plants are also marked by specific repertoires of genes related to the mutualistic or parasitic relationship established with the host plant. Genes encoding enzymes involved in the decomposition of plant cell wall components (e.g. carbohydrate-active enzymes, CAZymes), signal transduction, transport of nutrients and water, or effectors modulating the host immunity are essential for the success of colonization and acquisition of nutrients from the hosts. The content in CAZymes much likely reflects the fungal lifestyle. Mutualistic ectomycorrhizal fungi or obligate biotrophs such as rust fungi have a reduced or moderate ability to alter the plant cell wall, possibly to minimize triggering a plant response during colonization

(Duplessis, Cuomo, et al., 2011; Martin et al., 2008). In oomycetes and fungi interacting with plants, a large subset of genes code for proteins predicted to be secreted, the 'secretomes', and a subset of these consist of relatively small proteins that represent the effector repertoires necessary to achieve successful infection (Lowe & Howlett, 2012; Raffaele & Kamoun, 2012; Spanu, 2012). Those genomes are often marked by transposon invasions and eventually by the presence of multigene families, which explains their large size compared with those of other fungi and denotes a greater plasticity (Murat, Payen, Petitpierre, & Labbé, 2014; Raffaele & Kamoun, 2012). The association between candidate effector gene families and transposon-rich regions has been reported in the genomes of different plant pathogens (e.g. *Leptosphaeria maculans* and *P. infestans*) suggesting that transposon activity could have been central in the diversification of their effector repertoires (Haas et al., 2009; Rouxel et al., 2011). Indeed, effector genes found in transposon-rich regions show highly polymorphic profiles strongly supporting a possible adaptation to the host immune system (Raffaele & Kamoun, 2012).

2.2. Sequencing genomes of rust fungi

Progress has been made in the last decade, from sequencing rust fungi genomes with whole-genome shotgun strategies using the Sanger sequencing technology to NGS (Table 6.1) (Duplessis, Spanu, & Schirawski, 2014). This first generation of genome sequences has provided genome assemblies of great quality, despite a low-depth coverage due to the high cost of this technology compared with the more-recent next-generation sequencing techniques. The poplar leaf rust *M. larici-populina* and the wheat stem rust *P. graminis* f. sp. *tritici* were the first obligate biotrophic plant pathogens in the Basidiomycota phylum sequenced (Duplessis, Cuomo, et al., 2011; McDowell, 2011). The two genomes show a composition and organization illustrating not only striking similarities within the order Pucciniales but also clear differences with other obligate biotrophic pathogens in the Ascomycota and Basidiomycota phyla, such as powdery mildews and smut fungi. Indeed, both rust fungi have large genomes (80–101 Mb), a large number of repeats and TE (nearly 50% of the genome), and a large number of genes (>16,000), whereas other obligate biotrophic fungal pathogens only share one or two of these characteristics (see Duplessis et al., 2014, for details). The large genome size and the large number of genes observed for rust fungi might be related to the heteroecious and macrocyclic nature of the two species that were selected for genome sequencing. Sequencing the genomes of autoecious and microcyclic rust fungi is now required to

Table 6.1 Update on genome sequences statistics of Pucciniales

Rust species and isolate (reference)	***Melampsora larici-populina*** **98AG31 (Duplessis, Cuomo, et al., 2011[a])**	***Puccinia graminis*** **f. sp.** ***tritici*** **CDL75-36-700-3 (Duplessis, Cuomo, et al., 2011[a])**	***Puccinia striiformis*** **f. sp.** ***tritici*** **PST-130 (Cantu et al., 2011)**	***Puccinia striiformis*** **f. sp.** ***tritici*** **CY32 (Zheng et al., 2013)**	***Puccinia triticina*** **1-1 (BBBD) (C. Cuomo, J. Fellers, L. Szabo and G. Bakkeren unpublished[a])**
Sequencing approach	Sanger sequencing, whole-genome shotgun strategy	Sanger sequencing, whole-genome shotgun strategy, Illumina sequencing to validate SNPs between haplotypes	Illumina Genome Analyser II sequencing (53.8 million reads paired-end 85–101 pb reads); draft genome	Fosmid-to-fosmid strategy by Illumina GA paired-end sequencing (1920 pools of 19,200 36 Kb average fosmids; validation of 10 selected fosmids by end-to-end Sanger sequencing); whole genome	Illumina whole-genome shotgun strategy, paired-end 454 pyrosequencing, paired-end Sanger sequencing of 15,000 BAC clones (see Fellers et al., 2013)
Number of contigs/ scaffolds	3264/462	4557/392	29,178 (Contigs only)	12,833/4283	14,820
Total assembly size of contigs/ scaffolds	97.7/101.1 Mb	81.5/88.6 Mb	64.8 Mb (estimated genome size 68.2–78.8 Mb)	115.5 Mb/130.6 Mb (estimated size, 110 Mb)	135.34 Mb (estimated size, 135 Mb)
Gap (% of genome)	3.4%	8%	n.a.	n.a.	n.a.
Contigs N50/ L50	27/1.1 Mb	30/0.97 Mb	n.a./5.1 Kb	18 Kb/n.a.	10.37 Kb
Scaffolds N50/ L50	265/112.3 Kb	546/39.5 Kb	n.a.	125.3 Kb/n.a.	544.26 Kb

Sequencing depth (*x*-fold)	6.9 ×	12 ×	59 ×	26 ×	31 ×
Transposable and repetitive elements (% of genome)	45%	43.7%	17.8% (partial annotation)	48.9%	42–45%
Number of predicted genes	16,399	17,773 (revised to 15,800 by the Broad Institute upon RNA-seq analysis)	20,423 (lowered to 18,149 in Cantu et al., 2013)	25,288	14,880
Resequencing of isolates	Resequencing of 14 isolates from diverse populations collected over 20 years in France, at a 20–30 × depth by Illumina paired-end sequencing; mapping onto the 98AG31 reference genome, determination of SNPs, InDels and SV (A. Persoons & S. Duplessis, unpublished)		Resequencing of four isolates from the United States and the United Kingdom by Illumina, paired-end sequencing, determination of intra- and interisolates SNPs (Cantu et al., 2013 BMC Genomics): • PST-21 (66 × coverage, 43,106 contigs, 73 Mb total length, N50	Resequencing of five isolates from diverse origin worldwide by Illumina paired-end sequencing, mapping onto CY-32, determination of SNPs, InDels, and structural variants, determination of SNPs in PST-130: • CY23 (29 × coverage)	Resequencing of 120 isolates from diverse genetic lineages and race designations across North America, and Europe, by Illumina paired-end sequencing, mapping onto *Pt* 1-1, determination of SNPs, InDels, and structural variants (J. Fellers, J.A. Kolmer, C.A. Cuomo, G. Bakkeren, B.D. McCallum, & B. Saville, et al., unpublished)

Continued

Table 6.1 Update on genome sequences statistics of Pucciniales—cont'd

Rust species and isolate (reference)	*Melampsora larici-populina* 98AG31 (Duplessis, Cuomo, et al., 2011)	*Puccinia graminis* f. sp. *tritici* CDL75-36-700-3 (Duplessis, Cuomo, et al., 2011)	*Puccinia striiformis* f. sp. *tritici* PST-130 (Cantu et al., 2011)	*Puccinia striiformis* f. sp. *tritici* CY32 (Zheng et al., 2013)	*Puccinia triticina* 1-1 (BBBD) (C. Cuomo, J. Fellers, L. Szabo and G. Bakkeren unpublished)
			3.9 Kb, 20,653 genes), • PST-43 (26 × coverage, 49,784 contigs, 71 Mb total length, N50 3.3 Kb, 21,036 genes), • PST-87/7 (15 × coverage, 55,502 contigs, 53 Mb total length, N50 1.3 Kb, 20,688 genes), • PST-08/21 (21 × coverage, 50,898 contigs, 56 Mb total length, N50 1.6 Kb, 20,875 genes)	• PK-CDRD (27 × coverage) • Pst-78 (25 × coverage) • Hu09-2 (23 × coverage) • 104E137A- (22 × coverage)	

[a]*M. larici-populina*: http://genome.jgi.doe.gov/programs/fungi/index.jsf; *Puccinia* spp.: http://www.broadinstitute.org/annotation/genome/puccinia_group/GenomesIndex.html.

Selected genomic features from the rust fungi genomes sequenced to date. The table also includes NGS resequencing data. n.a., non available determine how far these biological features impact the genomic landscape of rust fungi. It is likely that TE invasion played a particularly important role in genome size expansion. The genome of the flax rust *Melampsora lini* is estimated to be close to 200 Mb (Peter Dodds, personal communication), and the genome of the microcyclic rust fungus *Phakopsora pachyrhizi*, responsible for the soybean rust disease, is estimated to be >800 Mb (Igor Grigoriev, personal communication). Analysis is ongoing and the number of predicted genes is not yet revealed, but it is very likely that repetitive DNA and TE will explain such genome size suggesting that genome expansion in rust fungi is a common trend that is not related to the ability to infect multiple host or to go through multiple differentiation stages in the biological cycles.

In the wake of these rust genome sequences, the draft genome of *Puccinia striiformis* f. sp. *tritici* (isolate PST-130) was obtained by Illumina-only sequencing (Cantu et al., 2011). A partial assembly of 65 Mb was gathered with a greater sequencing depth compared with the other rust genomes, but at a lower resolution (see number and size of contigs/scaffolds in Table 6.1). The striking differences in its assembly are likely due to variable genome sizes and differing content in repetitive sequences among rust fungi. Nonetheless, this pioneer sequencing of a rust genome with a NGS-only-based approach allowed the prediction of a draft catalogue of 20,423 putative genes, which was much larger than the number reported in the sister species *P. graminis* f. sp. *tritici* (Cantu et al., 2011). By combining different sequencing technologies and pools of large DNA fragments, a different isolate of the same wheat rust species originating from China (isolate CY32) was sequenced by Illumina and partially validated by Sanger sequencing (Zheng et al., 2013). The fosmid-to-fosmid sequencing approach used in this study helped to meet the challenging profile of rust genomes and to reach greater assembly quality at higher sequencing depth compared to the previous report for this species. After assembly, the genome of *P. striiformis* f. sp. *tritici* was estimated at a much larger size of 130 Mb, and the number of predicted genes was increased to 25,288. This places the wheat stripe rust fungus among the ones having the largest gene repertoire recorded so far in fungi, ahead of the mutualistic biotrophic fungus *Laccaria bicolor* (Martin et al., 2008). The repeat and TE content of this wheat leaf rust genome was estimated at nearly 50%, similar in range to previous observations in rust fungi. The strategy established by Zheng and collaborators greatly helped improving the assembly of this complex genome. The genome size was estimated to be twice the size of the draft sequence of the PST-130 isolate. Discrepancies observed between the two genomes could have reflected major divergence between isolates, but this was not supported by the resequencing of genomes of isolates collected worldwide (Zheng et al., 2013). The comparison more likely illustrates the limitations of a sequencing strategy based only on the sequencing of small reads from small DNA fragments as input. Although the depth of sequencing and the assembly quality of the PST-130 isolate differed to standards in genomic studies, the report confirmed both the large size of genome alluded to a similarly large genome and the gene complements of rust fungi. It also demonstrated the possibility to generate a large amount of relevant genetic information for a nonmodel species that is particularly economically relevant. Indeed, although only a portion of the genome was captured when compared with the genome of the CY-32 isolate, a closely similar number of genes was found (later reduced to 18,149 in Cantu et al., 2013; Table 6.1). Although the assembled draft genome did not properly illustrate the complexity of the genomic landscape of *P. striiformis* f. sp. *tritici*, it did succeed in capturing a significant portion of the functional genome, the gene space. On the one hand, the strategy defined by Zheng et al. (2013) could be very promising in the future for fungal genomes enriched in TE and for which difficulties in assembly are encountered (e.g. *Blumeria* spp.; Hacquard, Kracher, et al., 2013; Wicker et al., 2013); on the other hand, obtaining draft genomes at lower coverage and at lower cost based only on NGS technologies could be a faster way to gather relevant functional information in important pathogens. These studies are providing promising avenues for future sequencing of rust fungi with very large estimated genomes, that is, the soybean rust *P. pachyrhizi* or the coffee rust *Hemileia vastatrix* (Carvalho, Carvalho, Barreto, & Evans, 2014).

2.3. Major genomic features of rust fungi

The comparative analysis of gene complements and synteny in rust genomes highlights a complex evolution in the order Pucciniales with intense TE activity, rearrangements, insertions, and deletions. Specific gene families were particularly abundant compared with other basidiomycetes. Most of these families do not have functional information, and among those with annotations, helicases, leucine-rich repeat proteins, oligopeptide transporters, and different types of glycosyl hydrolases, lipases and peptidases were particularly overrepresented (Duplessis, Cuomo, et al., 2011; Zheng et al., 2013). Such expansions may correspond to specific processes related to host infection, as well as DNA repair and maintenance. Interestingly, more genes with gene ontology (GO) terms related to DNA recombination were found between the two wheat rust fungi than with *M. larici-populina* (Zheng et al., 2013). The expansion of genes with functional annotations related to DNA metabolism could also correlate with the invasion of the genomes of rust fungi by TEs. Whereas orthology was only evidenced for half of the genes between *M. larici-populina* and *P. graminis* f. sp. *tritici*, a greater conservation of the gene complements was observed within Pucciniaceae (Zheng et al., 2013). Few synteny blocks could be detected between the genomes of the poplar rust and the genomes of the wheat stem rust (Duplessis, Cuomo, et al., 2011). On the contrary, extensive evidence of microsynteny was found between the wheat rust fungi (Cantu et al., 2011; Fellers et al., 2013). Comparisons between *M. larici-populina, P. graminis* f. sp. *tritici*, and *P. striiformis* f. sp. *tritici* identified an important reshuffling of the assembled genomic scaffolds, consistent with ancient transposon activity (Zheng et al., 2013). The specificity of the gene catalogues in the Pucciniaceae and in the Melampsoraceae illustrates the old divergence between these two phylogenetic lineages (Aime et al., 2006). Future data from the *M. lini* genome analysis will help in determining with more precision the level of conservation of gene content at the taxonomic level of families within the order Pucciniales. Interestingly, the species-specific genes found in the three rust genomes published to date encode a majority of proteins of unknown function that belong to large expanded families, including secreted proteins (Duplessis, Cuomo, et al., 2011; Zheng et al., 2013).

2.4. NGS to assess genome-scale polymorphism in rust fungi

The facilitated access to NGS has allowed establishing programs for sequencing genomes of selected isolates for a given species and mapping onto reference genomes, opening the door to revealing population genomic and association genetic studies. Illumina sequencing was used to study polymorphisms within and between dikaryotic isolates of *P. striiformis* f. sp. *tritici* (Cantu et al., 2013; Zheng et al., 2013) (Table 6.1). Both studies relied on paired-end sequencing of small DNA fragments mapped to a reference genome at depth coverage of 15–66 × in order to determine single-nucleotide polymorphisms (SNPs) and in one case other events such as insertions/deletions (InDels) and structure variants (Table 6.1). In the study by Cantu et al. (2013), the sequenced genomes of each isolate were *de novo* assembled and compared with the reference draft genome of PST-130 in order to assess completeness of sequencing, whereas in the other study, the reference genome was the Chinese isolate CY-32 (Zheng et al., 2013). Considering the large differences between the PST-130 and CY-32 genome assemblies, it will be particularly relevant and useful to compare the two genomic resources and to eventually refine their assemblies. All *P. striiformis* f. sp. *tritici* genomes were sequenced from dikaryotic spores, and different types of polymorphic sites were recorded, such as intra- and extra-isolate SNPs. A large portion of the SNPs identified by Zheng et al. (2013) were located in coding regions ($\sim 1/3$), and half of them were non-synonymous, indicating important genetic diversity. The two studies identified different levels of polymorphisms in the sequenced genomes. In total, between 81,000 and 109,000 SNPs were detected per isolate by Zheng et al. (2013) of which ~82–84% were heterozygous. Cantu et al. (2013) identified a much larger amount of SNPs (>350,000 per genome) with a similar high rate of heterozygous SNPs. Major differences in the final numbers of SNPs probably reflect the mapping strategy and the SNP calling parameters and programs used. Such differences in SNP detection of the same rust are quite disturbing and highlight the need to refine these analyses and define parameters that will allow cross study comparisons.

In comparison, 88,083 and 129,172 SNPs were detected in *M. larici-populina*, based on Sanger sequencing only (note the low sequencing depth at 6.9×), and in *P. graminis* f. sp. *tritici*, based on Illumina sequencing and comparison to the Sanger assembled genome, respectively (Duplessis, Cuomo, et al., 2011). The sequencing of 14 isolates of *M. larici-populina* by Illumina at a greater depth (between 20 and 30×) identified a larger number of polymorphic sites when all genomes were considered (>580,000); however, overall, an average of ~150,000 SNPs per genome was found indicating polymorphic rates in the range of those reported for *Puccinia* spp. genomes (A. Persoons & S. Duplessis, unpublished data).

Although the Illumina-based strategy chosen by Cantu et al. (2013) led to a smaller *de novo* genome assembly size for isolate PST-130 than for the CY-32 isolate, the number of predicted genes was in the same range with more than 20,000 gene models per genome. The predicted gene complements have a rather good support in term of complete copies of conserved eukaryotic genes (72.1% SD ± 14.6%). This indicates once more that an Illumina-only-based sequencing strategy applied at adequate depth can capture a significant portion of the functional genome despite the complexity of the genome per se. Of course, the critical limit is the quality of the manual curation performed after the automatic gene prediction.

3. RUST TRANSCRIPTOMICS

Before the availability of NGS technologies, most of the knowledge on molecular determinants of rust fungi was based on expressed sequence tags (ESTs) and cDNA-array-based expression profiling. Pioneer studies explored the molecular biology of a variety of Pucciniales mostly focusing on the process of host infection and on purified haustorial structures from infected plant tissues (for a complete review, see Duplessis, Joly, & Dodds, 2012; Fernandez, Talhinhas, & Duplessis, 2013). These gene expression studies particularly showed that the haustorium is a crucial structure both for nutrients uptake (Mendgen & Hahn, 2002; Voegele, Hahn, & Mendgen, 2009; Voegele, Struck, Hahn, & Mendgen, 2001) and effector release (Catanzariti, Dodds, Lawrence, Ayliffe, & Ellis, 2006; Hahn & Mendgen, 1997; Joly, Feau, Tanguay, & Hamelin, 2010; Link & Voegele, 2008). A few studies explored other stages of the rust fungi biology, providing insights into relevant functions expressed at various developmental steps of the rust life cycle (Warren & Covert, 2004; Xu et al., 2011). Most expression studies were focused on resting rust uredospores collected outside the host plant and during *in planta* biotrophic growth by directly isolating RNA from infected host tissues. In less than 20 years (1990–2010), a total of 168,199 Pucciniales ESTs have been deposited at the National Center for Biotechnology Information (as of January 2014). This number has hardly changed between 2010 and 2013, whereas in the same period, a significant number of publications appeared describing the use of genome-wide oligoarrays or of NGS approaches to determine the transcriptome of rust fungi at various stages and in different types of plant–rust interactions (for a detailed list, see Table 6.2). In total, nearly 100,000 unique genes or transcripts were reported by these studies for seven rust fungi, indicating that in only 4 years, significant knowledge has been generated compared to what was accumulated in the two decades before.

3.1. Genome oligoarray-based transcriptomics

The availability of the *M. larici-populina* and *P. graminis* f. sp. *tritici* genomes allowed the design of custom whole-genome oligoarrays to perform genome-wide expression surveys during the infection process of the telial hosts of these fungi, respectively, poplar and wheat (Duplessis, Cuomo, et al., 2011; Duplessis, Haquard, et al., 2011). These pioneer

Table 6.2 Recent genome-wide expression studies of Pucciniales

Species and isolate (reference)	Interaction, biological stage	Transcriptome approach	Number of genes covered	Number of genes detected
Hemileia vastatrix 178a, CIFC collection (Fernandez et al., 2012)	Coffee rust infected leaves at 21 dpi (3 days before uredinia formation)	454 pyrosequencing GS-FLX Titanium; 352,146 reads	Unknown, no reference genome	22,774 assembled contigs, 6763 assigned to the fungus
Melampsora larici-populina 98AG31(Hacquard et al., 2010)	Laser capture microdissection of uredinia (area 1); spongy mesophyll containing infection hyphae, haustoria, and sporogenous hyphae (area 2); and infected palisade mesophyll containing infection hyphae and haustoria (area 3) from infected poplar leaves (Beaupré cv.) at 7 dpi	Custom whole-genome oligoarrays	15,388 putative genes (published ahead of the release of expert genome annotation in Duplessis, Cuomo, et al., 2011)	8145, 7786, and 7288 expressed transcripts in area 1, 2, and 3, respectively, accounting for a total of 9650 unique expressed transcript (63%)
Melampsora larici-populina 98AG31 (Duplessis, Cuomo, et al., 2011)	Infected poplar leaves (Beaupré cv.) at 4 dpi, resting and germinating (3 h) uredospores	Custom whole-genome oligoarrays	13,093 genes assayed out of 16,399 genes included in the final genome annotation of isolate 98AG31	71% of the transcripts were detected in at least one situation; 6466 were expressed in all three situations; 7582, 7541, and 7656 transcripts were expressed at 4 dpi, in resting, and in germinating uredospores, respectively

Melampsora larici-populina 98AG31 (Duplessis, Haquard, et al., 2011)	Infected poplar leaves (Beaupré cv.) at 2, 6, 12, 24, 48, 96, and 168 hpi and comparison to resting and germinating (3 h) uredospores	Custom whole-genome oligoarrays	13,093 genes assayed out of 16,399 genes included in the final genome annotation of isolate 98AG31	<500 transcripts detected at 2, 6, or 12 hpi due to fungal transcript dilution; 4279, 6216, 7856, and 8326 transcripts detected at 24, 48, 96, and 168 hpi, respectively; 7735 and 7872 in resting and germinating uredospores, respectively, when normalized separately from *in planta* situations
Melampsora larici-populina 93ID6 and 98AG31 (Petre et al., 2012)	Infected poplar leaves (Beaupré cv.) at 18, 21, 24 hpi in incompatible interaction (isolate 93ID6) and at 18, 24, 48 hpi in compatible interaction (isolate 98AG31)	454 pyrosequencing GS-FLX Titanium; 713,505 reads	16,399 genes included in the final genome annotation of isolate 98AG31	90,398 contigs, from which only 649 were assigned to 280 fungal genes (isolate 98AG31 genome annotation)
Melampsora larici-populina 98AG31 (Hacquard, Delaruelle, et al., 2013)	Telia from early autumn-collected naturally infected poplar leaves (Beaupré cv.)	Custom whole-genome oligoarrays	13,093 genes assayed out of 16,399 genes included in the final genome	9588 transcripts expressed in telia, including 395 telia-specific transcripts

Continued

Table 6.2 Recent genome-wide expression studies of Pucciniales—cont'd

Species and isolate (reference)	Interaction, biological stage	Transcriptome approach	Number of genes covered	Number of genes detected
	comparison to resting and germinating uredospores as well as *in planta* 168 hpi sample from Duplessis, Haquard, et al. (2011)		annotation of isolate 98AG31	
Phakopsora pachyrhizi Thai1 (Link et al., 2014)	Purified haustoria from 12 dpi infected *Glycine max* leaves	454 pyrosequencing GS-FLX Titanium 1,051,753 reads	Unknown, no reference genome	11,872 assembled contigs, 4483 unique *P. pachyrhizi* contigs
Phakopsora pachyrhizi MS06-1 (Tremblay, Hosseini, Li, Alkharouf, & Matthews, 2012)	10 dpi infected *Glycine max* leaves	Illumina Genome Analyser II 5.96 million 36 bp reads	Unknown, no reference genome	~2.4 million reads assigned to *P. pachyrhizi*, 32,940 assembled *P. pachyrhizi* contigs
Phakopsora pachyrhizi MS06-1 (Tremblay, Hosseini, Li, Alkharouf, & Matthews, 2013)	15 spi, 7 hpi, 48 hpi, and 10 dpi in susceptible *Glycine max* leaves	Illumina Genome Analyser II 24.6 million single reads (from 3.5 to 9 million reads per time point)	Unknown, no reference genome	23–55% Reads were assigned to *P. pachyrhizi*; 6531, 4627, 4273, and 12,284 *de novo* assembled *P. pachyrhizi* transcripts at 15 spi, 7 hpi, 48 hpi, and 10 dpi, respectively, accounting for 27,715 fungal transcripts of which 19,000 represents new transcripts not previously reported

Puccinia graminis f. sp. *tritici* CDL75-36-700-3, race SCCL (Duplessis, Cuomo, et al., 2011)	Infected wheat leaves (cv. McNair 701) and infected barley leaves (cv. Hypana) at 7 and 8 dpi, respectively, and resting and germinating (24 h) uredospores	Custom whole-genome oligoarrays	20,228 putative genes defined ahead of the final genome annotation that comprises 17,773 genes	A total of 9818 transcripts were expressed, 6570 being expressed in all four situations
Puccinia striiformis f. sp. *tritici* isolate UK PST-08/21 (Cantu et al., 2013)	Infected wheat leaves (cv. Avocet 'S') at 6 and 14 dpi and purified haustoria at 7 dpi	66.7 and 200.4 million reads from infected leaves and haustoria, respectively, by Illumina-based RNA-seq Genome Analyser II (76 bp single reads)	19,073 predicted genes based on Cantu et al. (2011, 2013)	12 and 28.8 million reads from infected leaves and haustoria; comparison between haustoria and infected leaves, focus on secreted proteins, 57-induced/31-repressed in haustoria transcripts encoding secreted proteins; 411-induced and 333-repressed transcripts encoding non secreted proteins
Puccinia striiformis f. sp. *tritici* Pst-104E137A- (Garnica, Upadhyaya, Dodds, & Rathjen, 2013)	Purified haustoria and uredospores	454 pyrosequencing GS-FLX Titanium 729,036 (413 bp) and 457,071 (420 bp) reads and Illumina Genome	Unpublished reference genome of *P. striiformis* f. sp. *tritici* local isolate Pst-104E137A-	12,846 contigs in haustorial transcriptome, 12,282 assembled transcripts for combined haustoria and

Continued

Table 6.2 Recent genome-wide expression studies of Pucciniales—cont'd

Species and isolate (reference)	Interaction, biological stage	Transcriptome approach	Number of genes covered	Number of genes detected
		Analyser GX II 500 million of 100 bp paired-end reads		uredospores transcriptomes
Puccinia triticina, isolates MHDS, MLDS, MJBJ, TDBG, THBJ, and TNRJ (Bruce et al., 2013)	Infected wheat leaves (susceptible Thatcher cv.) at 6 dpi	Illumina 165 millions reads (60 bp, paired-end) in total, 26.4, 25.5, 23.4, 27.7, 33.2, 28.4 millions reads for isolates MHDS, MLDS, MJBJ, TDBG, THBJ, and TNRJ, respectively	Mapping to the unpublished reference genome V2 of *P. triticina* isolate BBBD race 1 (J. Fellers, C. Cuomo, L. Szabo, G. Bakkeren, B. McCallum, B. Saville, unpublished)	222,571 reads assigned to the fungus, focus on 1450 secreted proteins encoding transcripts in reference genome; 543 uniquely secreted protein-encoding transcripts identified
Uromyces appendiculatus SWBR1 (Link et al., 2014)	Purified haustoria from 10 dpi infected *Phaseolus vulgaris* leaves	454 pyrosequencing GS-FLX Titanium 894,873 reads	Unknown, no reference genome	14,581 assembled contigs, 7582 unique *U. appendiculatus* contigs

The table lists recently published transcriptome analyses of rust fungi using whole-genome oligoarrays or NGS.
spi, seconds postinoculation; hpi, hours postinoculation; dpi, days postinoculation.

transcriptomics studies delivered the first snapshots of genome-wide rust fungus gene expression. They particularly showed that about a third to a half of the genes were not expressed in uredospores or during the infection process, suggesting that they may have a role at other stages of the life cycle (Table 6.2). Among the genes expressed during host colonization (including uredospore germination) were a large number of genes of unknown function. This included rust-specific gene families, among which were many of the small secreted protein-encoding genes. The induced or upregulated expression of those secreted protein genes during host colonization strengthened their profile as rust candidate effectors with important functions during infection. This is particularly true for those that are specifically expressed only during the biotrophic growth *in planta*. Some of these putative effectors are significantly overexpressed at late stages of infection when the host tissues are showing an intense colonization by infection hyphae and haustoria, which is consistent with early molecular data obtained from purified haustoria of *U. fabae* or *M. lini* (Catanzariti et al., 2006; Hahn & Mendgen, 1997) and with recent data obtained by NGS approaches for *P. striiformis* f. sp. *tritici* (Cantu et al., 2013; Garnica et al., 2013), *P. pachyrhizi*, and *U. appendiculatus* (Link et al., 2014) by sequencing RNA isolated from purified haustoria. Interestingly, late stages of infection of poplar and wheat leaves by *M. larici-populina* and *P. graminis* f. sp. *tritici*, respectively, are marked by the expression of a large panel of genes related to the biotrophic growth such as CAZymes, transporters, proteases, and lipases (Duplessis, Cuomo, et al., 2011). A time-course infection study showed that most of the genes falling in the categories mentioned in earlier text were highly and significantly differentially expressed at late stages of infection. However, distinct sets of dozens to hundreds of small secreted protein-encoding genes were expressed sequentially, that is, they appeared to be expressed in distinct waves indicating that *M. larici-populina* possesses a highly dynamic secretome, which may be important for the interplay with components of the host plant immunity system (Duplessis, Haquard, et al., 2011; Hacquard et al., 2012). Such a dynamic pattern of expression *in planta* was further confirmed in other rust–plant interaction studies (Bruce et al., 2013; Cantu et al., 2013; Fernandez et al., 2012; Tremblay et al., 2013). Transcription profiles of fungal cell types were studied when the fungus releases uredospores from the host. At this stage, only the palisade mesophyll contains infection hyphae and haustoria, whereas the spongy mesophyll is filled with huge amounts of sporogenous hyphae and spores. The rust fungus expressed very different genetic programs in these two plant compartments and the most highly expressed fungal genes in the palisade compared to the spongy mesophyll coded for candidate effectors, suggesting that these might play a role in the maintenance of infection structures *in planta* during the later stages of biotrophy (Hacquard et al., 2010). These gene expression profiles were compared with those obtained from telia of *M. larici-populina* harvested early in autumn, allowing a direct comparison of expression profiles in another spore-forming structure in the poplar leaf (Hacquard, Delaruelle, et al., 2013). A larger number of genes specifically expressed in telia were found, most of which encode unknown functions. Similarly, a large number of specific ESTs with unknown functions were identified in *P. triticina* teliospores (Xu et al., 2011) highlighting that the biological processes associated with this type of rust spores remain mostly uncharacterized. In *M. larici-populina*, telia and uredinia have the most similar expression profiles when compared to fungal hyphae undergoing biotrophic growth in poplar leaves, indicating shared components in the genetic programs ongoing in these spore-forming structures. Among the most highly regulated genes reported in telia are several with functions possibly related to spore survival, that is, overwintering, such as those encoding thaumatin-like proteins and aquaporins that may help to prevent osmotic damage due to desiccation. Also, several meiotic-related transcripts were overexpressed in teliospores and showed temporal patterns of expression during karyogamy, an important biological process occurring in teliospores (Hacquard, Delaruelle, et al., 2013). These studies demonstrate that the use of common custom oligoarrays is an efficient approach to realize transcriptomic comparisons to gain a better understanding of the biology of the rust fungus at different life cycle stages. However, this method still has pitfalls such as the detection of pathogen transcripts at stages containing very low amounts of fungal biomass inside the host (e.g. early stages of infection) (Duplessis, Haquard, et al., 2011).

3.2. RNA-Seq-based transcriptomics

Since 2011, considerable progress has been made by applying NGS to the study of several rust fungi, including *H. vastatrix*, *M. larici-populina*, *P. pachyrhizi*, *P. graminis* f. sp. *tritici*, *P. striiformis* f. sp. *tritici*, *P. triticina*, and *U. appendiculatus* (see Table 6.2). The possibility to compare different stages of infection in the telial hosts of rust fungi can help reveal patterns of gene expression during the infection process.

One challenge when studying different infection stages of a biotrophic fungus is the effect of dilution of the pathogen/host RNA. Within 48 h of infection of poplar by *M. larici-populina*, only 649 of the one million sequences generated by 454 pyrosequencing were assigned to rust genes (Table 6.2), representing less than 1% of the total sequences (Petre et al., 2012). Still, these contigs corresponded only to 280 unique *M. larici-populina* genes, half of them encoding small secreted proteins representing early-expressed candidate effectors at a stage when the first haustoria are recorded in the host (Laurans & Pilate, 1999). The level of infection could be an important aspect of the success in obtaining fungal transcripts. At a late stage of infection of *Coffea arabica* leaves by the coffee leaf rust fungus *H. vastatrix*, but before uredinia formation, the plant tissue is heavily colonized. About 30% of the total contigs (6763) assembled from 352,146 reads were attributed to the fungus based on a predictive comparative analysis to rust fungi and *C. arabica* sequences in databases (Table 6.2) (Fernandez et al., 2012). This nonmodel rust fungus lacks genomic information support, but based on the report of gene complements in rust fungi so far, it can be estimated that about a third of the *H. vastatrix* genes have been revealed by this transcriptomics approach. The study identified different cellular categories that can relate to the fungal growth *in planta*. The presence of 382 transcripts encoding small secreted proteins, among which a small set shown to be specifically expressed in haustoria and conserved among other rust fungi, indicates that this approach is also powerful to reveal putative candidate effectors in nonmodel rust fungi.

Similarly, time-course infection of soybean leaves colonized by *P. pachyrhizi* identified more than 4000 stage-specific transcripts at four different time points and in total 27,715 expressed fungal transcripts including 19,000 unique transcripts not previously recorded for rust fungi (Tremblay et al., 2013). Those transcript numbers suggest that some rust fungi with large estimated genomes like *P. pachyrhizi* may have larger gene complements than other rust fungi, or it could mean that the assembly parameters used in the corresponding study left numerous alternate transcripts ungrouped. The comparison of transcript expression profiles at the different stages of host colonization by *P. pachyrhizi* confirms the dynamic temporal regulation of gene expression also reported for *M. larici-populina*. What we have gleaned from the various studies is that the genetic programs expressed by rust fungi are finely regulated and this likely reflects different and/or specific processes occurring in the vastly different fungal structures formed during infection (i.e. germ tubes and appressoria at the leaf surface and infection hyphae and haustoria in the leaf mesophyll) to produce uredospores at the leaf surface (i.e. spore-forming cells).

Application of NGS technologies to precisely dissect the infection process, that is, appressorium formation, direct penetration through cuticle for the soybean rust or other rust fungi when invading their aecial hosts, development of the substomatal vesicle and further infection hyphae and haustoria in the mesophyll, and then formation of uredospores and their release, still has limitations related to the depth of sequencing in order to reach a proper coverage of the transcriptome. For example, early steps in colonization of host tissue are impossible to capture at the moment, due to the small amount of fungal biomass in the collected host tissues. It is most likely that only a small fraction of the most highly expressed genes—that is, the tip of the iceberg—are detected as illustrated in the poplar rust fungus at the onset of haustoria formation (Petre et al., 2012). This issue also poses problems in terms of normalization between colonization stages. NGS-based fold-change levels calculated between stages in time-course studies should be considered cautiously wherever saturation is not reached in the cumulative curves for transcript coverage (see chapter by Kohler & Tisserant, 2014). However, even if the transcriptome is not complete, at comparable sequencing depth, it reflects a significant expression and may help to uncover genes commonly expressed between stages to unravel the common host infection toolkit of rust fungi. A problem with experimental rust fungus systems is a near impossibility to achieve synchronous infections to obtain sufficient material from a specific infection stage for RNA extraction. Single-cell analysis using micromanipulation techniques is however becoming feasible (Lin et al., 2014), and this may be applied to rust fungal pathosystems in the near future.

An alternative approach is to focus on specialized infection structures. Haustoria are crucial infection structures that are formed by rust fungi within host cells by breaching the cell wall but only penetrating beyond by invaginating the plasmalemma leaving the membrane intact. By isolating RNA from haustoria, it is therefore possible to enrich transcripts involved in host–pathogen interactions. Isolation of haustoria and recovery of RNA led to the description of key rust determinants involved in nutrient acquisition and delivery of effectors into host cells (Catanzariti et al., 2006; Hahn & Mendgen, 1997). RNA-Seq studies in rust fungi established the transcriptome profiles of purified haustoria in *P. pachyrhizi* and *U. appendiculatus* (Link et al., 2014) and in *P. striiformis* f. sp. *tritici* (Cantu et al., 2013; Garnica et al., 2013). In these studies, a particular focus was on transcripts coding for predicted secreted proteins in the rust infection structure. Altogether, these studies show that a plethora of putative candidate effectors are expressed in those infection structures. In *P. striiformis* f. sp. *tritici*, fungal transcripts of secreted proteins representing promising candidate effectors of the wheat stripe rust fungus were identified by comparing leaf and haustoria transcript (Table 6.2) (Cantu et al., 2013). In *P. pachyrhizi* and *U. appendiculatus*, more than 11,000 contigs were generated for each species, and among these, 4483 and 7532 contigs were assigned to the two rust fungi, respectively (Link et al., 2014). Annotation of gene families and comparison to other fungi identified conserved families and specific families of secreted proteins. The candidate effector selection pipeline applied by Cantu et al. (2013) and previously defined to analyse *M. larici-populina* and *P. graminis* f. sp. *tritici* predicted genes (Saunders et al., 2012) and also identified conserved and specific tribes of haustorially expressed secreted protein gene families in *P. striiformis* f. sp. *tritici*.

It is important to note that some of these genes, although highly expressed in the haustorium, are not specific for this structure and many are also expressed in infection hyphae, uredospores, and other cell types of rust fungi in different host tissues, illustrating very dynamic patterns of gene regulation (Duplessis et al., 2012). Also, the detailed analysis of genes, overexpressed in *P. striiformis* f. sp. *tritici* haustoria and uredospores, showed major differences in the fungus prior and during infection (Garnica et al., 2013). Particularly, the two stages are highly contrasted for many cellular categories, including cell cycle, DNA metabolic and lipid metabolic processes, signal transduction in uredospores, generation of precursor metabolites and energy, translation, and vitamin and carbohydrate metabolic processes in haustoria. Such detailed analyses are highly valuable and forthcoming cross comparison between different rust species at different stages should help to identify the core components in the infection machinery of rust fungi and to decipher more precisely the specialized mode of acquisition of nutrients from the host.

A particularly exciting prospect is that of identified race-specific determinants in rusts. The combination in a dedicated pipeline of transcriptomic and genomic data of polymorphisms in gene sequences and their comparison to corresponding genes in other rust fungus isolates differing in resistance gene interaction helped to pinpoint the most promising candidate effectors for further functional characterization; these could be candidates with a role in the interplay with the host immune system (Cantu et al., 2013). In *P. triticina*, Illumina-based RNA-Seq was used to identify potential avirulent genes by screening expression in distinct isolates representing six rust races (Bruce et al., 2013). In this case, the sequenced reads were compared to predicted transcripts in the *P. triticina* reference genome sequenced by the Broad Institute (J. Fellers, C.A. Cuomo, L. Szabo, G. Bakkeren, et al. unpublished data) and to a large collection of wheat ESTs. A particular focus was on small secreted protein-encoding genes. Among the 543 genes identified, 15 accumulated nonsynonymous mutations representing putative effector candidates that evolved under the pressure of the host immune system, and 11 showed positive correlation with reactions to multiple resistance genes in wheat. This pioneer study is the first step in the broad identification of determinants of interest for selection breeding programs (Bruce et al., 2013). It is envisaged that similar genome-wide associations can be made when comparing isolates differing in virulence (e.g. *P. striiformis* variants infecting differently at different temperatures) or even host range.

3.3. Comparison of transcriptome in different hosts

The comparison of transcriptomes during the interaction of a rust fungus with both its telial and aecial hosts promises to elucidate questions regarding the gene sets required for interacting with different hosts. The rust is confronted with very different host challenges in these different parts of its life cycle. In pine rusts, for example, monokaryotic spores infect the pine needles and grow within the host cambium and into the woody portion of the trees where it forms a hymenium. But dikaryotic aeciospores and uredospores infect the telial hosts and germinate and penetrate via stomata, followed by intercellular growth and formation of haustoria. It is likely that different genes are required for these interactions. Comparing infection of the two hosts of heteroecious rusts is complicated by the fact that it is not possible to conduct infection with the same genetic individual. The infection of the aecial host is via meiosis-derived basidiospores that are thus segregating (Doudrick, Nance, Nelson, Snow, & Hamelin, 1993), while infection of the telial host is via postspermatization dikaryotic aeciospores. Nevertheless, by using single-spore cultures derived from the telial host, it is possible to obtain genetic material that is relatively homogeneous. *Cronartium quercuum* f. sp. *fusiforme* transcriptomic profiles were strikingly different on the telial (oak) and aecial (loblolly pine) hosts. Half of the transcriptome sampled on a microarray was differentially expressed in these two hosts (J. Davis, et al., unpublished data). Similar outcomes were observed for the poplar rust fungus, when Illumina RNA-seq was applied to infection stages in the aecial host, the larch (S. Duplessis, et al., unpublished data), and compared with expression during infection of the telial host, poplar (Duplessis, Haquard, et al., 2011). Particularly, a core set of candidate effectors were expressed in both hosts, whereas different sets were specifically expressed in each of the two hosts. Given the divergence between angiosperms and gymnosperms, it is likely that different sets of rust effectors interact with their respective targets in the telial and the aecial hosts. Rather different gene sets are also expressed in *P. triticina* when comparing Illumina RNA-seq-generated transcriptome profiles from infected wheat samples at 5 days after infection (dikaryotic uredospore-derived) and from pycnia and aecia (haploid and mated fungal cell mixtures) from the alternate host (J. Fellers, C. Cuomo, L. Szabo, G. Bakkeren, et al., unpublished data).

4. RUST SECRETOME, EFFECTORS, AND AVIRULENT GENES

There is a long history of studying host–pathogen interactions in rust pathosystems. The gene-for-gene hypothesis was tested and verified for the first time in the flax rust pathosystem (Flor, 1959). This field has been transformed recently by the discovery of a large repertoire of 'effectors' in plant pathogens. These are mostly proteins that are secreted by pathogens into the plant apoplast or into the host cells. Effectors are believed to manipulate host cell processes and facilitate infection by suppressing host defences, or by modifying host metabolism. Several of the avirulent genes that have been studied using classical genetic approaches are now demonstrated to be effectors recognized by cognate host resistant gene products (Bozkurt, Schornack, Banfield, & Kamoun, 2012; Dodds et al., 2009; Gan et al., 2010; Kamoun, 2007). This makes effector discovery and characterization extremely promising for host resistance screening and in monitoring of pathogen avirulence.

Since effectors are secreted, several pipelines have been developed to identify and characterize the secretome of plant pathogens. Such pipelines rely on computer-assisted identification of the presence of signal peptides, which are indicators that the proteins likely are secreted. Additional criteria often include size (putative effectors are usually small proteins) and being rich in cysteine (for folding and stability). Depending on the pathogen under study, other criteria can be added to the screening pipeline. In rust fungi, for example, additional criteria that have been proposed include expression *in planta*, homology to haustorial proteins, presence of internal repeats, long intergenic regions, and absence of PFAM domains except those associated with pathogenicity (Hacquard et al., 2012; Saunders et al., 2012). Complete genome sequencing projects for rust fungi have revealed striking numbers of lineage-specific genes at the taxonomic level of either the order Pucciniales or the Melampsoraceae and Pucciniaceae families. Among those, several expanded gene families representing unknown functions encode secreted proteins and probably candidate rust effectors required to establish a successful infection in the plant host(s). In the three rust genomes sequenced and published so far, the secretome accounts for a large portion of the gene complement (~8%) encompassing nearly 2000 secreted proteins, and a majority are encoded by multigene families (Duplessis, Cuomo, et al., 2011; Zheng et al., 2013). Although the secretomes of the three rust fungi showed similar numbers, composition, and genomic organization, they greatly differ and exhibit a high specificity, reflecting a possible adaptation to the different hosts. The sequencing of more genomes in the order Pucciniales is much awaited to determine whether this assumption is correct. Also, comparative analyses between the secretomes of rust fungi have shown that a few genes and gene families are conserved across Pucciniales or even across Basidiomycota and showed an important diversity (Duplessis, Cuomo, et al., 2011; Hacquard et al., 2012; Saunders et al., 2012). However, up to now, only a few secreted proteins have been proven 'true effectors' (i.e. *M. lini* avirulent proteins and *U. fabae* RTP1; Catanzariti et al., 2006; Duplessis et al., 2012; Kemen et al., 2005), and it remains to be seen to which extent secretomes of rust fungi contribute to the expression of virulence.

In some cases, mapped populations segregating for avirulence can be used to map markers onto whole-genome sequences and identify avirulence. By taking advantage of the fact that haploid basidiospores of *Cronartium* spp. cause infections to pines, a bulk segregant analysis was used to map *Avr1* in *C. quercuum* f. sp. *fusiforme* (Kubisiak et al., 2011). These markers are now proving extremely useful. They were used to find the scaffold that comprises the putative *Avr1* gene in the *C. quercuum* f. sp. *fusiforme* genome (J. Davis et al., unpublished data). Combining this with a whole-genome outlier F_{st} analysis contrasting the bulk segregants, pycnial pools can provide a powerful approach to identify the *Avr1* locus. Among cereal-infecting rust fungi, 57 F2 progenies of a wheat leaf rust, *P. triticina*, sexual cross between two parental isolates (race 9 with virulence formula SBDG (Kolmer, 1996) and race 161 FBDJ) were sequenced using Illumina. At least, nine seedling and two adult plant avirulent genes were shown to segregate in this F2 population. In order to generate a high-resolution genetic linkage map, genome-wide single-nucleotide polymorphisms (SNPs) were identified by comparison to the *P. triticina* 1-1 reference genome. Using this information and employing the genome sequence information from the two parents and the F1 isolate, more than 25,000 SNPs were selected and used to generate a preliminary genetic linkage map. Correlation with the virulence profiles of the segregants yielded candidate avirulent genes, which are currently being investigated. In addition, although they were obtained from different isolates, genetic linkages allowed the creation of pseudomolecules, which were used to confirm physical linkage of many unassembled contigs in the reference genome (D.L. Joly, B.D. McCallum, B.S. Mulock, & G. Bakkeren, unpublished data). A similar effort is under way for a wheat stem rust, *P. graminis* f.sp. *tritici*, mapping population (Zambino, Kubelik, & Szabo, 2000).

5. POPULATION GENOMICS: FROM GENOMES TO LANDSCAPES

5.1. The rapidly evolving rust genomes

The ability of rust fungi to evolve and adapt has been widely reported. The appearance of rust races that can overcome resistance genes in crops and trees shortly after deployment has been one of the most enduring challenges for breeders (Kinloch, 2003; Kinloch, Sniezko, & Dupper, 2003; Kolmer, 2005; Pinon & Frey, 2005). There is a long history of characterizing rusts from a phenotypic and later from a population genetics perspective. However, the identification of the genes involved in adaptation has been challenging. The identification of genetic markers to track pathogen virulence has been complicated by the lack of adequate genetic maps, at the most yielding loose linkages representing large distances between genetic markers and virulence loci (Kubisiak et al., 2011; Zambino et al., 2000).

The genomic signature of this rapid evolution has yet to be elucidated in most pathogens. In some, the ability to evade host resistance response is related to genome structure. The *L. maculans* genome has a bipartite structure with AT-rich blocks containing effector genes and transposable elements that promote rapid diversification for evasion of host recognition (Rouxel et al., 2011). In *Phytophthora infestans*, gene-sparse, repeat-rich, and effector-rich regions that can rapidly generate new alleles and promote host jumps have been implicated (Raffaele et al., 2010). The genome sequences of rusts do not display such patterns although several *P. triticina*

BAC clones harboring effector homologues were surrounded by repeats and TE-like sequences (Fellers et al., 2013). However, there is evidence that some genome compartments are under positive selection and likely candidates in the host–pathogen arms race. By resequencing and comparing the genomes of multiple rust strains, it is possible to identify and map genomic regions where among-strain differentiation points to positive selection.

The study of population genome resequencing is in its infancy, but it can be expected to generate extremely useful results. Resequencing even a few isolates, especially of the dikaryotic phase, can be informative since two different nuclei are sampled and heterozygotes can be revealed. The genome of *P. striiformis* f. sp. *tritici* is highly heterozygous and possesses a diverse gene repertoire and more genes encoding secreted proteins compared with nonobligate fungal pathogens (Cantu et al., 2013; Zheng et al., 2013). The presence of a large number of genetic variants could be part of the explanation for the rapid evolution of this rust. The discovery that a large fraction of the SNPs are nonsynonymous and vary among regions could indicate regional adaptation. This high level of diversity in Asian *P. striiformis* f. sp. *tritici* could be interpreted as a result of sexual reproduction being important in generating regional races. The aecial host of this rust, barberry, is present in western China and central Asia, and therefore, completion of the sexual stage is possible and could drive this high level of diversity. However, sampling of more isolates in regions with and without aecial hosts would be required to compare observed and expected heterozygosity and determine if sexual reproduction is frequent or a rare occurrence.

An exciting prospect is to sequence and contrast rusts with different virulence profiles or host specificity. The comparison of genomes of four UK isolates of *P. striiformis* f. sp. *tritici* that differed in virulence to two wheat varieties highlighted the enrichment of nonsynonymous polymorphisms in haustoria-enriched proteins that were predicted to be secreted (Cantu et al., 2013). This allowed the narrowing of a complement of nearly 3000 predicted secreted proteins to five candidate effectors that differed between the isolates.

5.2. Host–pathogen adaptation in coevolved pathosystems

A better understanding of host–pathogen coevolution is important to devise better deployment strategies of resistance genotypes in crops and plantations. Antagonistic coevolution is believed to be a critical force driving the evolution of diversity, and knowledge of how this balance is achieved in nature along spatial and temporal scales could be gained via studies of natural pathosystems (Thrall et al., 2012). Some of the most insightful studies on host–pathogen coevolution have been conducted in the flax rust *M. lini* and its wild host *Linum marginale* pathosystem. Variation in host resistance and pathogen infectivity is presumed to be the hallmark of a balanced wild pathosystem and to be essential for persistence of biotrophic pathogens (Laine, Burdon, Dodds, & Thrall, 2011). Although much has been learned by studying virulence phenotypes, there are exciting new studies examining natural variation in effector genes in natural populations.

Population genomics approaches in these nonmodel pathosystems are now feasible given the low cost of genome sequencing. Processes such as adaptation and evolution and the reconstruction of introduction and colonization events can be studied at the genome level. Evolutionary changes in the adaption of *M. larici-populina* to new hosts and new environments in northeastern North America were studied by resequencing 44 strains from the source and the introduced populations. Genome-wide distribution of polymorphisms displayed the signature of a severe founder event into the North American population (N. Feau et al., unpublished data). A genome scan of the four largest scaffolds revealed 32 regions having a strong divergence between the founder and the source populations. These regions include 14 secreted protein-coding genes that are candidate for adaptation (Feau, Frey, Duplessis, & Hamelin, 2013).

An advantage of studying *M. lini* in its wild pathosystem is that effector loci have been characterized functionally (Catanzariti et al., 2006, 2010; Dodds, Lawrence, Catanzariti, Ayliffe, & Ellis, 2004). Extensive sequence variation and an excess of nonsynonymous to synonymous mutations at two effector loci (AvrP123 and AvrP4) were found in the flax rust in its native range in Australia (Barrett et al., 2009). These loci therefore display the hallmark of positive selection. Alleles at these loci were demonstrated functionally to be differentially recognized by *R*-genes in the host. This result is consistent with the presence of variants of both effector genes that evolve to evade recognition by *R*-genes in the host in a classic antagonistic coevolutionary scenario (Barrett et al., 2009; Thrall et al., 2012).

It may also be informative to study evolution of the secretome in closely related rust species that differ in host specificity. The hosts are likely imposing evolutionary constraints acting more strongly on secreted proteins or effectors compared to the rest of the genome (Joly et al., 2010). More than half of the predicted proteins discovered in EST libraries of poplar rust species specialized on different poplar hosts were lineage-specific. Secreted proteins had accelerated divergence and increased ratios of nonsynonymous to synonymous mutations when compared with nonsecreted proteins. In comparison of flax rusts attacking different hosts, the *avrP4* locus displayed 14 codons under significant positive selection. These mutations were present in the C-terminal 28-amino-acid region, suggesting that this region could be important in the interactions with host recognition proteins (Van der Merwe et al., 2009). Assessing allelic variation in putative effectors between distinct rust isolates will be important to identify and monitor avirulence alleles in populations and provide risk prediction and early warning for appearance of novel alleles. The rust *Melampsora allii-populina* shares poplar trees as the same telial hosts with *M. larici-populina*. The scheduled sequencing of this poplar rust fungus within the 1000 fungal genomes project at the JGI (http://genome.jgi.doe.gov/programs/fungi/index.jsf) will bring a great opportunity to compare at the genome scale the gene complements in the two genomes in order to determine the extent of specific genes related to poplar and to the distinct telial hosts (*Larix* spp. and *Allium* spp.). Whole-genome resequencing in those pathosystems promises to highlight additional genomic features associated with host specificity and host-pathogen equilibrium.

6. COMING UP NEXT IN RUST GENOMICS

The latest developments in rust fungal genomics have been providing exciting insights and tremendous amounts of new information increasing our understanding of the biology of these ecologically and economically relevant fungal pathogens. Since 2011, three genomic analyses and 14 genome-wide or NGS-based transcriptome studies have been published, and many more are expected in the next years. Pucciniales remains one of the orders with the lowest genome coverage in the fungal kingdom, and efforts are currently under way by the JGI and other international sequencing centres to address this paucity and add more species to this particularly important clade of pathogens (Grigoriev, 2014; Grigoriev et al., 2011, 2014). It is clear that the large sizes of rust genomes are challenging for such genome sequencing and assembly efforts. However, the most recent examples that were based on NGS technology to obtain draft or high-quality genome sequences are paving the way to future research in rust genomics to better understand the biology and ecology of these unique plant pathogens.

The different studies reported in this chapter have now provided a considerable level of information about the genomes, predicted secretomes, and correlated transcriptomes of different rust fungi. A detailed cross-species comparative analysis using these combined data is now the next challenge to precisely determine the extent of the core repertoire of rust effectors and the specificity of the secretome at different taxonomic levels. Correlating expression profiles in the various specific infection structures and comparing a variety of rust species should help to define a minimal set of functional effectors required for infection among the thousands of predicted secreted proteins. Although somewhat out of the scope of this review chapter, the generation of the genomic and transcriptomic data for rust fungi has made feasible interesting and complementary proteomic research. In particular, with respect to revealing potential effector suites produced in *P. triticina* haustoria, protein identification using generated proteomic spectra was greatly improved because of available comprehensive gene model predictions (Song et al., 2011). In this pathosystem, protein spectra derived from purified haustoria from various virulent and avirulent isolates are currently analysed against generated genomic resources from the corresponding isolates to identify specific effector variants (C. Rampitsch, B.D. McCallum, & G. Bakkeren, unpublished data). Similar approaches have led to successful effector identification in the barley powdery mildew fungus (Bindschedler et al., 2009; Godfrey, Zhang, Saalbach, & Thordal-Christensen, 2009) and in general, proteomic research greatly complements genomic and transcriptomic approaches, assists with genome annotations, and helps answer the various questions raised in this chapter (Tan, Ipcho, Trengove, Oliver, & Solomon, 2009).

We have highlighted some of the limitations that still hinder the analysis of important steps of the infection process by transcriptomics, such as during early stages of colonization *in planta*. Other stages are also lacking detailed information such as meiosis during teliospore germination, the colonization of the aecial host by basidiospores, and the subsequent mating during the sexual stage. NGS may help to capture relevant information about these obscure and sometimes ephemeral stages. However, this requires a very high sequencing depth, which at the moment still comes at a high cost. Micromanipulation and single-cell transcriptome analysis may soon be commonplace. Other future developments in sequencing technologies, particularly the possibility to sequence longer DNA fragments and improvements of the computer programs used to assemble short sequences, may help to cope with the richness in repetitive elements in the rust genomes. Further than the rust transcriptome per se, another advantage of using a NGS RNA-seq approach is the possibility to realize dual transcriptome sequencing, capturing both the host and the pathogen transcript expression profiles. This approach has so far been poorly explored, probably due to the lack of assembled and annotated plant host genomes. The availability of the poplar and wheat genome sequences will allow for such kind of analyses with *M. larici-populina* and the *Puccinia* cereal rust fungi (Petre et al., 2012). But even in the absence of a known host genome, it is possible to collect such information on the host based on the availability of extensive plant and fungal genome databases to ensure proper transcript designation in mixed infections (Fernandez et al., 2012). Beyond the identification of rust disease determinants during host infection, this will help to unravel the functions and the pathways that are triggered by the rust effectors to achieve proper biotrophic growth in the case of compatible plant–rust interactions and components of rust resistance and defence in incompatible interactions.

With the actual sequencing depth of NGS, it is now possible to pool DNA of various isolates and gather very large amounts of genome sequences that can be compared to a reference genome. This opens great perspectives for sampling and resequencing rust isolates from different landscapes and different hosts, across altitude and latitude gradients to identify adaptive genome regions. Population genomics approaches will allow the identification of genomic regions subjected to positive and purifying selection to target loci responsible for major traits such as virulence. The inclusion of very large sampling in populations will allow the comparison of rusts with different demographic histories (bottleneck and expansions) and the identification of genome regions under selective sweeps. Genome resequencing is still in its infancy for rust fungi, but the first reports already indicate a great genomic variability, which is a good basis to conduct population genomics of rust fungi.

ACKNOWLEDGEMENTS

S. D. would like to acknowledge his colleagues Pascal Frey, Fabien Halkett, and Stéphane De Mita at INRA Nancy for fruitful discussions on the *Populus–Melampsora* pathosystem; the ANR for supporting rust genomics projects in the frame of the young scientist grant POPRUST (ANR-2010-JCJC-1709-01) and the 'Investissements d'Avenir' program (ANR-11-LABX-0002-01, Lab of Excellence ARBRE); and the Joint Genome Institute (Office of Science of the U.S. Department of Energy under contract no. DE-Ac02-05cH11231) for the sequencing of the genome of the poplar rust fungus *Melampsora larici-populina*.

G. B. acknowledges the collaborations with Christina Cuomo, John Fellers, Les Szabo, Jim Kolmer, Brent McCallum, and Barry Saville and the sequencing work performed by the Broad Institute, Cambridge, MA (funded through grants from NSF-CSREES/USDA), and the Michael Smith Genome Sciences Centre, Vancouver, BC (funded through grants from Genome BC and the Ontario Research Fund), and the research funds from AAFC.

R.H. acknowledges the collaborations with John Davis, Nicolas Feau, Braham Dhillon, Amanda Pendleton and Igor Gregoriev and Genome BC's support through Strategic Opportunity Funds #131 and Genome Canada through the Large Scale Applied Genomic Program project #164.

REFERENCES

Aime, M. C., Matheny, P. B., Henk, D. A., Frieders, E. M., Nilsson, R. H., Piepenbring, M., et al. (2006). An overview of the higher level classification of Pucciniomycotina based on combined analyses of nuclear large and small subunit rDNA sequences. *Mycologia, 98*, 896–905.

Barrett, L. G., Thrall, P. H., Dodds, P. N., van der Merwe, M., Linde, C. C., Lawrence, G. J., et al. (2009). Diversity and evolution of effector loci in natural populations of the plant pathogen *Melampsora lini*. *Molecular Biology and Evolution, 26*, 2499–2513.

Bindschedler, L. V., Burgis, T. A., Mills, D. J., Ho, J. T., Cramer, R., & Spanu, P. D. (2009). In planta proteomics and proteogenomics of the biotrophic barley fungal pathogen *Blumeria graminis* f. sp. *hordei*. *Molecular and Cellular Proteomics, 8*, 2368–2381.

Bozkurt, T. O., Schornack, S., Banfield, M. J., & Kamoun, S. (2012). Oomycetes, effectors, and all that jazz. *Current Opinion in Plant Biology, 15*, 483–492.

Bruce, M., Neugebauer, K. A., Joly, D. L., Migeon, P., Cuomo, C. A., Wang, S., et al. (2013). Using transcription of six *Puccinia triticina* races to identify the effective secretome during infection of wheat. *Frontiers in Plant Science, 4*, 520.

Cantu, D., Govindarajulu, M., Kozik, A., Wang, M., Chen, X., Kojima, K. K., et al. (2011). Next generation sequencing provides rapid access to the genome of *Puccinia striiformis* f. sp. *tritici*, the causal agent of wheat stripe rust. *PLoS One, 6*, e24230.

Cantu, D., Segovia, V., MacLean, D., Bayles, R., Chen, X., Kamoun, S., et al. (2013). Genome analyses of the wheat yellow (stripe) rust pathogen *Puccinia striiformis* f. sp. *tritici* reveal polymorphic and haustorial expressed secreted proteins as candidate effectors. *BMC Genomics, 14*, 270.

Carvalho, G. M. A., Carvalho, C. R., Barreto, R. W., & Evans, H. C. (2014). Coffee rust genome measured using flow cytometry: Does size matter? *Plant Pathology*, http://dx.doi.org/10.1111/ppa.12175.

Catanzariti, A. M., Dodds, P. N., Lawrence, G. J., Ayliffe, M. A., & Ellis, J. G. (2006). Haustorially expressed secreted proteins from flax rust are highly enriched for avirulence elicitors. *Plant Cell, 18*, 243–256.

Catanzariti, A. M., Dodds, P. N., Ve, T., Kobe, B., Ellis, J. G., & Staskawicz, B. J. (2010). The AvrM effector from flax rust has a structured C-terminal domain and interacts directly with the M resistance protein. *Molecular Plant-Microbe Interactions, 23*, 49–57.

Cummins, G. B., & Hiratsuka, Y. (2003). *Illustrated genera of rust fungi* (3rd ed.). St Paul, MN: American Phytopathological Society.

Dodds, P. N., Lawrence, G. J., Catanzariti, A. M., Ayliffe, M. A., & Ellis, J. G. (2004). The *Melampsora lini* AvrL567 avirulence genes are expressed in haustoria and their products are recognized inside plant cells. *Plant Cell, 16*, 755–768.

Dodds, P. N., Rafiqi, M., Gan, P. H., Hardham, A. R., Jones, D. A., & Ellis, J. G. (2009). Effectors of biotrophic fungi and oomycetes: Pathogenicity factors and triggers of host resistance. *New Phytologist, 183*, 993–1000.

Doudrick, R., Nance, W., Nelson, C., Snow, G., & Hamelin, R. (1993). Detection of DNA polymorphisms in a single urediniospore-derived culture of *Cronartium quercuum* f. sp. *fusiforme*. *Phytopathology, 83*, 388–392.

Duplessis, S., Cuomo, C. A., Lin, Y. C., Aerts, A., Tisserant, E., Veneault-Fourrey, C., et al. (2011). Obligate biotrophy features unraveled by the genomic analysis of rust fungi. *Proceedings of the National Academy of Sciences of the United States of America, 108*, 9166–9171.

Duplessis, S., Haquard, S., Delaruelle, C., Tisserant, E., Frey, P., Martin, F., et al. (2011). *Melampsora larici-populina* transcript profiling during germination and time-course infection of poplar leaves reveals dynamic expression patterns associated with virulence and biotrophy. *Molecular Plant-Microbe Interactions, 24*, 808–818.

Duplessis, S., Joly, D. L., & Dodds, P. N. (2012). Rust effectors. In F. Martin & S. Kamoun (Eds.), *Effectors in plant–microbes interactions* (pp. 155–193). Oxford, UK: Wiley-Blackwell, Chapter 7.

Duplessis, S., Spanu, P. D., & Schirawski, J. (2014). Biotrophic fungi (powdery mildews, Rusts and Smuts). In F. Martin (Ed.), *Ecological genomics of the fungi* (pp. 149–168). Oxford, UK: Wiley-Blackwell.

Feau, N., Frey, P., Duplessis, S., & Hamelin, R. C. (2013). Genome-wide signature of adaptation in a recently introduced pathogen. *Phytopathology, 103*, 42.

Fellers, J., Soltani, B., Bruce, M., Linning, R., Cuomo, C., Szabo, L., et al. (2013). Conserved loci of leaf and stem rust fungi of wheat share synteny interrupted by lineage-specific influx of repeat elements. *BMC Genomics, 14*, 60.

Fernandez, D., Talhinhas, P., & Duplessis, S. (2013). Rust fungi achievements and future challenges on genomics and host-parasite interactions. In F. Kempken (Ed.), *The Mycota: Vol XI, Agricultural applications* (pp. 315–342). (2nd ed.). Heidelberg, Germany: Springer.

Fernandez, D., Tisserant, E., Talhinhas, P., Azinheira, H., Vieira, A., Petitot, A. S., et al. (2012). 454-pyrosequencing of *Coffea arabica* leaves infected by the rust fungus *Hemileia vastatrix* reveals in planta expressed pathogen-secreted proteins and plant functions expressed in a late compatible plant-rust interaction. *Molecular Plant Pathology, 13*, 17–37.

Flor, H. H. (1959). Differential host range of the monocaryon and the dicaryons of a eu-autoecious rust. *Phytopathology, 49*, 794–795.

Gan, P. H., Rafiqi, M., Ellis, J. G., Jones, D. A., Hardham, A. R., & Dodds, P. N. (2010). Lipid binding activities of flax rust AvrM and AvrL567 effectors. *Plant Signaling and Behaviour, 5*, 1272–1275.

Garnica, D. P., Upadhyaya, N. M., Dodds, P. N., & Rathjen, J. P. (2013). Strategies for wheat stripe rust pathogenicity identified by transcriptome sequencing. *PLoS One, 8*, e67150.

Godfrey, D., Zhang, Z., Saalbach, G., & Thordal-Christensen, H. (2009). A proteomics study of barley powdery mildew haustoria. *Proteomics, 9*, 3222–3232.

Grigoriev, I. V. (2014). A changing landscape of fungal genomics. In F. Martin (Ed.), *Ecological genomics of the fungi* (pp. 3–20). Oxford, UK: Wiley-Blackwell.

Grigoriev, I. V., Cullen, D., Goodwin, S. B., Hibbett, D., Jeffries, T. W., Kubicek, C. P., et al. (2011). Fueling the future with fungal genomics. *Mycology, 2*, 192–209.

Grigoriev, I. V., Nikitin, R., Haridas, S., Kuo, A., Ohm, R., Otillar, R., et al. (2014). MycoCosm portal: Gearing up for 1000 fungal genomes. *Nucleic Acids Research, 42*, D699–D704.

Haas, B. J., Kamoun, S., Zody, M. C., Jiang, R. H., Handsaker, R. E., Cano, L. M., et al. (2009). Genome sequence and analysis of the Irish potato famine pathogen *Phytophthora infestans*. *Nature, 461*, 393–398.

Hacquard, S., Delaruelle, C., Frey, P., Tisserant, E., Kohler, A., & Duplessis, S. (2013). Transcriptome analysis of poplar rust telia reveals overwintering adaptation and tightly coordinated karyogamy and meiosis processes. *Frontiers in Plant Science, 4*, 456.

Hacquard, S., Delaruelle, C., Legue, V., Tisserant, E., Kohler, A., Frey, P., et al. (2010). Laser capture microdissection of uredinia formed by *Melampsora larici-populina* revealed a transcriptional switch between biotrophy and sporulation. *Molecular Plant-Microbe Interactions, 23*, 1275–1286.

Hacquard, S., Joly, D. L., Lin, Y. C., Tisserant, E., Feau, N., Delaruelle, C., et al. (2012). A comprehensive analysis of genes encoding small secreted proteins identifies candidate effectors in *Melampsora larici-populina* (poplar leaf rust). *Molecular Plant-Microbe Interactions, 25*, 279–293.

Hacquard, S., Kracher, B., Maekawa, T., Vernaldi, S., Schulze-Lefert, P., Loren, Ver, et al. (2013). Mosaic genome structure of the barley powdery mildew pathogen and conservation of transcriptional programs in divergent hosts. *Proceedings of the National Academy of Sciences of the United States of America, 110*, E2219–E2228.

Hahn, M., & Mendgen, K. (1997). Characterization of in planta-induced rust genes isolated from a haustorium-specific cDNA library. *Molecular Plant-Microbe Interactions, 10*, 427–437.

Joly, D. L., Feau, N., Tanguay, P., & Hamelin, R. C. (2010). Comparative analysis of secreted protein evolution using expressed sequence tags from four poplar leaf rusts (*Melampsora* spp.). *BMC Genomics, 11*, 422.

Kamoun, S. (2007). Groovy times: Filamentous pathogen effectors revealed. *Current Opinion in Plant Biology, 10*, 358–365.

Kämper, J., Kahmann, R., Bölker, M., Ma, L. J., Brefort, T., Saville, B. J., et al. (2006). Insights from the genome of the biotrophic fungal plant pathogen *Ustilago maydis*. *Nature, 444*, 97–101.

Kemen, E., & Jones, J. D. (2012). Obligate biotroph parasitism: Can we link genomes to lifestyles? *Trends in Plant Science, 17*, 448–457.

Kemen, E., Kemen, A. C., Rafiqi, M., Hempel, U., Mendgen, K., Hahn, M., et al. (2005). Identification of a protein from rust fungi transferred from haustoria into infected plant cells. *Molecular Plant-Microbe Interactions, 18*, 1130–1139.

Kinloch, B. B. (2003). White pine blister rust in north america: Past and prognosis. *Phytopathology, 93*, 1044–1047.

Kinloch, B., Jr., Sniezko, R., & Dupper, G. (2003). Origin and distribution of Cr2, a gene for resistance to white pine blister rust in natural populations of western white pine. *Phytopathology, 93*, 691–694.

Kohler, A., & Tisserant, E. (2014). Exploring the transcriptome of mycorrhizal interactions. *Advances in Botanical Research, 70*.

Kolmer, J. A. (1996). Genetics of resistance to wheat leaf rust. *Annual Review of Phytopathology, 34*, 435–455.

Kolmer, J. A. (2005). Tracking wheat rust on a continental scale. *Current Opinion in Plant Biology, 8*, 441–449.

Kubisiak, T. L., Anderson, C. L., Amerson, H. V., Smith, J. A., Davis, J. M., & Nelson, C. D. (2011). A genomic map enriched for markers linked to Avr1 in *Cronartium quercuum* f. sp. *fusiforme*. *Fungal Genetics and Biology, 48*, 266–274.

Laine, A. L., Burdon, J. J., Dodds, P. N., & Thrall, P. H. (2011). Spatial variation in disease resistance: From molecules to metapopulations. *Journal of Ecology, 99*, 96–112.

Laurans, F., & Pilate, G. (1999). Histological aspects of a hypersensitive response in poplar to *Melampsora larici-populina*. *Phytopathology, 89*, 233–238.

Laurie, J. D., Ali, S., Linning, R., Mannhaupt, G., Wong, P., Güldener, U., et al. (2012). Genome comparison of barley and maize smut fungi reveals targeted loss of RNA silencing components and species-specific presence of transposable elements. *Plant Cell, 24*, 1733–1745.

Lin, K., Limpens, E., Zhang, Z., Ivanov, S., Saunders, D. G., Mu, D., et al. (2014). Single nucleus genome sequencing reveals high similarity among nuclei of an endomycorrhizal fungus. *PLoS Genetics*, *10*, e1004078.

Link, T. I., Lang, P., Scheffler, B. E., Duke, M. V., Graham, M. A., Cooper, B., et al. (2014). The haustorial transcriptomes of Uromyces appendiculatus and Phakopsora pachyrhizi and their candidate effector families. *Molecular Plant Pathology*, http://dx.doi.org/10.1111/mpp.12099.

Link, T. I., & Voegele, R. T. (2008). Secreted proteins of *Uromyces fabae*: Similarities and stage specificity. *Molecular Plant Pathology*, *9*, 59–66.

Lowe, R. G., & Howlett, B. J. (2012). Indifferent, affectionate, or deceitful: Lifestyles and secretomes of fungi. *PLoS Pathogens*, *8*, e1002515.

Martin, F. (2014). *The ecological genomics of fungi*. Oxford, UK: Wiley-Blackwell.

Martin, F., Aerts, A., Ahrén, D., Brun, A., Danchin, E. G., Duchaussoy, F., et al. (2008). The genome of *Laccaria bicolor* provides insights into mycorrhizal symbiosis. *Nature*, *452*, 88–92.

Martin, F., Kohler, A., Murat, C., Balestrini, R., Coutinho, P. M., Jaillon, O., et al. (2010). Périgord black truffle genome uncovers evolutionary origins and mechanisms of symbiosis. *Nature*, *464*, 1033–1038.

McDowell, J. M. (2011). Genomes of obligate plant pathogens reveal adaptations for obligate parasitism. *Proceedings of the National Academy of Sciences of the United States of America*, *108*, 8921–8922.

Mendgen, K. W., & Hahn, M. (2002). Plant infection and the establishment of fungal biotrophy. *Trends in Plant Science*, 7, 352–356.

Murat, C., Payen, T., Petitpierre, D., & Labbé, J. (2014). Repeated elements in filamentous fungi with a focus on wood-decay fungi. In F. Martin (Ed.), *Ecological genomics of the fungi* (pp. 21–40). Oxford, UK: Wiley-Blackwell.

Petre, B., Morin, E., Tisserant, E., Hacquard, S., Da Silva, C., Poulain, J., et al. (2012). RNA-Seq of early-infected poplar leaves by the rust pathogen *Melampsora larici-populina* uncovers PtSultr3;5, a fungal-induced host sulfate transporter. *PLoS One*, 7, e44408.

Pinon, J., & Frey, P. (2005). Interactions between poplar clones and *Melampsora* populations and their implications for breeding for durable resistance. In M. H. Pei, & A. R. McCracken (Eds.), *Rust diseases of willow and poplar* (pp. 139–154). Wallingford, UK: CAB International.

Raffaele, S., Farrer, R. A., Cano, L. M., Studholme, D. J., MacLean, D., Thines, M., et al. (2010). Genome evolution following host jumps in the irish potato famine pathogen lineage. *Science*, *330*, 1540–1543.

Raffaele, S., & Kamoun, S. (2012). Genome evolution in filamentous plant pathogens: Why bigger can be better. *Nature Reviews. Microbiology*, *10*, 417–430.

Rouxel, T., Grandaubert, J., Hane, J. K., Hoede, C., van de Wouw, A. P., Couloux, A., et al. (2011). Effector diversification within compartments of the *Leptosphaeria maculans* genome affected by Repeat-Induced Point mutations. *Nature Communications*, *2*, 202.

Saunders, D. G. O., Win, J., Cano, L. M., Szabo, L. J., Kamoun, S., & Raffaele, S. (2012). Using hierarchical clustering of secreted protein families to classify and rank candidate effectors of rust fungi. *PLoS One*, 7, e29847.

Schirawski, J., Mannhaupt, G., Münch, K., Brefort, T., Schipper, K., Doehlemann, G., et al. (2010). Pathogenicity determinants in smut fungi revealed by genome comparison. *Science*, *330*, 1546–1548.

Song, X., Rampitsch, C., Soltani, B., Mauthe, W., Linning, R., Banks, T., et al. (2011). Proteome analysis of wheat leaf rust fungus, *Puccinia triticina*, infection structures enriched for haustoria. *Proteomics*, *11*, 944–963.

Spanu, P. D. (2012). The genomics of obligate (and nonobligate) biotrophs. *Annual Review of Phytopathology*, *50*, 91–109.

Spanu, P. D., Abbott, J. C., Amselem, J., Burgis, T. A., Soanes, D. M., Stüber, K., et al. (2010). Genome expansion and gene loss in powdery mildew fungi reveal tradeoffs in extreme parasitism. *Science, 330*, 1543–1546.

Tan, K. C., Ipcho, S. V., Trengove, R. D., Oliver, R. P., & Solomon, P. S. (2009). Assessing the impact of transcriptomics, proteomics and metabolomics on fungal phytopathology. *Molecular Plant Pathology, 10*, 703–715.

Thrall, P. H., Laine, A. L., Ravensdale, M., Nemri, A., Dodds, P. N., Barrett, L. G., et al. (2012). Rapid genetic change underpins antagonistic coevolution in a natural host-pathogen metapopulation. *Ecology Letters, 15*, 425–435.

Toome, M., Ohm, R. A., Riley, R. W., James, T. Y., Lazarus, K. L., Henrissat, B., et al. (2014). Genome sequencing provides insight into the reproductive biology, nutritional mode and ploidy of the fern pathogen Mixia osmundae. *New Phytologist,* (in press).

Tremblay, A., Hosseini, P., Li, S., Alkharouf, N. W., & Matthews, B. F. (2012). Identification of genes expressed by *Phakopsora pachyrhizi*, the pathogen causing soybean rust, at a late stage of infection of susceptible soybean leaves. *Plant Pathology, 61*, 773–786.

Tremblay, A., Hosseini, P., Li, S., Alkharouf, N. W., & Matthews, B. F. (2013). Analysis of *Phakopsora pachyrhizi* transcript abundance in critical pathways at four time-points during infection of a susceptible soybean cultivar using deep sequencing. *BMC Genomics, 14*, 614.

Van der Merwe, M. M., Kinnear, M. W., Barrett, L. G., Dodds, P. N., Ericson, L., Thrall, P. H., et al. (2009). Positive selection in AvrP4 avirulence gene homologues across the genus *Melampsora. Proceedings. Biological sciences/The Royal Society, 276*, 2913–2922.

Voegele, R. T., Hahn, M., & Mendgen, K. (2009). The uredinales: Cytology, biochemistry, and molecular biology. In K. Esser, & H. B. Deising (Eds.), *The Mycota. A comprehensive treatise on fungi as experimental systems for basic and applied research* (pp. 69–98). Heidelberg, Germany: Springer.

Voegele, R. T., Struck, C., Hahn, M., & Mendgen, K. (2001). The role of haustoria in sugar supply during infection of broad bean by the rust fungus Uromyces fabae. *Proceedings of the National Academy of Sciences of the United States of America, 98*, 8133–8138.

Warren, J. M., & Covert, S. F. (2004). Differential expression of pine and *Cronartium quercuum* f. sp. *fusiforme* genes in fusiform rust galls. *Applied and Environmental Microbiology, 70*, 441–451.

Wicker, T., Oberhaensli, S., Parlange, F., Buchmann, J. P., Shatalina, M., Roffler, S., et al. (2013). The wheat powdery mildew genome shows the unique evolution of an obligate biotroph. *Nature Genetics, 45*, 1092–1096.

Xu, J., Linning, R., Fellers, J., Dickinson, M., Zhu, W., Antonov, I., et al. (2011). Gene discovery in EST sequences from the wheat leaf rust fungus *Puccinia triticina* sexual spores, asexual spores and haustoria, compared to other rust and corn smut fungi. *BMC Genomics, 12*, 161.

Zambino, P. J., Kubelik, A. R., & Szabo, L. J. (2000). Gene action and linkage of avirulence genes to DNA markers in the rust fungus Puccinia graminis. *Phytopathology, 90*, 819–826.

Zheng, W., Huang, L., Huang, J., Wang, X., Chen, X., Zhao, J., et al. (2013). High genome heterozygosity and endemic genetic recombination in the wheat stripe rust fungus. *Nature Communications, 4*, 1–10.

CHAPTER SEVEN

Truffle Phylogenomics: New Insights into Truffle Evolution and Truffle Life Cycle

Thibaut Payen*, Claude Murat*[,1], Gregory Bonito†
*Institut National de la Recherche Agronomique (INRA), UMR 1136 INRA/Lorraine University, Interactions Arbres/Micro-organismes, Centre de Nancy, Champenoux, France
†Royal Botanic Gardens, Melbourne, South Yarra, Australia
[1]Corresponding author: e-mail address: claude.murat@nancy.inra.fr

Contents

Abstract

Truffles are ectomycorrhizal filamentous fungi belonging to genus *Tuber* naturally found across a range of climates in the Northern Hemisphere. Some truffle species such as Périgord black truffle (*T. melanosporum*) and the white truffle (*T. magnatum*) in Europe, the black Chinese truffle (*T. indicum*) in Asia, and the pecan truffle (*T. lyonii*) in North America are particularly famous due to the organoleptic qualities of their fruiting bodies and are among the world's most precious culinary delicacies. In this chapter, we summarize the recent studies pertaining to the phylogeography and biogeographic history of the genus *Tuber*, genomic analysis of the black truffle *Tuber melanosporum*, and comparative genomics towards understanding the life cycle, ecology, and evolution of *Tuber* species. Such understanding is critical to the anticipated future of truffle production in light of global change.

Advances in Botanical Research, Volume 70
ISSN 0065-2296
http://dx.doi.org/10.1016/B978-0-12-397940-7.00007-0

1. INTRODUCTION

True truffles are ectomycorrhizal filamentous fungi belonging to genus *Tuber* and living in symbiosis with numerous trees and shrubs. Truffles are naturally found across the Northern Hemisphere under a range of climates. Some European species including the burgundy truffle (*Tuber aestivum*), the Périgord black truffle (*T. melanosporum*), and the white piedmont truffle (*T. magnatum*) are particularly famous due to their sought after fruiting bodies (called ascocarps or truffles), which are among the world's most precious culinary delicacies. Economically important truffle species also occur in North America (i.e. the pecan truffle *T. lyonii*) and in Asia (i.e. the Chinese truffles *T. indicum* and *T. sinoaestivum*).

The cultivation of truffles was born with Joseph Talon (1793–1873) who told "*Si vous voulez récolter des truffes, semez des glands*" (If you want to harvest truffles, sow acorns). However, the link between truffles and trees was not fully understood until the first description of ectomycorrhizae by Frank (1889), who described ectomycorrhizae formed by *T. aestivum* and hypothesized the function of these symbiotic structures. After the grape phylloxera crisis of at the end of the nineteenth century, (Hall, Brown, & Zambonelli, 2007) grape vineyards in France were replaced with truffle orchards and the golden age of truffle production began.

During the twentieth century, truffle production in France declined from more than 1000 t to less than 100 t. This decline of truffle production has been attributed to several causes. The first rural desertification led to a decrease in the surface area of land devoted to truffle cultivation. Further, remaining truffle orchards were no longer managed; as the trees matured their canopies gradually closed altering microhabitats and water availability and leading to a decline in truffle production. Climate change has also been evoked to explain declining truffle yields (Büntgen et al., 2012) through alterations in regional rainfall and climate patterns.

The success in the mass production of seedlings colonized with *T. melanosporum* ectomycorrhizae began in the 1970s and has led to an increase in the area planted with truffles and their hosts (Chevalier, 1998; Chevalier & Grente, 1979). In France alone, 800–1000 ha of plantations have been established annually over the past 20 years. However, even with these efforts, truffle production has not reached the levels achieved at the beginning of the twentieth century. The truffle industry is currently requesting innovative tools to intelligently develop truffle cultivation into a sustainable and fully profitable activity. In this chapter, we will present the contribution of comparative genomics towards understanding the truffle life cycle, ecology, and evolution. Such understanding is critical to the anticipated future of truffle production in light of global change. The most recent studies pertaining to the phylogeography and biogeographic history of the genus *Tuber* will also be presented.

2. TRUFFLE DIVERSITY AND PHYLOGEOGRAPHY

2.1. *Tuber* biodiversity

The genus *Tuber* is among the most diverse groups of fungi to produce belowground fruiting bodies known as truffles. *Tuber* is estimated to include a minimum of 180 species worldwide (Bonito, Gryganskyi, Trappe, & Vilgalys, 2010). Much of this diversity remains undescribed. The majority of newly described and undescribed diversity in *Tuber* resides in poorly studied clades. In particular, the spiny-spored Rufum clade and the aveolate-reticulate spored Maculatum and Puberulum clades are particularly speciose (Bonito, Gryganskyi, et al., 2010). Many (but not all) of the species in these groups produce smaller-sized fruiting bodies that are morphologically cryptic and lack any apparent culinary value, attributes that may explain why these species have been largely overlooked by mycologists. With renewed interest in *Tuber* diversity and with the aid of molecular tools for identification, an impressive number of new *Tuber* species have recently been described in Europe (Alvarado, Moreno, & Manjón, 2012; Božac, Širić, & Kos, 2012), North America (Bonito, Trappe, Rawlinson, & Vilgalys, 2010; Guevara et al., 2013), and Asia (Deng, Liu, Liu, & Wang, 2013; Fan, Cao, & Li, 2012a,2012b; Fan, Cao, & Yu, 2012; Fan, Hou, & Li, 2012). Some of these newly described species are likely to have value as culinary species.

2.2. Phylogeography and origin of *Tuber*

Recent studies aimed at assessing the phylogenetic diversity of *Tuber* and reconstructing the truffle family tree have given new understanding pertaining to the origin, evolution, and biogeographic history of *Tuber* (Bonito et al., 2013; Jeandroz, Murat, Wang, Bonfante, & Tacon, 2008). Based on extensive biogeographic sampling of *Tuber*, multigene datasets were generated for molecular dating and phylogenetic approaches. In these studies, *Tuber* was resolved as monophyletic, and a common ancestor that was most likely an aboveground cup fungus was inferred. Bonito et al. (2013) used a concatenated dataset of 4 loci to resolve 11 major clades in *Tuber* (Fig. 7.1) and found a high degree of species-level endemism in the genus. While species may show high levels of endemism, *Tuber* clades including the Rufum, Melanosporum, Puberulum, Maculatum, and Macrosporum are distributed across the entire Northern Hemisphere (Europe, Asia, North America, Central America, and Northern Africa), and the Aestivum and Excavatum clades are distributed in both Europe and Asia. A few clades show continental endemism including the Gennadii and Multimaculatum clades in Europe, the Japonicum clade in Asia, and the Gibbosum clade in North America. While some *Tuber* species appear to be generalists in regard to their ectomycorrhizal hosts (e.g. *T. aestivum* and *T. indicum*), host preferences show higher-level patterns. For instance, most of the species in the Rufum, Excavatum, Aestivum, Maculatum, and Gennadii clades tend to be associated with angiosperm hosts, while those in the Japanicum and Gibbosum clades appear to prefer hosts in the Pinaceae. The Puberulum clade exhibits the widest range of host associations and many species are host generalists.

Bonito et al. (2013) carried out divergence dating analyses in BEAST using a relaxed clock model with an uncorrelated exponential prior distribution. They estimated the divergence of *Tuber* from other genera occurred during the early Cretaceous Period 156 Mya (fig. 5—node 4 in Bonito et al., 2013), with most of the extant subgeneric lineages present 65 Mya and experiencing major radiations during the Paleogene (Bonito et al., 2013). Estimates for the mean ages of the most recent common ancestor (MRCA) of the 11 resolved *Tuber* clades are Multimaculatum (121 Mya), Aestivum (101 Mya), Rufum (86 Mya), Melanosporum (79 Mya), Puberulum (65 Mya), Japonicum (46 Mya), Excavatum (43 Mya), Maculatum (67 Mya), Macrosporum (43 Mya), Gennadii (48 Mya), and Gibbosum (27 Mya). While there is still much uncertainty whether the origin of *Tuber* was European or Asian (Bonito et al., 2013; Jeandroz et al., 2008), putative geographic origins of the most common ancestors for a majority of clades were resolved by Bonito et al. (2013) with North America as the ancestral area of the Gibbosum, Maculatum, Rufum, and Melanosporum clades; Europe as the ancestral area of the Aestivum, Excavatum, and Gennadii clades; and Asia as the ancestral area for the Japonicum clade.

Tuber is often considered to have a Laurasian distribution because species are naturally distributed across Northern latitudes in North America,

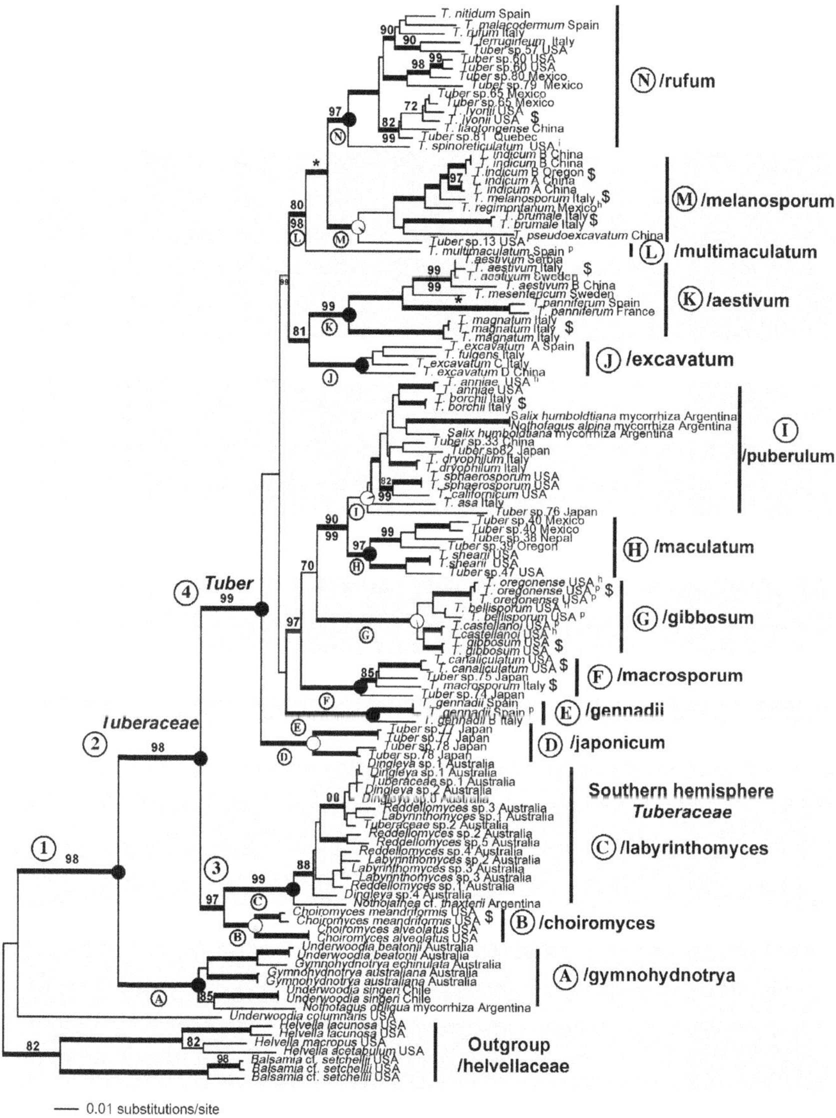

Figure 7.1 Maximum likelihood (ML) phylogenetic reconstruction of the *Tuberaceae* phylogeny based on ITS, 28S rDNA, EF1, and RPB2 genic loci. Thickened branches represent ML bootstrap support >70 and posterior probabilities of 100. ML bootstrap values above nodes are based on 1000 replicates. Posterior probabilities are presented below nodes. Thickened branches without numbers received maximum ML and Bayesian support values. Reconstructed ancestral host plant associations (based on maximum likelihood) are represented at internal nodes by circles, black for ancestors in symbiotic association with angiosperms, white for ancestors in symbiotic association with *Pinaceae*, and grey for ancestors in symbiotic association with angiosperms and *Pinaceae*.

(*Continued*)

Europe, Asia, and Northern Africa. However, species of *Tuber* belonging to the Puberulum clade were recently detected as ectomycorrhizae on *Nothofagus* in Argentina (Bonito et al., 2013). While humans have dispersed fungi to many continents, these South American species appear to be phylogenetically novel and are likely native species whose ancestors migrated into South America from the North, putatively in association with an ectomycorrhizal host such as *Salix*. In addition to having wide host preferences, the Puberulum clade appears to have the widest geographic distribution of all *Tuber* clades.

With the growing interest in cultivating truffles *Tuber* is now being introduced into areas outside their native range. For instance, the European black truffle *Tuber melanosporum*, which is the most important cultivated *Tuber* species, has been introduced globally, including Sweden, South Africa, Morocco, Israel, China, United States, Canada, Mexico, Chile, Argentina, Australia, and New Zealand. An understanding of the ecological and physiological constraints of *T. melanosporum* is necessary to ensuring that these efforts are successful. Unfortunately, in the act of introducing *Tuber melanosporum* into these countries, other *Tuber* species have also been introduced outside their range. Given that unintentional introductions of *T. indicum* or *T. brumale* have been documented in nearly every country where mycologists have carried out molecular sequence-based assessments (e.g. the United States, Italy, New Zealand, Australia, and Canada), it is likely that nontarget *Tuber* introductions are a frequent occurrence where efforts to cultivate *T. melanosporum* have been made (Bonito, Trappe, Donovan, & Vilgalys, 2011; Guerin-Laguette, Cummings, Hesom-Williams, Butler, & Wang, 2013; Murat, Zampieri, Vizzini, & Bonfante, 2008). Biosecurity measures to mitigate against further unintentional introductions of *Tuber* are needed and should include molecular testing of inoculum used for inoculations. While specific primers have been designed for rapid identification of target species (Amicucci, Guidi, Zambonelli,

Figure 7.1—Cont'd Nodes supported by transitions in spore ornamentation from alveolate-reticulate to spiny are shown with an asterisk *. Economically important species are denoted by the symbol $ after their name and geographic origin. The phylogeny is rooted with taxa from the *Helvellaceae* including species of epigeous *Helvella* and hypogeous *Balsamia*. Major lineages of *Tuber* and *Tuberaceae* are indicated to the right of the tree. The *Tuberaceae* form a monophyletic group, which is resolved as a sister group to a previously unrecognized Southern Hemisphere lineage (Gymnohydnotrya). Type specimens are denoted by the superscripts: h, holotype; i; isotype; and p, paratype. *Reprinted with permission from Bonito et al. (2013).*

Potenza, & Stocchi, 2000; Bonito, 2009; Paolocci, Rubini, Granetti, & Arcioni, 1999; Zampieri, Mello, Bonfante, & Murat, 2009), high-throughput sequenced-based assessments of inoculum slurries and root communities (i.e. amplicon sequencing by 454 pyrosequencing or Illumina) offer a more comprehensive strategy for ensuring inoculum quality for truffle tree nurseries and may allow for simultaneous assessments of overall genetic diversity levels or biogeographic origin of species in the inoculum.

3. HARNESSING GENOMES TO UNRAVEL THE BLACK PÉRIGORD TRUFFLE LIFE CYCLE

3.1. Black Périgord truffle life cycle: An intimate relation between the fungi, plant, and climate

The life cycle of the black Périgord truffle begins in late winter or early spring with the germination of haploid spores released by mature ascocarps. The growing haploid mycelium colonizes tree roots to form ectomycorrhizal symbiotic associations (Paolocci, Rubini, Riccioni, & Arcioni, 2006; Riccioni et al., 2008). In May or June, after plasmogamy (the fusion of haploid mycelia and the initial fertilization event), an ascogenous heterokaryotic tissue surrounded by homokaryotic maternal tissues gives birth to the ascocarp. This ascocarp primordium is linked to the ectomycorrhizae until the truffle fruiting body completely matures and requires carbon stored in the trunk or roots of the host plant for the truffle to grow and mature (Le Tacon et al., 2013). As recently demonstrated, truffles are heterothallic species and mating-type genes coding for sexual compatibility have been identified (see in the succeeding text). The young truffle, born in the late spring or beginning of summer, grows slowly during the summer and starts to mature in autumn. The crop in the Northern Hemisphere generally starts in mid-November and finishes mid-March.

The interaction between truffles and their host plant is critical since truffles need to be associated with a host plant to complete their life cycle by producing mature fruiting bodies. Indeed, truffle ascocarps need carbon to grow, and as demonstrated recently, nearly all of this carbon comes from the host plant photosynthate via ectomycorrhizae (Le Tacon et al., 2013). In this study, stable ^{13}C-labelling was used to demonstrate that the ascocarps accumulate host carbon until reaching complete maturity. *Tuber* ascocarps remained a carbon sink even several months after the end of photosynthetic carbon assimilation by the host and at low temperature. These results have important field implications. For instance, guidelines for tree pruning are not yet available. Winter pruning could reduce carbon availability for maturing truffles. More detailed studies on the interaction and dialogue between truffle and host tree has been investigated through transcriptomic analyses (Kohler & Tisserant, 2014).

The sequenced black truffle genome is an exceptional resource for investigating the development and physiology of the black truffle (Kües & Martin, 2011). Transcriptomic analyses of tissues under different conditions including vegetative hyphal growth (Amicucci et al., 2011), cold stress (Zampieri et al., 2011), different stages in the life cycle (Montanini et al., 2011), and metal homeostasis involvement (Bolchi et al., 2011) have been investigated. Splivallo and colleagues (Splivallo, Ottonello, Mello, & Karlovsky, 2011) also highlighted putative ecological roles of truffle volatiles as chemical signals between the fungus, other soil organisms, and plant roots.

The long duration of the truffle life cycle makes truffle harvests sensitive to climatic conditions of the entire year and particularly sensitive to the water balance of summer months, which is often critical in the Mediterranean climate. It is currently understood and well documented that black truffle production is tightly linked to climate (Büntgen et al., 2012). The role of drought in diminishing truffle production is well established. For instance, in 1914, Pradel demonstrated the importance of supplementing water to truffle-producing trees during drought summers to maintain truffle production. More recently, Le Tacon, Delmas, Gleyze, and Bouchard (1982) determined that maintaining soil water potential above 3.5 during the summer months with irrigation increased truffle production by a factor of 18 in 1979–1980 and by more than 40 in 1980–1981. Moreover, the interannual variability in black truffle production observed over the past 25 years is highly correlated with fluctuations in rainfall (Le Tacon et al., 2014). The consequence of drought for truffle production could be important. For example, the 2003 drought caused a drop in French production to about 80%; most of the truffle grounds affected, without water, did not produce any truffles as a result of this drought. This suggests that models and guidelines for sustainable watering of truffières are needed. Drought severely impacts truffle growers' incomes and consequently the whole truffle industry, leading to unpredictable truffle production and a fluctuating truffle market. Drought is expected to increase in both frequency and severity in most European regions as a result of changes in the temporal distribution of rainfall and increase in potential evapotranspiration driven by global warming (Sheffield, Wood, & Roderick, 2012). Truffles may have nonlinear and inconsistent responses to drought, depending on which phase of their life cycle is subjected to water stress. A better knowledge of climatic conditions and the effects of truffle growers' interventions (e.g. soil tilling, tree pruning, and watering) on the different phases of the truffle life cycle, such as the distribution of soil mycelium, is needed to anticipate the response of truffles to climate stresses.

In contrast to other ectomycorrhizal fungi (e.g. *Hebeloma cylindrosporum*) (Marmeisse et al., 2004), it is impossible to complete the truffle life cycle *in vitro*. Indeed, it is possible to obtain free-living mycelium of different *Tuber* spp., which grow slowly (Iotti, Amicucci, Stocchi, & Zambonelli, 2002), and ectomycorrhizae (Giomaro et al., 2002) can be produced, but to our knowledge, the sexual reproduction of *Tuber* has never been observed *in vitro*. The only documented production of ascocarps in control conditions was the fruiting of *T. maculatum* in vases associated with *Pinus strobus* (Fassi & Fontana, 1969). However, this was never replicated. The truffle life cycle is therefore challenging to study. Many unknowns regarding the basic biology and life cycle of truffles deserve further attention in order to increase truffle production. These include the following:

1. What factors promote and are responsible for spore germination?
2. How large are individual *Tuber* genets and what is their lifespan?
3. How can sexual reproduction of truffles be promoted?
4. How can we preserve and enhance ascocarp (fructification) growth?
5. Which is the role of microfauna and microbes on the truffle life cycle?
6. Are truffles able to adapt to climatic changes?
7. How can we enhance the truffle life cycle with management techniques?

A major advancement to our understanding of the truffle biology has come from the recently sequenced *T. melanosporum* genome (Martin et al., 2010). This effort has reaffirmed the importance of *T. melanosporum* as a model ectomycorrhizal ascomycete species. Molecular tools developed from this research are already filling many knowledge gaps, and more tools and insights pertaining to truffle biology and ecology are expected.

3.2. Intraspecific genetic diversity: From microsatellites to SNPs

The level of intraspecific genetic diversity of *T. melanosporum* has been debated since Bertault, Raymond, Berthomieu, Callot, and Fernandez (1998) argued that this species exhibits low levels of genetic diversity. More recently, Riccioni et al. (2008) used polymorphic microsatellites and questioned this conclusion. Thanks to the sequencing of the black truffle genome (Martin et al., 2010) and the development of highly polymorphic microsatellite makers, we now have a better view of the intraspecific genetic diversity of this species, which has a genotypic diversity index close to its theoretical maximum (Murat et al., 2011). Several of the microsatellites developed detected more than 10 alleles and had a value of expected heterozygosity (i.e. estimating the fraction of all individuals who would be heterozygous for any randomly chosen locus) greater than 0.5 indicating that these SSRs are highly polymorphic. These polymorphic markers were used to characterize small-scale genetic diversity in two truffle orchards in Rollainville (Lorraine, France; 48°21′45″ North and 5°44′23″ East) and in Montemartano (Umbria, Italy; 42°44′00″ North and 12°44′00″ East) (Murat et al., 2013). In both truffle orchards, a pronounced spatial genetic structure was found with numerous small-sized genets. Indeed, the maximum distances between ramets of the same genet were 2.35 and 4.70 m in the Montemartano and the Rollainville truffle orchards, respectively. Most genets had a size of less than 1 m, and many were limited to one sampling core. In Rollainville, the sampling was carried out in a 60-m^2 area during two consecutive years and only three genets were shared out of the 10 and 13 genets found during the 2010–2011 and 2011–2012 seasons, respectively. One other interesting result was that the spatial distribution of the *T. melanosporum* genets on their hosts was nonrandom, resulting in field patches colonized by genets that shared the same mating types (see in the succeeding text). These results suggest that *Tuber* invests mainly in sexual reproduction. Indeed, it seems that the genet size depends more on competition and the time elapsed since the last disturbance than on other factors such as forest age (Douhan, Vincenot, Gryta, & Selosse, 2011). The management of truffle orchards (e.g. tree pruning, tillage, grass cutting, and searching for truffles with dogs) causes disturbance to create an open woodlands effect, and it is well known that *T. melanosporum* fruiting bodies are generally found in such ecosystems, while the presence of competing fungi in closed canopies depresses its fructification (Hall et al., 2007).

Using intersimple sequence repeat (ISSR) fingerprints, García-Cunchillos and colleagues (2014) confirmed high levels of genetic diversity in Spanish black truffle populations. They analysed 190 ascocarps from 23 natural populations distributed throughout Spain. Each ascocarp presented a unique genotype and Nei, Shannon, and Bayesian heterozygosity genetic diversity indices showed high levels of genetic diversity throughout the Iberian populations (hN = 0.393, hS = 0.678, and HS = 0.418). Interestingly, they identified strong genetic structure among the 23 Spanish populations as well as within geographic groups. These results confirm that southern truffle populations in Spain include augmented levels' genetic diversity compared with northern truffle populations suggesting that the southern Iberian Peninsula may have served as a refuge for truffles during the last glaciation as proposed by Riccioni et al. (2008).

Classical population genetic analyses with microsatellites allowed the investigation on a very small proportion of the *Tuber* genome. With continual advances in high-throughput sequencing technologies, exhaustive investigation of the genomic variation is now feasible. Investigations of genetic variation in *Tuber* address questions pertaining to adaptation and genome plasticity of the species. Screening and identification of polymorphisms among strains of the same species is the most common way to identify genetic variation. Among all the polymorphisms, single nucleotide polymorphisms (SNPs) have gained much interest in the scientific community (Ganal, Altmann, & Röder, 2009). SNPs have been used for investigating the genetic origin of human diseases such as asthma and diabetes (Moffatt et al., 2010; Yamauchi et al., 2010) and also in crop plants (Ganal et al., 2009; Rafalski, 2002) to identify genomic regions involved in disease resistance in maize (Poland, Bradbury, Buckler, & Nelson, 2011). The identification of SNPs by genome resequencing was used for filamentous fungi (e.g. *Neurospora crassa* (Ellison et al., 2011), *Fusarium graminearum* (Cuomo et al., 2007), and *Coccidioides* spp. (Neafsey et al., 2010); see Stukenbrock, 2014, in this book) but to our knowledge not yet for mycorrhizal species. To assess the overall genetic diversity of *T. melanosporum* and factors driving genome plasticity of this species the genome of six *T. melanosporum* genotypes were sequenced using Illumina technology and were compared to the reference genome (Martin et al., 2010). By comparing seven genomes, we identified more than 400,000 SNPs confirming that the genetic diversity of *T. melanosporum* is similar to that of other filamentous fungi (Payen Thibaut, Murat Claude, Gigant Anaïs, Morin Emmanuelle, De Mita Stéphane and Martin Francis, in preparation). SNPs were more frequent in repeated sequences although some SNPs were also identified in coding regions of more than 2000 genes. The generated SNPs provide a valuable resource for future population genetic and genomics studies. In this analysis, we also highlighted the role of transposable elements, which represents about 60% of the black truffle genome, in its genome plasticity. These first results are promising to investigations into black truffle adaptive capacities to climate change. A new sampling of isolates covering all European populations is currently underway to perform a robust and more comprehensive population genomic analysis.

3.3. Sexual reproduction: Where are both mating-type strains?

The interest of scientists in sexual reproduction of fungi dates back to the nineteenth century when Ehrenberg first reported sex in a fungal species (Ehrenberg, 1829). In filamentous ascomycetes, in addition to the widespread potential for asexual (vegetative) reproduction, two main types of sexual reproduction can be found: homothallism and heterothallism. Whereas homothallic fungi are self-fertile, heterothallic species depend on the presence of a partner of the opposite mating type to reproduce and are therefore self-sterile (Kronstad, 2007; Murtagh, Dyer, & Crittenden, 2000; Paoletti et al., 2007). Concerning mycorrhizal fungal species, in addition to their critical role in establishing and maintaining the complexity in agroforest ecosystems worldwide (Smith & Read, 2008), the production of edible fruiting bodies (e.g. boletes and truffles) by some of them makes the understanding of their sexual reproduction mode very attractive for mycologists.

In 1998, Bertault and colleagues concluded that *T. melanosporum* had "a very closed mating system, such as homothallism or even exclusive selfing". This assertion was challenged by the data from microsatellites markers (Paolocci et al., 2006; Riccioni et al., 2008) and more recently by the identification of mating-type genes (Rubini et al., 2011, 2011). Microsatellite genotyping of ECM and asci, along with their surrounding sterile tissue in the ascocarps, the gleba, indicates that events of outcrossing are frequent in *T. melanosporum* (Riccioni et al., 2008). Conclusive evidence of sexual reproduction of *T. melanosporum* has been gained by looking at the structure and organization of the mating-type (*MAT*) locus in the sequenced genome of strain Mel28 (Martin et al., 2010). The *MAT* locus is a unique region of the fungal genome that governs the establishment of cell-type identity and orchestrates the sexual cycle (Fraser & Heitman, 2003). All known heterothallic ascomycetes have a single *MAT* locus with two alternative and dissimilar homologous sequences called idiomorphs (Metzenberg & Glass, 1990). One contains a *MAT* gene that encodes a protein with an α-box domain (*MAT1-1-1*), while the other contains an alternative *MAT* gene that encodes a high mobility group (HMG) protein (*MAT1-2-1*) (Debuchy, Berteaux-Lecellier, & Silar, 2010). Conversely, in homothallic filamentous ascomycetes, a single strain harbours both *MAT1-1-1* and *MAT1-2-1* genes. In Mel28 strain, only the *MAT1-2-1* gene was identified, whereas the second *MAT* gene was characterized in those strains that do not carry *MAT1-2-1*. Associated with outcrossing evidence produced with microsatellites, it was concluded that

T. melanosporum is a heterothallic species (Rubini, Belfiori, Riccioni, Arcioni, et al., 2011; Rubini, Belfiori, Riccioni, Tisserant, et al., 2011). Consequently, for *T. melanosporum* to complete its life cycle, the recognition of two truffle strains of opposite mating type is needed. Interestingly, both mating-type strains are not distributed randomly in the root systems of host plants in productive truffle orchards (Murat et al., 2013; Rubini, Belfiori, Riccioni, Arcioni, et al., 2011; Rubini, Belfiori, Riccioni, Tisserant, et al., 2011). Rather, patches colonized by strains of the same mating types are observed. This suggests that competition to colonize the root system may occur between the two mating types. Interestingly, both mating types can be detected in the soil in these patches. However, both mating types colonizing the same root systems were detected in some truffle orchards trees in Australia (Linde & Selmes, 2012). While the maternal strain producing the gleba can be detected in the surrounding ectomycorrhizae (Murat et al., 2013) and confirming the link between the ectomycorrhizae and ascocarps (Le Tacon et al., 2013), where does the paternal strain come from? To answer this question, studies are underway to identify the genotype of the paternal strain and investigate its distribution in the soil.

Recently, Belfiori, Riccioni, Paolocci, and Rubini (2013) identified mating-type genes in the Chinese truffle (*T. indicum*), a close relative to *T. melanosporum* (Fig. 7.1). Interestingly, they identified different sequences and the structure of the MAT idiomorphs between *T. indicum* strains belonging to different ITS classes suggested the presence of a complex of cryptic species. The mating-type idiomorphs of *T. indicum* and *T. melanosporum* are quite well conserved, and the sexual compatibility between the two species cannot be fully excluded. This is an important result since there is at least one documented case of the introduction of *T. indicum* into Italy (Murat et al., 2008). To limit possibilities of introgression between the two species, it is critical to avoid additional introductions of *T. indicum* into the natural range of *T. melanosporum*.

The characterization of mating-type genes identified by sequencing the black truffle genome is a very good example of a quick transfer from fundamental research to the truffle industry. Indeed, an international patent (no. WO2012/032098) was recently deposited for the characterization of the mating-type strains. This patent will soon be exploited to provide truffle industry with innovative tools. Indeed, promoting the initiation of sexual reproduction by truffle fungi is a critical point of interest, which should help to increase truffle production and yield. But as highlighted in the succeeding text, the technical management (e.g. soil tilling, tree pruning, and watering) of truffle orchards is also important to allow ascocarps in the soil to develop and ripen over a duration of several months. This is particularly true in the context of climate change.

4. COMPARATIVE GENOMIC IN PEZIZOMYCETES

Comparative genomics allows for the assessment of relationships between genomes to understand the function and evolutionary processes acting upon them. The link between fungal life strategies (wood decay, symbiosis, and parasites) and genome structure (size, repeat sequences, and gene catalogue) has been a hot topic for several years (Martin et al., 2011). For example, biotrophic pathogens (Spanu et al., 2010), ectomycorrhizal symbionts (Martin et al., 2008, 2010), and endomycorrhizal fungi *Rhizophagus irregularis* (Tisserant et al., 2013) have a decreased repertoire of genes involved in the degradation of plant cell wall polysaccharides and in the biosynthesis of secondary metabolite toxins.

Until recently, *T. melanosporum* was the sole representative of the Pezizomycotina, a speciose and diverse clade of fungi, to have a genome sequence available (Martin et al., 2010). In 2013, Traeger and colleagues published the genome of *Pyronema confluens* a saprotrophic pezizomycete. Interestingly, *P. confluens* has a small genome of 50 Mb compared to the 125 Mbp genome of *T. melanosporum*. In order to gain information on the evolution in Pezizomycetes, and more precisely the Tuberaceae, the genome of new species is currently being sequenced. In the succeeding text, we will provide an update of the different genomic projects that are currently realized in Pezizomycetes.

4.1. Sequencing the genome of *Tuber* spp.

Genome sequencing of five other *Tuber* spp. is currently underway (Table 7.1). These species were chosen because they are economically important species and represent different clades within the *Tuber* phylogeny (Fig. 7.1). They include the following:

1. *T. aestivum*, known as the burgundy truffle, is harvested in all the European countries and is sold for between 100 and 400 €/kg. This species belongs to the Aestivum clade and is becoming popular due to its broad range of host preference, large range of soil types and climates that it can fruit, and availability as mycorrhized plants by nurseryman.

Table 7.1 List of the current genomic projects in the Pezizomycetes

Species	Family	Genome size (Mb)	% of repeated sequences	Material	Status of the project	Natural habitat	Project
Ascobolus immersus	Ascobolaceae	59.53	NA	Mycelium	Complete	Worldwide	1000 Fungal Genomes Project (JGI-DOE)
Choiromyces venosus	Tuberaceae	124	54	Fruiting body	Complete	Europe	Mycorrhizal Genomics Initiative (JGI-DOE)
Pyronema confluens	Pyrenomycetaceae	50	12	Mycelium	Published	France	Traeger et al. (2013)
Terfezia boudieri	Pezizaceae	63	NA	Mycelium	Pending	North Africa and southern Europe	Mycorrhizal Genomics Initiative (JGI-DOE)
Tuber aestivum	Tuberaceae	135.6	50	Fruiting body	Annotation pending	Europe	TuberEvolve (Genoscope-INRA)
Tuber borchii	Tuberaceae	NA	NA	Mycelium	Pending	Europe	Metatranscriptomics of Soil Forest Ecosystems (JGI-DOE)
Tuber indicum	Tuberaceae	NA	NA	Fruiting body	Pending	China	INRA/Kunming University
Tuber lyonii	Tuberaceae	NA	NA	Mycelium	Pending	North America	Duke University
Tuber magnatum	Tuberaceae	192.8	58	Fruiting body	Annotation pending	Italy and South-Central Europe	INRA—UNITO
Tuber melanosporum	Tuberaceae	124.95	58	Mycelium	Published	Southwest Europe	Martin et al. (2010)

NA, not available.

2. *T. borchii*, called the bianchetto, is harvested in many European countries but is most famous in Italy where it is commercialized and marketed for between 50 and 100 €/kg. This species belongs to the Puberulum clade and is characterized by its broad range of host associates, aggressive growth, and wide edaphic tolerances. *Tuber borchii* is now being cultivated on numerous continents.
3. *T. indicum*, known as the Chinese truffle, is phylogenetically and morphologically very close to *T. melanosporum* and belongs to the Melanosporum clade. It is mainly harvested in China but is mostly exported to Europe and the United States. This species has been inadvertently introduced into Italy (Murat et al., 2008) and the United States (Bonito et al., 2011). *T. indicum* is sold in Europe for between 20 and 100 €/kg.
4. *T. lyonii*, known as the pecan truffle, is native to North America. This species fruits naturally in pecan orchards and under oak trees in the southern and eastern United States but is found as far north as Quebec. This species has spiny spores and belongs to the Rufum clade. Recent efforts to cultivate this species are underway and the first fruitings from artificially colonized seedlings were produced this past year (*personal communication*, Tim Brenneman). *Tuber lyonii* sells for between $200 and 300 per lb in the United States.
5. *T. magnatum*, known as the Piedmont white truffle, is harvested in Italy and central Europe (Croatia, Serbia, Slovenia, Romania, Bulgaria, and Hungary). It has also been found in southern France the past few years. This species has yet to be cultivated and is the most expensive truffle species sold at prices of 1000–3000 €/kg (but can reach more than 100,000 €/kg in auction sales).

The production of bulk mycelium for DNA isolation (or inoculation) is often difficult for *Tuber* spp. because they are fastidious to isolate and grow so slowly in pure culture. However, sufficient mycelium of *T. melanosporum* (Gérard Chevalier—INRA Clermont-Ferrand), *T. borchii* (Alessandra Zambonelli and Mirco Iotti—University of Bologna), and *T. lyonii* (Gregory Bonito—Duke University) was grown to produce several grams of mycelium, enough to allow DNA and RNA to be extracted for genome and transcriptome sequencing. For genome sequencing of *T. indicum*, *T. aestivum*, and *T. magnatum*, nucleic acids were isolated from fresh young fruiting bodies. The gleba (internal part) of the truffle ascocarp is rich in microorganisms such as bacteria (Antony-Babu et al., 2014) and therefore extracted nucleic acids usually contain a mix belonging to *Tuber* and the other microorganisms. When obtaining nucleic acids from fungal fruiting bodies, it is therefore necessary to exclude carry-over from associated bacteria prior to sequencing. In the first attempt to sequence the *T. magnatum* genome from a fruiting body, more than 90% of the 454 reads mapped to bacteria. More rigorous quality checks were necessary to avoid the huge bacterial contamination in further efforts. First, PCR with bacterial and fungal primers on serial dilutions of the DNA are done. Next, preliminary sequencing run is carried out with 454 GS Junior pyrosequencer to assess levels of bacterial contamination. Finally, samples deemed of sufficient quality for genome sequencing (i.e. with less than 10% of bacterial contamination) are sequenced and obtained reads are filtered using bacterial genomes as reference sequences to exclude bacterial reads.

Genome assemblies are currently available for *T. aestivum* and *T. magnatum* (Table 7.1). Interestingly, these two genomes are rather large with a size of 135 and 192 Mbp with 50% and 58% of repeated sequences, respectively. As for *T. melanosporum*, the gypsy-like retrotransposons are the most frequent repeat sequences in these two genomes. Preliminary analyses suggest that the same gypsy-like retrotransposon families are present in the three annotated *Tuber* genomes (*T. aestivum*, *T. magnatum*, and *T. melanosporum*) indicating that the invasion of these retrotransposons in *Tuber* genomes occurred early in the evolution of this genus. The structure of the *Tuber* and *Choiromyces venosus* (see in the succeeding text) genomes is conserved as shown also by large syntenic blocks (Fig. 7.2). The genome of *T. lyonii* has been sequenced but assemblies are not yet available. Comparisons between these four clades of *Tuber* should help address questions pertaining to host specificity, sexual and asexual reproduction, and development and growth of truffles.

4.2. Sequencing the genome of other Pezizomycetes

In the framework of the Mycorrhizal Genomics Initiative (Martin & Bonito, 2013; Martin et al., 2011), the genomes of two other truffles outside of the genus *Tuber* have been sequenced: *C. venosus* (http://genome.jgi.doe.gov/Chove1/Chove1.home.html) and *Terfezia boudieri* (http://genome.jgi.doe.gov/Terbo1/Terbo1.home.html). *Choiromyces venosus*, commonly called the pig truffle, belongs to the Tuberaceae and has a strong, distinctive odour. Gastronomic value of this whitish truffle varies throughout Europe, as it is consumed in a few countries (i.e. Germany, Hungary, and Sweden) but considered toxic in France and Italy. However, no records of *C. venosus*

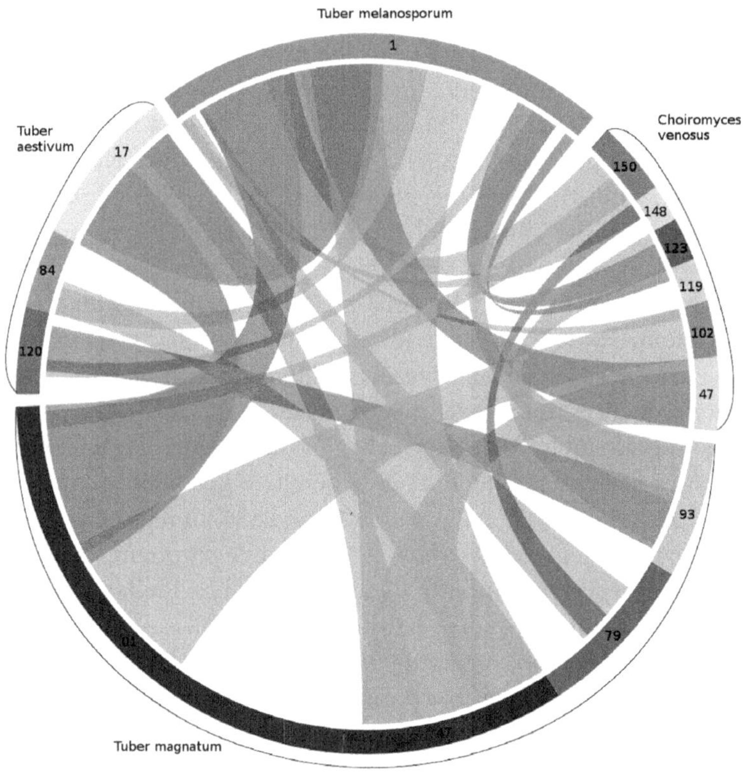

Figure 7.2 Example of synteny between the largest scaffold of *T. melanosporum* (scaffold 1) and its homologues in *T. aestivum*, *T. magnatum*, and *C. venosus*. The number of scaffold is indicated for each species. The synteny was realized using SyMAP version 4.0 (Soderlund, Bomhoff, & Nelson, 2011). (See the colour plate.)

poisoning have ever been reported in northern Europe where it has consumed for more than a century (Wedén, Larsson, Burman, & Backlund, 2009). This symbiotic species is associated with deciduous and coniferous trees and has a preference for clay soils. The genome of *C. venosus* is similar to that of *Tuber* spp. in size (125 Mbp) and repeat content (54%) (Table 7.1). There are also large syntenic block between *C. venosus* and *Tuber* spp. (Fig. 7.2). The phylogenetic position of *Choiromyces* was for a time ambiguous since Percudani, Trevisi, Zambonelli, and Ottonello (1999) included this genus inside *Tuber* spp. However, more recent phylogenetic analyses clearly separate these two genera with strong statistical support (Fig. 7.1).

Terfezia boudieri, known as desert truffles, are harvested around the Mediterranean Sea and the Middle East in areas associated with *Helianthemum* spp. It belongs to the Pezizaceae and represents an independent origin of the truffle lifestyle. Its genome has being sequenced. The comparison between *C. venosus*, *T. boudieri*, and *Tuber* spp. should provide new insights into the evolution of the truffle life strategy (e.g. hypogeous fruiting body formation).

Recently, through the "1000 Fungal Genomes Project" Initiative (JGI-DOE), the genome of *Ascobolus immersus*, belonging to the Ascobolaceae, has been sequenced. *Pyronema confluens* (Traeger et al., 2013), which belongs to the Pyronemataceae, has also been sequenced recently. Interestingly, two of these three last species have smaller-sized genomes than *Tuber* spp. and fewer repetitive elements (Table 7.1).

5. CONCLUDING REMARKS

The age of fungal genomics is well upon us. While mycorrhizal species such as *Tuber* may be fastidious to culture, various approaches have been made for obtaining sufficient quantities of nucleic acids for genome and transcriptome sequencing. These include growing cultures in dilute liquid media, across several agar plates, or by extracting nucleic acids directly from fresh young fruiting bodies. DNA from bacteria associated with truffle fruiting bodies can be problematic when obtaining genomic DNA from fruiting body materials; however, these bacteria are likely responsible for some of the organoleptic qualities of truffles and their development may be of interest to the study of the truffle microbiome. Several *Tuber* genomes are now sequenced including six species (*T. aestivum*, *T. borchii*, *T. indicum*, *T. lyonii*, *T. magnatum*, and *T. melanosporum*) and representing four of the 11 clades in *Tuber*. When fully assembled, these data should help resolve questions pertaining to truffle growth and development, ecological adaptability, origin, and evolution. Tools arising from these genomic projects should empower a new generation of truffle growers and researchers with the ability to simultaneously screen planted materials for contamination, diseases, genetic diversity, and origin of target species using high-throughput sequencing approaches. In addition, genomes of other Pezizomycetes are being sequenced and will offer new perspectives on the evolution of this particular class for which species with different life history strategies exist. We are convinced that this will lead to new insights and knowledge through the analysis of fruiting body development (hypogeous vs. epigeous), symbiosis-related genes, molecular crosstalk between fungus and host, the impact of repetitive elements on genome organization, and the consequence of intrahyphal and intraascocarp bacteria on fungal growth, function, and development.

ACKNOWLEDGEMENTS

We are grateful to Francis Martin and the Mycorrhizal Genomics Initiative consortium for access to unpublished genome data. Sequence data of *Choiromyces venosus* were produced by the U.S. Department of Energy Joint Genome Institute in collaboration with the user community. Sequence data of *Tuber aestivum* were produced by the Genoscope. Our research was financed by the "Institut National de la Recherche Agronomique" (INRA), the Région de Lorraine, the lab of excellence ARBRE (ANR-11-LABX-0002-01), and the ANR SYSTERRA SYSTRUF (ANR-09-STRA-10). The computations were performed at the INRA Nancy Ecogenomics facilities. Part of our research was sponsored by the Genomic Science Program of the U.S. Department of Energy, Office of Biological and Environmental Research (under contract DE-AC05-00OR22725).

REFERENCES

Alvarado, P., Moreno, G., & Manjón, J. L. (2012). Comparison between Tuber gennadii and *T. oligospermum* lineages reveals the existence of the new species *T. cistophilum* (Tuberaceae, Pezizales). *Mycologia, 104*(4), 894–910.

Amicucci, A., Balestrini, R., Kohler, A., Barbieri, E., Saltarelli, R., Faccio, A., et al. (2011). Hyphal and cytoskeleton polarization in Tuber melanosporum: A genomic and cellular analysis. *Fungal Genetics and Biology, 48*(6), 561–572.

Amicucci, A., Guidi, C., Zambonelli, A., Potenza, L., & Stocchi, V. (2000). Multiplex PCR for the identification of white Tuber species. *FEMS Microbiology Letters, 189*(2), 265–269.

Antony-Babu, S., Deveau, A., Van Nostrand, J. D., Zhou, J., Le Tacon, F., Robin, C., et al. (2014). Black truffle-associated bacterial communities during the development and maturation of Tuber melanosporum ascocarps and putative functional roles. *Environmental Microbiology*. http://dx.doi.org/10.1111/1462-2920.12294.

Belfiori, B., Riccioni, C., Paolocci, F., & Rubini, A. (2013). Mating type locus of Chinese black truffles reveals heterothallism and the presence of cryptic species within the *T. indicum* species complex. *PLoS One, 8*(12), e82353.

Bertault, G., Raymond, M., Berthomieu, A., Callot, G., & Fernandez, D. (1998). Trifling variation in truffles. *Nature, 394*(6695), 734.

Bolchi, A., Ruotolo, R., Marchini, G., Vurro, E., di Toppi, L. S., Kohler, A., et al. (2011). Genome-wide inventory of metal homeostasis-related gene products including a functional phytochelatin synthase in the hypogeous mycorrhizal fungus Tuber melanosporum. *Fungal Genetics and Biology, 48*(6), 573–584.

Bonito, G. (2009). Fast DNA-based identification of the black truffle Tuber melanosporum with direct PCR and species-specific primers. *FEMS Microbiology Letters, 301*(2), 171–175.

Bonito, G. M., Gryganskyi, A. P., Trappe, J. M., & Vilgalys, R. (2010). A global meta-analysis of Tuber ITS rDNA sequences: Species diversity, host associations and long-distance dispersal. *Molecular Ecology, 19*(22), 4994–5008.

Bonito, G., Smith, M. E., Nowak, M., Healy, R. A., Guevara, G., Cázares, E., et al. (2013). Historical biogeography and diversification of truffles in the tuberaceae and their newly identified southern hemisphere sister lineage. *PloS One, 8*(1), e52765.

Bonito, G., Trappe, J. M., Donovan, S., & Vilgalys, R. (2011). The Asian black truffle Tuber indicum can form ectomycorrhizas with North American host plants and complete its life cycle in non-native soils. *Fungal Ecology, 4*(1), 83–93.

Bonito, G., Trappe, J. M., Rawlinson, P., & Vilgalys, R. (2010). Improved resolution of major clades within Tuber and taxonomy of species within the Tuber gibbosum complex. *Mycologia, 102*(5), 1042–1057.

Božac, R., Širić, I., & Kos, I. (2012). Tuber decipiens: A new black Tuber species from Croatia. *Periodicum Biologorum, 114*(1), 125–128.

Büntgen, U., Egli, S., Camarero, J. J., Fischer, E. M., Stobbe, U., Kauserud, H., et al. (2012). Drought-induced decline in Mediterranean truffle harvest. *Nature Climate Change, 2*(12), 827–829.

Chevalier, G. (1998). The truffle cultivation in France: Assessment of the situation after 25 years of intensive use of mycorrhizal seedlings. In *Proceedings from the first international meeting on ecology, physiology and cultivation of edible mycorrhizal mushrooms* (pp. 3–4), Uppsala, Sweden.

Chevalier, G., & Grente, J. (1979). Application pratique de la symbiose ectomycorhizienne: Production à grande échelle de plants mycorhizés par la truffe (Tuber melanosporum Vitt.). *Mushroom Science, 10*, 483–505.

Cuomo, C. A., Güldener, U., Xu, J.-R., Trail, F., Turgeon, B. G., Di Pietro, A., et al. (2007). The Fusarium graminearum genome reveals a link between localized polymorphism and pathogen specialization. *Science, 317*(5843), 1400–1402.

Debuchy, R., Berteaux-Lecellier, V., & Silar, P. (2010). Mating systems and sexual morphogenesis in ascomycetes. In K. A. Borkovich & D. J. Ebbole (Eds.), *Cellular and molecular biology of filamentous fungi* (pp. 501–536). Washington: ASM Press. Retrieved from, http://hal.archives-ouvertes.fr/hal-00480603.

Deng, X. J., Liu, P. G., Liu, C. Y., & Wang, Y. (2013). A new white truffle species, Tuber panzhihuanense from China. *Mycological Progress, 12*, 557–561.

Douhan, G. W., Vincenot, L., Gryta, H., & Selosse, M.-A. (2011). Population genetics of ectomycorrhizal fungi: From current knowledge to emerging directions. *Fungal Biology, 115*(7), 569–597.

Ehrenberg, C. G. (1829). Syzygites, eine neue Schimmelgattung: Nebst Beobachtungen über sichtbare Bewegung in Schimmeln. *Verhandlungen der Gesellschaft naturforschender Freunde zu Berlin, 1*, 98–109.

Ellison, C. E., Hall, C., Kowbel, D., Welch, J., Brem, R. B., Glass, N. L., et al. (2011). Population genomics and local adaptation in wild isolates of a model microbial eukaryote. *Proceedings of the National Academy of Sciences of the United States of America, 108*(7), 2831–2836.

Fan, L., Cao, J.-Z., & Li, Y. (2012a). Tuber microsphaerosporum and *Paradoxa sinensis* spp. nov. *Mycotaxon, 120*(1), 471–475.

Fan, L., Cao, J.-Z., & Li, Y. (2012b). Tuber sinosphaerosporum sp. nov. from China. *Mycotaxon, 122*(1), 347–353.

Fan, L., Cao, J.-Z., & Yu, J. (2012). Tuber in China: *T. sinopuberulum* and *T. vesicoperidium* spp. nov. *Mycotaxon, 121*(1), 255–263.

Fan, L., Hou, C.-L., & Li, Y. (2012). Tuber microverrucosum and *T. huizeanum* two new species from China with reticulate ascospores. *Mycotaxon, 122*(1), 161–169.

Fassi, B., & Fontana, A. (1969). Sintesi micorrizica tra Pinus strobus e Tuber maculatum. II. Sviluppo dei semenzali trapianti e produzione di ascocarpi. *Allionia, 15*, 115–120.

Frank, B. (1889). Über die auf Wurzelsymbiose beruhende Ernährung gewisser Bäume durch unterirdische Pilze. *Berichte der Deutschen Botanischen Gesellschaft, 3*, 128–145.

Fraser, J. A., & Heitman, J. (2003). Fungal mating-type loci. *Current Biology, 13*(20), R792–R795.

Ganal, M. W., Altmann, T., & Röder, M. S. (2009). SNP identification in crop plants. *Current Opinion in Plant Biology, 12*(2), 211–217.

García-Cunchillos, I., Sánchez, S., Barriuso, J. J., & Pérez-Collazos, E. (2014). Population genetics of the westernmost distribution of the glaciations-surviving black truffle *Tuber melanosporum*. *Mycorrhiza, 24*(1), 89–100.

Giomaro, G., Sisti, D., Zambonelli, A., Amicucci, A., Cecchini, M., Comandini, O., et al. (2002). Comparative study and molecular characterization of ectomycorrhizas in Tilia americana and Quercus pubescens with Tuber brumale. *FEMS Microbiology Letters, 216*(1), 9–14.

Guerin-Laguette, A., Cummings, N., Hesom-Williams, N., Butler, R., & Wang, Y. (2013). Mycorrhiza analyses in New Zealand truffières reveal frequent but variable persistence of Tuber melanosporum in co-existence with other truffle species. *Mycorrhiza, 23*(2), 87–98.

Guevara, G., Bonito, G., Trappe, J. M., Cázares, E., Williams, G., Healy, R. A., et al. (2013). New North American truffles (*Tuber* spp.) and their ectomycorrhizal associations. *Mycologia, 105*(1), 194–209.

Hall, I. R., Brown, G. T., & Zambonelli, A. (2007). In *Taming the truffle: The history, lore, and science of the ultimate mushroom: Vol. 304.* Oregon: Timber Press Portland.

Iotti, M., Amicucci, A., Stocchi, V., & Zambonelli, A. (2002). Morphological and molecular characterization of mycelia of some Tuber species in pure culture. *New Phytologist, 155*(3), 499–505.

Jeandroz, S., Murat, C., Wang, Y., Bonfante, P., & Tacon, F. L. (2008). Molecular phylogeny and historical biogeography of the genus Tuber, the "true truffles" *Journal of Biogeography, 35*(5), 815–829.

Kohler, A., & Tisserant, E. (2014). Exploring the transcriptome of mycorrhizal interactions. In F. Martin (Ed.), Genomics of plant-related fungi. *Advances in Botanical Research, 70,* 23–78.

Kronstad, J. W. (2007). Self-fertility: The genetics of sex in lonely fungi. *Current Biology, 17*(19), R843–R845.

Kües, U., & Martin, F. (2011). On the road to understanding truffles in the underground. *Fungal Genetics and Biology, 48*(6), 555–560.

Le Tacon, F., Delmas, J., Gleyze, R., & Bouchard, D. (1982). Influence du regime hydrique du sol et de la fertilisation sur la fructification de la truffe noire du Perigord (Tuber melanosporum Vitt.) dans le Sud-Est de la France [Vaucluse; irrigation, paillage]. *Acta Oecologia Oecologia Applicata, 3,* 291–306.

Le Tacon, F., Zeller, B., Plain, C., Hossann, C., Bréchet, C., & Robin, C. (2013). Carbon transfer from the host to Tuber melanosporum mycorrhizas and ascocarps followed using a 13C pulse-labeling technique. *PLoS One, 8*(5), e64626.

Le Tacon, F., Marçais, B., Courvoisier, M., Murat, C., Montpied, P., & Becker, M. (2014). Climatic variations explain annual fluctuations in French Périgord black truffle wholesale markets but do not explain the decrease in black truffle production over the last 48 years. *Mycorrhiza, 24*(1), 115–125.

Linde, C. C., & Selmes, H. (2012). Genetic diversity and mating type distribution of Tuber melanosporum and their significance to truffle cultivation in artificially planted truffieres in Australia. *Applied and Environmental Microbiology, 78*(18), 6534–6539.

Marmeisse, R., Guidot, A., Gay, G., Lambilliotte, R., Sentenac, H., Combier, J.-P., et al. (2004). Hebeloma cylindrosporum—a model species to study ectomycorrhizal symbiosis from gene to ecosystem. *New Phytologist, 163*(3), 481–498.

Martin, F., Aerts, A., Ahrén, D., Brun, A., Danchin, E. G. J., Duchaussoy, F., et al. (2008). The genome of Laccaria bicolor provides insights into mycorrhizal symbiosis. *Nature, 452*(7183), 88–92.

Martin, F., & Bonito, G. M. (2013). Ten years of genomics for ectomycorrhizal fungi: What have we achieved and where are we heading? In A. Zambonelli & B. Bonito (Eds.), *Soil Biology Series. Edible ectomycorrhizal mushrooms.* Berlin: Springer-Verlag.

Martin, F., Cullen, D., Hibbett, D., Pisabarro, A., Spatafora, J. W., Baker, S. E., et al. (2011). Sequencing the fungal tree of life. *New Phytologist, 190,* 818–821.

Martin, F., Kohler, A., Murat, C., Balestrini, R., Coutinho, P. M., Jaillon, O., et al. (2010). Périgord black truffle genome uncovers evolutionary origins and mechanisms of symbiosis. *Nature, 464*(7291), 1033–1038.

Metzenberg, R. L., & Glass, N. L. (1990). Mating type and mating strategies in Neurospora. *Bioessays, 12*(2), 53–59.

Moffatt, M. F., Gut, I. G., Demenais, F., Strachan, D. P., Bouzigon, E., Heath, S., et al. (2010). A large-scale, consortium-based genomewide association study of asthma. *New England Journal of Medicine*, *363*(13), 1211–1221.

Montanini, B., Levati, E., Bolchi, A., Kohler, A., Morin, E., Tisserant, E., et al. (2011). Genome-wide search and functional identification of transcription factors in the mycorrhizal fungus Tuber melanosporum. *New Phytologist*, *189*(3), 736–750.

Murat, C., Riccioni, C., Belfiori, B., Cichocki, N., Labbé, J., Morin, E., et al. (2011). Distribution and localization of microsatellites in the Perigord black truffle genome and identification of new molecular markers. *Fungal Genetics and Biology*, *48*(6), 592–601.

Murat, C., Rubini, A., Riccioni, C., De la Varga, H., Akroume, E., Belfiori, B., et al. (2013). Fine-scale spatial genetic structure of the black truffle (Tuber melanosporum) investigated with neutral microsatellites and functional mating type genes. *New Phytologist*, *199*(1), 176–187.

Murat, C., Zampieri, E., Vizzini, A., & Bonfante, P. (2008). Is the Perigord black truffle threatened by an invasive species? We dreaded it and it has happened! *New Phytologist*, *178*(4), 699–702.

Murtagh, G. J., Dyer, P. S., & Crittenden, P. D. (2000). Reproductive systems: Sex and the single lichen. *Nature*, *404*(6778), 564.

Neafsey, D. E., Barker, B. M., Sharpton, T. J., Stajich, J. E., Park, D. J., Whiston, E., et al. (2010). Population genomic sequencing of Coccidioides fungi reveals recent hybridization and transposon control. *Genome Research*, *20*(7), 938–946.

Paoletti, M., Seymour, F. A., Alcocer, M. J. C., Kaur, N., Calvo, A. M., Archer, D. B., et al. (2007). Mating type and the genetic basis of self-fertility in the model fungus Aspergillus nidulans. *Current Biology*, *17*(16), 1384–1389.

Paolocci, F., Rubini, A., Granetti, B., & Arcioni, S. (1999). Rapid molecular approach for a reliable identification of *Tuber* spp. ectomycorrhizae. *FEMS Microbiology Ecology*, *28*(1), 23–30.

Paolocci, F., Rubini, A., Riccioni, C., & Arcioni, S. (2006). Reevaluation of the life cycle of Tuber magnatum. *Applied and Environmental Microbiology*, *72*(4), 2390–2393.

Percudani, R., Trevisi, A., Zambonelli, A., & Ottonello, S. (1999). Molecular phylogeny of truffles (Pezizales: Terfeziaceae, Tuberaceae) derived from nuclear rDNA sequence analysis. *Molecular Phylogenetics and Evolution*, *13*(1), 169–180.

Poland, J. A., Bradbury, P. J., Buckler, E. S., & Nelson, R. J. (2011). Genome-wide nested association mapping of quantitative resistance to northern leaf blight in maize. *Proceedings of the National Academy of Sciences of the United States of America*, *108*(17), 6893–6898.

Rafalski, A. (2002). Applications of single nucleotide polymorphisms in crop genetics. *Current Opinion in Plant Biology*, *5*(2), 94–100.

Riccioni, C., Belfiori, B., Rubini, A., Passeri, V., Arcioni, S., & Paolocci, F. (2008). Tuber melanosporum outcrosses: Analysis of the genetic diversity within and among its natural populations under this new scenario. *New Phytologist*, *180*(2), 466–478.

Rubini, A., Belfiori, B., Riccioni, C., Arcioni, S., Martin, F., & Paolocci, F. (2011). *Tuber melanosporum*: Mating type distribution in a natural plantation and dynamics of strains of different mating types on the roots of nursery-inoculated host plants. *New Phytologist*, *189*(3), 723–735.

Rubini, A., Belfiori, B., Riccioni, C., Tisserant, E., Arcioni, S., Martin, F., et al. (2011). Isolation and characterization of MAT genes in the symbiotic ascomycete Tuber melanosporum. *New Phytologist*, *189*(3), 710–722.

Sheffield, J., Wood, E. F., & Roderick, M. L. (2012). Little change in global drought over the past 60 years. *Nature*, *491*(7424), 435–438.

Smith, S. E., & Read, D. J. (2008). *Mycorrhizal symbiosis* (3rd ed.). San Diego, CA: Academic Press.

Soderlund, C., Bomhoff, M., & Nelson, W. M. (2011). SyMAP v3.4: A turnkey synteny system with application to plant genomes. *Nucleic Acids Research*, *39*(10), e68. http://dx.doi.org/10.1093/nar/gkr123.

Spanu, P. D., Abbott, J. C., Amselem, J., Burgis, T. A., Soanes, D. M., Stüber, K., et al. (2010). Genome expansion and gene loss in powdery mildew fungi reveal tradeoffs in extreme parasitism. *Science*, *330*(6010), 1543–1546.

Splivallo, R., Ottonello, S., Mello, A., & Karlovsky, P. (2011). Truffle volatiles: From chemical ecology to aroma biosynthesis. *New Phytologist*, *189*(3), 688–699.

Stukenbrock, E. H. (2014). Speciation genomics of fungal plant pathogens. In F. Martin (Ed.), Genomics of plant-related fungi. *Advances in Botanical Research, 70*, 397–423

Traeger, S., Altegoer, F., Freitag, M., Gabaldon, T., Kempken, F., et al. (2013). The genome and development-dependent transcriptomes of *Pyronema confluens*: A window into fungal evolution. *PLoS Genet*, *9*(9), e1003820. http://dx.doi.org/10.1371/journal.pgen.1003820.

Tisserant, E., Malbreil, M., Kuo, A., Kohler, A., Symeonidi, A., Balestrini, R., et al. (2013). Genome of an arbuscular mycorrhizal fungus provides insight into the oldest plant symbiosis. *Proceedings of the National Academy of Sciences of the United States of America.* http://dx.doi.org/10.1073/pnas.1313452110.

Wedén, C., Larsson, S., Burman, R., & Backlund, A. (2009). The edible truffle Choiromyces venosus and its use in Sweden. *Acta Botanica Yunnanica*, *31*(S16), 94–96.

Yamauchi, T., Hara, K., Maeda, S., Yasuda, K., Takahashi, A., Horikoshi, M., et al. (2010). A genome-wide association study in the Japanese population identifies susceptibility loci for type 2 diabetes at UBE2E2 and C2CD4A-C2CD4B. *Nature Genetics*, *42*(10), 864–868.

Zampieri, E., Balestrini, R., Kohler, A., Abbà, S., Martin, F., & Bonfante, P. (2011). The Perigord black truffle responds to cold temperature with an extensive reprogramming of its transcriptional activity. *Fungal Genetics and Biology*, *48*(6), 585–591.

Zampieri, E., Mello, A., Bonfante, P., & Murat, C. (2009). PCR primers specific for the genus Tuber reveal the presence of several truffle species in a truffle-ground. *FEMS Microbiology Letters*, *297*(1), 67–72.

CHAPTER EIGHT

The Natural Histories of Species and Their Genomes: Asymbiotic and Ectomycorrhizal *Amanita* Fungi

Jaqueline Hess[*,1], **Anne Pringle**[†]
[*]Section for Genetics and Evolutionary Biology, University of Oslo, Oslo, Norway
[†]Organismic and Evolutionary Biology, Harvard University, Cambridge, Massachusetts, USA
[1]Corresponding author: e-mail address: jaqueline.hess@ibv.uio.no

Contents

Abstract

Genome architectures are likely shaped by species' ecologies, but genomes are rarely discussed in ecological contexts. A major force in evolution is symbiosis, and a symbiotic niche may shape a genome's size, gene order and gene content. The mycorrhizal symbiosis is ubiquitous and critical to the function of diverse ecosystems. Although there are tens of thousands of mycorrhizal fungi, there are no general principles defining the genetic architectures of these fungi. General principles may not exist, perhaps because of the multiple, independent origins of the symbiosis. But research with pathogenic

Advances in Botanical Research, Volume 70
ISSN 0065-2296
http://dx.doi.org/10.1016/B978-0-12-397940-7.00008-2

fungi suggests general principles at work in the evolution of pathogen genomes, and to enable a more holistic understanding of the forces shaping genomes of mutualists, we focus on the genus *Amanita* and the role of ecology in genome evolution. *Amanita* is an emerging model for the ecology and evolution of symbiosis, and to date, our laboratory has sequenced the genomes of six species with diverse niches. We describe the natural histories of these species and current research on genomics. We offer novel analyses targeting two questions: did the evolution of the ectomycorrhizal symbiosis facilitate an adaptive radiation of symbiotic *Amanita* and how are the genomes of asymbiotic fungi different from the derived genomes of ectomycorrhizal fungi? We also discuss the role transposable elements may have had in generating genomic variation and a potential link between transposable element proliferation and patterns of speciation. Our descriptions of the genus identify as yet unexplored questions connecting genomics to the ecology of species' ranges and range expansions.

1. INTRODUCTION

How does symbiosis shape genetic architectures? In this chapter, we consider symbiosis as a close physical association between two individuals of different species and mutualism as any interaction benefitting both individuals. Symbioses may also be parasitisms and mutualisms may or may not be symbioses.

Mutualisms are often asymbiotic, for example, plants and pollinators or insects associated in Müllerian mimicry. Asymbiotic mutualisms are rarely considered as an explicit force shaping the genetic architectures of mutualists. However, floral morphology is critical to pollination and clearly evolves in response to pollinator choice (Venail, Dell'Olivo, & Kuhlemeier, 2010), and Müllerian mimics evolve to look like each other. It seems likely asymbiotic mutualisms will influence the spectrum of genes and their arrangement in a genome (Heliconius Genome Consortium, 2012). Fungi seem more often involved in symbiotic mutualisms, and we will not consider asymbiotic associations further.

Research on the genetic architectures of symbionts has focused on bacteria (McCutcheon & Moran, 2012; Moran, 1996; Moran, McLaughlin, & Sorek, 2009; Moran & Wernegreen, 2000), but how these data translate to fungi is unclear. Bacteria enabled paradigms defining the evolution of endosymbiosis (Martin & Schnarrenberger, 1997; Tamas et al., 2002), for example, endosymbiotic bacteria lose many of the genes found in closely related asymbiotic lineages and have reduced genome sizes (Mira, Ochman, & Moran, 2001; van Ham et al., 2003). However, fungal symbioses often involve individuals that live both inside and outside of organisms. In mycorrhizal symbioses, some parts of the fungus are physically associated with a plant, while others are in soil and exposed. The evolutionary trajectory of an organism that maintains itself outside of a host may be very different from the trajectory of an endosymbiont (Law & Lewis, 1983; Zoller & Lutzoni, 2003), and in contrast to an endosymbiotic bacterium, an ectosymbiotic fungus may maintain the genes necessary for independent growth. The dynamics of genome size evolution may also be quite different, because mycorrhizal symbioses are not vertically transmitted (Smith & Read, 2010).

The ectomycorrhizal (ECM) symbiosis has multiple, independent origins (James et al., 2006), and because the ancestor of these symbionts was a saprotroph, extant ECM species represent different gains of the symbiotic niche (James et al., 2006; Matheny et al., 2006). Comparative genomics of published ECM genomes identifies coarse genomic differences between saprotrophic and biotrophic fungi (Martin et al., 2008, 2010; Martin & Selosse, 2008; Nagendran, Hallen-Adams, Paper, Aslam, & Walton, 2009; Plett & Martin, 2011), and intriguing patterns of evolution are suggested from the two genomes published to date: *Laccaria bicolor* and *Tuber melanosporum*. Both genomes are characterized by elevated transposable element (TE) content (Martin et al., 2008, 2010). The basidiomycete *L. bicolor* has a total genome size of 61 MB and an estimated TE content around 20–25%, while the ascomycete *T. melanosporum* genome is of 125 MB and reaches to as much as 60% TEs. However, the gene content in these two species is radically different. While *L. bicolor* encodes around 23,000 genes, many more than most free-living basidiomycetes, *T. melanosporum* only harbours about 7500 genes, which only barely exceeds the numbers found in a larger yeast genome (Jones et al., 2004). Only 19% of *T. melanosporum* genes are part of larger gene families, compared to 55% of the genes found in *L. bicolor* (Martin et al., 2010), revealing very different genome architectures that both enable the same kind of symbiosis. But there are also commonalities, including the loss of plant cell wall degrading enzymes (PCWDEs) and expansions in gene families involved in signalling (Veneault-Fourrey & Martin, 2011).

Parasitisms also shape the genetic architectures of fungi; although in ecological contexts disease may seem very different from mutualism, the interactions may share genetic pathways, for example, oomycete pathogens and arbuscular mycorrhizal fungi use a signal expressed from the same plant gene to colonize plants (Wang et al., 2012). Parasitisms and mutualisms may impose common selective forces on genomes, perhaps including changes in genome size, but even when selective forces diverge the mechanisms enabling change may be similar, for example, changes in gene family copy number or the proportions of TEs.

Like *L. bicolor* and *T. melanosporum*, many plant pathogens maintain expanded genomes densely populated by TEs (Grandaubert, Balesdent, & Rouxel, 2014). Plant pathogens show differing patterns of expansions and contractions in the gene families involved in pathogenicity, including effectors and PCWDEs (Raffaele & Kamoun, 2012). Although the direct impact of TEs on these genomes is only discussed in a few cases, their effects are striking and include accelerated evolutionary rates of effectors caused by repeat-induced point mutation of nearby TEs (Grandaubert et al., 2014; Rouxel et al., 2011) and a fusion of an effector family with a TE resulting in joint proliferation (Sacristán et al., 2009). Moreover, simulations suggest rearrangements mediated by TEs may contribute towards the compartmentalization of genomes into slower- and faster-evolving regions and so aid the generation of genomic plasticity underpinning adaptation to new environments (Crombach & Hogeweg, 2007).

TEs appear as a common theme in research on the architectures of fungal genomes, but the discovery of TEs and their evolutionary potential is not restricted to the fungi; the impact of TEs on genomes is widespread and their significance as a mechanism generating heritable variation is widely appreciated (Kidwell & Lisch, 2001; Levin & Moran, 2011; Raffaele & Kamoun, 2012; Werren, 2011). TEs can generate large amounts of genetic diversity, for example, by facilitating chromosomal rearrangements, and besides their immediate consequences on gene content and gene order, the rearrangements mediated by TEs may also play a role in reproductive isolation, accelerating the process of speciation (Böhne, Brunet, Galiana-Arnoux, Schultheis, & Volff, 2008; Oliver & Greene, 2009, 2011, 2012; Oliver, McComb, & Greene, 2013; Zeh, Zeh, & Ishida, 2009).

To begin dissecting the variety of mechanisms shaping the genomes of ECM fungi, and maybe identify causal changes, comparisons of more closely related species are necessary. There are great evolutionary distances among species with sequenced genomes, and identifying the causes of genomic differences is difficult: genomic differences may be correlated with differences in ecological niche but may be caused by the unique evolutionary trajectories taken by distantly related species.

The Amanita *are a novel model for understanding the changes in genetic architecture associated with an evolution of a mutualistic symbiosis*: In this genus, the evolution of the ECM symbiosis occurred once, and saprotrophic *Amanita* form a strongly supported clade basal to a monophyletic clade of ECM species (Wolfe, Tulloss, & Pringle, 2012; Fig. 8.1). *Amanita* houses over 500 described species, and the majority are ECM. Approximately 30 species of *Amanita* appear to be asymbiotic. But the full diversity of asymbiotic *Amanita* may be unknown; at least a handful of species remain undescribed (e.g. "sp-C13", http://www.amanitaceae.org/?Amanita+sp-C13), while others are known only from a single collection. Asymbiotic *Amanita* are often found in Africa, Asia or South America, at sites far away from traditional centres of mycology.

The single origin of symbiosis and its species richness render the genus *Amanita* a great model system to investigate the genomic changes around the evolution of symbiosis and the subsequent evolutionary trajectories of individual ECM species at a finer scale. We have sequenced the genomes of five *Amanita* and an out-group species (Fig. 8.1 and Table 8.1). We chose to sequence three symbiotic *Amanita* (*A. brunnescens*, *A. polypyramis* and *A. muscaria* var. *guessowii*, sampling from each of the major ECM clades) and two asymbiotic *Amanita* (the closely related *A. thiersii* and *A. inopinata*). The saprotrophic fungus *Volvariella volvacea* was sequenced as an out-group.

To facilitate thinking about the *Amanita* as a model, we briefly describe the natural histories of sequenced species and then describe current analyses of genomes, focusing on TEs. Species descriptions focus on ecology; useful information about morphology and taxonomy are provided by both mushroomexpert.com and amanitaceae.org. At least three salient questions emerge from the descriptions of species and their genomes, and we concentrate on two: does symbiosis influence the pace of speciation in ECM *Amanita*, and are TEs causing apparent changes in synteny among the different species? We relate the second question to patterns of speciation. We conclude by briefly discussing the third, unanswered question of whether or how genome evolution may enable range expansions.

2. THE FUNGI AND THEIR GENOMES

2.1. The out-group *Volvariella volvacea*, an edible mushroom and decomposer of agricultural waste

An extensive literature on *V. volvacea* focuses on its use as a crop (Bao et al., 2013; Chang, 1977; Date & Mizuno, 1997). The fungus is cultivated throughout Asia, and especially in China, where it has been grown since the eighteenth century. Protocols to grow the mushroom were developed at least in part by Buddhist monks (Chang, 1977). The mushrooms are

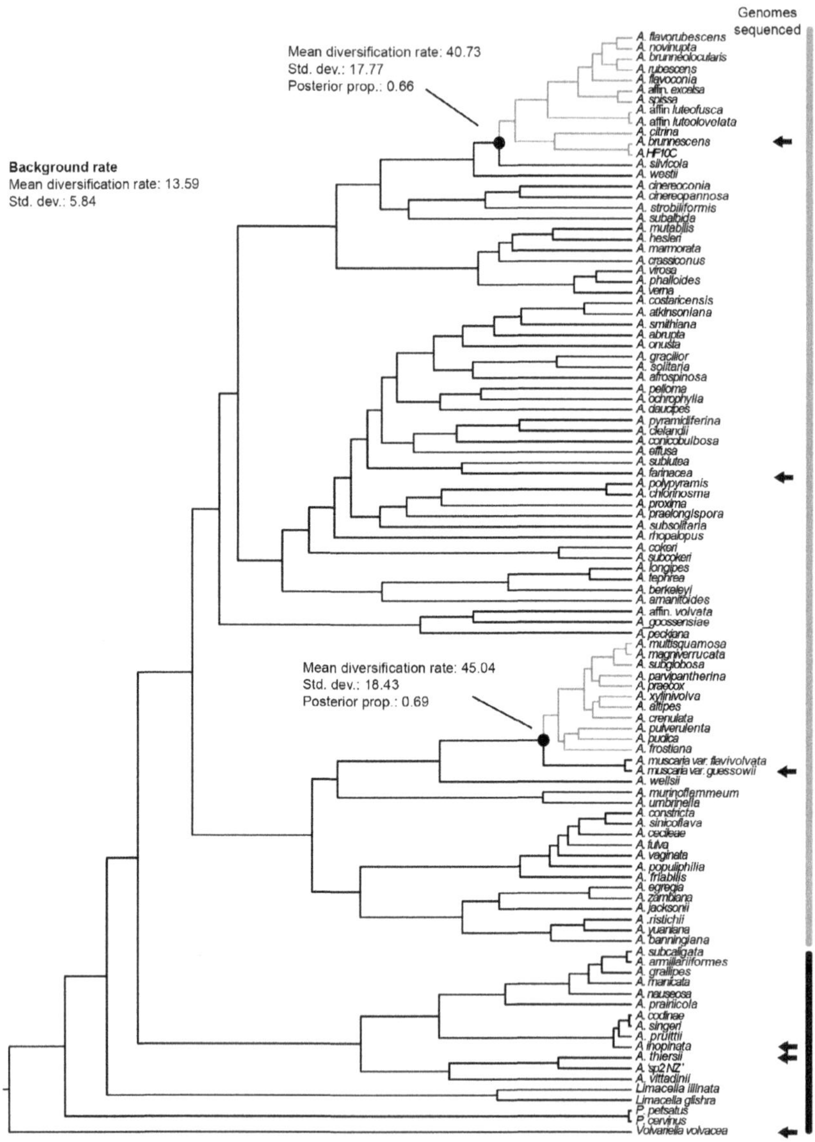

Figure 8.1 Phylogeny of the genus *Amanita*, based on the analysis of Wolfe, Tulloss, et al. (2012). Branch lengths were recalculated using the nucLSU gene and a relaxed molecular clock model in BEAST (Drummond & Rambaut, 2007). Rates of diversification were estimated with MEDUSA (Harmon, Weir, Brock, Glor, & Challenger, 2008) on a sample of 1000 trees from the BEAST posterior distribution, and inferred shifts recovered in >50% of trees are highlighted. Arrows mark the species for which genomic data are available.

Table 8.1 Basic statistics of the assemblies of sequenced *Amanita* genomes

Species	Sequencing centre	Assembly size (bp)	N50 (kB)	CEGMA genes (%)	CEGMA redundancy	TE (assembled) (%)	TE (coverage corrected) (%)
A. brunnescens	Harvard	57,556,770	11	94.35	1.81	17.9	36.4
A. polypyramis	Harvard	23,557,560	54	95.56	1.28	11.6	59.6
A. muscaria guess.	JGI	40,699,759	17	92.34	1.10	8.9	21.6
A. thiersii	JGI	33,689,220	77	95.97	1.11	26.4	36.6
A. inopinata	Harvard	22,122,871	156	95.97	1.11	4.8	8.9
V. volvacea	Harvard	52,426,718	55	95.56	1.57	4.6	5.2

Percentages of CEGMA (Parra, Bradnam, & Korf, 2007) genes recovered in each assembly were used as estimates of gene space completeness. CEGMA redundancy is the average copy number of single copy CEGMA genes detected in each genome. Assembly size is given as an approximation to genome size but may be a poor estimator. CEGMA redundancy and the proportion of unassembled TEs (as reflected by the difference between corrected and assembled TE content; Hess et al., in review) may give an indication of how assembly size related to true genome size. Higher redundancy values mean the true genome size is smaller than assembly size, while a high unassembled portion of TEs means that true genome size is larger than assembly size.

considered a health food. The species is relatively inefficient at converting substrates to mushrooms, with yields described as between 10% and 13% when the fungus is grown on rice "straw" (plant stalks leftover when rice is harvested) and 30–40% when grown on cotton wastes, for example, old clothes (Date & Mizuno, 1997). The fungus is tropical and requires temperatures greater than 25 °C to fruit but is introduced to North America where it can be found in woodchips, compost piles, greenhouses and gardens (Kuo, 2011). The mycelia of *V. volvacea* have no clamp connections. The species appears to be homothallic and capable of mating with itself (Bao et al., 2013).

2.2. *Amanita thiersii*, a fungus of lawns undergoing a range expansion

An extensive account of the natural history of *A. thiersii* is provided by Wolfe, Kuo, and Pringle (2012). The native range of the fungus may or may not include North America; although it was originally described from Texas in 1952, over the last decades, *A. thiersii* has moved from North to southern Illinois (Kuo, 2013a), and the fungus may be an invasive species originally introduced to Texas from an as yet unidentified home range. Alternatively, the fungus may be native and moving in response to environmental change. Recently, the fungus was found near Baltimore, MD, on the East Coast of the United States (www.mushroomobserver.org; Tulloss personal communication). Mushrooms are generally found in lawns, where the species decomposes grass litter. Although *A. thiersii* is not mycorrhizal, it stimulates plant growth (Wolfe, Kuo, et al., 2012), perhaps because decomposition releases limiting nutrients to soil. Nothing is known about the mating system of the fungus, but the genetic diversity of populations across North America is low.

2.3. *Amanita inopinata*, an *Amanita* known only from introduced ranges

The "unexpected" *Amanita* is an enigmatic fungus originally described from scattered localities in the southeast corner of England (Reid, 1987). After the description was published, a New Zealand mycologist recognized it as a rarely collected species found in both the North and South Islands (Ridley, 2000). The fungus is considered an introduction to both England and New Zealand and has also appeared in the Netherlands (Bas, 2001). Perhaps because the fungus is an *Amanita* and *Amanita* is typically an ECM genus, careful notes of the trees around collections are available (FRDBI, 2013; Ridley, 2000). Many are not hosts of ECM fungi, for example,

Chamaecyparis lawsoniana and *Taxus baccata*, and the fungus is currently assumed to be asymbiotic (Kibby, 2005; Wolfe, Tulloss, et al., 2012). However, molecular probes testing for a specific cellulose decomposition pathway found no evidence of these genes in *A. inopinata*, and its ecological niche remains unknown (Wolfe, Tulloss, et al., 2012).

2.4. *Amanita muscaria*, a species complex of ECM fungi with different ecologies

A. muscaria is the charismatic, widely recognized red-and-white-spotted mushroom of fairy tales and video games. However, it is very clearly a species complex of cryptic genetic species (Geml, Laursen, O'Neill, Nusbaum, & Taylor, 2006; Geml, Tulloss, Laursen, Sazanova, & Taylor, 2008), and these species look different, have different ranges and associate with different hosts. For example, the European mushroom (which keeps the name *A. muscaria*) is red with white spots and is often associated with oak, while the eastern North American mushroom (*A. muscaria* var. *guessowii*) is yellow with white spots and is primarily associated with conifers. A more southern North American mushroom (*A. muscaria* var. *persicina*) has a peach coloured cap and associates with both oak and pine. The genome sequenced to date is an isolate of *A. muscaria* var. *guessowii* collected in Pennsylvania; however, as additional genomes are sequenced, the complex will provide an opportunity to compare the genomes of very closely related genetic species with different habitats. For simplicity sake, we discuss the sequenced genome as "*A. muscaria*" and not "*A. muscaria* var. *guessowii*".

2.5. *Amanita polypyramis*, an ECM fungus

Relatively little is known about the natural history of *A. polypyramis*. The species is found in the United States from New Jersey south to Texas and Florida and in Mexico and Central America (Kuo, 2013b), including in the Guanacaste Conservation Area of Costa Rica (Tulloss, 2013). In contrast to *A. muscaria* but like *V. volvacea*, *A. polypyramis* grows in the tropics. The fungus associates with oaks and perhaps pines as well. Mushrooms are very large, with caps reaching to 20 cm across.

2.6. *Amanita brunnescens*, another ECM fungus about which relatively little is known

The species is found in eastern North America and associates with various hardwoods and conifers. The mushrooms are very common.

2.7. Genomics to date, and comparisons to *L. bicolor* and *T. melanosporum*

In the *Amanita*, decomposition pathways are lost by ECM species (Nagendran et al., 2009; Wolfe, Tulloss, et al., 2012). Preliminary analyses of the *A. muscaria* and *A. thiersii* genomes show a large reduction in many carbohydrate active enzyme (CAZyme) families in *A. muscaria*. The genome of *A. muscaria* encodes 279 CAZymes, while *A. thiersii* encodes 370. Losses are generally concentrated in the families involved in the degradation of plant cell wall material (Chaib de Mares, 2013). This pattern seems to be a common feature of crown group ECM species and a basic strategy used by biotrophic fungi to escape detection by the plant immune system (Veneault-Fourrey & Martin, 2011; MGI http://mycor.nancy.inra.fr/IMGC/MycoGenomes/).

Like *L. bicolor*, *A. muscaria* has an amplified genome encoding 18,153 genes, almost twice as many genes as *A. thiersii*, which houses 10,354 (http://genome.jgi-psf.org/Amamu1/Amamu1.info.html; http://genome.jgi.doe.gov/Amath1/Amath1.info.html). The types of amplified gene families show close similarities to gene families amplified in *L. bicolor*. Among the five largest gene clusters in *A. muscaria*, two contain protein–protein interaction domains (e.g. NACHT and WD40), while another two appear to be tyrosine kinases (Martin et al., 2008; http://genome.jgi-psf.org/clustering/pages/cluster/clusters.jsf?runId=1898&organism=Amamu1). Analysis of the secretome reveals an overall decrease in numbers of secreted proteins in *A. muscaria* compared to *A. thiersii*, but with a clear shift towards a higher percentage of small secreted proteins in *A. muscaria* (Chaib de Mares, 2013).

TE distributions across the six sequenced genomes show no simple pattern with respect to ecological niche, although we find evidence for changes in TE dynamics following the evolution of the ECM symbiosis (Hess et al., in review). Abundant numbers of TEs are found in two of the three ECM species (36% genomic content in *A. brunnescens* and 59% in *A. polypyramis*), as well as the asymbiotic species *A. thiersii* (37%). The third mycorrhizal species, *A. muscaria*, houses a moderate proportion of TEs (21%), while both *A. inopinata* and *V. volvacea* possess few TEs (less than 10% in both cases). TE repertoires across the *Amanita* are dominated by RNA-based elements from the Gypsy, Copia and LINE superfamilies, and together, these make up over 80% of TE diversity (Hess et al., in review). Phylogenetic analysis of these three most abundant retrotransposon superfamilies mirrors the patterns found in assemblies and reveals large numbers of recently diverged elements in the three TE-rich species. While *A. thiersii* houses amplifications of all three superfamilies, amplifications in *A. brunnescens* and *A. polypyramis* are concentrated in the LINE and Gypsy superfamilies but are especially prominent among LINE elements where 84% of TEs are from either *A. brunnescens* or *A. polypyramis*. Although *A. muscaria* amplifications are smaller than those found in *A. brunnescens*, *A. polypyramis* or *A. thiersii*, they outnumber those found in *A. inopinata* and *V. volvacea*.

Despite the lack of a simple pattern of elevated TE content in ECM genomes as compared to asymbiotic genomes, the presence of TE amplifications among ECM lineages suggests the evolution of the ECM lineages was accompanied by a period of either increased rates of TE proliferation or lower rates of TE removal, arguably with the same potential for TE-mediated chromosomal rearrangements, duplications and deletions. The patterns of TE content evolution in *A. thiersii* appear different to those in the ECM species. Individual families are amplified among the different ECM species, but all three types of retrotransposons are amplified within *A. thiersii*, suggesting that different mechanisms are at work to elevate TE content.

3. ECOLOGICAL GENOMICS OF ASYMBIOTIC AND ECM *AMANITA* SPECIES

3.1. Does symbiosis influence the pace of speciation in ECM *Amanita*?

Symbioses may facilitate evolutionary radiations by enabling new ecological opportunities, and the literature offers many examples from different domains and kingdoms. Often, mutualisms are also correlated with increased rates of speciation, for example, pollinators appear to drive plant diversification (Johnson, 2010), a mutualism with anemones may have triggered the adaptive radiation of clownfishes (Litsios et al., 2012), and microbes may mediate the diversification of phytophagous insects (Janson, Stireman, Singer, & Abbot, 2008).

However, modern theory offers conflicting opinions on the role of mutualism in speciation. Early theory suggested stochasticity in geographically isolated populations of mutualists would spur diversification, as forces like genetic drift caused isolated populations to diverge (Kiester, Lande, & Schemske, 1984). The geographic mosaic theory of coevolution (Thompson, 1999) also suggests that mutualisms can promote speciation when interacting species are divided into metapopulations. In contrast, a more recent model of two coevolving species suggests that mutualisms can slow diversification when phenotypes "match", for example, when an aspect of a pollinator's shape matches the shape of a pollinated flower (Yoder & Nuismer, 2010). While plants and ECM fungi often grow as metapopulations, it is not clear what phenotype matching would mean in these symbioses.

It seems logical to ask whether the transition to symbiosis enabled an adaptive radiation of ECM *Amanita* species. Ryberg and Matheny (2012) compared diversification rates and times among genera of ECM in the Agaricales and found constant but different rates of diversification across all clades: the *Amanita* possessed intermediate rates of diversification. However, their analysis did not include asymbiotic species of the genus and does not explicitly test the hypothesis of differences in the rates of diversification between asymbiotic and symbiotic *Amanita*.

To test for changes in rates of diversification after the evolution of the ECM niche in *Amanita*, we used phylogenetic methods to infer a linearized phylogenetic tree and models of rate diversification. We collected nuclear large subunit ribosomal DNA (nucLSU) sequences from the 108 species analysed in Wolfe, Tulloss, et al. (2012), realigned them using PRANK (Löytynoja & Goldman, 2008) and manually removed regions of uncertainty, resulting in a 1598 bp alignment. We then ran BEAST v.1.75 (Drummond & Rambaut, 2007) under the GTR + Γ model with four discrete rate categories and a lognormal uncorrelated relaxed molecular clock to estimate rates of evolution for the nucLSU alignment on the fixed species tree from Wolfe, Tulloss, et al. (2012). The results of three parallel runs were combined after discarding 10% of the estimates as burn-in. A random sample of 1000 trees from the combined posterior set of trees served as the input data for MEDUSA (Harmon et al., 2008). MEDUSA fits a series of birth–death processes, modelling species diversification by using an increasing number of distinct rate partitions until the model improvement becomes insignificant. A summary of the MEDUSA results is shown in Fig. 8.1.

We recover two rate shifts that are well represented among the posterior trees; both are within ECM clades, one in subgenus Lepidella and the other in subgenus Amanita. In both cases, the estimated mean diversification rates are three times as fast as the background rate, although the estimates also have large standard deviations. Because we did not calibrate the molecular clock analysis, the units of the diversification rates are arbitrary, but rates can be compared to each other. The respective rate shifts are found in 66% and 69% of the posterior sample.

The interpretation of our results is complicated by the large confidence intervals on estimated node heights in the backbone of the tree. Large confidence intervals will affect diversification rate estimates in deeper clades and our ability to test for a shift in diversification rate directly after the evolution of the ECM niche. Increasing the size of our dataset (adding additional loci) might narrow confidence intervals by improving branch length estimates, although rates on long branches are generally difficult to estimate (Schwartz & Mueller, 2010).

The accurate estimation of diversification rates also depends on taxon sampling; a general assumption of most methods developed for this purpose is a completely sampled phylogeny (e.g. Heath, Hedtke, & Hillis, 2008; Pybus & Harvey, 2000). If species sampling is even and so, by corollary, all deep lineages have been sampled, any new species added will likely be near the tips of the tree. The underrepresentation of nodes with shallow branch lengths in "evenly incompletely" sampled trees may lead to underestimation of diversification rates near the tips of the tree (Heath et al., 2008; Pybus & Harvey, 2000). Biases in unevenly sampled phylogenies are obviously more severe and would incorrectly increase or decrease rates in subsets of the tree.

Given the relative overrepresentation of asymbiotic *Amanita* in our phylogeny (slightly less than half of the described species, compared to about a fifth of the ECM species), we may be underestimating relative rates of diversification for the ECM species. We are also likely to be underestimating diversification rates near the tips of the tree, due to the relatively sparse sample of our dataset. A more comprehensive analysis including additional species and a better understanding of biogeography and natural history of the genus will be key to detangling the biases that may complicate our analysis. Nevertheless, there is at least some support for an increase in diversification rates following the evolution of the ECM symbiosis, and the large number of extant ECM *Amanita* proves that these species have been very successful.

The genus *Amanita* is not the only ECM clade proposed to have undergone an adaptive radiation, and multiple other clades also contain specious ECM lineages (Ryberg & Matheny, 2012; Smith & Read, 2010). Two competing hypotheses are proposed to explain potential radiations: the "dual origins" hypothesis, in which ECM species are proposed to have radiated at the same time as the diversifications of major ECM plant hosts (with the Pinaceae during the Jurassic and the Angiosperms during the Cretaceous (Halling, 2001)), and the "convergent radiation" hypothesis (Bruns, Szaro, Gardes, & Cullings, 1998), in which ECM lineages are proposed to have radiated more recently, as ECM plant communities expanded ranges into cooling climates. Ryberg and Matheny's (2012) study offers support for the second hypothesis but with the caveat that radiations may not be specific to ECM clades and may therefore be unrelated to ECM niche (Bruns et al.,

1998; Ryberg & Matheny, 2012). Our results suggest that ECM clades of *Amanita* have higher rates of diversification than asymbiotic clades of *Amanita*, and give some weight to the idea that radiations of ECM lineages were indeed adaptive and facilitated by symbiosis.

3.2. Does symbiosis reshape the ECM genome?

A variety of mutational mechanisms shape genes and genomes. In addition to single nucleotide substitutions and short insertions or deletions in protein-coding genes and regulatory regions, larger-scale events may involve duplications, losses or rearrangements. The size of the genomic regions involved in these events will vary widely and may encompass anything from segments of a single gene up to an entire genome but in all cases will provide evolutionary novelty (Kondrashov, 2012; Ramos & Ferrier, 2012; Zhang, 2003).

The mechanisms generating duplications, deletions and chromosomal rearrangements include nonallelic homologous recombination, involving either neighbouring stretches of DNA on the same chromosome or dispersed genomic regions, and nonhomologous end joining of double-strand breaks (Lynch, 2007; Ramos & Ferrier, 2012). TEs play a critical role in facilitating nonhomologous recombination events, by providing dispersed stretches of identical sequence that can act as seeds for recombination (Fiston-Lavier, Anxolabéhère, & Quesneville, 2007; Ponce, Martinsen, Vicente, & Hartl, 2012; Ramos & Ferrier, 2012); Small sequence repeats have been shown to be sites of chromosomal rearrangement in fungi (Ohm et al., 2012). In theory, the younger the TE family and the more abundant it is across the genome, the higher the probability it will facilitate a recombination event. An analysis of TEs found within a genome and the extent of gene order conservation, or "synteny", among related genomes may suggest (i) the extent to which chromosomal rearrangements mediated by TEs have influenced the evolution of gene content within the clade and (ii) whether and when TE dispersal within the genome has shaped synteny; if TEs influence chromosomal rearrangements, they may colocalize with synteny breakpoints.

Because gene content is very different between *A. thiersii and A. muscaria* (Chaib de Mares, 2013; J. Hess et al., unpublished; Wolfe, Tulloss, et al., 2012) and because TEs are found in abundance in two of the three ECM genomes (Table 8.1; Hess et al., in review), we designed an analysis to explore synteny conservation between our canonical saprotroph *A. thiersii*, the closely related *A. inopinata* and the ECM *Amanita* genomes.

We first identified *A. thiersii* scaffolds containing the key extracellular cellulases of CAZy families GH6 and GH7 (www.cazy.org), because the loss of PCWDEs appears to have been an important event in the early evolution of the *Amanita* ECM symbiosis (Wolfe, Tulloss, et al., 2012). Draft genome assemblies were aligned using PROmer from the MUMmer package (Kurtz et al., 2004) and filtered for matches to the three *A. thiersii* target scaffolds. Matching segments shorter than 1000 bp were removed, because we are primarily interested in visualizing long-range synteny of gene-sized segments. Adjacent regions were combined if there was no intervening segment from a different scaffold, internal duplication or change in directionality of the match.

Figure 8.2 illustrates the conservation of synteny between the *A. thiersii* scaffolds 2, 4 and 18, containing the predicted GH6 (scaffold 2) and the two predicted GH7 (scaffolds 4 and 18) genes, and homologous scaffolds in other *Amanita* genomes. Comparisons reveal different amounts of synteny conservation. As expected, *A. inopinata*, the closest relative (Fig. 8.1), shows the strongest amount of synteny conservation. All three *A. thiersii* scaffolds house long segments in synteny with scaffolds in *A. inopinata*, and two of these segments span cellulase loci (see radial bars, Fig. 8.2). Nevertheless, we find evidence for chromosomal rearrangements: The third cellulase locus, on scaffold 2 (the first *A. thiersii* scaffold, as you move in a clockwise direction), appears to be adjacent to a large chromosomal inversion and missing from *A. inopinata*, and the locus on scaffold 4 (the second *A. thiersii* scaffold), which contains a fragmented GH7 gene, consists of a complex segmental duplication. The different parts of scaffold 4 are superimposed onto the same scaffold in *A. inopinata*. Despite the conserved syntenic segments spanning two of the three target cellulases, all three enzymes are absent from *A. inopinata* and the scaffolds are not alignable in these regions, suggesting gene loss was independent of chromosomal rearrangements. As discussed previously, the ecological niche of *A. inopinata* remains undefined, and although it appears asymbiotic, all of the key PCWDEs are missing (Wolfe, Tulloss, et al., 2012).

The three ECM species display variable amounts of synteny conservation, with *A. muscaria* showing long contiguous matches to *A. thiersii* scaffolds, followed by *A. polypyramis* and finally *A. brunnescens*, which displays the largest breakdown in synteny. None of the cellulase loci in are syntenic. Phylogenetically, all three ECM species are equally distant to *A. thiersii* (cf. Fig. 8.1), raising questions as to what might cause differences in observed patterns of conservation.

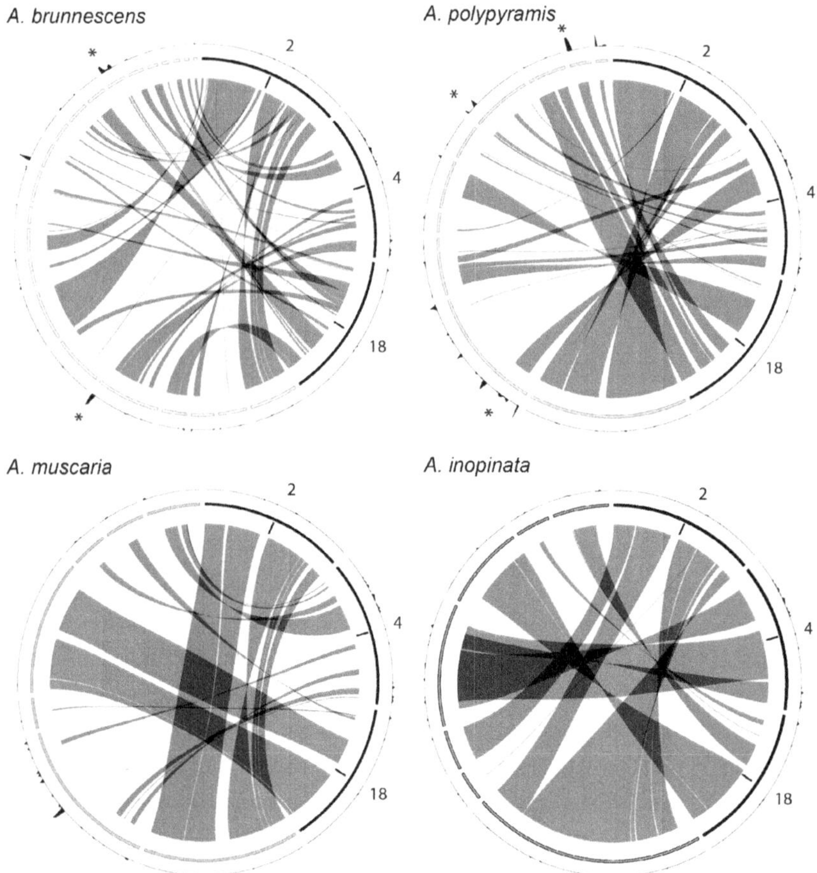

Figure 8.2 Conservation of synteny between the *Amanita thiersii* scaffolds containing key cellulases (GH6 and GH7), one on each scaffold, and matching scaffolds in other *Amanita* species. *A. thiersii* scaffolds are numbered and shown in black, and for the purpose of visualization, they are cropped to a total length of 100 kb surrounding the cellulase genes, the positions of which are indicated by the radial bars. Matching syntenic scaffolds are grey and cropped to the aligned positions plus a buffer of 10 kb on either side, if available. The line graph on the outer ring indicates TE density per 1000 bp window at the equivalent genomic coordinates. Asterisks mark the sites of potential TE-mediated synteny breakpoints.

The full interpretation of our data will first require a brief discussion of technical issues. The *A. muscaria* genome was sequenced by the US DOE Joint Genome Institute (http://genome.jgi-psf.org/Amamu1/Amamu1.home.html) using multiple Illumina libraries, including a 3.5 kb mate pair library, while the *A. brunnescens*, *A. polypyramis* and *A. inopinata* genomes were assembled from a single 0.3 kb paired end library. We expect the *A. muscaria* assembly to be of higher contiguity than the other assemblies because repeat regions will have been more easily resolved. Nevertheless, the *A. inopinata*, *A. polypyramis* and *A. brunnescens* assemblies are directly comparable and prove that assembly contiguity and synteny conservation are not strictly determined by the sequencing approach.

Relative TE abundance and distribution may also explain the degree of synteny conservation among the different ECM species. Among the ECM species, the *A. muscaria* genome houses the lowest proportion of TEs (Table 8.1). Technically, *A. polypyramis* houses a much larger proportion of TEs than *A. brunnescens*, but the higher assembly contiguity in *A. polypyramis* (N50 of 61 kb, compared to 11 kb in *A. brunnescens*; Table 8.1) suggests TEs are concentrated outside of gene-rich regions and may therefore be less of an influence on our predominantly genic target scaffolds. The *A. brunnescens* assembly is much more fragmented than the *A. polypyramis* assembly; the fragmentation may be caused by a more randomly distributed population of TEs. TEs and other repeated regions frequently form breakpoints in genome assemblies (Alkan, Sajjadian, & Eichler, 2011). The pattern of synteny conservation among ECM species may reflect positive relationships between the abundance of TEs, their distribution in a genome and synteny degradation.

Additional support for an influence of TEs on the degradation of synteny conservation is seen in the localization of TE-dense regions near synteny breakpoints (asterisks, Fig. 8.2). However, the many small scaffolds in the *A. brunnescens* may also reflect "interruptions" caused by repeats, rather than chromosomal rearrangements.

3.3. Conclusions

Emerging theory explicitly connects the questions we have explored: advances in genome sequencing and comparative genomics are enabling mechanistic frameworks synthesizing molecular patterns, including chromosomal rearrangements, with evolutionary phenomena, for example, adaptive radiations (Böhne et al., 2008; Jurka, Bao, & Kojima, 2011; Oliver et al., 2013; Oliver & Greene, 2009, 2011, 2012; Zeh et al., 2009). TEs are central to the new ideas proposed to explain adaptive radiations.

Adaptive radiations are defined by evolutionary innovation and increased rates of speciation, and TEs may facilitate both processes. Active transposition and ectopic recombination between young TE copies reshuffle the functional content of a genome. Domestication or "exaptation" of TE sequences and changes in gene regulation in the neighbourhood of TEs may also influence functional variation. An exhaustive list of examples and their evolutionary significance is found in Böhne et al. (2008), Oliver and Greene (2009, 2011, 2012) and Oliver et al. (2013). Moreover, the karyotypic variation resulting from chromosomal rearrangements can create reproductive barriers and cause reduced recombination and gene flow between chromosomal variants, accelerating the path to speciation (Böhne et al., 2008; Rieseberg, 2001). These ideas are encapsulated by the "TE-Thrust" model (Oliver & Greene, 2009, 2011, 2012), which proposes that lineages with large quantities of young TEs may be especially prone to speciation.

Based on our tentative evidence for an adaptive radiation following the evolution of the ECM symbiosis and our knowledge of genome architecture evolution in the *Amanita*, we can begin to ask whether and how TEs shaped the success of the ECM lineage. Two of the three ECM genomes we sequenced, *A. brunnescens* and *A. polypyramis* (Fig. 8.1), are rich in TEs, and the presence of closely related TE families in *A. polypyramis* and *A. muscaria* suggests a period of increased TE activity early in the ECM lineage (Hess et al., in review). Increased TE content, especially in *A. brunnescens*, and less so in *A. polypyramis*, coincides with a breakdown in long-range synteny (Fig. 8.2), suggesting that these ECM genomes may have undergone chromosomal rearrangements. The ECM fungus *A. brunnescens* falls within a lineage of increased speciation rates, as determined by our MEDUSA analysis (Fig. 8.1), while *A. polypyramis* and *A. muscaria* do not. Aggregate evidence strengthens the notion of a link between increased rates of speciation and changes in genome architecture.

The TE-Thrust model is focused on the age and abundance of TE families, but not on their distribution within the genome; on this point, the *Amanita* may provide a novel perspective. Our *A. polypyramis* results suggest that despite a high TE content, genomic rearrangements are less than in a species with lower TE content, *A. brunnescens*, perhaps because TE insertions are concentrated in regions without genes. A more nuanced model of adaptive radiations and TEs might consider patterns of TE distributions within genomes as an additional variable to explain the relationships between TEs and speciation rates.

Nonetheless, whether or not TEs influenced the radiation of ECM *Amanita* remains to be determined. Genomes from additional species, long-range sequencing libraries collected to improve existing assemblies and the sequencing of more individuals from different populations of already sequenced species may distinguish the potential influence of natural selection from stochasticity and distinguish between TE amplification as an active driver of speciation and patterns of TE content as a by-product of population genetics and demographic histories (Lynch, 2007). The *Amanita* belong to a charismatic genus, but even so, we also have remarkably little knowledge of the mating strategies, life cycles and demographic histories of many species. As we learn more about the genus, disentangling the various forces shaping speciation will be an ever more exciting and fruitful field of inquiry.

4. UNANSWERED QUESTIONS: RANGE EXPANSIONS AND GENOMIC ARCHITECTURES

An obvious feature of the natural histories of *Amanita* is range expansion; *A. thiersii* is moving north from Texas, while *A. inopinata* and *A. muscaria* (Vellinga, Wolfe, et al., 2009) seem to be establishing on several continents at once. In New Zealand, *A. muscaria* is invading in association with invasive pines (Dickie, Bolstridge, Cooper, & Peltzer, 2010). Other unsequenced *Amanita* are also invading novel habitats, for example, the ECM *A. phalloides* in California (Pringle, Adams, Cross, & Bruns, 2009). Although invasions appear as idiosyncratic phenomena, relatively little is known about the genomes of invasive species, and research on the ecological genomics of invasions may provide novel tools and discoveries (Suarez & Tsutsui, 2008). The genus *Amanita* is unique because it encompasses multiple introductions and invasions by species with both asymbiotic and symbiotic niches. Beyond obvious comparisons between decomposer and ECM fungi, salient questions will focus on what features of genomes enable dispersal, establishment or spread (Vellinga et al., 2009). The theory suggests targeting mating systems (is selfing an advantage?), genes involved in enabling associations with novel hosts (are generalists more likely to establish or spread? Pringle, Bever, et al., 2009), and genome plasticity perhaps mediated by the diversity of TEs in introduced populations (what do TE populations look like in native and invasive ranges?). The *Amanita* offer an exciting opportunity to push invasion biology in new directions.

ACKNOWLEDGEMENTS

Funding was provided by the National Science Foundation, Award Number 1021606. We are grateful to F. Martin's Laboratory and the Mycorrhizal Genomics Initiative consortium for access to unpublished genome data. Sequence data of *A. thiersii* and *A. muscaria* were produced by the US Department of Energy Joint Genome Institute, supported by the Office of Science of the US Department of Energy under Contract No. DE-AC02-05CH11231.

REFERENCES

Alkan, C., Sajjadian, S., & Eichler, E. E. (2011). Limitations of next-generation genome sequence assembly. *Nature Methods*, *8*, 61–65.

Bao, D., Gong, M., Zheng, H., Chen, M., Zhang, L., Wang, H., et al. (2013). Sequencing and comparative analysis of the straw mushroom (*Volvariella volvacea*) genome. *PLoS One*, *8*, e58294.

Bas, K. (2001). The unexpected one jumped the North Sea. *Field Mycology*, *2*, 40–41.

Böhne, A., Brunet, F., Galiana-Arnoux, D., Schultheis, C., & Volff, J. N. (2008). Transposable elements as drivers of genomic and biological diversity in vertebrates. *Chromosome Research*, *16*, 203–215.

Bruns, T. D., Szaro, T. M., Gardes, M., & Cullings, K. W. (1998). A sequence database for the identification of ectomycorrhizal basidiomycetes by phylogenetic analysis. *Molecular Ecology*, 7, 257–272.

Chaib de Mares, M. (2013). A genomic perspective on a transition to symbiosis in ectomycorrhizal *Amanita*. Masters thesis. Retrieved from http://irs.ub.rug.nl/dbi/521b6392a3b10.

Chang, S.-T. (1977). The origin and early development of straw mushroom cultivation. *Economic Botany*, *31*, 374–376.

Crombach, A., & Hogeweg, P. (2007). Chromosome rearrangements and the evolution of genome structuring and adaptability. *Molecular Biology and Evolution*, *24*, 1130–1139.

Date, K., & Mizuno, T. (1997). VII. Cultivation of fukurotake (*Volvariella* sp.) mushrooms. *Food Reviews International*, *13*, 397–400.

Dickie, I. A., Bolstridge, N., Cooper, J. A., & Peltzer, D. A. (2010). Co-invasion by *Pinus* and its mycorrhizal fungi. *New Phytologist*, *187*, 475–484.

Drummond, A. J., & Rambaut, A. (2007). BEAST: Bayesian evolutionary analysis by sampling trees. *BMC Evolutionary Biology*, 7, 214.

Fiston-Lavier, A. S., Anxolabéhère, D., & Quesneville, H. (2007). A model of segmental duplication formation in *Drosophila melanogaster*. *Genome Research*, *17*, 1458–1470.

FRDBI. (2013). *Fungal records database of Britain and Ireland*. September, Web site, http://www.fieldmycology.net/FRDBI/FRDBI.asp.

Geml, J., Laursen, G. A., O'Neill, K., Nusbaum, H. C., & Taylor, D. L. (2006). Beringian origins and cryptic speciation events in the fly agaric (*Amanita muscaria*). *Molecular Ecology*, *15*, 225–239.

Geml, J., Tulloss, R. E., Laursen, G. A., Sazanova, N. A., & Taylor, D. L. (2008). Evidence for strong inter- and intracontinental phylogeographic structure in *Amanita muscaria*, a wind-dispersed ectomycorrhizal basidiomycete. *Molecular Phylogenetics and Evolution*, *48*, 694–701.

Grandaubert, J., Balesdent, M. H., & Rouxel, T. (2014). Evolutionary and adaptive role of transposable elements in fungal genomes. *Advances in Botanical Research*, *70*, 79–107.

Halling, R. E. (2001). Ectomycorrhizae: Co-evolution, significance, and biogeography. *Annals of the Missouri Botanical Garden*, *88*, 5–13.

Harmon, L. J., Weir, J. T., Brock, C. D., Glor, R. E., & Challenger, W. (2008). GEIGER: Investigating evolutionary radiations. *Bioinformatics*, *24*, 129–131.

Heath, T. A., Hedtke, S. M., & Hillis, D. M. (2008). Taxon sampling and the accuracy of phylogenetic analyses. *Journal of Systematics and Evolution*, *46*, 239–257.

Heliconius Genome Consortium. (2012). Butterfly genome reveals promiscuous exchange of mimicry adaptations among species. *Nature*, *487*, 94–98.

Hess, J., Skrede, I., Wolfe, B.E., LaButti, K., Ohm, R.A., Grigoriev, I.V., et al., (in review). Transposable element dynamics among asymbiotic and ectomycorrhizal Amanita fungi.

James, T. Y., Kauff, F., Schoch, C. L., Matheny, P. B., Hofstetter, V., Cox, C. J., et al. (2006). Reconstructing the early evolution of Fungi using a six-gene phylogeny. *Nature, 443*, 818–822.

Janson, E. M., Stireman, J. O., Singer, M. S., & Abbot, P. (2008). Phytophagous insect-microbe mutualisms and adaptive evolutionary diversification. *Evolution, 62*, 997–1012.

Johnson, S. D. (2010). The pollination niche and its role in the diversification and maintenance of the southern African flora. *Philosophical Transactions of the Royal Society of London Series B, Biological Sciences, 365*, 499–516.

Jones, T., Federspiel, N. A., Chibana, H., Dungan, J., Kalman, S., Magee, B. B., et al. (2004). The diploid genome sequence of *Candida albicans*. *Proceedings of the National Academy of Sciences of the United States of America, 101*, 7329–7334.

Jurka, J., Bao, W., & Kojima, K. K. (2011). Families of transposable elements, population structure and the origin of species. *Biology Direct, 6*, 44.

Kibby, G. (2005). The invasion of *Amanita inopinata* continues! *Field Mycology, 6*, 31.

Kidwell, M. G., & Lisch, D. R. (2001). Perspective: Transposable elements, parasitic DNA, and genome evolution. *Evolution, 55*, 1–24.

Kiester, A. R., Lande, R., & Schemske, D. W. (1984). Models of coevolution and speciation in plants and their pollinators. *The American Naturalist, 124*, 220–243.

Kondrashov, F. A. (2012). Gene duplication as a mechanism of genomic adaptation to a changing environment. *Proceedings of the Royal Society B, 279*, 5048–5057.

Kuo, M. (2011). *Volvariella volvacea*. Retrieved from the MushroomExpert.Com Web site: http://www.mushroomexpert.com/volvariella_volvacea.html.

Kuo, M. (2013a). *Amanita thiersii*. Retrieved from the MushroomExpert.Com Web site: http://www.mushroomexpert.com/amanita_thiersii.html.

Kuo, M. (2013b). *Amanita polypyramis*. Retrieved from the MushroomExpert.Com Web site:http://www.mushroomexpert.com/amanita_polypyramis.html.

Kurtz, S., Phillippy, A., Delcher, A. L., Smoot, M., Shumway, M., Antonescu, C., et al. (2004). Versatile and open software for comparing large genomes. *Genome Biology, 5*, R12.

Law, R., & Lewis, D. H. (1983). Biotic environments and the maintenance of sex—Some evidence from mutualistic symbioses. *Biological Journal of the Linnean Society, 20*, 249–276.

Levin, H. L., & Moran, J. V. (2011). Dynamic interactions between transposable elements and their hosts. *Nature Reviews. Genetics, 12*, 615–627.

Litsios, G., Sims, C. A., Wüest, R. O., Pearman, P. B., Zimmermann, N. E., & Salamin, N. (2012). Mutualism with sea anemones triggered the adaptive radiation of clownfishes. *BMC Evolutionary Biology, 12*, 212.

Löytynoja, A., & Goldman, N. (2008). Phylogeny-aware gap placement prevents errors in sequence alignment and evolutionary analysis. *Science, 320*, 1632–1635.

Lynch, M. (2007). *The origins of genome architecture*. Sunderland, MA: Sinauer Associates Inc.

Martin, F., Aerts, A., Ahrén, D., Brun, A., Danchin, E. G. J., Duchaussoy, F., et al. (2008). The genome of *Laccaria bicolor* provides insights into mycorrhizal symbiosis. *Nature, 452*, 88–92.

Martin, F., Kohler, A., Murat, C., Balestrini, R., Coutinho, P. M., Jaillon, O., et al. (2010). Périgord black truffle genome uncovers evolutionary origins and mechanisms of symbiosis. *Nature, 464*, 1033–1038.

Martin, W., & Schnarrenberger, C. (1997). The evolution of the Calvin cycle from prokaryotic to eukaryotic chromosomes: A case study of functional redundancy in ancient pathways through endosymbiosis. *Current Genetics, 32*, 1–18.

Martin, F., & Selosse, M. A. (2008). The *Laccaria* genome: A symbiont blueprint decoded. *New Phytologist, 180*, 296–310.

Matheny, P. B., Curtis, J. M., Hofstetter, V., Aime, M. C., Moncalvo, J. M., Ge, Z. W., et al. (2006). Major clades of Agaricales: A multilocus phylogenetic overview. *Mycologia, 98*, 982–995.

McCutcheon, J. P., & Moran, N. A. (2012). Extreme genome reduction in symbiotic bacteria. *Nature Reviews. Microbiology, 10*, 13–26.
Mira, A., Ochman, H., & Moran, N. A. (2001). Deletional bias and the evolution of bacterial genomes. *Trends in Genetics, 17*, 589–596.
Moran, N. A. (1996). Accelerated evolution and Muller's rachet in endosymbiotic bacteria. *Proceedings of the National Academy of Sciences of the United States of America, 93*, 2873–2878.
Moran, N. A., McLaughlin, H. J., & Sorek, R. (2009). The dynamics and time scale of ongoing genomic erosion in symbiotic bacteria. *Science, 323*, 379–382.
Moran, N. A., & Wernegreen, J. J. (2000). Lifestyle evolution in symbiotic bacteria: Insights from genomics. *Trends in Ecology & Evolution, 15*, 321–326.
Nagendran, S., Hallen-Adams, H. E., Paper, J. M., Aslam, N., & Walton, J. D. (2009). Reduced genomic potential for secreted plant cell-wall-degrading enzymes in the ectomycorrhizal fungus *Amanita bisporigera*, based on the secretome of *Trichoderma reesei*. *Fungal Genetics and Biology, 46*, 427–435.
Ohm, R. A., Feau, N., Henrissat, B., Schoch, C. L., Horwitz, B. A., Barry, K. W., et al. (2012). Diverse lifestyles and strategies of plant pathogenesis encoded in the genomes of eighteen Dothideomycetes fungi. *PLoS Pathogens, 8*, e1003037.
Oliver, K. R., & Greene, W. K. (2009). Transposable elements: Powerful facilitators of evolution. *BioEssays, 31*, 703–714.
Oliver, K. R., & Greene, W. K. (2011). Mobile DNA and the TE-thrust hypothesis: Supporting evidence from the primates. *Mobile DNA, 2*, 8.
Oliver, K. R., & Greene, W. K. (2012). Transposable elements and viruses as factors in adaptation and evolution: An expansion and strengthening of the TE-Thrust hypothesis. *Ecology and Evolution, 2*, 2912–2933.
Oliver, K. R., McComb, J. A., & Greene, W. K. (2013). Transposable elements: Powerful contributors to angiosperm evolution and diversity. *Genome Biology and Evolution, 5*, 1886–1901.
Parra, G., Bradnam, K., & Korf, I. (2007). CEGMA: A pipeline to accurately annotate core genes in eukaryotic genomes. *Bioinformatics, 23*, 1061–1067.
Plett, J. M., & Martin, F. (2011). Blurred boundaries: Lifestyle lessons from ectomycorrhizal fungal genomes. *Trends in Genetics, 27*, 14–22.
Ponce, R., Martinsen, L., Vicente, L. M., & Hartl, D. L. (2012). Novel genes from formation to function. *International Journal of Evolutionary Biology, 2012*, 821645.
Pringle, A., Adams, R. I., Cross, H. B., & Bruns, T. D. (2009). The ectomycorrhizal fungus *Amanita phalloides* was introduced and is expanding its range on the west coast of North America. *Molecular Ecology, 18*, 817–833.
Pringle, A., Bever, J. D., Gardes, M., Parrent, J. L., Rillig, M. C., & Klironomos, J. N. (2009). Mycorrhizal symbioses and plant invasions. *Annual Review of Ecology, Evolution, and Systematics, 40*, 699–715.
Pybus, O. G., & Harvey, P. H. (2000). Testing macro-evolutionary models using incomplete molecular phylogenies. *Proceedings of the Royal Society B, 267*, 2267–2272.
Raffaele, S., & Kamoun, S. (2012). Genome evolution in filamentous plant pathogens: Why bigger can be better. *Nature Reviews. Microbiology, 10*, 417–430.
Ramos, O. M., & Ferrier, D. E. K. (2012). Mechanisms of gene duplication and translocation and progress towards understanding their relative contributions to animal genome evolution. *International Journal of Evolutionary Biology, 2012*, 846421.
Reid, D. A. (1987). New or interesting records of British hymenomycetes, VII. *Notes from the Royal Botanic Garden Edinburgh, 44*, 503–540.
Ridley, G. (2000). The New Zealand connection—*Amanita inopinata*—The mystery deepens. *Field Mycology, 1*, 117–118.
Rieseberg, L. H. (2001). Chromosomal rearrangements and speciation. *Trends in Ecology & Evolution, 16*, 351–358.

Rouxel, T., Grandaubert, J., Hane, J. K., Hoede, C., van de Wouw, A. P., Couloux, A., et al. (2011). Effector diversification within compartments of the Leptosphaeria maculans genome affected by repeat-induced point mutations. *Nature Communications, 2*, 202.

Ryberg, M., & Matheny, P. B. (2012). Asynchronous origins of ectomycorrhizal clades of Agaricales. *Proceedings of the Royal Society B, 279*, 2003–2011.

Sacristán, S., Vigouroux, M., Pedersen, C., Skamnioti, P., Thordal-Christensen, H., Micali, C., et al. (2009). Coevolution between a family of parasite virulence effectors and a class of LINE-1 retrotransposons. *PLoS One, 4*, e7463.

Schwartz, R. S., & Mueller, R. L. (2010). Branch length estimation and divergence dating: Estimates of error in Bayesian and maximum likelihood frameworks. *BMC Evolutionary Biology, 10*, 5.

Smith, S. E., & Read, D. J. (2010). *Mycorrhizal symbiosis* (3rd ed.). New York: Academic Press.

Suarez, A. V., & Tsutsui, N. D. (2008). The evolutionary consequences of biological invasions. *Molecular Ecology, 17*, 351–360.

Tamas, I., Klasson, L., Canbäck, B., Näslund, A. K., Eriksson, A. S., Wernegreen, J. J., et al. (2002). 50 million years of genomic stasis in endosymbiotic bacteria. *Science, 296*, 2376–2379.

Thompson, J. N. (1999). Specific hypotheses on the geographic mosaic of coevolution. *The American Naturalist, 153*(S5).

Tulloss, R. E. (2013). Amanita polypyramis. In R. E. Tulloss, & Z. L. Yang (Eds.), *Amanitaceae studies*. http://www.amanitaceae.org?Amanita+polypyramis.

van Ham, R. C. H. J., Kamerbeek, J., Palacios, C., Rausell, C., Abascal, F., Bastolla, U., et al. (2003). Reductive genome evolution in Buchnera aphidicola. *Proceedings of the National Academy of Sciences of the United States of America, 100*, 581–586.

Vellinga, E. C., Wolfe, B. E., & Pringle, A. (2009). Global patterns of ectomycorrhizal introductions. *New Phytologist, 181*, 960–973.

Venail, J., Dell'Olivo, A., & Kuhlemeier, C. (2010). Speciation genes in the genus Petunia. *Philosophical Transactions of the Royal Society of London B, 365*, 461–468.

Veneault-Fourrey, C., & Martin, F. (2011). Mutualistic interactions on a knife-edge between saprotrophy and pathogenesis. *Current Opinion in Plant Biology, 14*(444), 450.

Wang, E., Schornack, S., Marsh, J. F., Gobbato, E., Schwessinger, B., Eastmond, P., et al. (2012). A common signaling process that promotes mycorrhizal and oomycete colonization of plants. *Current Biology, 22*, 2242–2246.

Werren, J. H. (2011). Colloquium paper: Selfish genetic elements, genetic conflict, and evolutionary innovation. *Proceedings of the National Academy of Sciences of the United States of America, 108*, 10863–10870.

Wolfe, B. E., Kuo, M., & Pringle, A. (2012). *Amanita thiersii* is a saprotrophic fungus expanding its range in the United States. *Mycologia, 104*, 22–33.

Wolfe, B. E., Tulloss, R. E., & Pringle, A. (2012). The irreversible loss of a decomposition pathway marks the single origin of an ectomycorrhizal symbiosis. *PLoS One*, 7, e39597.

Yoder, J. B., & Nuismer, S. L. (2010). When does coevolution promote diversification? *The American Naturalist, 176*, 802–817.

Zeh, D. W., Zeh, J. A., & Ishida, Y. (2009). Transposable elements and an epigenetic basis for punctuated equilibria. *BioEssays, 31*, 715–726.

Zhang, J. (2003). Evolution by gene duplication: An update. *Trends in Ecology & Evolution, 18*, 292–298.

Zoller, S., & Lutzoni, F. (2003). Slow algae, fast fungi: Exceptionally high nucleotide substitution rate differences between lichenized fungi *Omphalina* and their symbiotic green algae *Coccomyxa*. *Molecular Phylogenetics and Evolution, 29*, 629–640.

CHAPTER NINE

Genomics of Arbuscular Mycorrhizal Fungi: Out of the Shadows

Mathilde Malbreil*[,†], Emilie Tisserant[‡], Francis Martin[‡], Christophe Roux*[,†,1]

*Université de Toulouse, UPS, UMR5546, Laboratoire de recherche en Sciences Végétales, BP 42617, F-31326 Castanet-Tolosan Cedex, France

[†]CNRS, UMR5546, BP 42617, F-31326, Castanet-Tolosan Cedex, France

[‡]Institut National de la Recherche Agronomique (INRA), UMR 1136 INRA/Lorraine University, Interactions Arbres/Micro-organismes, Centre de Nancy, Champenoux, France

[1]Corresponding author: e-mail address: roux@lrsv.ups-tlse.fr

Contents

Abstract

Arbuscular mycorrhizal (AM) symbiosis is the most widespread mutualistic association. It concerns 80% of land plants and involves fungi belonging to the phylum Glomeromycota. Benefits to the host plants due to this symbiosis range from nutrient supply to protection against pathogens. AM fungi are important components of the soil microbiome and are of great interest for managing sustainable agriculture, provided that their life cycle is better understood. Recently, major advances in the genomics of the model AM fungus *Rhizophagus irregularis* DAOM197198 have been published, offering new tools to investigate the biology of this symbiosis. In this chapter, we provide an overview of the efforts that were necessary to reach these results, from the discovery of these fungi and the description of their mutualistic incidence to their *in vitro* cultivation and on to genomics. The genome of DAOM197198 is estimated at *ca.*

Advances in Botanical Research, Volume 70
ISSN 0065-2296
http://dx.doi.org/10.1016/B978-0-12-397940-7.00009-4

150 Mb. It is haploid and less polymorphic than expected. Although it is an obligate biotrophic fungus, very little gene loss was observed. We put the *Rhizophagus* gene repertoire in perspective with previous investigations performed on the physiology of AM fungi: germination, early signalling with host plants, plant invasion, metabolism (phosphorous, carbon and nitrogen) and sexuality. Clearly, the publication of the genome of *R. irregularis* DAOM197198 is a turning point in the study of AM fungi, and large areas of their biology that still remain to be elucidated will now become accessible for investigation.

1. INTRODUCTION

Arbuscular mycorrhizal (AM) symbiosis is an association between plant roots and a specific fungal group, the glomeromycetes. This mutualistic association is ubiquist, concerning a vast majority of plant species. It is considered as the most widespread plant–fungus association, leading to the aphorism taken up by several mycorrhizologists: "plant do not have roots, they have mycorrhizae". In spite of their importance, more than a century was needed to outline the ecological role and agronomic interest of these fungi. One explanation is that this discrete microscopic underground association is formed by obligate biotrophic fungi that are difficult to manipulate, with no known sexual reproduction, and thus presenting no way to perform direct genetic studies. In a manuscript entitled *History of Research on Arbuscular Mycorrhiza*, Koide and Mosse (2004) traced the steps that had to be overcome to gain evidence about the importance of this symbiosis. The first description was reported in the middle of the nineteenth century, but these fungi were poorly investigated since Barbara Mosse first obtained *de novo* mycorrhized plants in 1953 (Mosse, 1953) and later produced evidence that the association is mutualistic. She also pioneered *in vitro* cultivation of the fungus she isolated, using roots and then the hairy root system, finding a way to cultivate the fungus in axenic conditions (see in the succeeding text). During the 1990s, *in vitro* culture conditions were optimized leading to the description of the fungal developmental steps, particularly thanks to the aggressive and well-sporulating isolate DAOM197198 of *Rhizophagus irregularis*. These last 15 years have seen an explosion of works on AM fungi covering the whole spectrum of biological studies: phylogeny, taxonomy, ecology, agronomy, genetics of plant–AM fungus interaction and physiology of the AM symbiosis. Since reports that different pea, bean and barrel medic mutants are defective for both mycorrhizal symbiosis and rhizobial symbiosis (Catoira et al., 2000; Duc, Trouvelot, Gianinazzi-Pearson, & Gianinazzi, 1989), the scientific advances performed on nodulation mapped out the way to investigate the mechanism of how AM symbiosis becomes established. This led to the identification of early signals produced by the plant hosts (strigolactones—Akiyama, Matsuzaki, & Hayashi, 2005; Besserer et al., 2006) and the fungal symbiont (lipochitooligosaccharides—Maillet et al., 2011). Symbiotic physiology has also been widely investigated, showing that phosphorus is not only a nutrient recruited by extraradical hyphae in soil and translocated to the host plant as described years earlier (Graham, Leonard, & Menge, 1981) but also a major regulator of the establishment and function of AM symbiosis (Balzergue, Puech-Pagès, Bécard, & Rochange, 2011; Breuillin et al., 2010; Harrison, Dewbre, & Liu, 2002; Javot, Penmetsa, Terzaghi, Cook, & Harrison, 2007). In the last decade, numerous transcriptomic studies have provided a good knowledge of the physiological metabolisms and cell regulation modified in the host plants through AM symbiosis.

This introductive panorama roughly and briefly shows that numerous aspects of AM symbiosis have been investigated *except* the genomics of AM fungi themselves. In this chapter, we will present the efforts made since 2004—a *long hard road* as prophetically announced (Martin et al., 2008)—to obtain the first version of the AM fungal genome, released in May 2013 (http://genome.jgi.doe.gov/Gloin1/Gloin1.home.html) before publication (Tisserant et al., 2013). In the meantime, an original approach based on the isolation and sequencing of independent nuclei was also conducted (Lin et al., 2014). These genomic data open a new era in the study of AM fungi.

1.1. The evolutionary and ecological success of AM symbiosis

Many works have pointed out that AM symbiosis is the most widespread plant–fungus association. First, the ability of plants to host AM fungi is shared by the vast majority of plant species. Although it is not possible to exhaustively assess the ability of all plants to participate in a mycorrhizal relationship, convergent estimations argue that more than 80% of plant species and 90% of plant families (Spermatophyta) are mycorrhizal (Smith & Read, 2010; Wang & Qiu, 2006). Secondly, these fungi are present in all continents, from sub-Arctic islands to Antarctic Peninsula as defined by microscopic observations on numerous sampling (Newsham, Upson, & Read, 2009; Smith & Read, 2010). Ongoing metagenomic studies will give deeper information about their global distribution (Moora et al., 2011; Opik et al., 2010). The ecological success of AM symbiosis is strengthened by the incidence that these fungi have on plant diversity and productivity (Van der Heijden et al., 1998; van der Heijden & Scheublin, 2007). One question is as follows: how could the genetic characters involved in AM symbiosis establishment be shared by the very great majority of plant species? A hypothesis to this wide distribution rose from data showing that AM symbiosis already existed in early land plants. Fossil records show that AM symbiosis was present in Rhynie chert, that is, as early as 400 million years ago (Dotzler, Krings, Taylor, & Agerer, 2006; Redecker, Kodner, & Graham, 2000). It has even been suggested that AM symbiosis has driven the evolution of the green lineage species from aquatic to land life by promoting the development of roots (Brundrett, 2002). According to its ancestral origin and the dispersion of nonmycorrhizal species across plant phylogeny (Wang & Qiu, 2006), the nonmycorrhizal status is considered as a loss of mycorrhizal ability that could be linked to the adaptation to specific environments, like aquatic plant species (Limnocharitaceae) or plants growing in nutrient-rich environments (Brassicaceae). Plants developed strategies to make up for the loss of nutrient availability, particularly phosphorous, showing a highly branched root system and dense and well-developed root hairs (Bucher, 2007; Koide, 1991).

Consequently, studies focusing on the fitness of plant interactions with AM fungi are of great interest for sustainable agriculture based on the conservation of microbial soil diversity.

1.2. The biology of AM fungi

1.2.1 Phylogeny and taxonomy

Since the first description of *Glomus microcarpum* (formerly *Endogone macrocarpus*) by the Tulasne & Tulasne, 1845, the phylogeny and taxonomy of AM fungi were totally reorganized following the availability of molecular data (see Stürmer, 2012, for an historical review of taxonomy). The main reason is that very few morphological characters are available to describe these nonseptated and apparently nonsexual fungi. As morphological characters, descriptors could only use mycelium aspect and spore organization: shape, colour, size and cell wall layers. Major phylogenetic breakthroughs, thanks to molecular data, were the repositioning of this fungus in an independent basal fungal group, the Glomeromycota (Schüßler, Schwarzott, & Walker, 2001) and the total revision of the taxonomy (Schüßler & Walker, 2010) taking into account the numerous papers published during a decade of molecular systematics of AM fungi by different authors. The delimitation of Glomeromycota is now stabilized and concerns 18 genera and 250 species (http://schuessler.userweb.mwn.de/amphylo/), some taxa still being revisited (Redecker et al., 2013). However, this phylogenetic architecture will probably evolve in the coming years since the repertoire of species studied has until now been based on spore-forming isolates. Metagenomic approaches based on next-generation sequencing methodology applied to soil and root samples, that is, without isolating spores, will lead to the identification of new operational taxonomic units (OTUs): species, genera and even higher taxonomic levels (Öpik et al., 2013).

In the particular case of the DAOM197198 isolate, the taxonomy evolved from *Glomus intraradices* N.C. Schenck and G.S. Sm to *Glomus irregulare* Błaszk., Wubet, Renker and Buscot, sp. nov. (Blaszkowski et al., 2008; Stockinger, Walker, & Schüssler, 2009) based on rRNA gene comparison and later *Rhizophagus irregularis* (N.C. Schenck and G.S. Sm.) C. Walker and A. Schüßler (Schüßler & Walker, 2010). This name was attributed because the rRNA genes of the *Glomus* genus-type species *G. macrocarpum* do not cluster with those of DAOM197198 and allied (Krüger, Krüger, Walker, Stockinger, & Schüßler, 2012; Schwarzott, Walker, & Schüßler, 2001) and since *Rhizophagus* was the first genus name historically attributed to this organism (Dangeard, 1896) before being synonymized with *Glomus* by Gerdemann and Trappe (1974). In fact, Dangeard had made a wrong pathological diagnosis as he thought he was describing a root pathogen causing poplar disease—hence the name Rhizo/phagus, etymologically root/eater—but this does not seem to alter the principle of the priority rule for taxonomists and DAOM197198 is now named *Rhizophagus irregularis*.

1.2.2 Life cycle

Since the works of Mosse and Hepper (1975) and later Bécard and Piché (1989a), the establishment of the symbiosis is usually described as occurring in three steps: (i) asymbiotic hyphal growth, where the spores can germinate and develop hyphae autonomously but during a limited period; (ii) presymbiotic growth, where hyphal growth is stimulated by host signal perception; and (iii) symbiotic life, where the fungus has penetrated the plant root and develops both intraradical mycelium (to exchange nutrients) and extraradical hyphae (to recruit nutrients in the soil and form new spores). These three steps consist in drastic morphological and physiological changes within the AM fungi, including discrete but essential steps of fungal development. In the vicinity of roots, the hyphae form numerous branches (Mosse & Hepper, 1975; reviewed by Juge, Coughlan, Fortin, & Piché, 2009).

This invasive growth in the rhizosphere enhances hyphal contact with the roots and then the formation of appressorium-like structures—hyphopodia (Giovannetti, Avio, Sbrana, & Citernesi, 1993). After penetration, the fungus grows biotrophically in the root as intercellular or intracellular mycelium in the cortical zone and then invades some cells with highly digitated haustorium-like structures, called arbuscules (Fig. 9.1). These structures

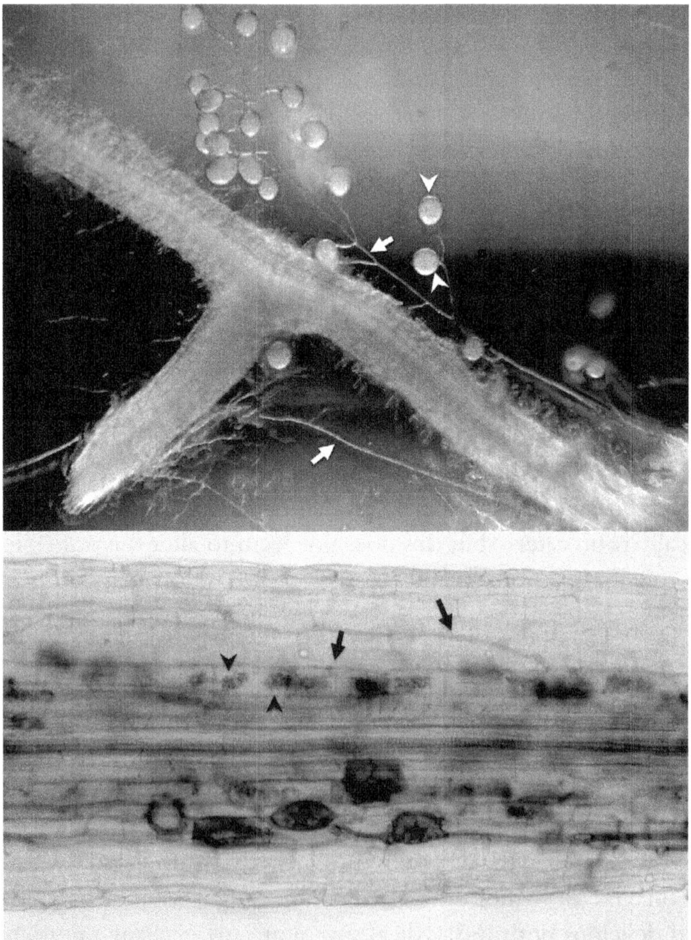

Figure 9.1 *In vitro* culture of *Rhizophagus irregularis* on carrot roots. On top, white arrows point to extraradical mycelium (ERM) that grows in the medium and forms spores (white arrowheads). Below, intraradical mycelium (IRM—black arrows) is visible in carrot roots after bleaching and staining. The fungus forms cleared intercellular vesicles (black stars) and arbuscules in host cortical cells (black arrowheads). (See the colour plate.)

are assumed to be the preferential zone of exchange between the two symbionts. As the intraradical mycelium (IRM) grows, a dense extraradical mycelium (ERM) net is formed in the soil. This hyphal net is involved not only in the supply ofboth water and nutrient to the host plant but also in fungus propagation as many spores can be formed. To date, no sexual mechanisms have been identified for the formation of such spores. They are dormancy structures that enable the invasion of new plants when climatic conditions are favourable.

2. TOWARD THE GENOME OF *RHIZOPHAGUS IRREGULARIS* DAOM197198

2.1. The first brick in the wall: Choosing a model organism

AM fungi are obligate biotrophic fungi difficult to produce for laboratory purposes, with no known sexuality, no genetic background and an unknown ploidy level. It was then a challenge to propose a "model" organism to launch a genome sequencing program.

2.1.1 *Rhizophagus irregularis* as a useful organism in laboratory studies

Several AM fungal species have been studied around the world. Considering the bibliography, the most popular are *Funneliformis mosseae* (formerly *Glomus mosseae*), *Gigaspora* sp. (*G. rosea*, *G. margarita* and *G. gigantea*) and *Rhizophagus irregularis* (formerly *Glomus intraradices*). All these species are easy to propagate in pot culture in association with different host plants (leek, parsley, Bahia grass, sorghum, etc.). However, genomic analyses need biological material in abundance, that is, mycelium and spores produced in axenic conditions. This step was overcome following the work of Mosse and Hepper (1975) who first succeeded in cultivating *F. mosseae* on *in vitro* root organ cultures from tomato and red clover. Later, the use of *Convolvulus sepium* hairy roots allowed *in vitro* cultivation of this species (Mugnier & Mosse, 1987). Unfortunately, it was unable to sporulate *in vitro* and then did not produce enough biomass for performing experiments on the fungus. Following improvements of *in vitro* culture conditions for *Gigaspora* sp. (Bécard & Fortin, 1988; Bécard & Piché, 1989b), an *in vitro* culture of one isolate of *Rhizophagus irregularis* was obtained (Chabot, Bécard, & Piché, 1992). Using the previously designed two-compartment system (Mugnier & Mosse, 1987), enhanced spore production was reached (St-Arnaud, Hamel, Vimard, Caron, & Fortin, 1996). This was the first

AM fungal isolate to be easily propagated *in vitro*. The fungal sample used was collected by Plenchette in 1978 (*Glomus* sp. #3 from *Fraxinus americana* L.—Pont-Rouge, Québec, Canada; Plenchette, Furian, & Fortin, 1982). White ash roots were used to inoculate leek plants, and then, the fungus was subcultured on the same host until it was deposited in 1981 at the Biosystematics Research Institute (Ottawa, Ontario) under the voucher DAOM181602. Subcultures propagated independently by a Canadian company—Premier Tech—were later deposited in 1987 under the voucher number DAOM197198 (C. Plenchette, personal communication). Chabot and coauthors used, in 1992, spore batches from pots that led to the voucher DAOM197198 for *in vitro* culture. To our knowledge, monosporal isolation has never been achieved since the collection of this isolate, suggesting that the monoclonal status of DAOM197198 depends on the genetic diversity of the initial isolate and on the genetic bottleneck due to the successive subcultures on leek from 1978 to 1992 and then on hairy roots since that time. Stockinger et al. (2009), analysing *in vitro* subcultures of DAOM197198 propagated in different laboratories, showed up to 32 allelic variants of ribosomal internal transcribed regions (ITS). Although intraindividual polymorphism is often observed among fungal species (Nilsson, Kristiansson, Ryberg, Hallenberg, & Larsson, 2008; Simon & Weiß, 2008), *R. irregularis* reaches the highest ITS variation rate compared with other fungi (Schoch et al., 2012). By analysing sequences from single nuclei, Lin et al. (2014) demonstrated that these variations occur within each single nucleus. According to Vankuren, den Bakker, Morton, and Pawlowska (2013), the high intraindividual polymorphism of ribosomal RNA genes could be linked to the asexual reproduction of AM fungus.

2.1.2 Organization and size of AM fungal genomes

The genetic status of Glomeromycetes is conflicting. As previously mentioned for ITS, multiple variants of ribosomal genes and other gene markers have been identified within each spore (Corradi et al., 2007; Sanders et al., 1995), suggesting that a single isolate in fact has a population of different nuclei (Hijri & Sanders, 2005; Kuhn, Hijri, & Sanders, 2001). Recent works suggest that allelic frequency of markers measured in daughter spores is modified according to the host species (Angelard et al., 2013). In opposition to this heterokaryotic hypothesis, other authors have suggested that polymorphism could be due to polyploidy (Pawlowska, 2005; Pawlowska & Taylor, 2004, read also Bever & Wang, 2005 and the reply of Pawlowska and Taylor in the same issue) or to gene duplication, each nucleus in spores being genetically identical. It must be pointed out that *R. irregularis* DAOM197198 strain was never used for such analyses.

Glomeromycota genome size estimations showed great variations according to species, ranging from 0.18 to 1.08 pg of DNA per nucleus (Bianciotto & Bonfante, 1992; Hosny, Gianinazzi-Pearson, & Dulieu, 1998), that is, 176 Mb to over 1 Gb using the conversion formula of Doležel, Bartoš, Voglmayr, and Greilhuber (2003). Glomeromycota hence present the largest genome sizes among fungi (37 Mb on average—Gregory et al., 2007). Flow cytometry assays performed on isolate DAOM197198 first led to a genome size estimation of around 15 Mb (Hijri & Sanders, 2004), but later measurements using different standards indicated that the genome size could be in fact 10 times higher (154.8 ± 6.2 Mb—Sedzielewska et al., 2011), although still remaining among the lowest sizes encountered across AM fungi.

In conclusion, isolate DAOM197198 of *Rhizophagus irregularis* was used for genome sequencing as (i) it is the AM fungus most studied in laboratories; (ii) it can sporulate heavily *in vitro*, providing large quantities of biological material; (iii) its genome size is lower than that of many other AM fungal species.

2.1.3 The genome of Rhizophagus Irregularis *DAOM197198: A cold case*

Considering that spores and coenocytic hyphae contain hundreds of putatively polymorphic nuclei and that the genome size estimations range can vary 10-fold, it was a real challenge to sequence the genome of *R. irregularis* DAOM197198. The sequencing program was announced in 2004, in the framework of the poplar mesocosm analysis (Martin et al., 2004). This program was supported by the Joint Genome Initiative and founded by the US Department of Energy. Several contributors, forming the Glomus Genome Consortium, participated to provide biological material and sequences. Five years later, a report stating the numerous difficulties encountered and strategies planned was published (Martin et al., 2008). These difficulties were resolved through the use of next-generation sequencing techniques, particularly suitable for discrete organisms producing little biomass, and new bioinformatic tools. First, a set of 25906 non-redundant virtual transcripts (773 nt avg. length) were obtained from 454 sequencing of cDNA libraries (Mirav2 assembly available at http://mycor.nancy.inra.fr/IMGC/GlomusGenome/index3.html; Tisserant et al., 2012). In April 2013, the first version of the genome was released

(Gloin1 at http://genome.jgi.doe.gov/Gloin1/Gloin1.home.html) using a hybrid assembly of 3.781 Gb of genomic sequences obtained by Sanger, 454 and Illumina sequencing procedures. The actual genome assembly, improved by using 766 Mb of PacBio sequences, consists in 12,421 scaffolds (N50 = 15.16 kb) for a total of 101 Mb (Tisserant et al., 2013).

From these data, the genome size was estimated to be 153 Mb, in accordance with the value of 154.8 ± 6.2 Mb measured by flow cytometry. The genome of *R. irregularis* is haploid and does not show evidence of recent duplication, although portions of genome were formerly duplicated. As expected for such a large fungal genome, transposable elements (TE) are strongly represented: up to 55 Mb of the genome is formed by repeated TE. This feature, combined with an A + T content of 72%—higher than the previous record among fungi that was observed on *Candida albicans* with an A + T content of 67%—explains the difficulties encountered in assembling the genomic data.

Intriguingly, the procedures developed to check polymorphism in the genome did not provide evidence of different haplotypes. The density of single-nucleotide polymorphism (SNP) was estimated at 0.43 per kb over the whole genome and 0.4 SNP per kb in the exome. Intraindividual genomic variations resulting in SNPs, although far less documented than interindividual SNPs used for population analyses, are not rare. Since NGS techniques allow this question to be tackled, several papers have reported such variations in multicellular organisms. In humans, for instance, transposon activity and gene copy number variation lead to clonal mosaicism (Ewing & Kazazian, 2010; Huallachain, Karczewski, Weissman, & Eckehart, 2012). In fungi, intraindividual SNP density is highly variable according to fungal species (0.06 SNP per kb in *Tuber melanosporum* and 0.78 SNP per kb in *Laccaria bicolor*), meaning that the value observed in *R. irregularis* is not consistent with the occurrence of multiple highly divergent genomes. Same low level of polymorphism was also found when comparing genomic sequences from four nuclei isolated from strain DAOM197198 (Lin et al., 2014). To sum up, on one side, DAOM197198 presents a classical haploid fungal genome, although atypically A + T-rich and invaded by TE. On the other side, works using targeted genomic markers showed a high and rapid genotypic plasticity driven by the plant host (Angelard et al., 2013), supporting a heterokayotic hypothesis. Ongoing works will help to understand the origin of this discrepancy. It can be speculated that *in vitro* cultivation of isolate DAOM197198 for 20 years on carrot roots has resulted in a significant loss of polymorphism. An alternative speculation is that the genome of AM fungi could evolve rapidly according to environmental conditions due to AT richness and TE activity. In *Fusarium*, a broad host spectrum fungal pathogen formed by host-adapted subpopulations, it was observed that regions with high SNP density could be linked to host specialization (Cuomo et al., 2007). The genomic assembly Gloin1 will be a powerful tool to investigate SNP distribution and to check for such putative variable regions.

Gene prediction based on transcriptomic data mapped on Gloin1 provided 28,232 protein-coding genes. Considering the conserved core eukaryotic set gene (Parra, Bradnam, & Korf, 2007), 98% of the genes are present in Gloin1. The genes have an average length of 1188 nt (890 nt as avg. transcript size) and an exon density of 3.5 exons per gene with an average intron length of 123 nt. Such organization is standard among fungi as exon density ranges from 2.5 for filamentous ascomycetes (e.g. *Neurospora* and *Magnaporthe*) to 5.5 for basidiomycetes such as *Cryptococcus neoformans*. The *Rhizophagus* gene repertoire is unusual among other obligate biotrophic organisms. Although *R. irregularis* is unable to grow axenically, as are all AM fungi, no large gene loss was observed. The conservation of the gene repertoire could be the result of the specific biology of AM fungi. Associated to their host, they form a metabolic dipole by recruiting water and minerals from the extraradical hyphae in soil while performing metabolic exchanges in intraradical mycelium, specifically in the arbuscules. Although fuelled by the carbon photosynthates provided by the plant host, the ERM has metabolic abilities close to those of a saprotrophic fungus (M. Malbreil, C. Roux & P.M. Delaux, unpublished data; Tisserant et al., 2012). Hence, one unexpected conclusion is that *R. irregularis* is physiologically closer to a hemibiotrophic fungus than to a strictly biotrophic one (Duplessis et al., 2011).

On the fringe of the nuclear genomic sequencing program, several works have dealt with the sequencing of the mitochondrial (mt) genome of AM fungi. As for other fungi, mitochondrial genomes of AM fungi are of great interest to investigate fungal ecology. Marker genes from mt genomes have long been used since no intraindividual polymorphism was observed, allowing strains to be tracked in natural conditions (Raab, Brennwald, & Redecker, 2005). These ecological aspects support the investigation of mt genomes. Taking advantage of pyrosequencing techniques, the first genome was obtained from *Rhizophagus irregularis* strain #FACE494 (FJ648425.1, deposited in January 2009—Lee & Young, 2009), before the publication of the mt genome of DAOM197198 by M. Hijri and B.F. Lang (HQ189519.1, deposited in September 2010—Lang & Hijri, 2009). AM fungal mt genome sizes range from 70.6 kbp for *R. irregularis* #FACE494 to 97 kb for *G. margarita* BEG34 (Pelin et al., 2012). The variability analyses of mt genomes within *R. irregularis* strains (Formey et al., 2012) or *Rhizophagus* species (Beaudet, Nadimi, Iffis, & Hijri, 2013) showed highly variable regions (Formey et al., 2012) that could lead to rapid mt genome evolution (Beaudet et al., 2013). One point of great interest is the homoplasmy observed through these works, whereas transient heteroplasmic stages were observed that could lead to genetic exchanges (De la Providencia, Nadimi, Beaudet, Rodriguez Morales, & Hijri, 2013).

3. THE BIOLOGY OF *RHIZOPHAGUS IRREGULARIS* FROM ITS GENE REPERTOIRE

Investigating the metabolism and gene regulation of AM fungi is constrained by the specific features of these fungi previously described: obligate biotrophy, multinucleate hyphae, no known sexuality and no unicellular stage. Moreover, it is still not possible to obtain stable transformant or mutant lines of these fungi; also, transient expression was reported (Helber & Requena, 2008). Therefore, forward and reverse genetics strategies are not applicable as yet. Metabolic approaches—enzyme activity, using isotope labelling—were the only way the biology of these organisms could be elucidated. Several studies have managed to link the physiological activity recorded to candidate genes, but as mutagenesis protocols are not available, gene expression monitoring and heterologous gene analyses were the only tools that are useable to validate protein function. Recently, functional validation of a candidate *Rhizophagus* gene was performed by using host-induced gene silencing (Helber et al., 2011). The approach is based on a host overexpressed antisense RNA strategy previously used to silence gene expression of pathogenic obligate biotrophic fungi (Nowara et al., 2010). Combined with the release of the *Rhizophagus* gene repertoire, these tools will enable the identification of new fungal genes involved in symbiosis, the upstream master regulator genes and the molecules that are involved in these signalling cascades. We describe later some of the physiological traits that will be deeper analysed in the coming years thanks to these new tools.

3.1. Spore germination and early signal perception

Hyphae from germinating spores grow slowly, consuming as little energy as possible, and form few hyphal branches. After a couple of weeks, in the absence of a host in their vicinity, the growth stops, the cytoplasm retracts into the spore and some species form new dormancy spores. This mechanism can be reiterated a *few to several times*, depending on the fungal species (Koske, 1981). The physiology of this survival strategy remains to be described. Only one gene, GmGIN1 from *G. mosseae*, putatively involved in the cell cycle has been identified (Requena, Mann, Hampp, & Franken, 2002).

In the vicinity of roots, perception of host signals trigger drastic physiological and morphological changes: plasmalemma ATPase activity is enhanced (Lei, Beard, Catford, & Piche, 1991), mitochondrial shape is modified and cell respiration and energy production increase via lipid catabolism activation, leading to hyphal branching (see Bécard et al., 2004). These physiological and morphological responses favour the contact of the fungus to a root and hence the possibilities of infection. The plant hormone strigolactone (SL) has been identified in root exudates as a host signal perceived by AM fungi (Akiyama et al., 2005) necessary for the establishment of the symbiosis (Gomez-Roldan et al., 2008). These molecules induce a significant and rapid increase in respiration, which is visible through mitochondrial activity and mitochondriome organization (Besserer, Becard, Jauneau, Roux, & Sejalon-Delmas, 2008; Besserer et al., 2006; Tamasloukht et al., 2003), fatty acid β-oxidation and ATP production, leading to hyphal branching. These responses are conserved in AM fungi as *R. irregularis* and *G. rosea* both responded in the same way. It was shown that these responses are dependent on an NADH dehydrogenase as well as an alternative oxidase in the respiratory chain (Besserer et al., 2008, Besserer, Bécard, Roux, & Séjalon-Delmas, 2009). Both enzymes are involved in GR24 response, while germination relies on the second only. Over the 138 genes dedicated to energy metabolism (KEGG metabolic pathway), very few have been investigated in response to SL. A pyruvate carboxylase (allowing CO_2 dark fixation) and a mitochondrial ADP/ATP translocase, involved in respiratory functions, are rapidly upregulated in *G. rosea*, while a cytochrome-c oxidase, an ATP synthase, ketoacyl thiolase, CuZnSOD, α-tubulin and sphingosine-1P lyase are induced later (Besserer et al., 2008; Tamasloukht et al., 2003). While gene regulation in response to SLs has been documented, nothing is known about the proteins involved in perception. A plant gene was identified, coding for an α/β hydrolase and able to bind and hydrolyse SL, strongly supporting a role in perception of the hormone (Hamiaux et al., 2012; Nakamura et al., 2013). No clear fungal homologue has been identified so far, indicating either that the current *Rhizophagus* genome assembly is not complete or that SL perception follows a different pathway in AM fungi.

Other signalling processes involved in the early steps of interaction have to be identified, either in root exudates or in contact with the roots. For instance, hyphopodial formation is linked to the contact of root epidermal cells (Nagahashi & Douds, 1997) where cutin monomers play a crucial role (Wang et al., 2012), but nothing is known about the genes responsible in such perception.

On the fungal side, recent studies have shown that AM fungi produce chitin-based compounds as signals able to elicit a plant response dependent on the SYM pathway. Lipochitooligosaccharides (Myc-LCOs) enhance root colonization and induce lateral root formation. Their structure is very close to that of Nod factors, signals produced by nitrogen-fixing rhizobia to prepare the legume host for symbiosis (Maillet et al., 2011). It was hypothesized that bacteria acquired genes involved in Nod factor synthesis by horizontal transfer from AM fungi. Germinated spore exudates of *R. irregularis*, *G. margarita* and *G. rosea* trigger calcium spiking in plant epidermal cells, a response also observed in hyphopodium-contacted cells (Chabaud et al., 2011; Genre et al., 2013). Molecules responsible for this plant response have been identified by Chabaud and colleagues and consist in short-chain chitin oligomers (CO) of 4 or 5 residues. Their production is stimulated by GR24. In both cases, the genes involved in their biosynthesis remain elusive. The production of chitin derivatives raises the question of their origin: are they by-products of hyphal growth (consistent with the increase of hyphal development and CO release in response to GR24) or specifically produced by chitin synthase or chitinase? Several chitin synthases are found in *R. irregularis* genome Gloin1, but no orthologues of the nodC gene, involved in the synthesis of the chitin backbone of Nod factor, have been identified yet. Expression pattern analyses in response to plant signals will be helpful to identify such genes.

3.2. Plant invasion

As far as we know, AM fungi do not have host specificity, and one species can interact with all mycotrophic plants. The analysis of the gene repertoire highlights that AM fungi have developed a specific symbiosis-associated gene pattern for furtive growth in their host, avoiding the plant immune detection.

First, it is worth noting that there is no gene coding for mycotoxin or host cell wall-degrading enzymes in the genome of *R. irregularis* (Tisserant et al., 2013). It can be assumed that the loss of such genes prevents the release of pathogenesis-associated and damage-associated molecular patterns that induce host defence mechanisms. It raises the question of the mechanism involved in the penetration of the fungus through host cell wall when growing in plants. The observation that the plant cell hosts the fungus during the first step of colonization (Genre, Chabaud, Timmers, Bonfante, & Barker, 2005) and that cell wall-remodelling plant genes are highly expressed in symbiotic root tissues (Guether et al., 2009) suggests that plant cell processes could be involved in cell wall loosening.

In addition to invasion by stealth, AM fungi have developed host defence-suppressive strategies. Pathogens (bacteria, fungi or oomycetes) secrete or inject "effector" proteins that interfere with plant response and promote virulence (Abramovitch, Anderson, & Martin, 2006; Kamoun, 2007; Valent & Khang, 2010; Wawra et al., 2012). A decrease in plant defence during AM fungus infection is usually observed after transient induction (García-Garrido & Ocampo, 2002; Güimil et al., 2005, Kapulnik et al., 1996; Pozo, Azcón-Aguilar, Dumas-Gaudot, & Barea, 1998; Pozo, Loon, & Pieterse, 2005; Zamioudis & Pieterse, 2012). Recently, an *R. irregularis* secreted protein 7 (SP7) was characterized thanks to a modified yeast secretion sequence trap method (Kloppholz, Kuhn, & Requena, 2011). This protein is secreted by the fungus into the plant cell where it interacts with ethylene response factor 19 (ERF19, putatively involved in plant immune system) and plays a role in modulating accommodation in the root. In the *R. irregularis* transcriptome (Tisserant et al., 2012), several putative small secreted proteins (SSPs) and small proteins (SPs) were detected, some among the most highly upregulated *in planta* and many being expressed specifically during symbiosis. The Gloin1 analysis revealed 376 proteins with a peptide signal, with 20% upregulated in *M. truncatula*, and among 153 SSPs detected, 19% are also upregulated. Moreover, most highly upregulated genes are species-specific, probably related to establishment of symbiosis (Tisserant et al., 2013).

Following the furtive strategy and defence suppression, a third level of defence host control consists in counteracting the remaining defence mechanisms. ROS have been shown to play a defensive and signalling role during plant invasion by pathogens or by rhizobia (O'Brien, Daudi, Butt, & Bolwell, 2012; Pauly et al., 2006). In AM symbiosis, an increase of ROS in mycorrhizal roots is also observed (Fester & Hause, 2005). A functional superoxide dismutase has been characterized in *G. margarita*. This enzyme turns superoxide into less damaging molecules (hydrogen peroxide and oxygen), actively participating in fungal resistance during the oxidative burst (Lanfranco, Novero, Bonfante, & Torino, 2005). Consistent with this function, the authors showed that this gene is highly induced *in planta* and only slightly expressed in germinating spores. The same features are observed in *R. irregularis* where superoxide dismutases form a gene family in the Gloin1 assembly.

3.3. Fungal metabolism during symbiotic life

Mutualism is a strategy for organisms to survive and should be considered in terms of costs and benefits for each partner. In AM symbiosis, up to 20% of the carbon fixed via photosynthesis is transferred to the fungal partner (Jakobsen & Rosendahl, 1990). To prevent cheating, strict control of nutrient flow must be settled to reach a fair trade (Kiers et al., 2011). However, many aspects of mutualism are still blurred and benefits are not always easy to discern (Walder et al., 2012). The most widely studied aspect of AM fungi is their role in mineral nutrient supply for the plants. This requires different metabolic machinery: import from soil to extraradical hyphae (including transporters and enzymes to facilitate nutrient accessibility), transport along the hyphae and finally export to the plant host. The coenocytic nature of AM fungal hyphae allows easier, faster and low-energy-demanding transport, and it is easy to observe in extraradical hyphae rapid and intense cross trafficking of vesicles. Linked to this double-exchange metabolism, it is noteworthy that *R. irregularis* devotes a large part of its metabolism to deliver nutrients to its host. *R. irregularis* possesses 276 gene models involved in inorganic ion transport and metabolism (Gloin1), and it has been shown that 2.6% of the upregulated transcripts *in planta* are major facilitator superfamily proteins and ABC transporters (Tisserant et al., 2013). Identification of the genes involved in these pathways has been a challenge over the past years, and models are proposed for the major nutrients.

3.3.1 Phosphate transport and metabolism

Phosphate is an essential nutrient and is involved in energy production and photosynthesis. It is preferentially taken up as orthophosphate (Pi) by plants, but unfortunately, this form occurs at low concentrations in soils, around 10 μM (Bieleski, 1973), due to its low solubility and low mobility, leading to a rapid depletion zone around the roots. The ability to uptake this nutrient and its availability significantly affects plant growth. Phosphate is considered as the main benefit that plants obtain by associating with AM fungi. Efficient phosphate acquisition by the fungus is partly due to the wider network of extraradical mycelium that explores a larger volume of substrate, going much further than the plant root depletion zone. Alkaline phosphatase activity and candidate genes have been identified in AM fungi (Gianinazzi, Gianinazzi-Pearson, & Dexheimer, 1979; Gianinazzi-Pearson & Gianinazzi, 1978; Liu, Parsons, Xue, Jones, & Rasmussen, 2013). *R. irregularis* possesses at least four different kinds of putative phosphatases expressed in intraradical mycelium that can cleave a broad range of substrates to release Pi (Tisserant et al., 2012). The uptake itself is carried out by transporters that are strongly expressed in the ERM. The first one to be identified was from *G. versiforme*, (Harrison & van Buuren, 1995), and since then, several others have been characterized (Benedetto, Magurno, Bonfante, & Lanfranco, 2005; Maldonado-Mendoza, Dewbre, & Harrison, 2001; Tisserant et al., 2012). To achieve the long-distance translocation from ERM to IRM, Pi is rapidly converted to polyphosphate (polyP) (Ezawa, Cavagnaro, Smith, Smith, & Ohtomo, 2004; Ezawa, Smith, & Smith, 2002; Viereck, Hansen, & Jakobsen, 2004), a phosphate chain composed of three to thousands of molecules linked by phosphoanhydride bonds (Kornberg, Rao, & Ault-Riché, 1999). PolyP synthesis activity has been detected in vacuolar membrane after fractionation of the cellular compartment of IRM (Tani, Ohtomo, Osaki, Kuga, & Ezawa, 2009), and genes coding for protein involved in the synthesis of polyP and a putative vacuolar transporter chaperone complex Vtc4p were found in *R. irregularis* (Tisserant et al., 2012). In IRM, polyP has to be hydrolyzed to free Pi that will be delivered in the apoplast where a specialized host plant transporter takes care of importation (Pumplin, Zhang, Noar, & Harrison, 2012). Several endopolyphosphatases are strongly upregulated in IRM, consistent with the model proposed. Interestingly, phosphate transporters expressed in ERM were also found in IRM (Benedetto et al., 2005; Fiorilli, Lanfranco, & Bonfante, 2013; Tisserant et al., 2012). No phytase-encoding genes were detected, indicating that the fungus cannot mobilize phosphorus of plant origin although it was hypothesized that reabsorption of Pi from the apoplastic space might take place as a mechanism to control the amount of nutrient exported. The fine-tuning of these exchanges remains to be described.

3.3.2 Nitrogen transport and metabolism

Although often underestimated, AM fungi supply significant amounts of the total N taken up by plants (Govindarajulu et al., 2005; Tanaka & Yano, 2005). Two high-affinity N transporters have been partially characterized in *R. irregularis*: GinAMT1 (López-Pedrosa, González-Guerrero, Valderas,

Azcón-Aguilar, & Ferrol, 2006) and GinAMT2 (Pérez-Tienda et al., 2011). Both are expressed in ERM, but GinAMT2 transcript levels are higher in IRM and GinAMT1 is induced in ERM at low N concentrations. A putative high-affinity nitrate transporter was also identified (Tian et al., 2010), and a transcriptomic approach carried out by Tisserant and colleagues (2012) showed that this gene was expressed more in IRM. Their study also revealed another nitrate transporter, expressed in all fungal compartments. Recently, three ammonium transporters from *Geosiphon pyriformis* were characterized. GpyrAMT1 and GpyrAMT2 are plasma membrane proteins, whereas GpyrAMT3 is localized in the vacuolar membrane. A functional amino acid (AA) permease was characterized in *G. mosseae* (GmosAAP, Cappellazzo, Lanfranco, Fitz, Wipf, & Bonfante, 2008) and several in *R. irregularis*, expressed in IRM, ERM and germinating spores. This enlarges the panel of means available not only to obtain N from the soil but also to obtain AA from the host (Tisserant et al., 2012). To be further assimilated via the glutamine synthetase/glutamate synthase (GS/GOGAT) cycle (Marzluf, 1996), nitrate has to be converted into ammonium by nitrate reductase and nitrite reductase. Transcripts corresponding to these genes were identified in *R. irregularis*: one nitrate reductase and two nitrite reductases, all expressed in ERM. N is then translocated along the hyphae to the IRM as arginine. The following step involves glutamine synthetase (GS), and 3 different transcripts have been identified in *R. irregularis* and *G. mosseae* (Breuninger, Trujillo, Serrano, Fischer, & Requena, 2004; Govindarajulu et al., 2005; Tian et al., 2010; Tisserant et al., 2012). Breuninger and colleagues showed that the two GSs identified in their study were expressed in all conditions and their activity was regulated but not their expression level. Only one putative glutamate synthase (GOGAT), the enzyme catabolizing the next step, has been identified (Tian et al., 2010) and confirmed by genomic data. Transcripts coding for proteins involved in further steps to synthesize arginine (a functional argininosuccinate synthetase (ASS), argininosuccinate lyase (AL) and carbamoyl-P-synthetase (CSP)) were identified and highly expressed in germinating spores, ERM and IRM, confirming intense N cycling in this fungus (Tian et al., 2010; Tisserant et al., 2012). It has been hypothesized that Arg is transported along the hyphae from ERM to IRM in vacuoles associated with polyP, coupling P and N translocation toward the host (Bago, Pfeffer, & Shachar-Hill, 2001). Once the Arg has reached the mycelium inside the host roots, it is hydrolyzed by arginase (functional in *R. irregularis*, GiCAR) releasing urea and ornithine. Partial sequences corresponding to arginase regulatory proteins (ArgRI,

ArgRIII and Mcm1) are present in Gloin1 (BLASTP cut-off E-value e-30, e-23 and e-24, respectively). Urea is further transformed into ammonium by urease (GiURE), while ornithine can be hydrolyzed either by ornithine aminotransferase (OAT, release glutamate) or by ornithine decarboxylase (OCD, releasing putrescine). *R. irregularis* possesses 2 OAT of which one is functional and 4 ODC of which again one is functional (Tian et al., 2010; Tisserant et al., 2012). Conversion of arginine back to ammonium might seem unrequired and an energy loss, but it saves 2 precious C units for the fungus (Govindarajulu et al., 2005). Finally, in Gloin1, 36 genes were identified in the KEGG pathway dedicated to arginine and proline metabolism.

3.3.3 Sugar transport and metabolism

Carbon flux occurs mainly from the plant to the fungus. *R. irregularis* presents 445 gene models involved in carbohydrate metabolism. They include transporters to obtain sugars in IRM and a machinery to convert them into more easily transportable C form to the ERM. It was shown long ago that C is mainly provided by the plant host as hexoses and preferentially as glucose (Pfeffer, Douds, Becard, & Shachar-Hill, 1999; Shachar-Hill et al., 1995; Solaiman & Saito, 1997). One study suggests that host plants deliver sucrose into the apoplast, converted into hexoses by a secreted plant acid invertase (Schaarschmidt, Roitsch, & Hause, 2006), confirmed by the lack of fungal secreted invertase (Tisserant et al., 2013). It is interesting to note that this model is also valid for ectomycorrhizal fungi (discussed in Plett & Martin, 2011). Hexoses are then imported via fungal transporters. The first AM fungal monosaccharide transporter (MST) was isolated from *G. pyriformis* (Schüssler, Martin, Cohen, Fitz, & Wipf, 2006) followed years later by the identification of 3 MSTs from *R. irregularis* as well as a sucrose transporter (Helber et al., 2011). Mst2 is specifically induced *in planta*, both in arbuscules and in hyphae, following the plant phosphate transporter expression pattern. When its expression is reduced, symbiosis is strongly impaired, presenting abnormal arbuscules. When incorporated, hexoses are then converted into trehalose, glycogen and lipids (Bago, Pfeffer, & Shachar-Hill, 2000; Pfeffer et al., 1999; Shachar-Hill et al., 1995). Trehalose and glycogen synthases are present in the *R. irregularis* transcript collection (Tisserant et al., 2012, 2013).

3.3.4 Lipid metabolism

Glomeromycetes can be qualified as "oleogenic" fungi: 25% of their dry weight consists of lipids (Bago et al., 2002; Jabaji-Hare, 1988; Murphy, 1990). Important lipid body trafficking is visible by microscopy along the coenocytic hyphae, mainly from IRM to ERM and from spore to germinating tip. A set of 432 lipid related genes were identified in Gloin1, for either transport or metabolism. Several experiments have shown that lipid metabolism has an unexpected and specific regulation mechanism: carbon is obtained from plants as hexose but mainly stored as triacylglycerol (a compact form of C storage, allowing long-distance translocation) in hyphae and more particularly in spores. Labelling experiments have revealed that palmitic acid biosynthesis (the first produced in fatty acid synthesis and precursor to longer ones) takes place in IRM only and is used in IRM, ERM or germinating spores (Pfeffer et al., 1999; Trépanier et al., 2005). This feature assumes fine regulation to distribute the synthesis or storage pool to where it is needed. Tisserant and colleagues (2012) showed that all the genes involved in fatty acid synthesis are present in *R. irregularis* in agreement with biochemical studies showing that the fungus did not obtain its FA from the plant but was able to synthesize them. Acetyl-CoA carboxylase is the main regulator of FA synthesis as it is responsible for the synthesis of malonyl CoA, the 2-C unit used for FA priming and further elongations to synthesize palmitate and longer C chains. Unexpectedly, microarray experiments invalidated the hypothesis that the spatial gap observed by isotope labelling approaches was due to differential expression of FA synthesis genes in ERM, IRM and germinating spores. Indeed, genes identified as being involved in this metabolism were detected in ERM in the same range as the expression level found in IRM (Tisserant et al., 2012). This seems unlikely and might imply posttranscriptional regulation. Several genes with an InterPro domain (Apweiler et al., 2000) related to fatty acid desaturase and lipase were upregulated (five- and fourfold changes respectively), and among all the genes that are upregulated *in planta*, 7% belong to the lipid transport and metabolism category (Tisserant et al., 2013).

3.4. Sexual reproduction of AM fungi

The ability of AM fungi to have a stage of sexual reproduction is an old debate. Sexual reproduction in fungi consists in finding a compatible mating partner, going through nuclear fusion and meiosis and shuffling genetic information and then clearing accumulation of deleterious mutations. No such cell events have been observed in AM fungi. As anastomosis is a well-described process among Glomeromycota (den Bakker, Vankuren, Morton, & Pawlowska, 2010; Vandenkoornhuyse, Leyval, & Bonnin, 2001), a consensus was reached that AM fungi reproduce asexually, exchanging genes by hyphal fusion and forming asexual spores. However, genes have been found that are related either to meiosis or to mating with the so-called mating type (MAT) loci, responsible for cell identity and thus compatibility. Meiosis is achieved, thanks to not only meiosis-specific genes but also genes involved in DNA repair and recombination. In four Glomus strains, a set of 51 homologous genes were identified, including meiosis-specific ones that altogether would be sufficient for meiosis (Halary et al., 2011). MAT loci include a homeodomain, α-box or high-mobility group (HMG) domain-containing protein (Fraser & Heitman, 2004). In Tisserant et al. (2012), several homologues of SexP and SexM, two HMG domain-containing ESTs involved in sex compatibility in *Phycomyces* (Idnurm, Walton, Floyd, & Heitman, 2008), were identified. In a study focused on HMG domain-containing genes in *R. irregularis* SwiC2 and DAOM197198 strains, a surprisingly elevated number of transcripts (76) was found, far exceeding what had been observed so far in other fungi (Riley et al., 2014). Gene expression patterns during crossing experiments were ambiguous and did not help to propose a model or a hypothesis. In Gloin1 assembly, prediction of MATA-HMG domain-containing proteins gave an even higher number with 146 genes predicted, 12 of them being upregulated *in planta* (Tisserant et al., 2013). In spite of all this evidence (recombination events, sufficient meiosis machinery and MAT loci), sexual events in AM fungi remain cryptic and no formal proof exists to give a proper answer yet (Corradi & Lildhar, 2012; Riley & Corradi, 2013). Mucoromycotina, the fungal group closest to AM fungi, produce trisporic acids, fungal hormones involved in the first step of recognition for mating (reviewed by Schimek & Wöstemeyer, 2009). Few genes have been characterized so far, and no clear homologous genes were found in the *R. irregularis* genome, except for TSP2, a 4-dihydromethyl-trisporate dehydrogenase that converts 4-dihydrotrisporin into trisporin (Wetzel, Scheibner, Burmester, Schimek, & Wöstemeyer, 2009). Blast of TSP2 from *Mucor mucedo*, accession number Q01213, matches on scaffold_28269 of Gloin1 with an E-value e-64. This is intriguing and it should open new perspectives about any sexual mechanisms occurring in *R. irregularis*.

4. CONCLUSION AND PERSPECTIVES

The work that led to the release of the genome assemblies of *R. irregularis* DAOM197198 was marked by a series of hurdles to be overcome. Expansion of transposable elements—one-third of the genome,

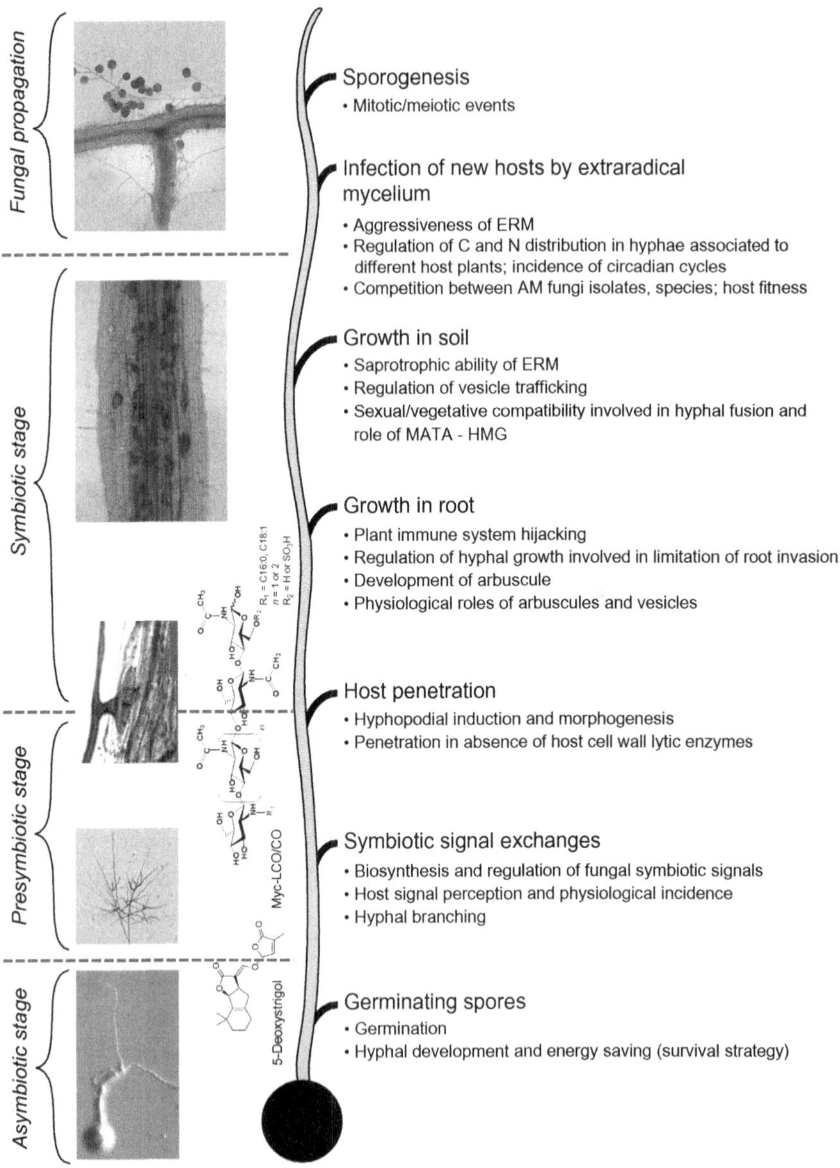

Figure 9.2 Sampling of fungal cell mechanisms and developmental steps that will be accessible for investigation using the gene repertoire of *R. irregularis* DAOM197198.

leading to a mosaic gene/transposon organization—the high AT content and the large size of this fungal genome have impaired sequencing and assembly. Sequencing technical improvement will fill the gaps within a few years. Finally, the organization as a unique haploid nuclear haplotype with limited polymorphism is surprising. The question of polykaryotism is as yet not fully resolved. Isolate DAOM197198, cultivated for 14 years on leek followed by more than 20 years on carrot hairy roots, might have lost a great part of its genetic polymorphism and thus have become the exception to the rule. Sequencing other *R. irregularis* strains will help to conclude whether DAOM197198 does indeed correspond to an artificially genetically stabilized line, the "white mouse" for AM fungi. Pending this final conclusion, the current assembly is a cornerstone in the study of AM fungi. Indeed, as summed up in Fig. 9.2, AM fungi remain elusive organisms, and the gene repertoire obtained will allow deeper investigation of their biology, ecology and genetics necessary to promote their rational use in sustainable agriculture.

ACKNOWLEDGEMENTS

The authors thank Dr Peter Winterton for the critical chapter reading and Marie Aizpuru for her help. M. M. is granted by the French Ministry of Higher Education and Research.

REFERENCES

Abramovitch, R. B., Anderson, J. C., & Martin, G. B. (2006). Bacterial elicitation and evasion of plant innate immunity. *Nature Reviews. Molecular Cell Biology*, *7*, 601–611.

Akiyama, K., Matsuzaki, K. I., & Hayashi, H. (2005). Plant sesquiterpenes induce hyphal branching in arbuscular mycorrhizal fungi. *Nature*, *435*, 824–827.

Angelard, C., Tanner, C. J., Fontanillas, P., Niculita-Hirzel, H., Masclaux, F., & Sanders, I. R. (2013). Rapid genotypic change and plasticity in arbuscular mycorrhizal fungi is caused by a host shift and enhanced by segregation. *The ISME Journal*, *11*, 2137–2146.

Apweiler, R., Attwood, T. K., Bairoch, A., Bateman, A., Birney, E., Biswas, M., et al. (2000). InterPro—An integrated documentation resource for protein families, domains and functional sites. *Bioinformatics*, *16*, 1145–1150.

Bago, B., Pfeffer, P. E., & Shachar-Hill, Y. (2000). Carbon metabolism and transport in arbuscular mycorrhizas. *Plant Physiology*, *124*, 949–958.

Bago, B., Pfeffer, P. E., & Shachar-Hill, Y. (2001). Could the urea cycle be translocating nitrogen in the arbuscular mycorrhizal symbiosis? *New Phytologist*, *149*, 4–8.

Bago, B., Zipfel, W., Williams, R. M., Jun, J., Arreola, R., Lammers, P. J., et al. (2002). Translocation and utilization of fungal storage lipid in the arbuscular mycorrhizal symbiosis. *Plant Physiology*, *128*, 108–124.

Balzergue, C., Puech-Pagès, V., Bécard, G., & Rochange, S. F. (2011). The regulation of arbuscular mycorrhizal symbiosis by phosphate in pea involves early and systemic signalling events. *Journal of Experimental Botany*, *62*, 1049–1060.

Beaudet, D., Nadimi, M., Iffis, B., & Hijri, M. (2013). Rapid mitochondrial genome evolution through invasion of mobile elements in two closely related species of arbuscular mycorrhizal fungi. *PloS One, 8*, e60768.

Bécard, G., & Fortin, J. (1988). Early events of vesicular–arbuscular mycorrhiza formation on Ri T-DNA transformed roots. *New Phytologist, 108*, 211–218.

Bécard, G., Kosuta, S., Tamasloukht, Mb, Séjalon-Delmas, N., & Roux, C. (2004). Partner communication in the arbuscular mycorrhizal interaction. *Canadian Journal of Botany, 82*, 1186–1197.

Bécard, G., & Piché, Y. (1989a). Fungal growth stimulation by CO_2 and root exudates in vesicular-arbuscular mycorrhizal symbiosis. *Applied and Environmental Microbiology, 55*, 2320–2325.

Bécard, G., & Piché, Y. (1989b). New aspects on the acquisition of biotrophic status by a vesicular—Arbuscular mycorrhizal fungus, *Gigaspora Margarita. New Phytologist, 112*, 77–83.

Benedetto, A., Magurno, F., Bonfante, P., & Lanfranco, L. (2005). Expression profiles of a phosphate transporter gene (GmosPT) from the endomycorrhizal fungus *Glomus mosseae. Mycorrhiza, 15*, 620–627.

Besserer, A., Becard, G., Jauneau, A., Roux, C., & Sejalon-Delmas, N. (2008). GR24, a synthetic analog of strigolactones, stimulates the mitosis and growth of the arbuscular mycorrhizal fungus *Gigaspora rosea* by boosting its energy metabolism. *Plant Physiology, 148*, 402–413.

Besserer, A., Bécard, G., Roux, C., & Séjalon-Delmas, N. (2009). Role of mitochondria in the response of arbuscular mycorrhizal fungi to strigolactones. *Plant Signaling Behavior, 4*, 75–77.

Besserer, A., Puech-Pagès, V., Kiefer, P., Gomez-Roldan, V., Jauneau, A., Roy, S., et al. (2006). Strigolactones stimulate arbuscular mycorrhizal fungi by activating mitochondria. *PLoS Biology, 4*, e226.

Bever, J. D., & Wang, M. (2005). Arbuscular mycorrhizal fungi, hyphal fusion and multigenomic structure. *Nature, 433*, E3–E4, Discussion E4.

Bianciotto, V., & Bonfante, P. (1992). Quantification of the nuclear DNA content of two arbuscular mycorrhizal fungi. *Mycological Research, 96*, 1071–1076.

Bieleski, R. L. (1973). Phosphate pools, phosphate transport, and phosphate. *Annual Review of Plant Physiology, 24*, 225–252.

Blaszkowski, J., Czerniawska, B., Wubet, T., Schuessler, T., Buscot, F., & Renker, C. (2008). *Glomus irregulare*, a new arbuscular mycorrhizal fungus in the Glomeromycota. *Mycotaxon, 106*, 247–267.

Breuillin, F., Schramm, J., Hajirezaei, M., Ahkami, A., Favre, P., Druege, U., et al. (2010). Phosphate systemically inhibits development of arbuscular mycorrhiza in *Petunia hybrida* and represses genes involved in mycorrhizal functioning. *The Plant Journal, 64*, 1002–1017.

Breuninger, M., Trujillo, C. G., Serrano, E., Fischer, R., & Requena, N. (2004). Different nitrogen sources modulate activity but not expression of glutamine synthetase in arbuscular mycorrhizal fungi. *Fungal Genetics and Biology, 41*, 542–552.

Brundrett, M. C. (2002). Coevolution of roots and mycorrhizas of land plants. *New Phytologist, 154*, 275–304.

Bucher, M. (2007). Functional biology of plant phosphate uptake at root and mycorrhiza interfaces. *New Phytologist, 173*, 11–26.

Cappellazzo, G., Lanfranco, L., Fitz, M., Wipf, D., & Bonfante, P. (2008). Characterization of an amino acid permease from the endomycorrhizal fungus *Glomus mosseae. Plant Physiology, 147*, 429–437.

Catoira, R., Galera, C., de Billy, F., Penmetsa, R. V., Journet, E. P., Maillet, F., et al. (2000). Four genes of *Medicago truncatula* controlling components of a nod factor transduction pathway. *Plant Cell, 12*, 1647–1666.

Chabaud, M., Genre, A., Sieberer, B. J., Faccio, A., Fournier, J., Novero, M., et al. (2011). Arbuscular mycorrhizal hyphopodia and germinated spore exudates trigger Ca2+ spiking in the legume and nonlegume root epidermis. *The New Phytologist*, *189*, 347–355.

Chabot, S., Bécard, G., & Piché, Y. (1992). Life cycle of *Glomus intraradix* in root organ culture. *Mycologia*, *84*, 315–321.

Corradi, N., Croll, D., Colard, A., Kuhn, G., Ehinger, M., & Sanders, I. R. (2007). Gene copy number polymorphisms in an arbuscular mycorrhizal fungal population. *Applied and Environmental Microbiology*, *73*, 366–375.

Corradi, N., & Lildhar, L. (2012). Meiotic genes in the arbuscular mycorrhizal fungi: What for? *Communicative and Integrative Biology*, *5*, 187–189.

Cuomo, C. A., Guldener, U., Xu, J. R., Trail, F., Turgeon, B. G., Di Pietro, A., et al. (2007). The *Fusarium graminearum* genome reveals a link between localized polymorphism and pathogen specialization. *Science*, *317*, 1400–1402.

Dangeard, P. A. (1896). Une maladie du peuplier dans l'ouest de la France. *Le Botaniste*, *5 série*, 38–43.

De la Providencia, I. E., Nadimi, M., Beaudet, D., Rodriguez Morales, G., & Hijri, M. (2013). Detection of a transient mitochondrial DNA heteroplasmy in the progeny of crossed genetically divergent isolates of arbuscular mycorrhizal fungi. *New Phytologist*, *200*, 211–221.

den Bakker, H. C., Vankuren, N. W., Morton, J. B., & Pawlowska, T. E. (2010). Clonality and recombination in the life history of an asexual arbuscular mycorrhizal fungus. *Molecular Biology and Evolution*, *27*, 2474–2486.

Doležel, J., Bartoš, J., Voglmayr, H., & Greilhuber, J. (2003). Nuclear DNA content and genome size of trout and human. *Cytometry. Part A*, *51*, 127–128.

Dotzler, N., Krings, M., Taylor, T. N., & Agerer, R. (2006). Germination shields in Scutellospora (Glomeromycota, Diversisporales, Gigasporaceae) from the 400 million-year-old Rhynie chert. *Mycological Progress*, *5*(3), 178–184.

Duc, G., Trouvelot, A., Gianinazzi-Pearson, V., & Gianinazzi, S. (1989). First report of non-mycorrhizal plant mutants (Myc-) obtained in pea (*Pisum sativum* L.) and faba bean (*Vicia faba* L.). *Plant Science*, *60*, 215–222.

Duplessis, S., Cuomo, C. A., Lin, Y. C., Aerts, A., Tisserant, E., Veneault-Fourrey, C., et al. (2011). Obligate biotrophy features unraveled by the genomic analysis of rust fungi. *Proceedings of the National Academy of Sciences*, *108*, 9166–9171.

Ewing, A. D., & Kazazian, H. H. (2010). High-throughput sequencing reveals extensive variation in human-specific L1 content in individual human genomes. *Genome Research*, *20*, 1262–1270.

Ezawa, T., Cavagnaro, T. R., Smith, S. E., Smith, F. A., & Ohtomo, R. (2004). Rapid accumulation of polyphosphate in extraradical hyphae of an arbuscular mycorrhizal fungus as revealed by histochemistry and a polyphosphate kinase/luciferase system. *New Phytologist*, *161*, 387–392.

Ezawa, T., Smith, S. E., & Smith, F. A. (2002). P metabolism and transport in AM fungi. *Plant and Soil*, *244*, 221–230.

Fester, T., & Hause, G. (2005). Accumulation of reactive oxygen species in arbuscular mycorrhizal roots. *Mycorrhiza*, *15*, 373–379.

Fiorilli, V., Lanfranco, L., & Bonfante, P. (2013). The expression of GintPT, the phosphate transporter of *Rhizophagus irregularis*, depends on the symbiotic status and phosphate availability. *Planta*, *237*, 1267–1277.

Formey, D., Molès, M., Haouy, A., Savelli, B., Bouchez, O., Bécard, G., et al. (2012). Comparative analysis of mitochondrial genomes of *Rhizophagus irregularis* - syn. *Glomus irregulare* - eveals a polymorphism induced by variability generating elements. *The New Phytologist*, *196*, 1217–1227.

Fraser, J. A., & Heitman, J. (2004). Evolution of fungal sex chromosomes. *Molecular Microbiology, 51*, 299–306.

García-Garrido, J. M., & Ocampo, J. A. (2002). Regulation of the plant defence response in arbuscular mycorrhizal symbiosis. *Journal of Experimental Botany, 53*, 1377–1386.

Genre, A., Chabaud, M., Balzergue, C., Puech-Pages, V., Novero, M., Rey, T., et al. (2013). Short-chain chitin oligomers from arbuscular mycorrhizal fungi trigger nuclear Ca2+ spiking in Medicago truncatula roots and their production is enhanced by strigolactone. *The New Phytologist, 198*, 190–202.

Genre, A., Chabaud, M., Timmers, T., Bonfante, P., & Barker, D. G. (2005). Arbuscular mycorrhizal fungi elicit a novel intracellular apparatus in Medicago truncatula root epidermal cells before infection. *Plant Cell, 17*, 3489–3499.

Gerdemann, J. W., & Trappe, J. M. (1974). The endogonales in the pacific northwest. *Mycologia Memoirs, 5*, 9–30.

Gianinazzi, S., Gianinazzi-Pearson, V., & Dexheimer, J. (1979). Enzymatic studies on the metabolism of vesicular–arbuscular mycorrhiza. III. Ultrastructural localization of acid and alkaline phosphatase in onion roots infected by *Glomus mosseae* (Nicol. & Gerd.). *New Phytologist, 82*, 127–132.

Gianinazzi-Pearson, V., & Gianinazzi, S. (1978). Enzymatic studies on the metabolism of vesicular-arbuscular mycorrhiza II. Soluble alkaline phosphatase specific to mycorrhizal infection in onion roots. *Physiological Plant Pathology, 12*, 45–53.

Giovannetti, M., Avio, L., Sbrana, C., & Citernesi, A. (1993). Factors affecting appressorium development in the vesicular-arbuscular mycorrhizal fungus *Glomus mosseae* (Nicol. & Gerd.) Gerd. and Trappe. *New Phytologist, 123*, 115–122.

Gomez-Roldan, V., Fermas, S., Brewer, P. B., Puech-Pages, V., Dun, E. A., Pillot, J.-P., et al. (2008). Strigolactone inhibition of shoot branching. *Nature, 455*, 189–194.

Govindarajulu, M., Pfeffer, P. E., Jin, H., Abubaker, J., Douds, D. D., Allen, J. W., et al. (2005). Nitrogen transfer in the arbuscular mycorrhizal symbiosis. *Nature, 435*, 819–823.

Graham, J. H., Leonard, R. T., & Menge, J. A. (1981). Membrane-mediated decrease in root exudation responsible for phosphorus inhibition of vesicular-arbuscular mycorrhiza formation. *Plant Physiology, 68*, 548–552.

Gregory, T., Nicol, J., Tamm, H., Kullman, B., Kullman, K., Leitch, I., et al. (2007). Eukaryotic genome size databases. *Nucleic Acids Research, 35*, D332–D338.

Guether, M., Balestrini, R., Hannah, M., He, J., Udvardi, M. K., & Bonfante, P. (2009). Genome-wide reprogramming of regulatory networks, transport, cell wall and membrane biogenesis during arbuscular mycorrhizal symbiosis in *Lotus japonicus*. *New Phytologist, 182*, 200–212.

Güimil, S., Chang, H.-S., Zhu, T., Sesma, A., Osbourn, A., Roux, C., et al. (2005). Comparative transcriptomics of rice reveals an ancient pattern of response to microbial colonization. *Proceedings of the National Academy of Sciences of the United States of America, 102*, 8066–8070.

Halary, S., Malik, S.-B., Lildhar, L., Slamovits, C. H., Hijri, M., & Corradi, N. (2011). Conserved meiotic machinery in Glomus spp., a putatively ancient asexual fungal lineage. *Genome Biology and Evolution, 3*, 950–958.

Hamiaux, C., Drummond, R. S. M., Janssen, B. J., Ledger, S. E. E., Cooney, J. M., Newcomb, R. D. D., et al. (2012). DAD2 is an α/β hydrolase likely to be involved in the perception of the plant branching hormone, strigolactone. *Current Biology, 22*, 2032–2036.

Harrison, M. J., Dewbre, G. R., & Liu, J. Y. (2002). A phosphate transporter from *Medicago truncatula* involved in the acquisition of phosphate released by arbuscular mycorrhizal fungi. *Plant Cell, 14*, 2413–2429.

Harrison, M. J., & van Buuren, M. L. (1995). A phosphate transporter from the mycorrhizal fungus *Glomus versiforme*. *Nature, 378*, 626–629.

Helber, N., & Requena, N. (2008). Expression of the fluorescence markers DsRed and GFP fused to a nuclear localization signal in the arbuscular mycorrhizal fungus *Glomus intraradices*. *New Phytologist, 177*, 537–548.

Helber, N., Wippel, K., Sauer, N., Schaarschmidt, S., Hause, B., & Requena, N. (2011). A versatile monosaccharide transporter that operates in the arbuscular mycorrhizal fungus *Glomus* sp is crucial for the symbiotic relationship with plants. *The Plant Cell, 23*, 3812–3823.

Hijri, M., & Sanders, I. R. (2004). The arbuscular mycorrhizal fungus *Glomus intraradices* is haploid and has a small genome size in the lower limit of eukaryotes. *Fungal Genetics and Biology, 41*, 253–261.

Hijri, M., & Sanders, I. R. (2005). Low gene copy number shows that arbuscular mycorrhizal fungi inherit genetically different nuclei. *Nature, 433*, 160–163.

Hosny, M., Gianinazzi-Pearson, V., & Dulieu, H. (1998). Nuclear DNA content of 11 fungal species in glomales. *Genome, 41*, 422–428.

Huallachain, M. O., Karczewski, K. J., Weissman, S. M., & Eckehart, A. (2012). Extensive genetic variation in somatic human tissues. *Proceedings of the National Academy of Sciences, 109*(44), 18018–18023.

Idnurm, A., Walton, F. J., Floyd, A., & Heitman, J. (2008). Identification of the sex genes in an early diverged fungus. *Nature, 451*, 193–196.

Jabaji-Hare, S. (1988). Lipid and fatty acid profiles of some vesicular-arbuscular mycorrhizal fungi, contribution to taxonomy. *Mycologia, 80*, 622–629.

Jakobsen, I., & Rosendahl, L. (1990). Carbon flow into soil and external hyphae from roots of mycorrhizal cucumber plants. *New Phytologist, 115*, 77–83.

Javot, H., Penmetsa, R. V., Terzaghi, N., Cook, D. R., & Harrison, M. J. (2007). A Medicago truncatula phosphate transporter indispensable for the arbuscular mycorrhizal symbiosis. *Proceedings of the National Academy of Sciences, 104*, 1720–1725.

Juge, C., Coughlan, P., Fortin, A., & Piché, Y. (2009). Growth and branching of asymbiotic, presymbiotic and extraradical AM fungal hyphae, clarification of concept and terminology. In D. Khasa, Y. Piché, & Andrew P. Coughlan (Eds.), *Advances in mycorrhizal science and technology* (pp. 39–50). Ottawa, ON: NRC Research Press.

Kamoun, S. (2007). Groovy times, filamentous pathogen effectors revealed. *Current Opinion in Plant Biology, 10*, 358–365.

Kapulnik, Y., Volpin, H., Itzhaki, H., Ganon, D., Galili, S., David, R., et al. (1996). Suppression of defence responses in mycorrhizal alfalfa and tobacco roots. *New Phytologist, 133*, 59–64.

Kiers, E. T., Duhamel, M., Beesetty, Y., Mensah, J. A., Franken, O., Verbruggen, E., et al. (2011). Reciprocal rewards stabilize cooperation in the mycorrhizal symbiosis. *Science, 333*, 880–882.

Kloppholz, S., Kuhn, H., & Requena, N. (2011). A secreted fungal effector of Glomus intraradices promotes symbiotic biotrophy. *Current Biology, 21*, 1204–1209.

Koide, R. T. (1991). Tansley review no. 29. Nutrient supply, nutrient demand and plant response to mycorrhizal infection. *New Phytologist, 117*, 365–386.

Koide, R. T., & Mosse, B. (2004). A history of research on arbuscular mycorrhiza. *Mycorrhiza, 14*, 145–163.

Kornberg, A., Rao, N. N., & Ault-Riché, D. (1999). Inorganic polyphosphate, a molecule of many functions. *Annual Review of Biochemistry, 68*, 89–125.

Koske, R. E. (1981). Multiple germination by spores of *Gigaspora gigantea*. *Transactions of the British Mycological Society, 76*, 328–330.

Krüger, M., Krüger, C., Walker, C., Stockinger, H., & Schüßler, A. (2012). Phylogenetic reference data for systematics and phylotaxonomy of arbuscular mycorrhizal fungi from phylum to species level. *New Phytologist, 193*, 970–984.

Kuhn, G., Hijri, M., & Sanders, I. R. (2001). Evidence for the evolution of multiple genomes in arbuscular mycorrhizal fungi. *Nature, 414*, 745–748.

Lanfranco, L., Novero, M., Bonfante, P., & Torino, S. (2005). The mycorrhizal fungus *Gigaspora margarita* possesses a CuZn superoxide dismutase that is up-regulated during symbiosis with legume hosts. *Plant Physiology, 137*, 1319–1330.

Lang, F. B., & Hijri, M. (2009). The complete *Glomus intraradices* mitochondrial genome sequence—A milestone in mycorrhizal research. *New Phytologist, 183*, 3–6.

Lee, J., & Young, J. P. W. (2009). The mitochondrial genome sequence of the arbuscular mycorrhizal fungus *Glomus intraradices* isolate 494 and implications for the phylogenetic placement of Glomus. *New Phytologist, 183*, 200–211.

Lei, J., Beard, G., Catford, J. G., & Piche, Y. (1991). Root factors stimulate 32P uptake and plasmalemma ATPase activity in vesicular-arbuscular mycorrhizal fungus, *Gigaspora margarita*. *New Phytologist, 118*, 289–294.

Lin, K., Limpens, E., Zhang, Z., Ivanov, S., Saunders, D. G. O., Mu, D., et al. (2014). Single nucleus genome sequencing reveals high similarity among nuclei of an endomycorrhizal Fungus. *PLoS Genetics, 10*, e1004078.

Liu, Q., Parsons, A. J., Xue, H., Jones, C. S., & Rasmussen, S. (2013). Functional characterisation and transcript analysis of an alkaline phosphatase from the arbuscular mycorrhizal fungus *Funneliformis mosseae*. *Fungal Genetics and Biology, 54*, 52–59.

López-Pedrosa, A., González-Guerrero, M., Valderas, A., Azcón-Aguilar, C., & Ferrol, N. (2006). GintAMT1 encodes a functional high-affinity ammonium transporter that is expressed in the extraradical mycelium of *Glomus intraradices*. *Fungal Genetics and Biology, 43*, 102–110.

Maillet, F., Poinsot, V., André, O., Puech-Pagès, V., Haouy, A., Gueunier, M., et al. (2011). Fungal lipochitooligosaccharide symbiotic signals in arbuscular mycorrhiza. *Nature, 469*, 58–63.

Maldonado-Mendoza, I. E., Dewbre, G. R., & Harrison, M. J. (2001). A phosphate transporter gene from the extra-radical mycelium of an arbuscular mycorrhizal fungus *Glomus intraradices* is regulated in response to phosphate in the environment. *Molecular Plant-Microbe Interactions, 14*, 1140–1148.

Martin, F., Gianinazzi-Pearson, V., Hijri, M., Lammers, P., Requena, N., Sanders, I. R., et al. (2008). The long hard road to a completed *Glomus intraradices* genome. *New Phytologist, 180*, 747–750.

Martin, F., Tuskan, G. A., DiFazio, S. P., Lammers, P., Newcombe, G., & Podila, G. K. (2004). Symbiotic sequencing for the Populus mesocosm. *New Phytologist, 161*, 330–335.

Marzluf, G. A. (1996). Regulation of nitrogen metabolism in mycelial fungi. In R. Brambl & G. A. Marzluf (Eds.), Berlin: Springer.

Moora, M., Berger, S., Davison, J., Öpik, M., Bommarco, R., Bruelheide, H., et al. (2011). Alien plants associate with widespread generalist arbuscular mycorrhizal fungal taxa, evidence from a continental-scale study using massively parallel 454 sequencing. *Journal of Biogeography, 38*, 1305–1317.

Mosse, B. (1953). Fructifications associated with mycorrhizal strawberry roots. *Nature, 171*, 974.

Mosse, B., & Hepper, C. (1975). Vesicular-arbuscular mycorrhizal infections in root organ cultures. *Physiological Plant Pathology, 5*(3), 215–223.

Mugnier, J., & Mosse, B. (1987). Spore germination and viability of a vesicular arbuscular mycorrhizal fungus, *Glomus mosseae*. *Transactions of the British Mycological Society, 88*(3), 411–413.

Murphy, D. J. (1990). Storage lipid bodies in plants and other organisms. *Progress in Lipid Research, 29*, 299–324.

Nagahashi, G., & Douds, David D., Jr. (1997). Appressorium formation by AM fungi on isolated cell walls of carrot. *New Phytologist, 136*, 299–304.

Nakamura, H., Xue, Y.-L., Miyakawa, T., Hou, F., Qin, H.-M., Fukui, K., et al. (2013). Molecular mechanism of strigolactone perception by DWARF14. *Nature Communications, 4*, 2613.

Newsham, K.-K., Upson, R., & Read, D.-J. (2009). Mycorrhizas and dark septate root endophytes in polar regions. *Fungal Ecology, 2*, 10–20.

Nilsson, R. H., Kristiansson, E., Ryberg, M., Hallenberg, N., & Larsson, K. H. (2008). Intraspecific ITS variability in the kingdom fungi as expressed in the international sequence databases and its implications for molecular species identification. *Evolutionary Bioinformatics Online, 4*, 193.

Nowara, D., Gay, A., Lacomme, C., Shaw, J., Ridout, C., Douchkov, D., et al. (2010). HIGS, host-induced gene silencing in the obligate biotrophic fungal pathogen *Blumeria graminis*. *The Plant Cell, 22*, 3130–3141.

O'Brien, J. A., Daudi, A., Butt, V. S., & Bolwell, G. P. (2012). Reactive oxygen species and their role in plant defence and cell wall metabolism. *Planta, 236*, 765–779.

Opik, M., Vanatoa, A., Vanatoa, E., Moora, M., Davison, J., Kalwij, J. M., et al. (2010). The online database MaarjAM reveals global and ecosystemic distribution patterns in arbuscular mycorrhizal fungi (Glomeromycota). *The New Phytologist, 188*, 223–241.

Öpik, M., Zobel, M., Cantero, J. J., Davison, J., Facelli, J. M., Hiiesalu, I., et al. (2013). Global sampling of plant roots expands the described molecular diversity of arbuscular mycorrhizal fungi. *Mycorrhiza, 23*, 411–430.

Parra, G., Bradnam, K., & Korf, I. (2007). CEGMA, a pipeline to accurately annotate core genes in eukaryotic genomes. *Bioinformatics, 23*(9), 1061–1067.

Pauly, N., Pucciariello, C., Mandon, K., Innocenti, G., Jamet, A., Baudouin, E., et al. (2006). Reactive oxygen and nitrogen species and glutathione, key players in the legume-Rhizobium symbiosis. *Journal of Experimental Botany, 57*, 1769–1776.

Pawlowska, T. E. (2005). Genetic processes in arbuscular mycorrhizal fungi. *FEMS Microbiology Letters, 251*, 185–192.

Pawlowska, T. E., & Taylor, J. W. (2004). Organization of genetic variation in individuals of arbuscular mycorrhizal fungi. *Nature, 427*, 733–737.

Pelin, A., Pombert, J.-F., Salvioli, A., Bonen, L., Bonfante, P., & Corradi, N. (2012). The mitochondrial genome of the arbuscular mycorrhizal fungus *Gigaspora margarita* reveals two unsuspected trans-splicing events of group I introns. *The New Phytologist, 194*, 836–845.

Pérez-Tienda, J., Testillano, P. S., Balestrini, R., Fiorilli, V., Azcón-Aguilar, C., & Ferrol, N. (2011). GintAMT2, a new member of the ammonium transporter family in the arbuscular mycorrhizal fungus *Glomus intraradices*. *Fungal Genetics and Biology, 48*, 1044–1055.

Pfeffer, P., Douds, D. D., Becard, G., & Shachar-Hill, Y. (1999). Carbon uptake and the metabolism and transport of lipids in an arbuscular mycorrhiza. *Plant Physiology, 120*, 587–598.

Plenchette, C., Furian, V., & Fortin, J. A. (1982). Effects of different endomycorrhizal fungi on five host plants grown on calcined montmorillonite clay [apple, asparagus, leek, strawberry, oats]. *Journal of the American Society for Horticultural Science, 107*, 535–538.

Plett, J. M., & Martin, F. (2011). Blurred boundaries, lifestyle lessons from ectomycorrhizal fungal genomes. *Trends in Genetics, 27*, 14–22.

Pozo, M. J., Azcón-Aguilar, C., Dumas-Gaudot, E., & Barea, J. M. (1998). Chitosanase and chitinase activities in tomato roots during interactions with arbuscular mycorrhizal fungi or *Phytophthora parasitica*. *Journal of Experimental Botany, 49*, 1729–1739.

Pozo, M. J., Loon, L. C., & Pieterse, C. M. J. (2005). Jasmonates—Signals in plant-microbe interactions. *Journal of Plant Growth Regulation, 23*, 211–222.

Pumplin, N., Zhang, X., Noar, R. D., & Harrison, M. J. (2012). Polar localization of a symbiosis-specific phosphate transporter is mediated by a transient reorientation of

secretion. *Proceedings of the National Academy of Sciences of the United States of America*, *109*, E665–E672.
Raab, P. A., Brennwald, A., & Redecker, D. (2005). Mitochondrial large ribosomal subunit sequences are homogeneous within isolates of Glomus (arbuscular mycorrhizal fungi, Glomeromycota). *Mycological Research*, *109*, 1315–1322.
Redecker, D., Kodner, R., & Graham, L. E. (2000). Glomalean fungi from the Ordovician. *Science*, *289*, 1920–1921.
Redecker, D., Schüssler, A., Stockinger, H., Stürmer, S. L., Morton, J. B., & Walker, C. (2013). An evidence-based consensus for the classification of arbuscular mycorrhizal fungi (Glomeromycota). *Mycorrhiza*, *23*, 515–531.
Requena, N., Mann, P., Hampp, R., & Franken, P. (2002). Early developmentally regulated genes in the arbuscular mycorrhizal fungus Glomus mosseae, identification of GmGIN1, a novel gene with homology to the C-terminus of metazoan hedgehog proteins. *Plant and Soil*, *244*, 129–139.
Riley, R., Charron, P., Idnurm, A., Farinelli, L., Dalpé, Y., Martin, F., et al. (2014). Extreme diversification of the mating type-high-mobility group (MATA-HMG) gene family in a plant-associated arbuscular mycorrhizal fungus. *New Phytologist*, *201*, 254–268.
Riley, R., & Corradi, N. (2013). Searching for clues of sexual reproduction in the genomes of arbuscular mycorrhizal fungi. *Fungal Ecology*, *6*, 44–49.
Sanders, I. R., Alt, M., Groppe, K., Boller, T., & Wiemken, A. (1995). Identification of ribosomal DNA polymorphisms among and within spores of the Glomales: Application to studies on the genetic diversity of arbuscular mycorrhizal fungal communities. *New Phytologist*, *130*(3), 419–427.
Schaarschmidt, S., Roitsch, T., & Hause, B. (2006). Arbuscular mycorrhiza induces gene expression of the apoplastic invertase LIN6 in tomato (*Lycopersicon esculentum*) roots. *Journal of Experimental Botany*, *57*, 4015–4023.
Schimek, C., & Wöstemeyer, J. (2009). Carotene derivatives in sexual communication of zygomycete fungi. *Phytochemistry*, *70*, 1867–1875.
Schoch, C. L., Seifert, K. A., Huhndorf, S., Robert, V., Spouge, J. L., Levesque, C. A., et al. (2012). Nuclear ribosomal internal transcribed spacer (ITS) region as a universal DNA barcode marker for Fungi. *Proceedings of the National Academy of Sciences*, *109*, 6241–6246.
Schüssler, A., Martin, H., Cohen, D., Fitz, M., & Wipf, D. (2006). Characterization of a carbohydrate transporter from symbiotic glomeromycotan fungi. *Nature*, *444*, 933–936.
Schüßler, A., Schwarzott, D., & Walker, C. (2001). A new fungal phylum, the Glomeromycota, phylogeny and evolution. *Mycological Research*, *105*, 1413–1421.
Schüßler, A., & Walker, C. (2010). *The glomeromycota: a species list with new families and new genera*. Libraries at The Royal Botanic Garden Edinburgh, The Royal Botanic Garden Kew, Botanische Staatssammlung Munich, and Oregon State University. Available at: www.amf-phylogeny.com.
Schwarzott, D., Walker, C., & Schüßler, A. (2001). Glomus, the largest genus of the arbuscular mycorrhizal fungi (Glomales), is nonmonophyletic. *Molecular Phylogenetics and Evolution*, *21*, 190–197.
Sedzielewska, K. A., Fuchs, J., Temsch, E. M., Baronian, K., Watzke, R., & Kunze, G. (2011). Estimation of the *Glomus intraradices* nuclear DNA content. *New Phytologist*, *192*, 794–797.
Shachar-Hill, Y., Pfeffer, P. E., Douds, D. D., Osman, S. F., Doner, L. W., & Ratcliffe, R. G. (1995). Partitioning of intermediary carbon metabolism in vesicular-arbuscular mycorrhizal leek. *Plant Physiology*, *108*, 7–15.
Simon, U. K., & Weiß, M. (2008). Intragenomic variation of fungal ribosomal genes is higher than previously thought. *Molecular Biology and Evolution*, *25*, 2251–2254.
Smith, S. E., & Read, D. J. (2010). *Mycorrhizal symbiosis*. London: Academic Press, Access Online via Elsevier.

Solaiman, Z. M., & Saito, M. (1997). Use of sugars by intraradical hyphae of arbuscular mycorrhizal fungi revealed by radiorespirometry. *New Phytologist, 136*, 533–538.

St-Arnaud, M., Hamel, C., Vimard, B., Caron, M., & Fortin, J. (1996). Enhanced hyphal growth and spore production of the arbuscular mycorrhizal fungus *Glomus intraradices* in an *in vitro* system in the absence of host roots. *Mycological Research, 100*, 328–332.

Stockinger, H., Walker, C., & Schüssler, A. (2009). '*Glomus intraradices* DAOM197198', a model fungus in arbuscular mycorrhiza research, is not *Glomus intraradices*. *New Phytologist, 183*(4), 1176–1187.

Stürmer, S. L. (2012). A history of the taxonomy and systematics of arbuscular mycorrhizal fungi belonging to the phylum Glomeromycota. *Mycorrhiza, 22*, 247–258.

Tamasloukht, M. B., Se, N., Kluever, A., Jauneau, A., Roux, C., Bécard, G., et al. (2003). Root factors induce mitochondrial-related gene expression and fungal respiration during the developmental switch from asymbiosis to presymbiosis in the arbuscular mycorrhizal fungus *Gigaspora rosea*. *Plant Physiology, 131*, 1468–1478.

Tanaka, Y., & Yano, K. (2005). Nitrogen delivery to maize via mycorrhizal hyphae depends on the form of N supplied. *Plant, Cell and Environment, 28*, 1247–1254.

Tani, C., Ohtomo, R., Osaki, M., Kuga, Y., & Ezawa, T. (2009). ATP-dependent but proton gradient-independent polyphosphate-synthesizing activity in extraradical hyphae of an arbuscular mycorrhizal fungus. *Applied and Environmental Microbiology, 75*, 7044–7050.

Tian, C., Kasiborski, B., Koul, R., Lammers, P. J., Bücking, H., & Shachar-Hill, Y. (2010). Regulation of the nitrogen transfer pathway in the arbuscular mycorrhizal symbiosis, gene characterization and the coordination of expression with nitrogen flux. *Plant Physiology, 153*, 1175–1187.

Tisserant, E., Kohler, A., Dozolme-Seddas, P., Balestrini, R., Benabdellah, K., Colard, A., et al. (2012). The transcriptome of the arbuscular mycorrhizal fungus *Glomus intraradices* (DAOM 197198) reveals functional tradeoffs in an obligate symbiont. *New Phytologist, 193*, 755–769.

Tisserant, E., Malbreil, M., Kuo, A., Kohler, A., Symeonidi, A., Balestrini, R., et al. (2013). Genome of an arbuscular mycorrhizal fungus provides insight into the oldest plant symbiosis. *Proceedings of the National Academy of Sciences, 110*, 20117–20122.

Trépanier, M., Bécard, G., Moutoglis, P., Willemot, C., Gagné, S., Avis, T. J., et al. (2005). Dependence of arbuscular-mycorrhizal fungi on their plant host for palmitic acid synthesis. *Applied and Environmental Microbiology, 71*, 5341–5347.

Tulasne, L. R., & Tulasne, C. (1845). Fungi nonnulli hypogaei, novi minus cogniti. *Giornale Botanico Italiano, 2*, 35–63.

Valent, B., & Khang, C. H. (2010). Recent advances in rice blast effector research. *Current Opinion in Plant Biology, 13*, 434–441.

Vandenkoornhuyse, P., Leyval, C., & Bonnin, I. (2001). High genetic diversity in arbuscular mycorrhizal fungi, evidence for recombination events. *Heredity, 87*, 243–253.

Van der Heijden, M., Klironomos, J., Ursic, M., Moutouglis, P., Streitwolf-Engel, R., Bollet, T., et al. (1998). Mycorrhizal fungal diversity determines plant biodiversity, ecosystem variability and productivity. *Nature, 74*, 69–72.

van der Heijden, M., & Scheublin, T. R. (2007). Functional traits in mycorrhizal ecology, their use for predicting the impact of arbuscular mycorrhizal fungal communities on plant growth and ecosystem functioning. *New Phytologist, 174*, 244–250.

Vankuren, N. W., den Bakker, H. C., Morton, J. B., & Pawlowska, T. E. (2013). Ribosomal RNA gene diversity, effective population size, and evolutionary longevity in asexual glomeromycota. *Evolution, 67*, 207–224.

Viereck, N., Hansen, P. E., & Jakobsen, I. (2004). Phosphate pool dynamics in the arbuscular mycorrhizal fungus *Glomus intraradices* studied by *in vivo* ^{31}P NMR spectroscopy. *New Phytologist, 162*, 783–794.

Walder, F., Niemann, H., Natarajan, M., Lehmann, M. F., Boller, T., & Wiemken, A. (2012). Mycorrhizal networks, common goods of plants shared under unequal terms of trade. *Plant Physiology*, *159*, 789–797.

Wang, B., & Qiu, Y. (2006). Phylogenetic distribution and evolution of mycorrhizas in land plants. *Mycorrhiza*, *16*, 299–363.

Wang, E., Schornack, S., Marsh, J. F., Gobbato, E., Schwessinger, B., Eastmond, P., et al. (2012). A common signaling process that promotes mycorrhizal and oomycete colonization of plants. *Current Biology*, *22*, 2242–2246.

Wawra, S., Belmonte, R., Löbach, L., Saraiva, M., Willems, A., & van West, P. (2012). Secretion, delivery and function of oomycete effector proteins. *Current Opinion in Microbiology*, *15*, 685–691.

Wetzel, J., Scheibner, O., Burmester, A., Schimek, C., & Wöstemeyer, J. (2009). 4-dihydrotrisporin-dehydrogenase, an enzyme of the sex hormone pathway of *Mucor mucedo*, purification, cloning of the corresponding gene, and developmental expression. *Eukaryotic Cell*, *8*, 88–95.

Zamioudis, C., & Pieterse, C. M. J. (2012). Modulation of host immunity by beneficial microbes. *Molecular Plant-Microbe Interactions*, *25*, 139–150.

CHAPTER TEN

Genomes of Plant-Associated Clavicipitaceae

Christopher L. Schardl[*,1], Carolyn A. Young[†], Neil Moore[‡], Nicholas Krom[†], Pierre-Yves Dupont[§], Juan Pan[*], Simona Florea[*], Jennifer S. Webb[*], Jolanta Jaromczyk[*], Jerzy W. Jaromczyk[‡], Murray P. Cox[§], Mark L. Farman[*]
[*]Department of Plant Pathology, University of Kentucky, Lexington, Kentucky, USA
[†]The Samuel Roberts Noble Foundation, Ardmore, Oklahoma, USA
[‡]Department of Computer Science, University of Kentucky, Lexington, Kentucky, USA
[§]Institute of Fundamental Sciences, Massey University, Palmerston North, New Zealand
[1]Corresponding author: e-mail address: schardl@uky.edu

Contents

Abstract

Fungi of family Clavicipitaceae serve as models for evolution on the symbiotic continuum from pathogenic to mutualistic. Clavicipitaceous fungi associate with plants, invertebrates, and other fungi. Most plant-associated Clavicipitaceae systemically colonize shoots, but the fungal fruiting structures are localized to inflorescences, florets, buds, leaves, or nodes. Many Clavicipitaceae decrease or eliminate host seed production, but some have evolved such intimate symbioses with plant hosts that they disseminate

Advances in Botanical Research, Volume 70
ISSN 0065-2296
http://dx.doi.org/10.1016/B978-0-12-397940-7.00010-0

clonally in seeds (vertical transmission) without damage or any reduction in plant fertility. In such cases, the fungi dramatically enhance host fitness by producing defensive alkaloids and through other mechanisms. To date, sequences have been assembled for 26 Clavicipitaceae representing 21 species in seven genera. These include three *Claviceps* species that fruit on and replace host ovaries, two *Metarhizium* species that parasitize insects and associate with plant roots, and 21 strains of systemic plant parasites or symbionts. Of the latter, 14 are capable of vertical transmission, and of those, 7 are strictly seed-borne mutualists in genera *Epichloë* and *Periglandula*. Alkaloid biosynthetic genes are widely distributed among these fungi. Gene clusters for ergot alkaloids and indole-diterpenes, both of which are neurotoxins in vertebrates and invertebrates, are present in members of all seven genera. The genes for anti-insect loline alkaloids and peramine have a more restricted distribution, but are present in many of the vertically transmissible *Epichloë* species. The availability of these genome sequences will facilitate studies of the evolution and mechanisms underlying the diversity of metabolism, host interactions, and niche adaptation of plant-associated Clavicipitaceae.

1. INTRODUCTION

1.1. Biology of plant-associated Clavicipitaceae

The family Clavicipitaceae, order Hypocreales, is composed of fungi that interact with a broad range of invertebrate animals and plants and occasionally with other fungi. The associations with plants span a symbiotic continuum including parasitism, mutualism, and pleiotropic symbioses where the relative benefits to host and symbiont depend on the developmental stages and modes of transmission and may be mediated by environmental variables. Diversity among the plant-associated Clavicipitaceae and their potential to protect host plants are also starkly evident in the variety of specialized (secondary) metabolites (SMs) that they produce, including antiherbivore alkaloids belonging to four different chemical classes and exhibiting considerable structural variants within those classes. Recent sequencing of genomes from a wide diversity of plant-associated Clavicipitaceae (Schardl, Young, Hesse, et al., 2013; Schardl, Young, Pan, et al., 2013) should facilitate comparative genomic analyses that can help address the mechanistic basis for mutualism and parasitism; the basis for variation in expression, location, and structure of fruiting bodies; variation in the ability and efficiency of vertical transmission in host seeds; and metabolic diversity. For example, such a comparative analysis has already revealed that the alkaloid gene loci in *Epichloë* species are highly dynamic and tend to have much more abundant repeat sequence than in representatives of genera *Aciculosporium*, *Claviceps*, *Metarhizium*, and

Periglandula (Schardl, Young, Hesse, et al., 2013; Schardl, Young, Pan, et al., 2013). The dynamics of SM genes and loci, as revealed through comparative genomics, is the main focus of this chapter.

1.1.1 Symbiosis and transmission strategies of plant-associated Clavicipitaceae

The interactions of Clavicipitaceae with plant hosts range from pathogenic to highly mutualistic (Table 10.1). Some (e.g. *Claviceps* species and *Neoclaviceps monostipa*) replace host seeds with their fruiting structures (Pažoutová, Kolarik, & Kolinska, 2004; Tudzynski & Scheffer, 2004), while others (*Metarhizium* species) are plant root associates and insect pathogens (Gao et al., 2011). Many form systemic (endophytic) associations throughout host shoots but fruit in a highly localized manner on leaves, nodes, buds, or inflorescences. Such systemic associations characterize *Aciculosporium* (Tanaka & Tanaka, 2008), *Atkinsonella* (Leuchtmann & Clay, 1989), *Balansia* (Diehl, 1950), *Ephelis* (Tanaka & Tanaka, 2008), *Epichloë* (Leuchtmann, Bacon, Schardl, White, & Tadych, 2014), *Heteroepichloë* (Tanaka & Tanaka, 2008), *Myriogenospora* (Glenn, Rykard, Bacon, & Hanlin, 1998), and *Parepichloë* (White & Reddy, 1998) species and possibly also *Cepsiclava phalaridis* (Walker, 2004), *Corallocytostroma* species (Pažoutová et al., 2004), *Neoclaviceps monostipa* (White & Reddy, 1998), and *Nigrocornus scleroticus* (Ryley, 2003). Some species grow in intercellular spaces (endobiotic growth), whereas others are restricted to surfaces of, and spaces between, plant tissue layers (epibiotic growth). We consider "endophytes" to encompass all species that grow in asymptomatic plant parts, whether endobiotically or epibiotically.

Remarkably, some plants have established symbioses with Clavicipitaceae that are heritable, being vertically transmitted via host seeds. In particular, many cool-season grasses (Poaceae subfamily Poöideae) possess seed-borne *Epichloë* species (including *Neotyphodium* species) (Schardl, 2010), and many morning glories (Convolvulaceae tribe Ipomoeeae) have seed-borne *Periglandula* species (Steiner, Leibner, Schardl, Leuchtmann, & Leistner, 2011). These same plant groups can possess other heritable symbionts (An et al., 1993; Cook et al., 2013), but the associations with Clavicipitaceae are particularly common and are the best-documented defensive mutualisms. Symbiotic and parasitic Clavicipitaceae often produce a wide array of alkaloids that antagonize invertebrate and sometimes vertebrate herbivores (Schardl, Young, Faulkner, Florea, & Pan, 2012; Schardl, Young, Hesse, et al., 2013).

Table 10.1 Characteristics of genera in plant-associated Clavicipitaceae[a]

Genus	Hosts	Seed transmission	Systemic growth	Fruiting type	Fruiting location
Aciculosporium	Poaceae, Bambusoideae	No	Endobiotic	Stroma	Bud
Atkinsonella	Poaceae	Yes	Epibiotic	Stroma	Inflorescence
Balansia	Poaceae or Cyperaceae	No	Endobiotic or epibiotic	Stroma	Bud, leaf, node, or inflorescence
Cepsiclava	Poaceae	No	Endobiotic	Sclerotium	Floret
Claviceps	Poaceae	No	None	Sclerotium	Floret
Corallocytostroma	Poaceae	No	ND	Sclerotium	Inflorescence
Epichloë	Poaceae, Poöideae	Yes	Endobiotic	Stroma	Inflorescence
Heteroepichloë	Poaceae, Bambusoideae	No	Epibiotic	Stroma	Leaf
Metarhizium	Plants and insects	No	In insect	Stroma	Insect
Myriogenospora	Poaceae, Panicoideae	No	Epibiotic	Stroma	Leaf
Neoclaviceps	Poaceae	No	ND	Hypothallus	Floret
Nigrocornus	Poaceae	No	ND	Sclerotium	Bud
Parepichloë	Poaceae, Panicoideae	No	Epibiotic	Stroma	Inflorescence
Periglandula	Convolvulaceae, Ipomoeeae	Yes	Epibiotic	None	None
Villosiclava	Poaceae, Oryzoideae	No	ND	Sclerotium	Floret

[a]Only genera described from sexual structures (teleomorphs) are listed.
ND, not determined.

1.1.2 *Vertically transmitted symbionts, including asexual* Epichloë *species*

The process of vertical transmission has best been described for *Epichloë* species in grasses (Freeman, 1904; Philipson & Christey, 1986; Sampson, 1937). These fungi can colonize most aerial portions of the host plant, including meristematic zones. In most cases, even the ovary and eventually the embryo are infected but undamaged, leading to vertical transmission. Among the Clavicipitaceae, this heritability is known only for *Epichloë* and *Periglandula* species and for *Atkinsonella hypoxylon* (Clay, 1994). It seems somewhat ironic, then, that sexual *Epichloë* species fruit on immature host inflorescences and arrest their development, thereby preventing seed production from the symptomatic tillers. This phenomenon, known as "choke disease", also occurs on most or all reproductive tillers of grasses infected with *Atkinsonella* or *Corallocytostroma* species (Pažoutová et al., 2004) and on grasses and sedges (Cyperaceae) infected with certain *Balansia* and *Ephelis* species (Diehl, 1950; Tanaka & Tanaka, 2008). What is remarkable about the sexual *Epichloë* species is that, in most infected hosts, the majority of reproductive tillers are completely asymptomatic and give rise to normal seeds bearing the fungal symbiont. In such symbioses, both the grass plant and its symbiotic *Epichloë* species have the benefit of a complete range of reproductive capabilities.

Asexual *Epichloë* (*Neotyphodium*) species primarily or exclusively use vertical transmission in host seeds as their means of dissemination (Schardl, Leuchtmann, & Spiering, 2004). The vast majority of the asexual *Epichloë* species have not been reported to produce any external hyphae or spores under natural conditions, but a few have been observed to produce hyphal nets or stromata bearing mitotic spores (conidia) (Tadych, Ambrose, Bergen, Belanger, & White, 2012). Though claims have been made that conidia can mediate the horizontal transmission of *Epichloë typhina* ssp. *poae* (Tadych et al., 2012), direct infection by germinating conidia (rather than from growing cultures) has not been demonstrated. However, hyphal proliferation enhanced by arthropod activities and host wounds, as observed for *E. typhina* (Alderman, 2013), may well provide the means for horizontal transmission. An overwhelming body of evidence nevertheless supports vertical transmission in seeds as the primary and often exclusive means of transmission for the asexual *Epichloë* species.

Most asexual *Epichloë* species produce several antiherbivore alkaloids, often in high abundance (Schardl et al., 2012; Schardl, Young, Pan, et al., 2013), as would be expected when such systems are subject to selection due to reliance on vertical transmission (Clay & Schardl, 2002; Selosse & Schardl, 2007). However, asexual reproduction can also exact a significant genetic cost in the long term, and it may be for that reason that most asexual *Epichloë* species are recently derived interspecific hybrids with two or even three genomes traceable to sexual ancestors (Schardl, 2010; Selosse & Schardl, 2007).

1.1.3 Biology and life history of ergot fungi

Although alkaloid production and vertical transmissibility can be a basis for mutualistic symbiosis, alkaloid production is not limited to the mutualistic symbionts, but is widespread in the Clavicipitaceae. Particularly important sources of alkaloids are the *Claviceps* species, the infamous "ergot" fungi that can contaminate grain supplies. These fungi infect individual host florets via the stigma, ultimately engulfing the host ovary and developing into dense sclerotia (resting structures) that are also called ergots (from old French, "argot", "cock's spur", in reference to their shape) (Pažoutová et al., 2004; Tudzynski & Scheffer, 2004). Ergots resemble seeds in density and sometimes size and shape, are not easily removed with the chaff, and were a dangerous source of mycotoxins until the advent of modern mechanical techniques to remove them (Schardl, Panaccione, & Tudzynski, 2006). One reason that wheat (*Triticum aestivum*) is preferable among the grains is that it self-pollinates before florets open, largely eliminating opportunities for *Claviceps* spores to access susceptible young stigmata. Open-pollinated grains such as rye (*Secale cereale*) and seed heads of forage grasses such as tall fescue (*Lolium arundinaceum*) are much more susceptible to infection by *Claviceps* species. Compounds of the ergot alkaloid chemical class have been purified from ergot fungi and sometimes chemically modified into important pharmaceuticals or into the illicit drug, lysergic acid diethylamide (LSD) (Hofmann, 1978; Schardl et al., 2006). Ergots can also contain tremorgenic indole-diterpenes, which have received less consideration in human health but are a recognized cause of livestock poisoning (Schardl, Young, Hesse, et al., 2013; Uhlig, Botha, Vrålstad, Rolén, & Miles, 2009).

1.2. Phylogenetic relationships

1.2.1 Gene trees

Phylogenetic studies of the order Hypocreales have recently resulted in the description of two new families, Cordycipitaceae and Ophiocordycipitaceae, comprising species previously aligned with Clavicipitaceae (Kepler et al.,

2012; Sung, Sung, Hywel Jones, & Spatafora, 2007). Whereas all three families are dominated by pathogens of insects, the Clavicipitaceae *sensu stricto* include a very large number of plant parasites and plant symbionts. We investigated relationships among the plant-associated Clavicipitaceae by maximum likelihood analysis of aligned sequences from public databases and sequenced genomes. Phylograms were generated based on partial coding sequences of genes for the largest subunit of RNA polymerase II (*rpbA*), β-tubulin (*tubB*), and aldehyde dehydrogenase I (*aldA*) (Fig. 10.1). Results were consistent for all three phylograms. Plant-associated Clavicipitaceae are grouped into clades associated with genera *Epichloë*, *Claviceps*, *Aciculosporium*, *Balansia*, *Periglandula*, and *Villosiclava*. Included in some clades were anamorphs linked with respective teleomorphic genera: *Ephelis* with *Balansia* and *Ustilaginoidea* with *Villosiclava*. Additionally, the anamorph genus *Neotyphodium* has recently been aligned with *Epichloë* (Leuchtmann et al., 2014), and the new names are used here (e.g. *E. coenophiala* = *Neotyphodium coenophialum*). Some clades had multiple teleomorphic genera as well. The *Balansia* clade (*B*) included *Atkinsonella* species, *Heteroepichloë* species, *Myriogenospora atramentosa*, and *Parepichloë cinerea*, and the *Aciculosporium* clade included *Cepsiclava phalaridis*, *Corallocytostroma ornithocopreoides*, and *Neoclaviceps monostipa*. Particularly interesting was that the *rpbA* gene indicated a close relationship of *Periglandula* species—which are seed-transmitted symbionts of morning glories—with *Verticillium epiphytum* and *Tyrranicordyceps fratricida*, both of which are fungal parasites of other fungi (Kepler et al., 2012).

1.2.2 Phylogenetic relationships of pathogenic and symbiotic life histories

The ergot fungi (*Claviceps* species; clade *C*; Fig. 10.1) cause highly localized infections of their hosts. They are restricted to the host florets, where they engulf the ovaries, access nutrients, and develop sclerotia, which then drop to the ground to germinate later and complete their development and infection cycle (Pažoutová, Olšovská, Linka, Kolínská, & Flieger, 2000). Thus, the ergot infections are more transient and localized compared to the long-term (i.e. constitutive) systemic infections of plant hosts that typify members of clades *A*, *B*, *E*, and *P*. Taken together, these clades form a group paraphyletic to clade *C*, implying that the ability to form constitutive localized infections was ancestral and was lost in the evolution of *Claviceps* species, which may have instead developed a much more efficient process of contagious spread.

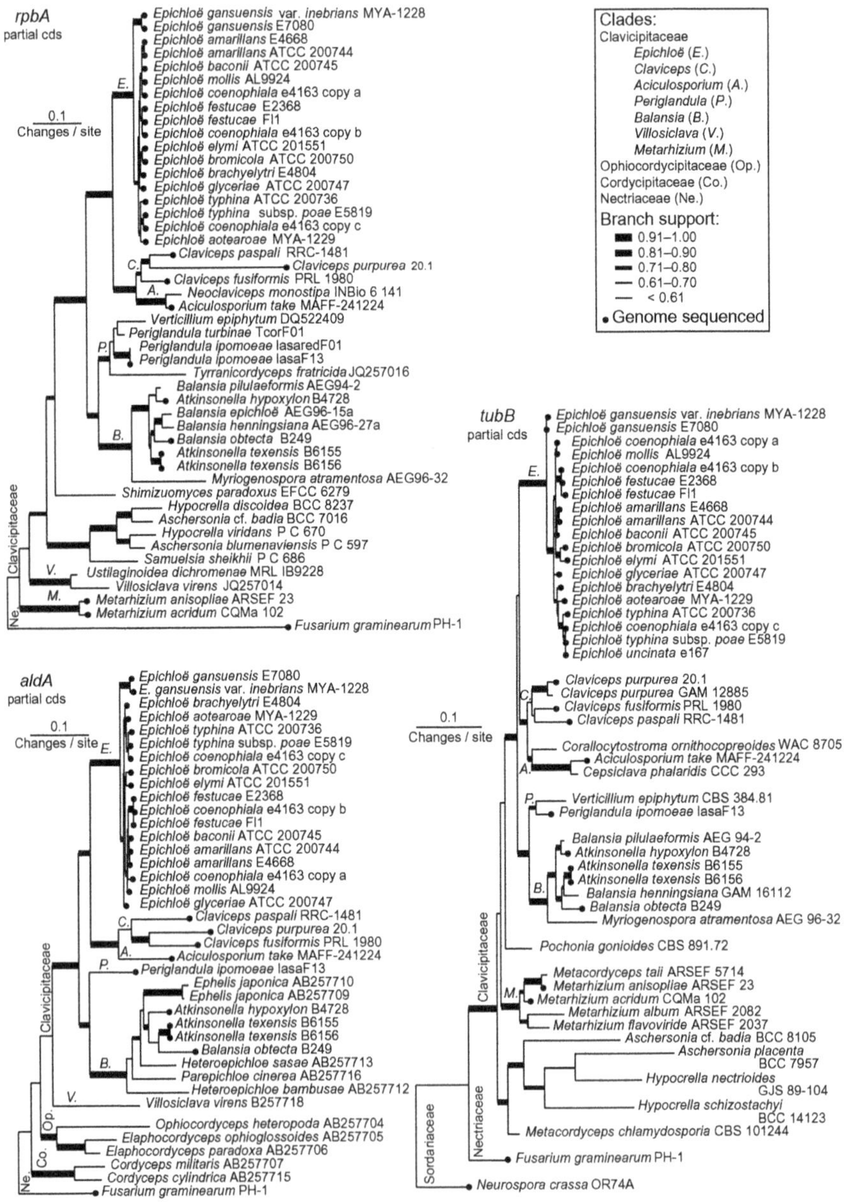

Figure 10.1 Phylogenies inferred from genes for RNA polymerase II largest subunit (*rpbA*), aldehyde dehydrogenase I (*aldA*), and β-tubulin (*tubB*). Partial coding sequences were obtained from sequenced genomes and public databases, aligned with MUSCLE (Edgar, 2004), and trees were inferred by maximum likelihood with PhyML implemented by Phylogeny.fr (Dereeper et al., 2008). Species names are followed by isolate identifiers or, if those were unavailable, the GenBank accession numbers.

Coding sequences of *aldA*, *rpbA*, and *tubB* were very similar among *Epichloë* species (clade *E*) (Fig. 10.1). Nevertheless, there was sufficient phylogenetic signal in all three sequence alignments to indicate a root to this clade. The two taxa symbiotic with *Achnatherum inebrians*, *Epichloë gansuensis*, and *E. gansuensis* var. *inebrians* formed a subclade separated by the root from the rest of the sequenced *Epichloë* species.

The characteristics and life cycles of *Epichloë* species are not particularly unique to that genus. The formation of stromata on inflorescences is common in clades *E* and *B* (Fig. 10.1), and highly efficient vertical transmission, which is so important for their mutualistic symbiotic associations, is shared with *Periglandula* species of clade *P*. However, the capability for a delicately balanced symbiosis that mixes stroma production on some tillers and seed transmission on others is common among sexual *Epichloë* species, but so far not characteristic of any other Clavicipitaceae or, for that matter, any other fungi described to date. Therefore, it appears that the features that characterize the *Epichloë* life history and host interactions, as well as most of their alkaloid biosynthetic capability, evolved well before the emergence of that fungal genus, whereas the particular combination characteristics that are crucial to their widespread and often mutualistic symbioses in grasses (Schardl et al., 2008) were selected in the origin of the genus *Epichloë*.

2. SEQUENCED GENOMES OF THE CLAVICIPITACEAE

Recent publications (Gao et al., 2011; Schardl, Young, Hesse, et al., 2013; Schardl, Young, Pan, et al., 2013) report genome sequences for a total of 17 genomes of Clavicipitaceae. Here, we report additional sequences, bringing the total to 26 (Table 10.2), including representatives of seven genera (considering *Epichloë* and *Neotyphodium* as a single genus), and 21 species. These, together with a large number of genome sequences from the other families of Hypocreales (Bushley et al., 2013; Cuomo et al., 2007; Güldener et al., 2006; Martinez et al., 2008; Wiemann et al., 2013; Zheng et al., 2011), make this order one of the most intensively studied groups of eukaryotes at the genomic level.

All of the published genome sequences in the Clavicipitaceae are haploids, although the genus *Epichloë* contains a very large number of polyploid hybrids (Leuchtmann et al., 2014; Moon, Craven, Leuchtmann, Clement, & Schardl, 2004). Comprehensive genome sequencing of the polyploid hybrids has only just begun (Schardl, Young, Pan, et al., 2013) because it requires much more sequence coverage. Nevertheless, a reasonable assembly

Table 10.2 Statistics for sequenced genomes of plant-symbiotic and plant-parasitic Clavicipitaceae

Organism	Strain	MT	Genome assembly length (bp)[a]	Genes	Total genic (Mb)	Total CDS (Mb)	% CDS	% Non-Rpt-IG	% Rpt	GC proportions: Genome	CDS	Non-Rpt-IG	Rpt
Aciculosporium take	MAFF-241224	B	58,707,902	8863	14.0	9.4	16.0	16.9	60.9	0.40	0.59	0.52	0.31
Atkinsonella hypoxylon	B4728	A	35,553,953	9763	15.6	12.7	35.7	32.8	23.4	0.44	0.52	0.48	0.31
Atkinsonella texensis	B6155	A	28,241,655	8878	13.8	12.5	44.2	45.4	8.2	0.50	0.53	0.49	0.36
At. texensis	B6156	B	28,146,212	8890	13.8	12.5	44.4	45.6	8.2	0.50	0.53	0.49	0.36
Balansia obtecta	B249	B	30,388,664	9779	14.9	12.9	42.4	48.0	3.5	0.48	0.53	0.47	0.27
Claviceps fusiformis	PRL 1980	B[b]	52,335,178	9784	19.2	11.9	22.7	30.4	42.3	0.37	0.55	0.41	0.23
Claviceps paspali	RRC-1481	B	28,922,829	8631	14.4	10.0	34.7	37.6	16.9	0.48	0.58	0.48	0.23
Claviceps purpurea	20.1	A	32,108,429	9452	17.5	12.2	39.4	41.6	7.7	0.52	0.55	0.50	0.49
Epichloë amarillans	ATCC 200744	B	37,962,913	10198	15.9	10.9	28.7	23.3	36.8	0.44	0.55	0.49	0.33
E. amarillans	E4668	B	40,693,583	11154	16.9	13.4	32.9	20.6	38.6	0.44	0.53	0.50	0.35

Epichloë aotearoae	MYA-1229	A	34,338,902	11045	17.9	13.1	38.1	28.1	20.4	0.44	0.54	0.49	0.27
Epichloë baconii	ATCC 200745	A	38,004,484	12669	20.5	13.5	35.5	26.9	19.5	0.42	0.53	0.47	0.26
Epichloë brachyelytri	E4804	B	44,051,370	12772	16.5	10.9	24.7	21.0	33.4	0.40	0.54	0.48	0.29
Epichloë coenophiala[c]	e4163	AAA	97,712,391	30268	51.3	36.1	36.9	12.6	35.5	0.43	0.53	0.48	0.36
Epichloë elymi	ATCC 201551	A	31,756,217	8426	15.4	10.8	33.9	29.9	30.4	0.47	0.55	0.50	0.34
Epichloë festucae	E2368	A	34,661,749	8306	15.9	11.1	32.5	24.4	30.6	0.44	0.55	0.48	0.28
E. festucae	Fl1	B	34,904,508	8649	15.6	10.9	31.9	28.7	26.1	0.44	0.55	0.48	0.28
Epichloë gansuensis	E7080	B	39,525,795	9030	16.6	11.7	29.7	25.7	38.6	0.44	0.54	0.49	0.33
E. gansuensis var. *inebrians*	MYA-1228	A	29,794,493	9823	16.1	11.2	38.5	35.9	15.0	0.47	0.54	0.49	0.27
Epichloë glyceriae	ATCC 200747	A	49,319,608	11483	20.4	14.9	30.3	24.4	41.0	0.45	0.54	0.49	0.36
Epichloë mollis	AL9924	B	36,111,612	10406	17.0	13.1	36.2	26.1	27.2	0.44	0.54	0.49	0.30

Continued

Table 10.2 Statistics for sequenced genomes of plant-symbiotic and plant-parasitic Clavicipitaceae—cont'd

Organism	Strain	MT	Genome assembly length (bp)	Genes	Total genic (Mb)	Total CDS (Mb)	% CDS	% Non-Rpt-IG	% Rpt	GC proportions			
										Genome	CDS	Non-Rpt-IG	Rpt
Epichloë typhina	ATCC 200736	A	41,288,070	8584	15.5	10.3	30.9	22.5	44.5	0.42	0.55	0.49	0.28
E. typhina ssp. *poae*	E5819	A	34,036,313	8770	15.2	10.5	25.1	27.9	32.1	0.43	0.55	0.48	0.24
Periglandula ipomoeae[d]	IasaF13	A[b]	35,301,553	12102	22.5	15.9	45.0	46.4	0.2	0.51	0.53	0.49	0.44

[a]Based on total of scaffolds (supercontigs) or contigs $\geq$500 bp.
[b]*C. fusiformis* PRL 1980 mating-type genes include *mtBA* and *mtAC*. *P. ipomoeae* IasaF13 mating-type genes *mtAA* and *mtAC* appear to have premature stop codons.
[c]Repeat statistics for the hybrid *E. coenophiala* were determined by masking the genetic regions prior to determining the repeat content.
[d]Statistics for *P. ipomoeae* are tentative because the assembly was filtered by selecting only contigs containing tBLASTx matches to genome sequences from other Clavicipitaceae.
Abbreviations: CDS, coding sequence; MT, mating type; non-Rpt-IG, nonrepetitive intergenic DNA; Rpt, repetitive DNA; GC, proportion of sequence that is G or C.

has been obtained for *Epichloë coenophiala*, an economically important hybrid endophyte with three ancestral genomes. This particular fungus has considerable economic and ecological importance as the common endophyte of tall fescue, which is widely distributed on several continents, and the most widely planted forage grass in the United States (Rudgers, Holah, Orr, & Clay, 2007; Schardl, Scott, Florea, & Zhang, 2009). Haploid genome sizes (Table 10.2) vary roughly twofold from just under 30 Mb for *Atkinsonella texensis*, *Claviceps paspali*, and *Epichloë gansuensis* var. *inebrians* to nearly 60 Mb for *Aciculosporium take* and *Claviceps fusiformis*. However, genetic content does not correlate with genome size in this group of fungi, which all have approximately the same number of genes. Genome size differences are almost entirely due to repeat DNA content. There is no obvious taxonomic or phylogenetic pattern associated with variation in genome size or repeat content. However, in the *Epichloë* species, alkaloid loci tend to have large blocks of repeat sequences interspersed among the genes, whereas the members of other genera have far less repeat sequence within those loci (Schardl, Young, Hesse, et al., 2013).

2.1. Repeat content

The estimated proportion of repetitive sequences for each genome was found to vary, ranging from the fewest in the *P. ipomoeae* (0.2%) genome, which could be a feature of the assembly, to the most in *A. take* (60.9%) (Table 10.2). Only five genomes, *P. ipomoeae*, *B. obtecta*, *At. texensis* B6155, *At. texensis* B6156, and *C. purpurea*, contained repeat content less than 10% of the total genome. The GC proportion of the repetitive sequences was low, ranging from 0.24 to 0.36 (average of 0.3 GC), apart from *C. purpurea* (0.49) and *P. ipomoeae* (0.44). The comparison of repeat sequences within a genome has indicated those with low GC content contain many C to T and G to A transitions that are likely due to repeat-induced point mutations (RIPs) (Fleetwood, Scott, Lane, Tanaka, & Johnson, 2007; Schardl, Young, Hesse, et al., 2013; Young et al., 2005). The evaluation of repeat sequences within the *Epichloë* species indicates there are more class I retrotransposon elements than class II DNA transposons. However, it is unlikely that any of the transposable elements are still functional since they have been rendered highly degenerate due to their extensive mutations.

The fragmented nature of the assemblies (largely due to AT-rich repeats) makes it difficult to determine the repeat distribution and integration bias patterns. However, it is very apparent that repeat sequences are overrepresented in the alkaloid biosynthesis clusters of *Epichloë* species compared to most other SM clusters and to alkaloid clusters in most members of the other genera (Schardl, Young, Hesse, et al., 2013). The extensive repeat blocks are likely to have impacted alkaloid cluster stability (Schardl, Young, Hesse, et al., 2013) (discussed later). Also prevalent in the genomes of *Epichloë* species, including their alkaloid biosynthesis clusters, are miniature inverted-repeat transposable elements (MITEs), particularly in gene promoter regions (Fleetwood et al., 2011; Schardl, Young, Hesse, et al., 2013). It appears that gene clusters have rearranged due to repetitive elements and genes being rendered nonfunctional because of integrating elements. Interestingly, although *At. texensis* has markedly fewer repeats than the *Epichloë* species, the *IDT* clusters found in the *At. texensis* isolates appear to also have been fragmented by repeat sequences.

2.2. Telomeric regions

The telomere repeat sequence, $(TTAGGG)_n$ when read towards the chromosome end, was identified as tandem repeats at the start or end of contigs in many of the sequenced genomes. Contigs that contained at least two copies of a telomere repeat were counted to determine the likely number of chromosome ends. The average chromosome number of nonhybrid and hybrid *Epichloë* species was 4.5 and 8.75, respectively. The predicted chromosome numbers for the two *E. festucae* isolates and *E. coenophiala* are consistent with data generated from chromosome separations using pulsed-field gel electrophoresis and from Southern-blot hybridizations to telomere repeat probes (Kuldau, Tsai, & Schardl, 1999; Schardl, C. L., Young, C. A., & Andreeva, K., unpublished data). Unfortunately, telomere repeat sequences were underrepresented in the genomes of *C. purpurea* 20.1, *Fusarium graminearum* PH-1, the two *Metarhizium* species, and some *Epichloë* species (namely, *E. typhina* E8, *E. elymi* E56, and *E. bromicola* E502).

Many of the predicted telomeres were contained on contigs with large subtelomeric AT-rich repeat regions. The telomere-linked *recQ* helicase (*TLH*) genes, commonly associated within the telomeres of *Magnaporthe oryzae* (Rehmeyer et al., 2006), did not appear to be associated with subtelomeric regions in the Clavicipitaceae. In fact, the total number of *recQ* genes identified in clavicipitaceous species was limited to one to four copies. Subterminal genetic regions did not appear to be shared across the Clavicipitaceae, but among the *Epichloë* species, the *EAS* and *IDT* clusters were often identified as subterminal (Schardl, Young, Hesse, et al., 2013).

2.3. Mating-type loci

By far, the most common genetic system governing mating types in Pezizomycotina (filamentous ascomycetes) involves one locus with two idiomorphs designated *MAT*1-1 and *MAT*1-2 (Turgeon & Yoder, 2000). Finding that nomenclature a bit cumbersome, we refer to the locus as *MT*, the idiomorphs as *MTA* and *MTB*, and the genes as *mtAA*, *mtAB*, and *mtAC* for the three genes of the *MTA* idiomorph and *mtBA* for the sole gene of *MTB*. With some exceptions (Vaillancourt, Du, Wang, Rollins, & Hanau, 2000), obligately outcrossing (heterothallic) species have a haploid genome with either *MTA* or *MTB* at the *MT* locus, whereas self-compatible (homothallic) species have linked *MTA* and *MTB* (Turgeon & Yoder, 2000).

Inspection of the sequenced genomes suggests that most Clavicipitaceae are heterothallic (obligately outcrossing), because their *MT* loci have either one idiomorph (*MTA*) or the other (*MTB*) (Table 10.2). Two strains seem exceptional: *Claviceps fusiformis* PRL 1980 has an *mtBA* gene and an *mtAC* gene. In contrast, *Periglandula ipomoeae* IasaF13 has premature stops in *mtAA* and *mtAC*. It is possible that neither of these strains is competent for sexual crosses. The *C. fusiformis* strain has been propagated clonally in culture for many decades, as a model for studying ergot alkaloid biosynthesis (Gröger & Floss, 1998), and no teleomorph (sexual state) is known for *P. ipomoeae*. *Balansia obtecta* B249 also appeared exceptional in that the *apnB* (DNA lyase) gene adjacent to *MTB* is a pseudogene (Schardl, C. L., unpublished data). Apparently, functional *apnB* genes are almost always located adjacent to *MT* loci in Pezizomycotina.

Hybrid *Epichloë* species are generally asexual (Moon et al., 2004), with the reported exception of *E. liyangensis* (Kang et al., 2011), yet they generally possess *MT* idiomorphs. There is no general pattern of the *MT* idiomorphs in hybrids. *Epichloë coenophiala* has an *MTA* idiomorph from each of its three ancestors (Takach & Young, 2014), the *E. festucae* x *E. typhina* hybrid strain Lp1 has *MTB* idiomorphs from both of its ancestors (unpubl. data of the authors), and *MTA* and *MTB* genotypes are found in *E. canadensis* (Charlton, Craven, Mittal, Hopkins, & Young, 2012), *E. uncinata*, and *Epichloë* sp. FaTG-4 from decaploid *Lolium arundinaceum* (Takach & Young, 2014).

2.4. Mitochondrial genomes

Complete mitochondrial genomes (mtDNA) were identified in the genome assemblies of *E. festucae* E2368 and *E. bromicola* E502 (Young, C. A. &

Schardl, C. L., unpublished data), and the mtDNA sequence was nearly complete in the *C. purpurea* 20.1 assembly (accession number FO082257). The A+T content in mtDNA of the two *Epichloë* strains and *C. purpurea* was 73% and 65%, respectively, consistent with other published sequences (Pantou, Kouvelis, & Typas, 2008). In the mtDNA of both *E. festucae* E2368 and *E. bromicola* E502, 15 protein-coding genes were identified, as were genes for the small- and large-subunit rRNAs of the mitochondrial ribosomes and for 27 tRNAs (predicted by tRNAscan-SE; Lowe & Eddy, 1997). The genes were ordered consistently with the mtDNA of Hypocreales when compared to the other members of Sordariomycetes (Pantou et al., 2008). The assembled *C. purpurea* mtDNA contained the same genes as the *Epichloë* isolates, except the *atp9* gene was absent, and although *nad4* was present, it was no longer adjacent to *nad1*. It is yet to be determined if these are real differences or an artefact of incomplete assembly of the *C. purpurea* mtDNA.

Mitochondrial genome sizes are variable in fungi, and this is also reflected in the Clavicipitaceae where *M. anisopliae* mtDNA (accession number NC_008068; Ghikas, Kouvelis, & Typas, 2006) is 24,673 bp, *C. purpurea* mtDNA is estimated at 55,537 bp, *E. bromicola* E502 mtDNA is 69,466 bp, and *E. festucae* mtDNA is the largest at 72,701 bp. The size differences between these genomes are largely due to variations of intron number and size and the presence of a large variable region found in E2368 and E502 associated with the tRNA cluster containing 12 tRNA genes flanked by *rnl/rps3* and *nad2*. The variable region in E2368 and E502 extends over 10 kb and appears to contain sequence similarity to *dpoA*, a DNA-directed RNA polymerase gene encoded on the opposite strand to the other mitochondrial genes. Multiple stop codons present throughout the coding region of *dpoA* suggest the gene is nonfunctional. In many fungi, *dpoA* genes not only are often found on a linear plasmid but also can be stably integrated into the mitochondrial genome in single or multiples sites (Formighieri et al., 2008). *Fusarium* species also contain a large variable coding region between *rnl/rps3* and *nad2*, but its encoded function is yet to be elucidated and it does not appear to be related to *dpoA* (Al-Reedy, Malireddy, Dillman, & Kennell, 2012).

2.5. Gene ontology categories

We compared the representation of genes across higher-order gene ontology (GO) terms for four fungi with annotated genome sequences, the symbiont *E. festucae* E2368, the grass pathogens *Fusarium graminearum* PH-1 and

Magnaporthe oryzae 70-15, and the saprophyte *Neurospora crassa* OR74A. All GO annotations were downloaded from Ensembl using BioMart. Then, the GO hierarchy was determined for each term. Level 1 nodes in the hierarchy are the three roots, biological process, molecular function, and cellular component. Level 2 nodes are the first-generation children of the level 1 nodes. Table 10.3 lists the proportion and the number of genes having a GO annotation linked to each level 2 GO term for the molecular function and biological process roots. Only the 10 most frequent categories are displayed.

Despite their different ecological niches and host interactions and differences in total gene numbers, the four fungi showed little difference in

Table 10.3 Gene ontology (GO) categories of inferred genes from genomes of *Epichloë festucae* E2368 (*Ef*), *Fusarium graminearum* PH-1 (*Fg*), *Magnaporthe oryzae* 70-15 (*Mo*), and *Neurospora crassa* OR74A (*Nc*)[a]

GO id and term	***Ef***		***Fg***		***Mo***		***Nc***	
Biological process	**No.**	**%**	**No.**	**%**	**No.**	**%**	**No.**	**%**
GO:0008152 metabolic process	4768	58	4437	59	4686	61	2929	58
GO:0051234 establishment of localization	1220	15	1341	18	1314	17	852	17
GO:0065007 biological regulation	813	10	706	9	586	8	445	9
GO:0044699 single-organism process	596	7	566	8	500	6	483	10
GO:0009987 cellular process	543	7	222	3	407	5	144	3
GO:0071840 cellular component organization or biogenesis	150	2	138	2	136	2	117	2
GO:0050896 response to stimulus	75	1	73	1	64	1	75	1
GO:0051704 multiorganism process	14	<1	9	<1	7	<1	2	<1
GO:0048511 rhythmic process	2	<1	0	0	0	0	1	<1
GO:0040007 growth	1	<1	1	<1	1	<1	1	<1
Total	8182	100	7493	100	7701	100	5049	100
GO id and term	***Ef***		***Fg***		***Mo***		***Nc***	
Molecular function	**No.**	**%**	**No.**	**%**	**No.**	**%**	**No.**	**%**
GO:0003824 catalytic activity	7302	54	5709	46	6876	48	2771	40

Continued

Table 10.3 Gene ontology (GO) categories of inferred genes from genomes of *Epichloë festucae* E2368 (*Ef*), *Fusarium graminearum* PH-1 (*Fg*), *Magnaporthe oryzae* 70-15 (*Mo*), and *Neurospora crassa* OR74A (*Nc*)—cont'd

GO id and term	***Ef***		***Fg***		***Mo***		***Nc***	
Molecular function	**No.**	**%**	**No.**	**%**	**No.**	**%**	**No.**	**%**
GO:0005488 binding	5273	39	5315	43	6309	44	3453	50
GO:0005215 transporter activity	344	3	458	4	420	3	242	3
GO:0001071 nucleic acid binding transcription factor activity	253	2	369	3	235	2	157	2
GO:0005198 structural molecule activity	149	1	141	1	143	1	143	2
GO:0009055 electron carrier activity	102	1	176	1	185	1	92	1
GO:0030234 enzyme regulator activity	62	<1	54	<1	53	<1	44	1
GO:0060089 molecular transducer activity	59	<1	59	<1	47	<1	40	1
GO:0004872 receptor activity [b]	52	<1	0	0	21	<1	8	<1
GO:0000988 protein binding transcription factor activity	26	<1	18	<1	20	<1	6	<1
Total	13622	100	12299	100	14309	100	6956	100

[a]Note that some genes may be assigned to multiple GO categories, and unassigned genes are not listed.
[b]The large differences in this category appear to be an artefact of differences in gene annotations by curators of the respective genomes.

proportions of genes represented by the 20 GO terms that we compiled. The apparent difference in receptor activity genes was likely an artefact of differences in the original annotations by curators. None of the other terms showed obvious differences in representation. Nevertheless, it will be worthwhile in the future to conduct analyses at other levels of GO terms and among a broader range of fungi.

2.6. Variation of SM clusters

The repertoire of genes and gene clusters associated with potential SM production were determined using the SMURF and antiSMASH pipelines

(Blin et al., 2013; Khaldi et al., 2010). Similar trends were seen with either analysis, and Fig. 10.2 presents antiSMASH results supplemented with manual identification of those alkaloid clusters missed by antiSMASH and SMURF. These programs failed to detect the *LOL* clusters due to a lack of key SM signature genes at this locus. They also failed to detect *idtG* from the *IDT* clusters, although antiSMASH often associated the *idtG* paralog, *ggsA* (required for primary metabolism production of geranylgeranyl diphosphate), with SM cluster prediction. The classification of the clusters is based on predicted functions that are considered signatures of SM (Bok et al., 2006), although sometimes orthologous clusters can be classified differently. In particular, *EAS* clusters for the production of the simple ergot alkaloids such as chanoclavine are placed in the prenyl class, while those clusters required for more complex ergot alkaloids are considered in the mixed class because both prenyl (DmaW) and NRPS (LpsA and LpsB) functionality are required. In addition, the fragmented nature of the *EAS* and *IDT* clusters also affected correct SM identification and class placement of these clusters.

Numbers of SM clusters identified within the *Epichloë* nonhybrid (1x ploidy) species appeared consistent, averaging 29 (range = 24–35), although more variation could be seen among individual classes. As expected, this number dramatically increased in the hybrid, *E. coenophiala*, whereby 69 clusters were identified, consistent with its higher (3 ×) ploidy level. Generally, the *Epichloë* species show an increase in SM clusters compared to members of related clades, *C*, *B*, and *A*, but not *P*. Based on fungal lifestyle, the fungal clades could be ranked fewest to most SM clusters, whereby saprophytes (e.g. *N. crassa*) contained fewest, followed by replacement parasites (both systemic and ergot; clades *C*, *B*, and *A*), then seed-transmitted endophytes (clades *E* and *P*), and finally "classic" pathogens of plants and animals (e.g. *Metarhizium* spp., *Fusarium graminearum*, and *Magnaporthe oryzae*). We speculate that in plant-associated Clavicipitaceae, an increase in SM clusters found in seed-transmitted endophytes represents selection for increased host protection in both *E* and *P*.

3. ALKALOID GENE LOCI

On a broad scale, there is considerable diversity of alkaloid profiles at all taxonomic levels from species on up. For example, different strains of *Epichloë festucae* can produce any of the four classes of alkaloids—ergot alkaloids, indole-diterpenes, lolines, and peramine—in various combinations, although only some strains of *E. coenophiala* (including e4163) are

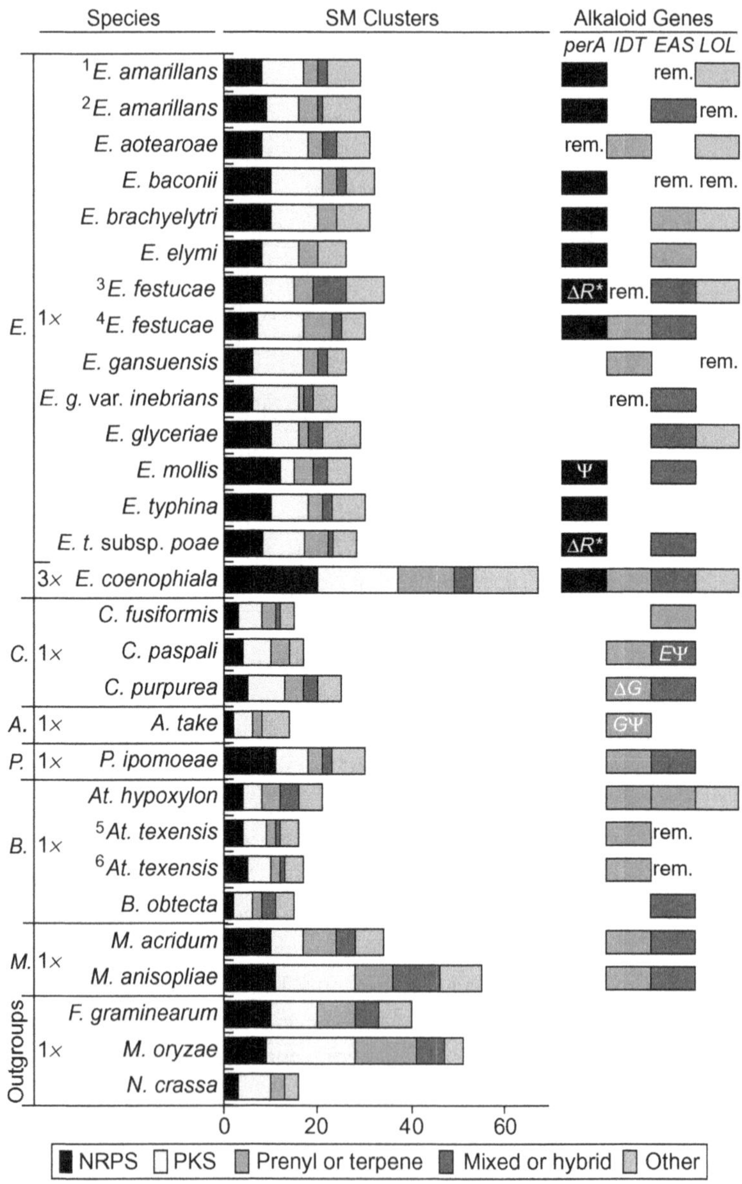

Figure 10.2 Numbers of predicted specialized metabolite (SM) clusters by class and fungal strain. SM classes are based on signature enzyme families as follows: NRPS, nonribosomal peptide synthase; PKS, polyketide synthase; prenyl or terpene, prenyl transferase or terpene synthase/cyclase; mixed or hybrid, clusters with a mixture of classes or hybrid NRPS/PKS; and others, known SM clusters lacking the aforementioned families. Cluster numbers were adjusted to account for each of the known alkaloid gene clusters. Specifically, the *LOL* clusters that were not identified by antiSMASH were added into the

known to produce all four (Schardl et al., 2012; Schardl, C. L., Young, C. A., Dinkins, R. D., Nagabhyru, P., & Panaccione, D. G., unpublished data). Comparing the syntenic loci among strains, it is apparent that those that lack any particular alkaloid class most often lack some or all of the genes at the corresponding locus. At least three of the four alkaloid classes (ergot alkaloids, indole-diterpenes, and lolines) exhibit considerable variation in structures, and such variation has implications for spectra and modes of activity (Pan et al., 2014; Schardl et al., 2006, 2012; Schardl, Young, Hesse, et al., 2013; Young et al., 2009).

3.1. Relationships between gene contents and alkaloid structures

3.1.1 Ergot alkaloids

Ergot alkaloids are known from a wide range of filamentous ascomycetes (Schardl et al., 2006). Simpler ergot alkaloids called clavines are known from several orders, but the Clavicipitaceae can produce an extraordinarily diverse range of ergot alkaloids from simple clavines such as chanoclavine I to much larger and more complex amides of lysergic acid (Fig. 10.3). The most complete ergot alkaloid biosynthesis (*EAS*) gene cluster has been identified in the morning glory symbiont, *Periglandula ipomoeae*, comprising 14 genes (Fig. 10.4). This endophyte produces the ergopeptine, ergobalansine, and the simpler lysergic acid amides, ergonovine and D-lysergic acid α-hydroxyethylamide (LAH). The latter two metabolites are also associated with *Claviceps paspali*, which shares with *P. ipomoeae* the genes designated *lpsC*, *easO*, and *easP*. (However, the sequenced *C. paspali* strain has an inactive *easE* gene, so it should produce only one or two early intermediates of the ergot alkaloid pathway.)

The presence of *lpsC* in genomes of *C. purpurea* and *Epichloë gansuensis* var. *inebrians* is in keeping with the finding that the LPS3 subunit encoded by this gene is required, together with the *lspB* gene product (LPS2), for ergonovine biosynthesis (Ortel & Keller, 2009). The LPS2 subunit also is required together with the *lpsA* gene product, LPS1, in the biosynthesis

others class, while the *EAS* and *IDT* clusters split by repeat sequences were condensed into their correct SM classes. The known alkaloid biosynthesis genes are indicated on the right with shaded boxes indicating SM class. Pseudogenes (Ψ), gene or domain deletions (Δ) that preclude production of the alkaloids, and cluster remnants (rem.) are also indicated. Genome ploidies are given as 1x or 3x. Where multiple strains of the same species were sequenced, they are distinguished by superscript numbers as follows: 1, ATCC 200744; 2, E4668; 3, E2468; 4, Fl1; 5, B6155; and 6, B6156.

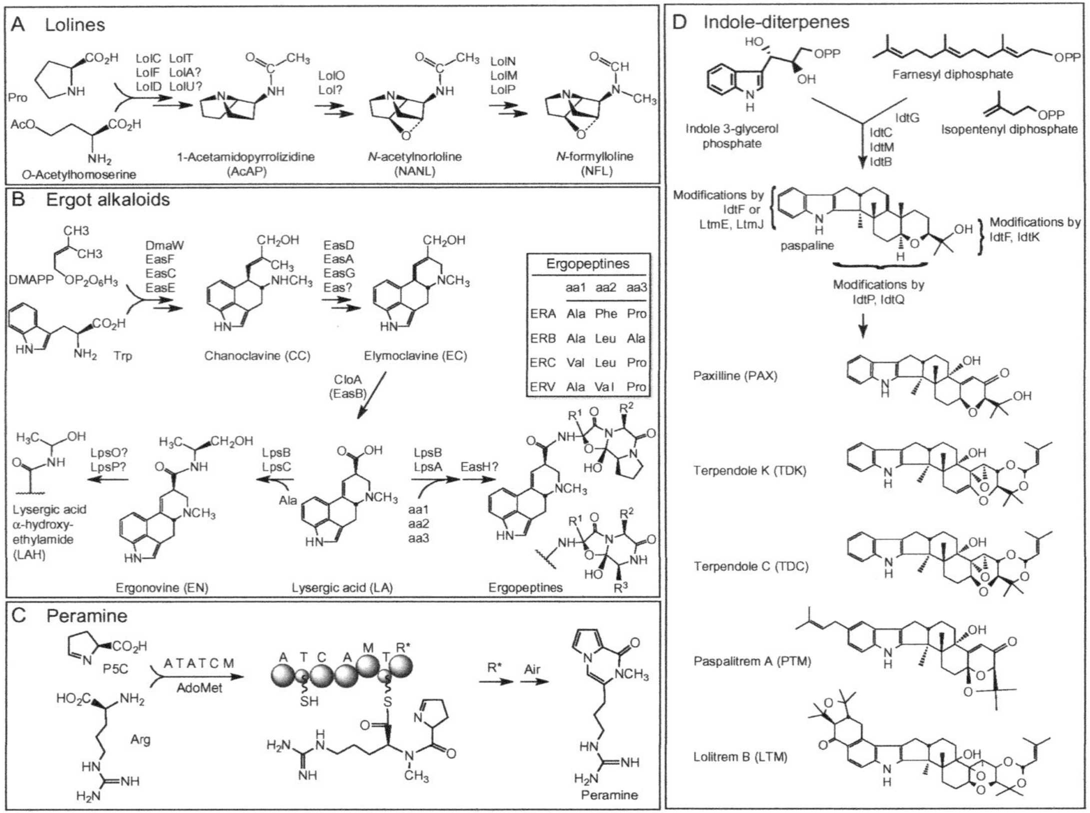

Figure 10.3 Alkaloid brief pathways. Summaries of biosynthetic pathways for four classes of alkaloids produced by Clavicipitaceae. (A) Loline alkaloids. (B) Ergot alkaloids. (C) Peramine. (D) Indole-diterpenes. Double arrows indicate multiple steps, and arrows are labelled according to the gene products (enzymes) that catalyse the steps or, in the case of peramine biosynthesis (panel C), the individual domains of the multifunctional peramine synthetase (*perA* gene product).

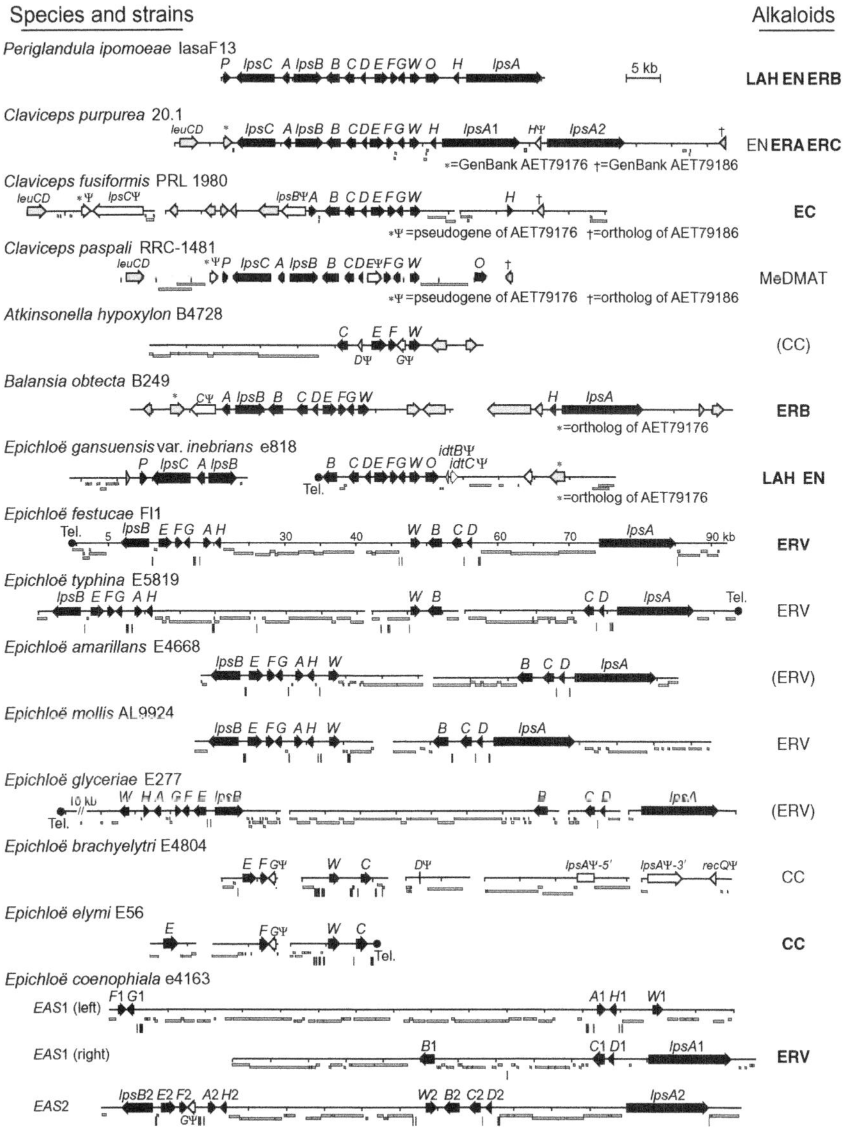

Figure 10.4 Structures of representative *EAS* loci and variation in flanking sequences. Associated ergot alkaloids are indicated at the right, abbreviated as indicated in Fig. 10.3 or as MeDMAT for *N*-methyl-dimethylallyltryptophan. The clusters have been ordered to show similarity of cluster gene order and orientation and common flanking genes. *EAS* genes are represented with black arrows and those labelled by single letters are as follows: *A*, *easA*; *B*, *cloA*; *C*, *easC*; *D*, *easD*; *E*, *easE*; *F*, *easF*; *G*, *easG*; *H*, *easH*; *O*, *easO*; *P*, *easP*; and *W*, *dmaW*. Other *EAS* genes, encoding nonribosomal peptide synthetase subunits, are *lpsA*, *lpsB*, and *lpsC*. Pseudogenes, white arrows; flanking genes, grey arrows; and telomeres (tel.), filled black circles. Depicted beneath each map are horizontal bars indicating transposon-derived repeats and vertical bars indicating miniature inverted-repeat transposable elements (MITEs).

of ergopeptine alkaloids. Variation in LPS1 determines which three amino acids are incorporated into the ergopeptide lactam (Riederer, Han, & Keller, 1996), which is then converted by EasH to the corresponding ergopeptine (Havemann, Vogel, Loll, & Keller 2014). Thus, *Periglandula* and *Balansia* species have been found to produce ergobalansine, and *Epichloë* species have been found to produce ergovaline (Schardl, Young, Hesse, et al., 2013) (Fig. 10.3). The two forms of *lpsA* in *C. purpurea* 20.1 are responsible for two different ergopeptines, ergotamine and ergocryptine (Haarmann, Lorenz, & Tudzynski, 2008). In all, 20 different amino acid combinations are evident in natural ergopeptines (some of which also are known in dihydroergopeptines) (Cheng, Coyle, Panaccione, & O'Connor, 2010; Schardl et al., 2006).

While *P. ipomoeae* is the only species so far found to have all 14 of the *EAS* genes known in Clavicipitaceae (although the fumigaclavine biosynthesis gene cluster of *Aspergillus fumigatus* has additional genes; Coyle & Panaccione, 2005), other ergot alkaloid producers lack functional copies of between 2 and 10 of these genes. So, in the extreme cases, only *dmaW*, *easF*, *easC*, and *easE* remain, and the fungus produces chanoclavine I as its pathway end product (Schardl, Young, Hesse, et al., 2013). Remnants of some of *EAS* genes in such reduced clusters indicate that gene loss is a major evolutionary process in ergot alkaloid diversification. Other processes that diversify these profiles include variation in *lpsA*, as discussed earlier in the text, and variation in *easA* associated with the production of either the classical 8–9 unsaturated ergot alkaloids or the dihydroergot alkaloids with a saturated D ring (Coyle, Cheng, O'Connor, & Panaccione, 2010).

3.1.2 Indole-diterpenes

Also produced by a range of filamentous ascomycetes, the indole-diterpenes represent a diverse array of compounds synthesized from indole-3-glycerol phosphate and geranylgeranyl diphosphate and followed by various modifications such as prenylations, epoxidations, and hydroxylations to provide structural diversity (Fig. 10.3) (Saikia et al., 2012). The reconstitution of the *Penicillium paxilli* paxilline biosynthetic pathway in both homologous and heterologous systems has clearly established the early pathway steps to the simple indole-diterpene, paspaline (Saikia, Parker, Koulman, & Scott, 2006; Tagami et al., 2013). Four gene products are required to produce paspaline: IdtG to form geranylgeranyl diphosphate, IdtC to form 3-geranylgeranyl indole, and followed by IdtM and IdtB for the epoxidation and cyclization steps, respectively. Beyond paspaline production, the indole-diterpene pathway can be more complex and pathway diversity is determined by both the presence and absence of *IDT* genes and variations in substrate and product specificity of pathway-specific gene products.

A key feature of *IDT* clusters (Fig. 10.5) is usually the presence of an *IDT* pathway-specific geranylgeranyl diphosphate synthase encoded by *idtG*, whereas a paralog, *ggsA*, is located elsewhere in the genome and specifies geranylgeranyl diphosphate synthase for primary metabolism. Conceivably, *ggsA* could serve both roles in some fungi, but genetic tests in *P. paxilli* demonstrate a requirement of *paxG* (ortholog of *idtG/ltmG*) for indole-diterpene biosynthesis (Saikia et al., 2006; Young, McMillan, Telfer, & Scott, 2001). Interestingly, an *IDT* gene cluster was recently sequenced from *Chaunopycnis alba* (which shares an ordinal relationship with the Clavicipitaceae), which is capable of producing terpendole E despite the absence of an identifiable *idtG* gene in the cluster (Motoyama, Hayashi, Hirota, Ueki, & Osada, 2012). However, without a complete sequence for *Ch. alba*, it remains unknown if an *idtG* ortholog may be present elsewhere in its genome. It is also conceivable that some fungi can utilize the housekeeping *ggsA* gene to direct the first step in the pathway. Sequenced genomes of some Clavicipitaceae have revealed *IDT* clusters that lack *idtG* (in *C. purpurea* 20.1 and *A. texensis* B6155 and B6156) or are predicted to contain a nonfunctional *idtG* (*A. take* MAFF-241224), but whether any of these clusters is functional is currently unknown.

Lolitrem B production is one of the most well-studied indole-diterpene pathways, elucidated based on the identification of intermediates present in perennial ryegrass (*Lolium perenne*) infected with *E. festucae* var. *festucae* and var. *lolii* (Munday-Finch, Wilkins, & Miles, 1998) and from gene deletion and heterologous expression experiments (Saikia et al., 2006, 2012; Young et al., 2006, 2009). In total, 11 genes are found at the *IDT/LTM* locus for *Epichloë* species capable of producing lolitrem B (Schardl, Young, Hesse, et al., 2013; Young et al., 2006). However, the production of lolitrem B occurs via a very complex metabolic grid apparently lacking any specific order of oxygenations (catalysed by IdtP and IdtQ) and prenylations (catalysed by IdtF, IdtK, IdtE, and IdtJ) of the geranylgeranyl and indole moieties. Such a metabolic grid gives a much larger number of intermediate products than would a direct linear pathway (Saikia et al., 2012). The *ltmE* and *ltmJ* genes required for the indole-ring prenylation steps seem to have arisen by duplications and neofunctionalizations of other *IDT* genes (Schardl, Young, Hesse, et al., 2013).

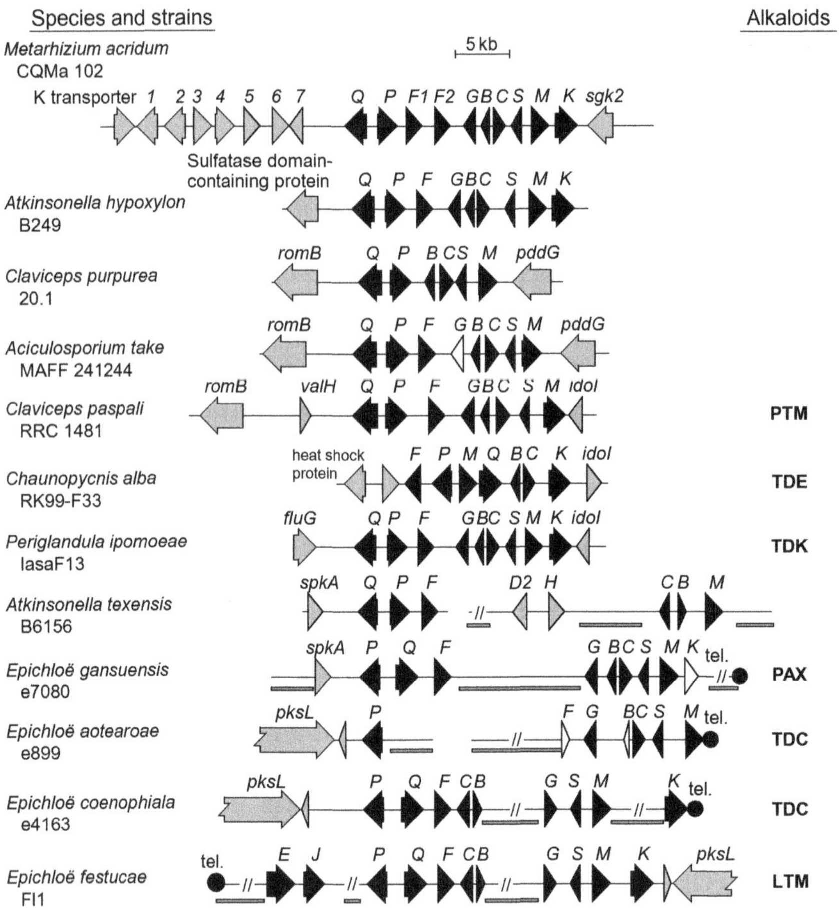

Figure 10.5 Structures of representative *IDT* loci and variation in flanking sequences. Associated indole-diterpenes are indicated at the right, abbreviated as indicated in Fig. 10.3 or as TDE for terpendole E (Motoyama et al., 2012). The clusters have been ordered to show similarity of cluster gene order and orientation and common flanking genes. *IDT* genes are represented with black arrows and genes are labelled by single letters, whereby *Q* = *idtQ*, *ltmQ* (*E. festucae*), or *terQ* (*Ch. Alba*; accession AB725916), etc. The *IDT* genes in the *M. acridum* cluster (accession GL698534) (representing a *Metarhizium IDT* example) have been renamed to represent homologs present in other Clavicipitaceae. The *M. acridum* genes labelled 1–7 have predicted functions possibly associated with SM. The *IDT* genes from B6156 were chosen as the *At. texensis* example as B6155 was identical but appeared to lack *idtF*. Additional *IDT* genes, likely required for TDC production, are present in e899 but each gene is embedded in AT-rich regions (Schardl, Young, Pan, et al., 2013). Pseudogenes, white arrows; flanking genes, grey arrows; telomeres (tel.), filled black circles; and *//*, regions of the cluster containing repetitive elements.

Other variations within the *IDT* clusters have revealed that *idtF* is sometimes nonfunctional and *idtK* can be absent or nonfunctional, leading to the accumulation of terpendoles (Schardl, Young, Hesse, et al., 2013; Schardl, Young, Pan, et al., 2013; Young et al., 2009). Additional indole-diterpene pathway variation can also stem from specificity differences of the P450 monooxygenases IdtP and IdtQ (Saikia et al., 2012). Recently, *IDT* gene clusters have been identified in genome sequences of a number of other Clavicipitaceae, namely, *M. acridum*, *M. anisopliae* (Schardl, Young, Hesse, et al., 2013), *At. hypoxylon*, and *At. texensis* (Young, C.A., and Schardl, C.L. unpublished data), but indole-diterpene products are yet to be identified from these species.

3.1.3 Lolines and aminopyrrolizidines

Loline alkaloids and related aminopyrrolizidines (Fig. 10.3A) are known from various cool-season grasses, legume genera *Adenocarpus* (Veen, Greinwald, Canto, Witte, & Czygan, 1992) and *Laburnum* (Tasso, Novelli, Sparatore, Fasoli, & Gotti, 2013), and a species of morning glory, *Argyreia mollis* (Tofern, Kaloga, Witte, Hartmann, & Eich, 1999). Orthologous clusters of genes, constituting the *LOL* locus, have been identified in several *Epichloë* species (including several former *Neotyphodium* species) (Schardl, Young, Hesse, et al., 2013) and recently in an *At. hypoxylon* isolate that was confirmed to produce 1-acetamidopyrrolizidine (Pan et al., 2014). The full cluster (Fig. 10.6) has 11 genes, but inactivating mutations or deletions of certain late-pathway genes are associated with natural variations in aminopyrrolizidine profiles (Pan et al., 2014; Schardl, Young, Hesse, et al., 2013). Those that accumulate 1-acetamidopyrrolizidine as the end product lack functional *lolO* genes; those that accumulate *N*-acetylnorloline lack *lolN*, *lolM*, and *lolP*; and those with the full cluster accumulate *N*-formylloline.

Four strains have been identified that, due to the mutation of *lolO*, accumulate 1-acetamidopyrrolizidine as an end product (Pan et al., 2014). These are strains of *E. elymi*, *E. brachyelytri*, *E. canadensis*, and *A. hypoxylon*. All possess only the *EAS* genes for chanoclavine I production. It is intriguing to speculate that this pattern results from positive selection for the combination of 1-acetamidopyrrolizidine and chanoclavine I, perhaps by synergistic action against herbivores.

3.1.4 Peramine

Among known SMs, peramine (Fig. 10.3) has a nearly unique pyrrolopyrazine structure bearing some similarity to the pyrazinone linkers in

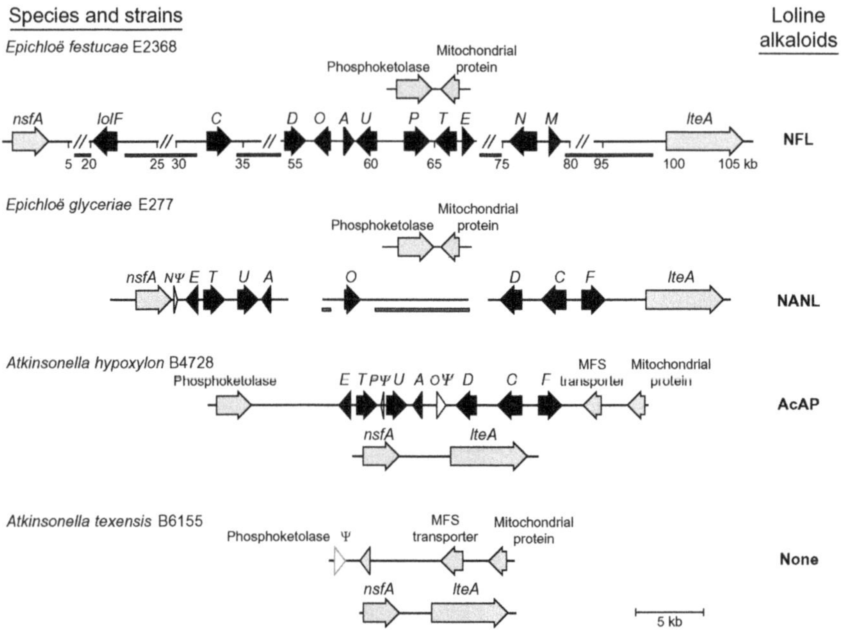

Figure 10.6 Evidence that the *LOL* cluster moved after being constituted during the evolution of plant-associated Clavicipitaceae. The *LOL* clusters in *Epichloë festucae* E2368 and *Epichloë glyceriae* E277 are located between housekeeping genes *nsfA* and *lteA*, but in opposite orientations. The *LOL* cluster in *Atkinsonella hypoxylon* B4728 is located between genes encoding a phosphoketolase and a mitochondrial protein, which are linked in the *Epichloë* species. Pseudogenes, white arrows; flanking genes, grey arrows; and //, regions of the cluster containing repetitive elements.

dragmacidins and proline-containing diketopiperazines (cyclic dipeptides) that contribute to flavours in beer (Gautschi et al., 1997). Interestingly, the ring system of peramine also resembles the ring system formed from amino acids II and III of ergopeptine alkaloids (Schardl et al., 2006). However, unlike these other natural products, one oxo group has been reduced to enamine, probably by the action of a C-terminal reductase domain (R*) of the multifunctional peramine synthase (Tanaka, Tapper, Popay, Parker, & Scott, 2005). The functions of peramine synthase have been predicted by bioinformatic analysis of the *perA* gene sequence. Curiously, variants of *perA* in several *Epichloë* species lack the R* domain but contain all of the others, and (at least in *E. festucae* E2368) this variant is expressed, suggesting that these *perA*-ΔR^* variants may actually determine synthesis of a related metabolite (Schardl, Young, Hesse, et al., 2013). If so, some condensation, esterase, reductase, or cyclase function needs to be provided in *trans* in order to release the chain from the enzyme, and until such a subunit is identified, the nature of the metabolite might remain unknown.

3.2. Variation in gene order

In the *LOL* gene clusters (Fig. 10.6), the gene orders are well conserved between species and even genera. For example, the order in *E. festucae* E2368 is *lolF, C, D, O, A, U, P, T, E, N, M,* and although the *LOL* cluster in *A. hypoxylon* lacks *lolP, N,* and *M,* all of the other *LOL* genes are in the same order. What is curious is that, while the gene order in the *LOL* cluster has been conserved, the placement and orientation of the *LOL* cluster in the genome have not. Among the *Epichloë* species, the cluster has flipped orientation at least once. In E2368, the *lolF* gene is nearer to the housekeeping gene *nsfA*, and *lolE, N,* and *M* are on the side nearer to *lteA*, whereas in *E. glyceriae* E277, *lolE* and a remnant of *lolN* are nearer to *nsfA*, and *lolF* is nearer to *lteA*. Comparing between genera, it is apparent that the entire cluster has moved while remaining intact. Genes for a predicted mitochondrial protein and a predicted phosphoketolase flank the *LOL* cluster (plus a predicted MSF transporter gene) in *A. hypoxylon* but are adjacent to each other in *Epichloë* species. The converse is true concerning the adjacent *nsfA* and *lteA* genes in *Atkinsonella* species, which are orthologous to the housekeeping genes that flank *LOL* in *Epichloë* species. It is not apparent which are the ancestral and which are derived positions of the *LOL* cluster, and it is even possible that the positions in *Epichloë* and *At. hypoxylon* are both derived from an ancestral condition where the locus was in yet another position in the genome.

Unlike the *LOL* gene clusters, more variation is found within the *IDT* gene order and cluster locations (Fig. 10.5). Based on observations of various *IDT* clusters across diverse species, it appears the most likely ancestral order of genes is *idtQ, P, F, G, B, C, S, M, K.* However, there is a variation among the *Epichloë* species, whereby the gene order has undergone a rearrangement that is now represented by *idtP, Q, F, C, B, G, S, M, K.* The two additional genes, *ltmE* and *ltmJ*, that are found in *E. festucae* Fl1 and other lolitrem B-producing isolates appear to be derived from gene duplication and neofunctionalization, whereby *ltmJ* has arisen from *ltmK*, and *ltmE* is the likely fusion of *ltmC* and *ltmF* (Schardl, Young, Hesse, et al., 2013). Gene duplication or acquisition is not limited to the *Epichloë* species but also can be found in other genera. The *IDT* clusters of *M. acridum* and *M. anisopliae* harbour what appears to be a fairly recent duplication of *idtF*. The *E. aotearoae* e899 *IDT* cluster harbours pseudogenes or remnants of *idtF* and *idtB* and lacks *idtQ* and *idtK* but has highly divergent paralogs evident in AT-rich regions of the genomes (Schardl, Young, Pan, et al., 2013).

Similar to the *LOL* clusters, it appears that the *IDT* clusters have also relocated (Fig. 10.5). It is clear that both *C. purpurea* and *A. take* share common flanking genes of *romB* and *ppdG*, but other species show more variation and tend to share only one flanking gene. Most commonly found in the *Epichloë* species is an association of the *IDT* clusters with a telomere, yet some of these clusters have flipped orientation. In Fl1, the telomere is most distal from *ltmK*, yet the *E. coenophiala* e4163 *IDT* cluster, which was derived from an *E. festucae* ancestor, has the telomere adjacent to *idtK*. Generally, a polyketide synthase gene or pseudogene and an associated hypothetical gene on one end and a telomere on the other end flank the *IDT* clusters of the *Epichloë* spp. However, *E. gansuensis* e7080 has a serine protein kinase adjacent to *idtP* that is also seen in the *At. texensis IDT* cluster, although the segment containing P450 monooxygenase genes *idtP* and *idtQ* has flipped in e7080. An *idoI* gene, so designated because it is similar to genes for indoleamine 2,3-dioxygenase, is adjacent to the *IDT* locus in *C. paspali* (flanks *idtM*), *Ch. alba* (flanks *idtK*), and *P. ipomoeae* (flanks *idtK*). The possibility has not been ruled out that the *idoI* gene product acts on indolediterpenes either for further modification or as part of a catabolic pathway.

The ergot alkaloid (*EAS*) clusters (Fig. 10.4), like the *IDT* clusters, are also telomeric in *Epichloë* species and internal in others. Like the other clusters, gene content varies among species, but unlike *LOL* and *IDT* clusters, the order of *EAS* genes also varies considerably. In *Claviceps*, *Atkinsonella*, and *Balansia* species, the clusters are internal, and whether this is also the case for *P. ipomoeae* is unknown. In the sequenced members of these genera, the gene order is conserved, though only *P. ipomoeae* has all of the known *EAS* genes associated with this family of fungi.

3.3. Variations in gene expression

The *EAS* locus seems exceptional in that, even though some strains also lack it, others seem to have a full complement of *EAS* genes but do not produce detectable levels of ergot alkaloids (Schardl, Young, Hesse, et al., 2013). Some such *E. festucae* strains were subjected to RNA-seq analysis, results of which indicated that their *EAS* genes were not expressed. Why those genes are silent in some strains, whereas other strains such as *E. festucae* Fl1 express the genes and the alkaloids, remains unknown. In both cases, the genes are interspersed with large blocks of AT-rich repeats and are located near telomeres, but there are some differences in the repeats and, surprisingly, the active locus in Fl1 is much closer to the telomere than is the inactive locus in strain E2368. It may be informative to conduct a detailed analysis of the effects of certain sequences in these regions as enhancers or silencers, as has been done for penicillin genes in *Aspergillus nidulans* (Shaaban et al., 2010).

Similar RNA-seq analysis indicated that genes of the partial *IDT* cluster in E2368 (which lacks key genes for indole-diterpene synthesis) are expressed very poorly compared to those in the complete *IDT* cluster of the lolitrem-producing strain, Fl1. In contrast, both the complete *perA* gene of Fl1 and the homologous *perA*-ΔR^* gene of E2368 are expressed well, even though the latter lacks the terminal domain (R^*) required for peramine biosynthesis. Conceivably, PerA-ΔR^* is involved in the synthesis of a related metabolite yet to be identified.

4. FUTURE PERSPECTIVES

To date, our focus on sequencing multiple plant-associated Clavicipitaceae has been to understand the dynamics of defensive alkaloid diversity, such as how variation within an alkaloid gene cluster is responsible for the accumulation of diverse major pathway end products. This research has provided insight into genes responsible for specific pathway steps (Pan et al., 2014) and has been especially important to decipher alkaloid diversity of mutualistic endophytes found in cool-season grasses (Charlton et al., 2012; Schardl et al., 2012; Schardl, Young, Hesse, et al., 2013; Schardl, Young, Pan, et al., 2013; Takach et al., 2012; Takach & Young, 2014). Comparative genomics studies among the haploid *Epichloë* species have the potential to unravel the genome dynamics and development of polyploid *Epichloë* species that are so prevalent in nature (Schardl, 2010). Understanding alkaloid diversity, along with the identification of additional SM, can provide the molecular basis behind some of the ecological fitness benefits symbionts provide their host.

In-depth studies on the genome dynamics of plant-associated Clavicipitaceae can also provide insight into the life history traits of these fungi. Signalling between the host and symbiont is required to establish and maintain systemic infections, whereby the balance of this interaction is tightly regulated and synchronized with the growth and development of the host, regardless of transmission mode. The development of genomic resources has already been employed to facilitate mapping signalling pathways and identifying physiological changes that occur in the interaction between symbiont and host (Schurmann, Buttermann, Herrmann, Giesbert, & Tudzynski, 2013; Tanaka, Takemoto, Chujo, & Scott, 2012). The availability of sequenced genomes for plant-associated Clavicipitaceae representing seven genera that span the symbiotic continuum will facilitate studies of the evolution and mechanisms underlying the diversity of metabolism, host interactions, and niche adaptation of plant-associated Clavicipitaceae.

ACKNOWLEDGMENTS

This work was supported by USDA-NIFA grant 2012-67013-19384, USDA-CSREES grant 2010-34457-21269, National Institutes of Health grants R01GM086888 and 2 P20 RR-16481, and the Samuel Roberts Noble Foundation. Genome sequence analysis was conducted in the University of Kentucky Advanced Genetic Technologies Center. This is publication number 14-12-025 of the Kentucky Agricultural Experiment Station, published with approval of the director.

REFERENCES

Alderman, S. C. (2013). Survival, germination, and growth of *Epichloë typhina* and significance of leaf wounds and insects in infection of orchardgrass. *Plant Disease*, *97*, 323–328.

Al-Reedy, R. M., Malireddy, R., Dillman, C. B., & Kennell, J. C. (2012). Comparative analysis of *Fusarium* mitochondrial genomes reveals a highly variable region that encodes an exceptionally large open reading frame. *Fungal Genetics and Biology*, *49*, 2–14.

An, Z.-q., Siegel, M. R., Hollin, W., Tsai, H.-F., Schmidt, D., & Schardl, C. L. (1993). Relationships among non-*Acremonium* sp. fungal endophytes in five grass species. *Applied and Environmental Microbiology*, *59*, 1540–1548.

Blin, K., Medema, M. H., Kazempour, D., Fischbach, M. A., Breitling, R., Takano, E., et al. (2013). antiSMASH 2.0—A versatile platform for genome mining of secondary metabolite producers. *Nucleic Acids Research*, *41*, W204–W212.

Bok, J. W., Hoffmeister, D., Maggio-Hall, L. A., Murillo, R., Glasner, J. D., & Keller, N. P. (2006). Genomic mining for *Aspergillus* natural products. *Chemistry & Biology*, *13*, 31–37.

Bushley, K. E., Raja, R., Jaiswal, P., Cumbie, J. S., Nonogaki, M., Boyd, A. E., et al. (2013). The genome of *Tolypocladium inflatum*: Evolution, organization, and expression of the cyclosporin biosynthetic gene cluster. *PLoS Genetics*, *9*, e1003496.

Charlton, N. D., Craven, K. D., Mittal, S., Hopkins, A. A., & Young, C. A. (2012). *Epichloë canadensis*, a new interspecific epichloid hybrid symbiotic with Canada wildrye (*Elymus canadensis*). *Mycologia*, *104*, 1187–1199.

Cheng, J. Z., Coyle, C. M., Panaccione, D. G., & O'Connor, S. E. (2010). A role for old yellow enzyme in ergot alkaloid biosynthesis. *Journal of the American Chemical Society*, *132*, 1776–1777.

Clay, K. (1994). Hereditary symbiosis in the grass genus *Danthonia*. *New Phytologist*, *126*, 223–231.

Clay, K., & Schardl, C. (2002). Evolutionary origins and ecological consequences of endophyte symbiosis with grasses. *American Naturalist*, *160*, S99–S127.

Cook, D., Beaulieu, W. T., Mott, I. W., Riet-Correa, F., Gardner, D. R., Grum, D., et al. (2013). Production of the alkaloid swainsonine by a fungal endosymbiont of the ascomycete order Chaetothyriales in the host *Ipomoea carnea*. *Journal of Agricultural and Food Chemistry*, *61*, 3797–3803.

Coyle, C. M., Cheng, J. Z., O'Connor, S. E., & Panaccione, D. G. (2010). An old yellow enzyme gene controls the branch point between *Aspergillus fumigatus* and *Claviceps purpurea* ergot alkaloid pathways. *Applied and Environmental Microbiology*, *76*, 3898–3903.

Coyle, C. M., & Panaccione, D. G. (2005). An ergot alkaloid biosynthesis gene and clustered hypothetical genes from *Aspergillus fumigatus*. *Applied and Environmental Microbiology*, *71*, 3112 3118.

Cuomo, C. A., Güldener, U., Xu, J.-R., Trail, F., Turgeon, B. G., Di Pietro, A., et al. (2007). The *Fusarium graminearum* genome reveals a link between localized polymorphism and pathogen specialization. *Science*, *317*, 1400–1402.

Dereeper, A., Guignon, V., Blanc, G., Audic, S., Buffet, S., Chevenet, F., et al. (2008). Phylogeny.fr: Robust phylogenetic analysis for the non-specialist. *Nucleic Acids Research*, *36*, W465–W469.

Diehl, W. W. (1950). *Balansia and the Balansiae in America*. Washington, D.C.: United States Department of Agriculture.

Edgar, R. (2004). MUSCLE: A multiple sequence alignment method with reduced time and space complexity. *BMC Bioinformatics*, *5*, 113.

Fleetwood, D. J., Khan, A. K., Johnson, R. D., Young, C. A., Mittal, S., Wrenn, R. E., et al. (2011). Abundant degenerate miniature inverted-repeat transposable elements in genomes of epichloid fungal endophytes of grasses. *Genome Biology and Evolution*, *3*, 1253–1264.

Fleetwood, D. J., Scott, B., Lane, G. A., Tanaka, A., & Johnson, R. D. (2007). A complex ergovaline gene cluster in epichloë endophytes of grasses. *Applied and Environmental Microbiology*, *73*, 2571–2579.

Formighieri, E. F., Tiburcio, R. A., Armas, E. D., Medrano, F. J., Shimo, H., Carels, N., et al. (2008). The mitochondrial genome of the phytopathogenic basidiomycete *Moniliophthora perniciosa* is 109 kb in size and contains a stable integrated plasmid. *Mycological Research*, *112*, 1136–1152.

Freeman, E. M. (1904). The seed fungus of *Lolium temulentum* L., the darnel. *Philosophical Transactions of the Royal Society of London, Series B*, *196*, 1–27.

Gao, Q., Jin, K., Ying, S.-H., Zhang, Y., Xiao, G., Shang, Y., et al. (2011). Genome sequencing and comparative transcriptomics of the model entomopathogenic fungi *Metarhizium anisopliae* and *M. acridum*. *PLoS Genetics*, 7, e1001264.

Gautschi, M., Schmid, J. P., Peppard, T. L., Ryan, T. P., Tuorto, R. M., & Yang, X. (1997). Chemical characterization of diketopiperazines in beer. *Journal of Agricultural and Food Chemistry*, *45*, 3183–3189.

Ghikas, D., Kouvelis, V., & Typas, M. (2006). The complete mitochondrial genome of the entomopathogenic fungus *Metarhizium anisopliae* var. *anisopliae*: Gene order and *trn* gene clusters reveal a common evolutionary course for all Sordariomycetes, while intergenic regions show variation. *Archives of Microbiology*, *185*, 393–401.

Glenn, A. E., Rykard, D. M., Bacon, C. W., & Hanlin, R. T. (1998). Molecular characterization of *Myriogenospora atramentosa* and its occurrence on some new hosts. *Mycological Research*, *102*, 483–490.

Gröger, D., & Floss, H. G. (1998). Biochemistry of ergot alkaloids—Achievements and challenges. *Alkaloids: Chemistry and Biology*, *50*, 171–218.

Güldener, U., Mannhaupt, G., Munsterkotter, M., Haase, D., Oesterheld, M., Stumpflen, V., et al. (2006). FGDB: A comprehensive fungal genome resource on the plant pathogen *Fusarium graminearum*. *Nucleic Acids Research*, *34*, D456–D458.

Haarmann, T., Lorenz, N., & Tudzynski, P. (2008). Use of a nonhomologous end joining deficient strain (Δku70) of the ergot fungus *Claviceps purpurea* for identification of a nonribosomal peptide synthetase gene involved in ergotamine biosynthesis. *Fungal Genetics and Biology, 45*, 35–44.

Hofmann, A. (1978). Historical view on ergot alkaloids. *Pharmacology, 16*, 1–11.

Havemann, J., Vogel, D., Loll, B., & Keller, U. (2014). Cyclolization of D-lysergic acid alkaloid peptides. *Chemistry & Biology*, in press. http://dx.doi.org/10.1016/j.chembiol.2013.11.008.

Kang, Y., Ji, Y. L., Zhu, K. R., Wang, H., Miao, H. M., & Wang, Z. W. (2011). A new *Epichloë* species with interspecific hybrid origins from *Poa pratensis* ssp *pratensis* in Liyang, China. *Mycologia, 103*, 1341–1350.

Kepler, R. M., Sung, G.-H., Harada, Y., Tanaka, K., Tanaka, E., Hosoya, T., et al. (2012). Host jumping onto close relatives and across kingdoms by *Tyrannicordyceps* (Clavicipitaceae) *gen. nov.* and *Ustilaginoidea* (Clavicipitaceae). *American Journal of Botany, 99*, 552–561.

Khaldi, N., Seifuddin, F. T., Turner, G., Haft, D., Nierman, W. C., Wolfe, K. H., et al. (2010). SMURF: Genomic mapping of fungal secondary metabolite clusters. *Fungal Genetics and Biology, 47*, 736–741.

Kuldau, G. A., Tsai, H.-F., & Schardl, C. L. (1999). Genome sizes of *Epichloë* species and anamorphic hybrids. *Mycologia, 91*, 776–782.

Leuchtmann, A., Bacon, C. W., Schardl, C. L., White, J. F., & Tadych, M. (2014). Nomenclatural realignment of *Neotyphodium* species with genus *Epichloë*. *Mycologia*, (in press).

Leuchtmann, A., & Clay, K. (1989). Morphological, cultural and mating studies on *Atkinsonella*, including *A. texensis*. *Mycologia, 81*, 692–701.

Lowe, T. M., & Eddy, S. R. (1997). tRNAscan-SE: A program for improved detection of transfer RNA genes in genomic sequence. *Nucleic Acids Research, 25*, 955–964.

Martinez, D., Berka, R. M., Henrissat, B., Saloheimo, M., Arvas, M., Baker, S. E., et al. (2008). Genome sequencing and analysis of the biomass-degrading fungus *Trichoderma reesei* (syn. *Hypocrea jecorina*). *Nature Biotechnology, 26*, 553–560.

Moon, C. D., Craven, K. D., Leuchtmann, A., Clement, S. L., & Schardl, C. L. (2004). Prevalence of interspecific hybrids amongst asexual fungal endophytes of grasses. *Molecular Ecology, 13*, 1455–1467.

Motoyama, T., Hayashi, T., Hirota, H., Ueki, M., & Osada, H. (2012). Terpendole E, a kinesin Eg5 inhibitor, is a key biosynthetic intermediate of indole-diterpenes in the producing fungus *Chaunopycnis alba*. *Chemistry & Biology, 19*, 1611–1619.

Munday-Finch, S. C., Wilkins, A. L., & Miles, C. O. (1998). Isolation of lolicine A, lolicine B, lolitriol, and lolitrem N from *Lolium perenne* infected with *Neotyphodium lolii* and evidence for the natural occurrence of 31-epilolitrem N and 31-epilolitrem F. *Journal of Agricultural and Food Chemistry, 46*, 590–598.

Ortel, I., & Keller, U. (2009). Combinatorial assembly of simple and complex d-lysergic acid alkaloid peptide classes in the ergot fungus *Claviceps purpurea*. *Journal of Biological Chemistry, 284*, 6650–6660.

Pan, J., Bhardwaj, M., Faulkner, J. R., Nagabhyru, P., Charlton, N. D., Higashi, R. M., et al. (2014). Ether bridge formation in loline alkaloid biosynthesis. *Phytochemistry, 98*, 60–68.

Pantou, M. P., Kouvelis, V. N., & Typas, M. A. (2008). The complete mitochondrial genome of *Fusarium oxysporum*: Insights into fungal mitochondrial evolution. *Gene, 419*, 7–15.

Pažoutová, S., Kolarik, M., & Kolinska, R. (2004). Pleomorphic conidiation in *Claviceps*. *Mycological Research, 108*, 126–135.

Pažoutová, S., Olšovská, J., Linka, M., Kolínská, R., & Flieger, M. (2000). Chemoraces and habitat specialization of *Claviceps purpurea* populations. *Applied and Environmental Microbiology, 66*, 5419–5425.

Philipson, M. N., & Christey, M. C. (1986). The relationship of host and endophyte during flowering, seed formation, and germination of *Lolium perenne*. *New Zealand Journal of Botany*, *24*, 125–134.
Rehmeyer, C., Li, W., Kusaba, M., Kim, Y.-S., Brown, D., Staben, C., et al. (2006). Organization of chromosome ends in the rice blast fungus, *Magnaporthe oryzae*. *Nucleic Acids Research*, *34*, 4685–4701.
Riederer, B., Han, M., & Keller, U. (1996). D-lysergyl peptide synthetase from the ergot fungus *Claviceps purpurea*. *Journal of Biological Chemistry*, *271*, 27524–27530.
Rudgers, J. A., Holah, J., Orr, S. P., & Clay, K. (2007). Forest succession suppressed by an introduced plant-fungal symbiosis. *Ecology*, *88*, 18–25.
Rylcy, M. J. (2003). *Nigrocornus scleroticus*, a common old world balansioid fungus. In J. F. Whiteet al. (Ed.)*Clavicipitalean fungi: Evolutionary biology, chemistry, biocontrol and cultural impacts* (pp. 247–271). New York, NY: Marcel-Dekker, Inc.
Saikia, S., Parker, E. J., Koulman, A., & Scott, B. (2006). Four gene products are required for the fungal synthesis of the indole-diterpene, paspaline. *FEBS Letters*, *580*, 1625–1630.
Saikia, S., Takemoto, D., Tapper, B. A., Lane, G. A., Frazer, K., & Scott, B. (2012). Functional analysis of an indole-diterpene gene cluster for lolitrem B biosynthesis in the grass endosymbiont *Epichloë festucae*. *FEBS Letters*, *586*, 2563–2569.
Sampson, K. (1937). Further observations on the systemic infection of *Lolium*. *Transactions of the British Mycological Society*, *21*, 84–97.
Schardl, C. L. (2010). The epichloae, symbionts of the grass subfamily Poöideae. *Annals of the Missouri Botanical Garden*, *97*, 646–665.
Schardl, C. L., Craven, K. D., Speakman, S., Stromberg, A., Lindstrom, A., & Yoshida, R. (2008). A novel test for host-symbiont codivergence indicates ancient origin of fungal endophytes in grasses. *Systematic Biology*, *57*, 483–498.
Schardl, C. L., Leuchtmann, A., & Spiering, M. J. (2004). Symbioses of grasses with seedborne fungal endophytes. *Annual Review of Plant Biology*, *55*, 315–340.
Schardl, C. L., Panaccione, D. G., & Tudzynski, P. (2006). Ergot alkaloids—Biology and molecular biology. *Alkaloids: Chemistry and Biology*, *63*, 45–86.
Schardl, C. L., Scott, B., Florea, S., & Zhang, D.-X. (2009). Epichloë endophytes: Clavicipitaceous symbionts of grasses. In H. B. Deising (Ed.), *The mycota V: Plant relationships* (pp. 275–306). Berlin: Springer-Verlag.
Schardl, C. L., Young, C. A., Faulkner, J. R., Florea, S., & Pan, J. (2012). Chemotypic diversity of epichloae, fungal symbionts of grasses. *Fungal Ecology*, *5*, 331–344.
Schardl, C. L., Young, C. A., Hesse, U., Amyotte, S. G., Andreeva, K., Calie, P. J., et al. (2013). Plant-symbiotic fungi as chemical engineers: Multi-genome analysis of the Clavicipitaceae reveals dynamics of alkaloid loci. *PLoS Genetics*, *9*, e1003323.
Schardl, C. L., Young, C. A., Pan, J., Florea, S., Takach, J. E., Panaccione, D. G., et al. (2013). Currencies of mutualisms: Sources of alkaloid genes in vertically transmitted epichloae. *Toxins*, *5*, 1064–1088.
Schurmann, J., Buttermann, D., Herrmann, A., Giesbert, S., & Tudzynski, P. (2013). Molecular characterization of the NADPH oxidase complex in the ergot fungus *Claviceps purpurea*: CpNox2 and CpPls1 are important for a balanced host-pathogen interaction. *Molecular Plant-Microbe Interactions*, *26*, 1151–1164.
Selosse, M. A., & Schardl, C. L. (2007). Fungal endophytes of grasses: Hybrids rescued by vertical transmission? An evolutionary perspective. *New Phytologist*, *173*, 452–458.
Shaaban, M., Palmer, J. M., El-Naggar, W. A., El-Sokkary, M. A., Habib, E. L. S. E., & Keller, N. P. (2010). Involvement of transposon-like elements in penicillin gene cluster regulation. *Fungal Genetics and Biology*, *47*, 423–432.
Steiner, U., Leibner, S., Schardl, C. L., Leuchtmann, A., & Leistner, E. (2011). *Periglandula*, a new fungal genus within the Clavicipitaceae and its association with Convolvulaceae. *Mycologia*, *103*, 1133–1145.

Sung, G. H., Sung, J. M., Hywel Jones, N. L., & Spatafora, J. W. (2007). A multi-gene phylogeny of Clavicipitaceae (Ascomycota, Fungi): Identification of localized incongruence using a combinational bootstrap approach. *Molecular Phylogenetics and Evolution, 44*, 1204–1223.

Tadych, M., Ambrose, K., Bergen, M., Belanger, F., & White, J. (2012). Taxonomic placement of *Epichloë poae* sp. nov. and horizontal dissemination to seedlings via conidia. *Fungal Diversity, 54*, 117–131.

Tagami, K., Liu, C., Minami, A., Noike, M., Isaka, T., Fueki, S., et al. (2013). Reconstitution of biosynthetic machinery for indole-diterpene paxilline in *Aspergillus oryzae*. *Journal of the American Chemical Society, 135*, 1260–1263.

Takach, J. E., Mittal, S., Swoboda, G. A., Bright, S. K., Trammell, M. A., Hopkins, A. A., et al. (2012). Genotypic and chemotypic diversity of *Neotyphodium* endophytes in tall fescue from Greece. *Applied and Environmental Microbiology, 78*, 5501–5510.

Takach, J. E., & Young, C. A. (2014). Alkaloid genotype diversity of tall fescue endophytes. *Crop Science* (in press).

Tanaka, A., Takemoto, D., Chujo, T., & Scott, B. (2012). Fungal endophytes of grasses. *Current Opinion in Plant Biology, 15*, 462–468.

Tanaka, E., & Tanaka, C. (2008). Phylogenetic study of clavicipitaceous fungi using acetaldehyde dehydrogenase gene sequences. *Mycoscience, 49*, 115–125.

Tanaka, A., Tapper, B. A., Popay, A., Parker, E. J., & Scott, B. (2005). A symbiosis expressed non-ribosomal peptide synthetase from a mutualistic fungal endophyte of perennial ryegrass confers protection to the symbiotum from insect herbivory. *Molecular Microbiology, 57*, 1036–1050.

Tasso, B., Novelli, F., Sparatore, F., Fasoli, F., & Gotti, C. (2013). (+)-Laburnamine, a natural selective ligand and partial agonist for the α4β2 nicotinic receptor subtype. *Journal of Natural Products, 76*, 727–731.

Tofern, B., Kaloga, M., Witte, L., Hartmann, T., & Eich, E. (1999). Phytochemistry and chemotaxonomy of the Convolvulaceae part 8—Occurrence of loline alkaloids in *Argyreia mollis* (Convolvulaceae). *Phytochemistry, 51*, 1177–1180.

Tudzynski, P., & Scheffer, J. (2004). *Claviceps purpurea*: Molecular aspects of a unique pathogenic lifestyle. *Molecular Plant Pathology, 5*, 377–388.

Turgeon, G., & Yoder, O. C. (2000). Proposed nomenclature for mating type genes of filamentous ascomycetes. *Fungal Genetics and Biology, 31*, 1–5.

Uhlig, S., Botha, C. J., Vrålstad, T., Rolén, E., & Miles, C. O. (2009). Indole-diterpenes and ergot alkaloids in *Cynodon dactylon* (Bermuda grass) infected with *Claviceps cynodontis* from an outbreak of tremors in cattle. *Journal of Agricultural and Food Chemistry, 57*, 11112–11119.

Vaillancourt, L., Du, M. Z., Wang, J., Rollins, J., & Hanau, R. (2000). Genetic analysis of cross fertility between two self-sterile strains of *Glomerella graminicola*. *Mycologia, 92*, 430–435.

Veen, G., Greinwald, R., Canto, P., Witte, L., & Czygan, F. C. (1992). Alkaloids of *Adenocarpus hispanicus* (Lam.) DC varieties. *Zeitschrift für Naturforschung C-A. Journal of Biosciences, 47*, 341–345.

Walker, J. (2004). *Claviceps phalaridis* in Australia: Biology, pathology and taxonomy with a description of the new genus *Cepsiclava* (Hypocreales, Clavicipitaceae). *Australasian Plant Pathology, 33*, 211–239.

White, J. F., Jr., & Reddy, P. V. (1998). Examination of structure and molecular phylogenetic relationships of some graminicolous symbionts in genera *Epichloë* and *Parepichloë*. *Mycologia, 90*, 226–234.

Wiemann, P., Sieber, C. M. K., von Bargen, K. W., Studt, L., Niehaus, E.-M., Espino, J. J., et al. (2013). Deciphering the cryptic genome: Genome-wide analyses of the rice pathogen *Fusarium fujikuroi* reveal complex regulation of secondary metabolism and novel metabolites. *PLoS Pathogens, 9*, e1003475.

Young, C., Bryant, M., Christensen, M., Tapper, B., Bryan, G., & Scott, B. (2005). Molecular cloning and genetic analysis of a symbiosis-expressed gene cluster for lolitrem biosynthesis from a mutualistic endophyte of perennial ryegrass. *Molecular Genetics and Genomics*, *274*, 13–29.

Young, C. A., Felitti, S., Shields, K., Spangenberg, G., Johnson, R. D., Bryan, G. T., et al. (2006). A complex gene cluster for indole-diterpene biosynthesis in the grass endophyte *Neotyphodium lolii*. *Fungal Genetics and Biology*, *43*, 679–693.

Young, C., McMillan, L., Telfer, E., & Scott, B. (2001). Molecular cloning and genetic analysis of an indole-diterpene gene cluster from *Penicillium paxilli*. *Molecular Microbiology*, *39*, 754–764.

Young, C. A., Tapper, B. A., May, K., Moon, C. D., Schardl, C. L., & Scott, B. (2009). Indole-diterpene biosynthetic capability of epichloë endophytes as predicted by *ltm* gene analysis. *Applied and Environmental Microbiology*, *75*, 2200 2211.

Zheng, P., Xia, Y. L., Xiao, G. H., Xiong, C. H., Hu, X., Zhang, S. W., et al. (2011). Genome sequence of the insect pathogenic fungus *Cordyceps militaris*, a valued traditional chinese medicine. *Genome Biology*, *12*, R116.

CHAPTER ELEVEN

Genomics, Lifestyles and Future Prospects of Wood-Decay and Litter-Decomposing Basidiomycota

Taina K. Lundell*[,1], Miia R. Mäkelä*, Ronald P. de Vries†, Kristiina S. Hildén*

*Division of Microbiology and Biotechnology, Department of Food and Environmental Sciences, University of Helsinki, Helsinki, Finland

†CBS-KNAW Fungal Biodiversity Centre, Utrecht, The Netherlands

[1]Corresponding author: e-mail address: taina.lundell@helsinki.fi

Contents

Advances in Botanical Research, Volume 70
ISSN 0065-2296
http://dx.doi.org/10.1016/B978-0-12-397940-7.00011-2

Abstract

Saprobic (*saprotrophic* and *saprophytic*) *wood-decay fungi* are in majority species belonging to the fungal phylum Basidiomycota, whereas *saprobic plant litter-decomposing fungi* are species of both the Basidiomycota and the second Dikarya phylum Ascomycota. Wood-colonizing white rot and brown rot fungi are principally polypore, gilled pleurotoid, or corticioid Basidiomycota species of the class Agaricomycetes, which also includes forest and grassland soil-inhabiting and litter-decomposing mushroom species. In this chapter, examples of lignocellulose degradation patterns are presented in the current view of genome sequencing and comparative genomics of fungal wood-decay enzymes. Specific attention is given to the model white rot fungus, lignin-degrading species *Phanerochaete chrysosporium* and its wood decay-related gene expression (transcriptomics) on lignocellulose substrates. Types of fungal decay patterns on wood and plant lignocellulose are discussed in the view of fungal lifestyle strategies. Potentiality of the plant biomass-decomposing Basidiomycota species, their secreted enzymes and respective lignocellulose-attacking genes is evaluated in regard to development of biotechnological and industrial applications.

1. INTRODUCTION: PLANT AND SOIL ORGANIC MATTER

Of the annual input of atmospheric carbon, the majority of the accumulation of CO_2 is generated by natural decomposition of organic matter. In terrestrial environments, one-third of the organic carbon pool is present in the structural components of living organisms, mainly plants and microbes, while two-thirds of the organic carbon is fixed in nonliving biomass. The dead organic substances are either compact plant lignocelluloses or soil organic matter of diverse composition. In order to initiate degradation and modification of these complex organic carbon pools, and thereby promote cycling of carbon and other nutrients in particular in the forest ecosystems, robust and versatile bioconversion reactions have evolved in the fungal lineages.

1.1. Organic matter and Earth's carbon cycle

Organic carbon pool reservoirs on Earth consist of fossil fuels, soil organic carbon, terrestrial biosphere and the ocean biosphere and dissolved organic compounds. Fossil fuels account for over 4000 Gtn (gigatons) by mass of the carbon pool, while terrestrial biomass—consisting of living and dead organisms and soil-stored organic carbon—accounts for 2000 Gtn (Falkowski et al., 2000). Most of the terrestrial biosphere-fixed organic carbon is stored in dead biomass (1200 Gtn), while living plants and other organisms sum up to 1/3 of the terrestrial organic carbon pool (Falkowski et al., 2000; Prentice et al., 2001).

Woody plants (trees, shrubs and lianas) account for 75% of stored terrestrial plant carbon (Horwarth, 2007). Tree-filled forests and wooded lands cover together 40% of the Earth's land area (over 9% of the planet's surface) making a sum of over 5×10^9 ha. This accounts for a global volume of ca. 540×10^9 m^3 of growing wood stock making a biomass of over 600 Gtn (in dry matter) of living wood (sum of above- and belowground wood) (FAO, 2010). Dead wood biomass reaches 67 Gtn in the manageable forests (FAO, 2010). It has been estimated that it takes about 10 years for plants to recycle their carbon, when their respiration (CO_2 formation) is subtracted from their gross primary production of organic carbon compounds by CO_2 fixation via photosynthesis (Horwarth, 2007). In forest ecosystems, turnover rates for complete degradation of the woody components and soil humified matter may reach up to thousands of years. In order to allow short-term carbon cycling in tree-based forest ecosystems, organisms specialized in conversion and degradation of all the components of wood lignocellulose, the wood-decay Basidiomycota fungi, play a crucial role.

More than 90% of the input of atmospheric carbon (in CO_2) is generated by natural decomposition of organic matter (Falkowski et al., 2000; Prentice et al., 2001). However, two-thirds of the terrestrial organic carbon is fixed in organic compounds of nonliving biomass. Of the dead organic carbon substances, the largest proportions are trapped in forest ecosystems in dead wood *lignocellulose* and *soil organic matter* (*SOM*). Wood lignocellulose is a compact composite of plant cell wall cellulose, hemicelluloses and lignin (see later). SOM on the contrary is a diverse term embracing both aboveground and belowground organic matter, such as forest tree and other plant-derived litter (leaves, needles, sticks, branches, hay and seeds), plant root exudates and rhizosphere microbes, dead plant roots, dead animals and microbes and humic substances composed of dead wood lignin-derived phenolic compounds, aliphatic chains and a substantial amount of nitrogen heterocyclic compounds such as pyrrole and pyridine rings (Horwarth, 2007).

Of the heterogeneous SOM, the humic substances are the most persistent reservoir of trapped carbon in the soil ecosystems and thereby generate an important carbon sink in the Earth's carbon cycle (Falkowski et al., 2000; Horwarth, 2007). SOM contains 3 × more carbon than what exists in the atmosphere or is fixed in the biomass of living plants. Degradation of mostly plant-derived organic matter is the key process to release other nutrients than carbon (nitrogen, phosphorous, mineral cations and sulphur) for use by soil microorganisms and plant roots. In these processes, soil-inhabiting litter-decomposing species play key functional roles.

1.2. Structure of plant biomass and lignocellulose

In forests ecosystems, the majority of SOM is derived from trees as dead wood lignocellulose derived from fallen trunks and branches, stumps and roots and littered leaves and needles. In grasslands, plant coverage is principally composed of nonwood species, and dead plant biomass is thereby dominated by grass-type litter lignocellulose. Most of the SOM consists of insoluble large polymers from the plant matter including polysaccharides (cellulose, hemicellulose, starch and pectin), lipids and waxes, proteins, nucleic acids, lignin and phenolic compounds (Horwarth, 2007). Of these macromolecules, lignin is the most recalcitrant for decomposition due to its aromatic and heterogeneous structure of various interlinkages and side groups and varying number of phenylpropanoid subunits.

The significance of the saprobic plant biomass-degrading Basidiomycota is seen in the evolution of their ability to completely decompose the tough woody (xylem) tissue lignocellulose. Lignocellulose composite forms the building material of the plant secondary cell walls and is composed of the polymeric polysaccharides cellulose, hemicelluloses and pectins and aromatic lignin heteropolymers (Bidlack, Malone, & Benson, 1992; Eriksson, Blanchette, & Ander, 1990; Sjöström, 1981). The long cellulose chains are arranged in packed microfibrils that are differentially orientated in the plant cell wall layers. Between these partially crystalline microfibril threads, hemicellulose units attach aromatic lignin moieties, thus generating a complex and rigid structure of the carbohydrate polymers (Fig. 11.1).

In trees, the amount of cellulose is 40–50% of the wood dry weight, and hemicelluloses and lignin each account for 15–30%, depending on the plant species and tissue (Eriksson et al., 1990; Sjöström, 1981). Cellulose is a homopolymer of β-1,4-glycosidic interlinked D-glucopyranoses forming up to 8000–10,000 unit long chains, which are tightly packed by hydrogen bonding in 40 chain bundles (microfibrils) (Bidlack et al., 1992; Sjöström, 1981). In plant cell walls, the orderly packed and crystalline cellulose microfibrils are embedded by hemicellulose–lignin matrix in several layers (Fig. 11.1). Angiosperm hemicelluloses are mainly xylans (β-1,4-D-xylopyranose chains), often acetylated methylglucurono-β-D-xylans, and some glucomannans (β-1,4-linked D-glucopyranose-D-mannopyranose

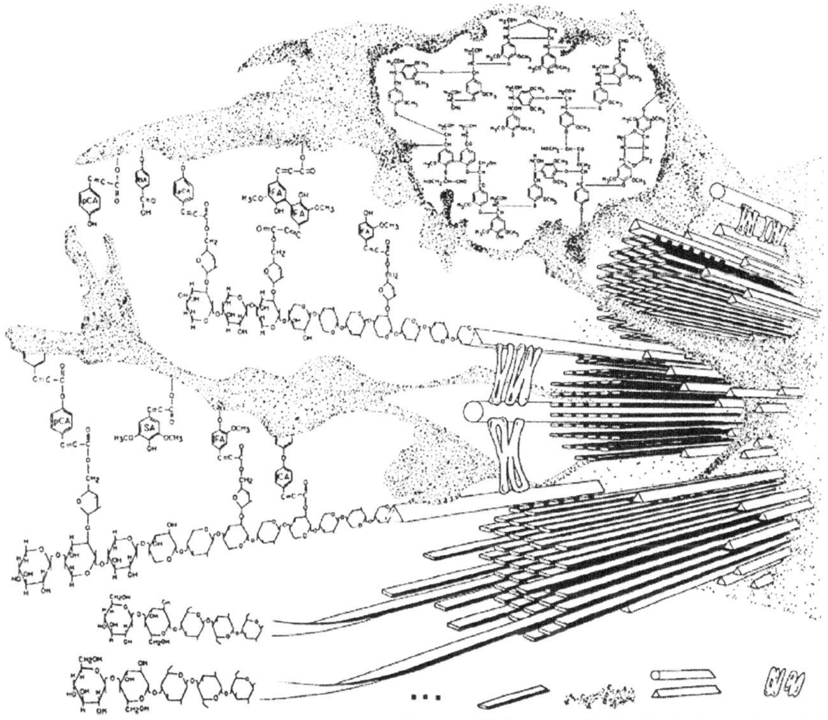

Figure 11.1 Plant lignocellulose. Schematic view of the lignified secondary cell wall molecular structure for grass plants. Polymeric cellulose microfibrils (linear polymers) are tightly packed in ordered layers surrounded by branched hemicellulose chains and aromatic lignin heteropolymers. Cross-links between hemicellulose chains and between hemicelluloses and lignin moieties (by ester and ether linkages) are depicted. *Picture according to the original illustration (Bidlack et al., 1992) is reprinted from* Current Opinion in Biotechnology, *Martínez, Ruiz-Dueñas, Martínez, del Río, and Gutiérrez (2009), Copyright (2009), with permission from Elsevier.*

chains). Coniferous gymnosperm hemicelluloses mainly comprise galactoglucomannans (glucomannan chains including α-1,6-linked galactose units). In addition to galactoglucomannans, coniferous softwood contains arabinoglucuronoxylan (Sjöström, 1981). Typical features for hemicelluloses are branching, partial acetylation of the sugar units and cross-linking also to other lignocellulose components. The hemicellulose backbones are substituted with a variety of carbohydrate monomers such as D-galactose, D-xylose, L-arabinose and D-glucuronic acid. Grass hemicelluloses are chemically diverse containing various pentose sugar units and more cross-linking, for example, by L-arabinofuranoses in arabinoxylans and to pectin (α-1,4-galacturonan and other D-galacturonic acid-containing branched polysaccharide chains) (Fig. 11.1). Grass hemicelluloses also contain covalent ester and ether bonds to ferulic and *p*-coumaric acid lignin monomers (Bidlack et al., 1992).

Lignin is an aromatic, non-polysaccharide heteropolymer of phenylpropanoid monolignol (*p*-coumaryl, feruloyl and sinapyl alcohol) subunits (Fig. 11.1). The degree of polymerization—number of interlinked phenylpropanoid units—as well as type of covalent linkages and amount of methyl-ether side groups is a highly variable characteristic of lignin depending on plant species and cell wall localization (Sjöström, 1981; Vanholme, Demedts, Morreel, Ralph, & Boerjan, 2010). The most condensed lignin is present in the middle lamellae interconnected to plant primary cell wall pectins and in gymnosperm softwood to xyloglucan hemicellulose. Xyloglucan is also synthesized in tension wood in angiosperm trees (Mellerowicz, Immerzeel, & Hayashi, 2008). The various ether and C–C bond interlinkages are formed via enzyme-catalysed radical coupling upon lignin biosynthesis (Vanholme et al., 2010). This chemical heterogeneity makes lignins difficult to degrade by hydrolytic enzyme attack. The most efficient degraders of this persistent component of the plant cell wall lignocellulose complex are the white rot wood decay Basidiomycota species of the systematic class Agaricomycetes, which are able to decompose not only cellulose and hemicelluloses but also the recalcitrant lignin polymers.

In addition to the lignocellulosic complex, plant cell walls contain proteins (enzymes and structural proteins), monomeric lignin precursors and other phenolic compounds and branched pectin polysaccharides (Eriksson et al., 1990; Sjöström, 1981). Pectin, lignin and hemicellulose fill the cell corners and interlayers forming the so-called middle lamellae in mature xylem cells (wood tissue). Thus, the compact wood composite structure is formed. Tree stems (trunks) include chemically distinct vertical compartments, starting from the bark outer layers, followed by living phloem and cambial wood tracheid-forming cells and then the innermost dead wood cell xylem compartments of sapwood and heartwood, including rays of living parenchyma cells. In conifers, resin ducts dissect the inner wood compartments horizontally.

2. WOOD AND LITTER DECAY AND FUNGAL LIFESTYLES

Four main patterns of fungal decay of wood may visually be recognized: white rot, brown rot, soft rot and blue stain. The wood white and brown rot-causing species belong to Basidiomycota, whereas soft rot and blue stain fungi represent species of Ascomycota. In the forest ecosystems, decomposition of plant litter is in turn due to diverse species of fungi representing both Basidiomycota and Ascomycota.

2.1. White rot

White rot-decayed wood is fibrous, soft and white to yellow in colour and often contains darks manganese deposits. The typical wood-decay white rot fungi are Basidiomycota Agaricomycetes species, mainly classified to the order Polyporales (Table 11.1), which colonize wood trunks forming protruding brackets or resupinate corticioid fruiting bodies (basidiocarps). A few gilled pleurotoid basidiocarp-forming species (e.g. species of *Pleurotus* and *Lentinula*) in other orders of Agaricomycetes are efficient colonizers of dead wood such as tree stumps resulting with white rot. A general feature of the white rot fungi is the production of an array of lignin-degrading peroxidases (PODs) together with multicopper oxidase (MCO) laccases and H_2O_2-generating enzymes (Table 11.2). Some wood-decay fungi are severe tree pathogens, in particular the conifer pathogenic species of the genus *Heterobasidion* (Table 11.1).

According to recent large fungal comparative genomics studies and molecular evolutionary calculations, the crucial ability to degrade lignin and cause wood white rot appeared to the ancestral Agaricomycetes fungal species at the same time when the large organic carbon deposits of the Carboniferous era, resulting from massive land plants of that time, started to decline about 290 million years ago in the beginning of the Devonian era (Floudas et al., 2012; Horwarth, 2007). Based on evolutionary analysis, the key genes for lignin-degrading ability, which are coding for the secreted class II peroxidase (POD) enzymes, emerged in the ancient Agaricomycetes at the same time. Apparently, the first lignin-degrading enzymes were manganese-oxidizing peroxidases (MnPs) evolved from a single, more simple generic peroxidase-encoding gene (Floudas et al., 2012). Since then, after several gene duplications, a multitude of lignin-degrading peroxidases divided to various subfamilies (short-MnPs, long-MnPs, atypical MnPs, VPs and LiPs; Table 11.2) and specific functions have appeared in the white rot Agaricomycetes species (Floudas et al., 2012; Ruiz-Dueñas et al., 2013).

2.2. Brown rot

Brown rot-decayed wood is dry, brown, powdery, and cracking to cubicles. Brown rot wood decay fungi are mainly Polyporales species, apparently

Table 11.1 Fungal decay of wood and lignocellulose: Basidiomycota model species and decay features

Decay type	Fungal species	Systematics	Growth substrate in nature	Specific features
Wood white rot	*Phanerochaete chrysosporium* G, S, B	Phanerochaetaceae, Polyporales, Agaricomycetes, Basidiomycota	Coniferous wood chips, lignocellulose composts	Model white rot fungus, lignin depolymerizing, cellulose decay, first Basidiomycota species genome sequenced; secretes PODs and CAZymes
	Gelatoporia (*Ceriporiopsis*) *subvermispora* G, S, B	Phanerochaetaceae, Polyporales, Agaricomycetes, Basidiomycota	Dead coniferous wood (esp. spruces), fallen trunks, burned wood	Selectively lignin-degrading, biopulping fungus, rare species found in boreal forests; secretes PODs and laccase
	Trametes versicolor G, S, B	Polyporaceae, Polyporales, Agaricomycetes, Basidiomycota	Dead deciduous and angiosperm tree wood, stumps and fallen trunks	Common species in temperate climate zones, model white rot fungus; secretes PODs, laccases, CAZymes; medicinal
	Heterobasidion annosum s.l., *H. parviporum*, *H. irregulare* G, P, S	Bondarzewiaceae, Russulales, Agaricomycetes, Basidiomycota	Conifers (pines, spruces, firs), deciduous trees, attacks living trees and seedlings	Species complex, tree pathogens, common in boreal and temperate zone forests; secretes laccases and PODs

Table 11.1 Fungal decay of wood and lignocellulose: Basidiomycota model species and decay features—cont'd

Decay type	Fungal species	Systematics	Growth substrate in nature	Specific features
Wood brown rot	*Fomitopsis pinicola* G, P, S	Fomitopsidaceae, Polyporales, Agaricomycetes, Basidiomycota	Norway spruce, Scots pine, silver birch	Typical brown rot-causing fungus, may be pathogenic and kill young trees, common in boreal forests
	Serpula lacrymans G, S	Serpulaceae, Boletales, Agaricomycetes, Basidiomycota	Wooden houses and constructs, moisture-damaged buildings, timber, planks	Causes destructive dry rot in buildings, more rare in nature, nonpathogenic in soil and tree roots
	Coniophora puteana G, S	Boletaceae, Boletales, Agaricomycetes, Basidiomycota	Stumps of conifers and deciduous trees, timber, wood constructs	Tolerates cold, grows well in humid constructs and cellars, may cause destruction
Wood soft rot	*Phialophora* sp., *Scytalidium* sp., *Paecilomyces* sp. S, P, B	Dothideomycetes, Leotiomycetes, Eurotiomycetes, Ascomycota	Unprotected wood constructs, moisture-damaged buildings, wood chip composts	Soften surface of timber and planks, some species generate microhyphae, cellulose
	Xylaria hypoxylon, *X. polymorpha*, *X. longipes* S	Xylariaceae, Xylariales, Sordariomycetes, Ascomycota	Stumps, wood chips and sticks on soil, mulch and detritus	Soften pieces of wood, polysaccharide decay, partial modification of wood lignin

Continued

Table 11.1 Fungal decay of wood and lignocellulose: Basidiomycota model species and decay features—cont'd

Decay type	Fungal species	Systematics	Growth substrate in nature	Specific features
Wood blue stain	*Ophiostoma* sp., *O. piceae* G, S, *Grosmannia clavigera* G, P	Ophiostomataceae, Ophiostomatales, Sordariomycetes, Ascomycota	Timber and felled trunks, live in resin ducts or phloem and parenchyma cells, pathogenic species attack living trees	Blue and black stain of trunks, no real decomposition of wood lignocellulose, also tree pathogenic species
Litter LDS	*Agaricus bisporus* (var. *bisporus*, var. *burnettii*) G, S, B	Agaricaceae, Agaricales, Agaricomycetes, Basidiomycota	Grasslands, cultivated on composted horse manure and straw, not on wood	Edible mushroom, commercial production worldwide, high nutritional and medicinal value; laccases and HTPs
	Volvariella volvacea G, S, B	Pluteaceae, Agaricales, Agaricomycetes, Basidiomycota	Cultivated on rice straw, slow decomposition of lignocellulose	Edible straw mushroom, commercial production in southern China and Asia; secretes laccases
	Coprinopsis cinerea G, S	Psathyrellaceae, Agaricales, Agaricomycetes, Basidiomycota	Grasslands, dung composts, high nitrogen demand, not on wood	Basidiomycota genetics model species; secretes laccases

LDS, litter-decomposing, soil-inhabiting; G, genome sequence available; S, saprotroph; P, pathogen; B, biotechnologically produced or applied species; CAZyme, carbohydrate-active enzyme; POD, class II lignin-modifying peroxidase; HTP, heme-thiolate peroxidase.
Modified from recent reviews on the topic (Lundell & Mäkelä, 2013; Lundell, Mäkelä, & Hildén, 2010).

Table 11.2 Oxidoreductase-encoding genes involved in lignin degradation in the genomes of plant biomass-decomposing Basidiomycota species

	Wood white rot										Litter LDS			EM	Wood brown rot						
AAs in CAZy database	*Pc*	*Gs*	*Ad*	*Ps*	*Fm*	*Ds*	*Tv*	*Sh*	*Hi*	*Sc*	*Ab*	*Vv*	*Cc*	*Lb*	*Rp*	*Fp*	*Cp*	*Gt*	*Sl*	*Wc*	*Da*
AA1 MCOs	5	9	10	13	11	13	10	20	18	6	na	na	17	15	7	7	8	4	6	5	5
AA1_1 laccase	0	7	7	12	10	11	7	15	16	2	11	11	17	9	7	5	6	3	4	3	0
AA2 PODs class II peroxidases	16	16	19	11	17	12	26	6	7	0	2	4	1	1	1	2	0	0	0	1	0
MnP (total)	5	13	6	10	16	9	13	5	6	0	2	4	0	0	0	0	0	0	0	0	0
MnP-short	0	1	0	2	3	5	13	0	6	0	2	4	0	0	0	0	0	0	0	0	0
Long	5	7	0	5	11	0	0	0	0	0	0	0	0	0	0	0	0	0	0	0	0
Extra long	0	5	0	3	0	4	0	0	0	0	0	0	0	0	0	0	0	0	0	0	0
Atypical	0	0	6	0	2	0	0	5	0	0	0	0	0	0	0	0	0	0	0	0	0
VP	0	0	0	0	0	3	3	0	0	0	0	0	0	0	0	0	0	0	0	0	0
LiP	10	2	0	0	0	0	10	0	0	0	0	0	0	0	0	0	0	0	0	0	0
Low-redox GP	1	1	11	1	1	0	0	1	1	0	0	na	1	1	1	2	0	0	0	1	0
AA3 GMC oxidoreductases																					
AA3_1 CDH	1	1	1	1	1	1	1	1	2	1	1	1	1	0	0	0	2	1	2	0	0
AA3_2 AAO/GO	4	7	0	4	2	9	3	14	12	5	10	8	18/27	7	3	1	0	2	0/6	0	0
AA3_3 AOX	1	4	3	3	2	4	4	7	16	3	na	na	0	0	1	4	2	1	1	4	1

Continued

Table 11.2 Oxidoreductase-encoding genes involved in lignin degradation in the genomes of plant biomass-decomposing Basidiomycota species—cont'd

	Wood white rot										Litter LDS			EM	Wood brown rot						
AAs in CAZy database	***Pc***	***Gs***	***Ad***	***Ps***	***Fm***	***Ds***	***Tv***	***Sh***	***Hi***	***Sc***	***Ab***	***Vv***	***Cc***	***Lb***	***Rp***	***Fp***	***Cp***	***Gt***	***Sl***	***Wc***	***Da***
AA3_4 POX	1	na	2	0	0	0	2	0	0	1	na	na	0	0	0	0	0	0	0	0	0
AA5 radical copper-oxidases																					
AA5_1 GLOX	7	3	7	9	4	9	9	8	5	2	3	3	6	11	2	4	6	2	2	4	3
AA6 1,4-benzo-quinone reductase	4	1	3	3	3	1	1	1	2	4	4	na	3	2	1	1	2	3	2	1	1
AA9 (GH61) LPMO	15	9	19	14	13	15	18	16	10	22	11	26	35	5	2	4	10	4	5	2	0
DyP	0	0	11	5	3	1	2	2	1	0	0	3	4	2	0	0	0	0	0	0	0
HTP	3	9	16	8	4	4	3	10	5	3	25	na	8	5	3	4	2	6	3	5	6

Enzyme classification is following the CAZy database (http://www.cazy.org/) auxiliary activities (AAs) description when possible. LDS, litter-decomposing, soil-inhabiting saprotroph; EM, ectomycorrhizal. *Pc*, *Phanerochaete chrysosporium; Gs*, *Gelatoporia* (*Ceriporiopsis*) *subvermispora; Ad*, *Auricularia delicata; Ps*, *Punctularia strigosozonata; Fm*, *Fomitiporia mediterranea; Ds*, *Dichomitus squalens; Tv*, *Trametes versicolor; Sh*, *Stereum hirsutum; Hi*, *Heterobasidion irregulare; Sc*, *Schizophyllum commune; Ab*, *Agaricus bisporus; Vv*, *Volvariella volvacea; Cc*, *Coprinopsis cinerea; Lb*, *Laccaria bicolor; Rp*, *Rhodonia* (*Postia*) *placenta; Fp*, *Fomitopsis pinicola; Cp*, *Coniophora puteana; Gt*, *Gloeophyllum trabeum; Sl*, *Serpula lacrymans; Wc*, *Wolfiporia cocos; Da*, *Dacryopinax* sp. MCOs, multicopper oxidases; PODs, class II heme-including peroxidases; LiP, lignin peroxidase; low-redox GP, general low-redox potential peroxidase; MnP, manganese peroxidase; VP, versatile peroxidase; GMC, glucose–methanol–choline oxidoreductase superfamily; CDH, cellobiose dehydrogenase; AAO, aryl alcohol oxidase; AOX, alcohol oxidase; POX, pyranose oxidase; GO, glucose oxidase; GLOX, glyoxal oxidase; LPMO, lytic polysaccharide monooxygenase; DyP, dye-decolorizing peroxidase, DyP-type peroxidase family; HTP, heme-thiolate peroxidase, peroxidase family 2; na, not annotated.

Data are derived from recent genomic studies (Bao et al., 2013; Chen et al., 2013; Fernandez-Fueyo et al., 2012; Floudas et al., 2012; Mäkelä et al., 2014; Martin et al., 2008; Martinez et al., 2004; Morin et al., 2012; Olson et al., 2012; Ruiz-Dueñas et al., 2013; Suzuki et al., 2012; Yakovlev et al., 2013).

having lost their lignin-degrading POD genes and possessing only one or two genes for nonligninolytic, low-redox potential general peroxidases (Ruiz-Dueñas et al., 2013). The loss of the ancient MnP-encoding genes has apparently occurred several times in the Agaricomycetes lineage (Floudas et al., 2012), thus leading to adaptation to other lifestyles than lignin degradation and white rot and resulting with either brown rot or ectomycorrhizal lifestyles (see Chapter 9). A few of the brown rot species are closely related to ectomycorrhizal species (e.g. *Serpula lacrymans* is classified to Boletales) (Eastwood et al., 2011). Brown rot fungi cause severe destruction of wood cellulose leading to brown and modified lignin residue, which also contains iron. The main mechanism for fungal brown rot destruction of wood polysaccharides is nonenzymatic attack, and oxidation and degradation proceed by Fenton chemistry producing highly reactive oxygen radicals (1) $Fe^{2+} + H_2O_2 + H^+ \rightarrow Fe^{3+} + HO^{\cdot} + H_2O$; (2) $Fe^{3+} + H_2O_2 \rightarrow Fe^{2+} + HOO^{\cdot} + H^+$ (Fenton, 1894; Jensen, Houtman, Ryan, & Hammel, 2001; Fig. 11.2).

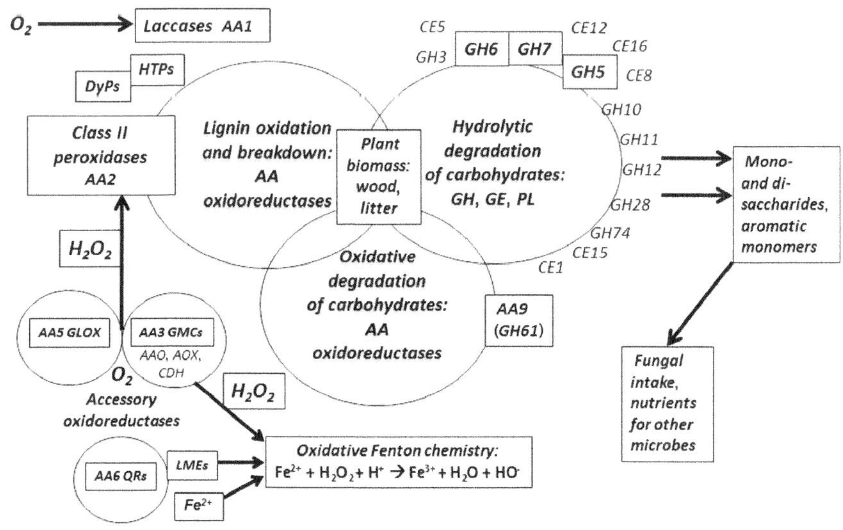

Figure 11.2 Participation of fungal-secreted CAZy and oxidoreductase enzymes in degradation of plant biomass. Enzyme abbreviations and classes are explained in the text. LMEs, low-molecular-weight compounds; QRs, quinone reductases. *Scheme is modified from original illustration given by Dimitrios Floudas (Clark University, Worcester, MA, United States), with kind permission from the author to be used here.*

2.3. Soft rot

Soft rot is a particular type of Ascomycota wood decay when the surface of wood material is colonized by green or dark "mould" fungi, mainly by species of the classes Sordariomycetes (genus *Trichoderma*), Dothideomycetes (*Phialophora*) and Eurotiomycetes (e.g. *Paecilomyces* and *Aspergillus*) (Table 11.1). In forest environments, pieces of dead wood in soil, dead tree roots, tree stumps and fallen branches may be colonized by Ascomycota species of the genus *Xylaria* causing a specific type of wood soft rot. *Xylaria* sp. decompose wood lignocellulose causing as well some degradation of wood lignin (Liers, Ullrich, Steffen, Hatakka, & Hofrichter, 2006). The "mould" Ascomycota are incapable to advance to the heartwood in dead tree trunks but may decompose cellulose and hemicelluloses of wounded or cut wood surfaces and timber. However, many wood surface soft rot fungi are efficient decomposers of plant litter (see later).

2.4. Blue stain

Blue stain fungi are Ascomycota species not destructing wood xylem lignocellulose. These fungi (e.g. species of the genera *Ophiostoma* and *Grosmannia*) are frequently disseminated as spores by wood-inhabiting beetles. Their hyphae extend in living wood trunks—thus either being pathogenic by growing in living ray parenchyma or phloem cells or being more saprobic by decomposing wood resins and waxes while growing in the resin ducts of conifers (Ballard, Walsh, & Cole, 1984; DiGuistini et al., 2011). Typical for blue stain species is the generation of dark-coloured melanins on their hyphal cell walls for protection against light, drought and host tree resistance factors. The wood-staining saprobic species *Ophiostoma piceae* is able to grow on triglycerides and oleic acid, and CAZy glycoside hydrolases are expressed on these substrates (Haridas et al., 2013). Since the blue stain species are not decomposing the main wood lignocellulose components (cellulose, hemicellulose and lignin), they are considered to be less important for wood organic carbon cycling.

2.5. Litter decomposition

Forest litter-decomposing fungi may roughly be divided into two subgroups according to their lifestyles: (i) species that colonize plant litter and debris on top of the humic organic O-layer and (ii) species that may attack smaller pieces of wood, bark, tree branches and dead tree roots. In the boreal and temperate forest ecosystems, the tree leaf litter and needle colonizers are mainly Ascomycota species. Litter quality, seasonal changes and variations in the top soil temperature are factors affecting litter biomass degradation rates (Prescott, 2010; Zhang, Hui, Luo, & Zhou, 2008). Extracellular hydrolytic enzyme activities are connected to soil pH, availability of nutrients (C, N and P) and microbial nutrient demand (Sinsabaugh et al., 2008). In a metaproteomic study on European forest oak leaf litter, secreted enzymes of representatives of the classes Leotiomycetes, Sordariomycetes and Eurotiomycetes dominated (80% frequency) (Schneider et al., 2012). A succession of fungi occurred to more Basidiomycota species in leaf litter sampled in spring compared to samples taken during autumn.

Fungal communities have been analysed in temperate soils of pine tree and mixed deciduous tree forests, demonstrating equal shares of Basidiomycota and Ascomycota taxonomic units (O'Brien et al., 2005). Similar results have been obtained in boreal spruce tree (*Picea abies*)-covered forest soils, where the main species identified were root endophytic Ascomycota and Basidiomycota species (Korkama-Rajala, Müller, & Pennanen, 2008). In studies using forest soil-embedded litter bags containing beech and spruce litter, it was demonstrated that Ascomycota species dominated the fungal communities (Aneja et al., 2006). The same was observed for decaying alder leaf litter in acidified streams (Clivot et al., 2013). Also in the decaying wood logs, the Ascomycota species are commonly found, together with a high variety of Basidiomycota and representatives of other fungal phyla, as is revealed by fungal community pyrosequencing metagenomic approach (Kubartova, Ottosson, Dahlberg, & Stenlid, 2012). It should be noted that soil fungal and bacterial communities are dynamic entities and variable during plant biomass degradation processes (Stursova, Zifcakova, Leigh, Burgess, & Baldrian, 2012).

3. FUNGAL ENZYMES IN DEGRADATION OF PLANT LIGNOCELLULOSE CARBOHYDRATES: GENOMIC VIEW

Due to the structural complexity of lignocellulosic plant biomass, a wide variety of carbohydrate-active enzymes and oxidoreductases are produced by fungal species to degrade the plant polysaccharides and lignin. Genomic studies have confirmed that both hydrolytic and oxidative enzyme activities are functional, together with chemical oxidation by Fenton reaction, in order to achieve efficient decay of the wood lignocellulose components (Fig. 11.2). However, depending on fungal species and lifestyles, the repertoire of enzymes and their gene numbers differs significantly (Tables 11.2 and 11.3). Genome sequences of plant biomass-degrading fungi have provided important molecular-level information on their strategies of biomass decay and also revealed that the fungal enzyme systems are much more diverse than was previously considered.

3.1. Carbohydrate-active enzymes

Carbohydrate-active enzymes (CAZymes) synthesize, modify or degrade plant biomass carbohydrates, such as polysaccharides and glycoconjugates. In the CAZy database (www.cazy.org; Lombard, Golaconda Ramulu, Drula, Coutinho, & Henrissat, 2013), these enzymes are classified based on their similarities in amino acid sequence, protein structure and enzymatic mechanism. Currently over 300 enzyme families can be grouped into four functional classes: glycoside hydrolases (GHs), glycosyltransferases (GTs), polysaccharide lyases (PLs) and carbohydrate esterases (CEs) based on their structural or functional domains. The role of GHs, PLs and CEs is to mediate the breakdown of polysaccharides, whereas GTs catalyse the formation of glycosidic linkages. Recently, oxidoreductases involved in lignin modification and degradation, such as laccases and class II peroxidases, have been included in the CAZy database as 10 auxiliary activity (AA) families (Levasseur, Drula, Lombard, Coutinho, & Henrissat, 2013). Comparisons of the genomic content of *wood- and litter-decomposing fungal species* have shown that there is a correlation between the number and type of CAZyme-encoding genes and the fungal capability to use various carbon sources and lignocellulosic substrates (Eastwood et al., 2011; Floudas et al., 2012).

3.2. Cellulose breakdown

Complete enzymatic hydrolysis and breakdown of cellulose chains require β-1,4-glycosidic bond-cleaving endoglucanase (EG), cellobiohydrolase (CBH) and β-glucosidase (BGL) activities. β-1,4-Endoglucanases hydrolyse noncrystalline regions of cellulose microfibrils, and the CBH enzymes attack the ends of the cellulose chains (Medie, Davies, Drancourt, & Henrissat, 2012). CBHs act in a unidirectional manner from either the reducing or nonreducing end of cellulose chains releasing disaccharide cellobiose units. Cellobiose is then cleaved to glucose by β-1,4-glucosidases, and this final step is crucial for complete hydrolysis of cellulose chains to glucose. β-Glucosidases are found in CAZy families GH1 and GH3, and in the wood-decay Basidiomycota genomes, GH3 enzymes are apparently more

Table 11.3 CAZy-encoding gene families in the genomes of plant biomass-decomposing Basidiomycota species

	Wood white rot										LDS			EM	Wood brown rot						
	Pc	***Gs***	***Ad***	***Ps***	***Fm***	***Ds***	***Tv***	***Sh***	***Hi***	***Sc***	***Ab***	***Vv***	***Cc***	***Lb***	***Rp***	***Fp***	***Cp***	***Gt***	***Sl***	***Wc***	***Da***
GHs																					
GH3	9	6	12	12	7	7	11	15	11	11	8	13	7	2	5	12	12	9	9	7	8
GH5	5	18	8	6	6	5	5	6	7	3	19	5	6	3	5	5	3	5	8	4	5
GH6	1	1	2	1	2	1	1	1	1	1	1	3	5	0	0	0	2	0	1	0	0
GH7	7	3	6	5	2	4	4	3	1	2	1	6	6	0	0	0	2	0	0	0	0
GH9	1	1	na	1	1	1	1	1	na	1	1	1	1	1	0	0	na	1	1	0	1
GH10	6	6	4	5	4	5	6	6	2	5	2	9	6	0	4	2	3	3	1	4	3
GH11	1	1	3	1	0	0	0	1	0	1	2	0	6	0	0	0	0	0	0	0	0
GH12	2	2	1	2	3	3	5	5	4	1	2	2	1	3	2	2	4	2	1	2	1
GH28	4	6	10	13	16	7	11	17	8	3	6	3	3	7	7	13	13	10	7	9	6
GH35	15	9	19	14	13	15	18	16	10	22	11	4	35	5	2	4	10	4	5	2	0
GH43	4	2	26	7	6	7	3	10	4	12	4	11	4	0	1	7	6	5	1	1	5
GH45	2	2	2	1	na	3	3	1	na	1	1	0	na	na	1	2	1	1	na	1	1
GH74	4	1	1	2	4	1	1	2	1	1	1	0	1	0	0	0	0	1	1	0	0
CEs																					

Continued

Table 11.3 CAZy-encoding gene families in the genomes of plant biomass-decomposing Basidiomycota species—cont'd

	Wood white rot										LDS			EM	Wood brown rot						
	Pc	*Gs*	*Ad*	*Ps*	*Fm*	*Ds*	*Tv*	*Sh*	*Hi*	*Sc*	*Ab*	*Vv*	*Cc*	*Lb*	*Rp*	*Fp*	*Cp*	*Gt*	*Sl*	*Wc*	*Da*
CE1	4	3	3	2	0	0	3	1	1	4	2	2	3	0	0	0	0	1	0	0	0
CE16	2	5	29	8	6	10	7	10	5	10	10	0	5	3	5	11	6	6	3	6	4
CE5	0	0	3	1	0	0	0	1	0	2	6	1	6	1	0	0	1	0	0	0	0
CE8	2	2	3	6	3	3	2	4	3	2	2	2	0	4	1	2	2	2	2	1	3
CE12	0	0	1	0	2	2	0	3	2	1	3	0	1	0	0	0	0	0	0	0	0
CE15	2	2	6	2	1	2	2	1	1	2	0	0	8	0	1	1	0	1	0	1	1

For enzyme functions, see CAZy database (http://www.cazy.org/). GHs, glycosyl hydrolases; CEs, carbohydrate esterases; na, not annotated. For other abbreviations, see Table 11.2 legend.

Data are derived from recent genomic studies (Chen et al., 2013; Fernandez-Fueyo et al., 2012; Floudas et al., 2012; Hori et al., 2013; Mäkelä et al., 2014; Morin et al., 2012; Olson et al., 2012; Vanden Wymelenberg et al., 2009).

abundant than GH1 gene models. It should be noted that β-glucosidases are also involved in the hydrolysis of other substrates (e.g. glycosides) and several of these enzymes are intracellular. It is therefore likely that only a subset of these enzymes is needed for lignocellulose breakdown.

Extracellular EG activity is common in plant biomass-converting Ascomycota species (de Vries & Visser, 2001; Kluczek-Turpeinen, Maijala, Hofrichter, & Hatakka, 2007) and is reported for wood-decay white and brown rot, litter-decomposing and ectomycorrhizal Basidiomycota Agaricomycetes species (Baldrian & Valaskova, 2008; Maijala, Fagerstedt, & Raudaskoski, 1991). β-1,4-Glycosidic bond-cleaving endoglucanase activities are present in several CAZy families (GH 5, 7, 12 and 45). One of the largest GH families in the wood-decay and litter-decomposing species is GH5, which is present in variant number of gene models (3–19) in all the species sequenced (Table 11.3).

In addition to endoglucanases, GH5 also includes internal β-1,4-glycosidic bond-cleaving activities on hemicellulose substrates (xylanase and mannanase activities) and exoglucanase activity at the cellulose chain ends (Medie et al., 2012; www.cazy.org). This lack of specificity towards the polysaccharide substrate is also found in other GH families. Therefore, the function of a protein of a particular GH family cannot always be predicted from its gene model or sequence structure. Potential endoglucanase activities are as well found in CAZy families GH45 and GH12, which are present in both white and brown rot fungi (Table 11.3), but the number of gene models is lower than for GH5 and their role in wood decay is not known (Morin et al., 2012). It has been suggested that fungal GH45 and GH12 proteins may cause swelling in the lignocellulose polysaccharide complexes, thereby acting synergistically with other cellulases (Igarashi, Ishida, Hori, & Samejima, 2008; Medie et al., 2012). Fungal GH12 enzymes from Ascomycota act on xyloglucan hemicellulose, which is primarily located in the primary cell wall of wood cells. Interestingly, a single copy of a putative GH9 endoglucanase with a unique transmembrane-anchored domain is almost constitutively identified in the wood-decay white and brown rot Basidiomycota genomes (Table 11.3).

CBH activities needed for cleavage of chain ends of crystalline cellulose are found in CAZy families GH6 and GH7. Typically, only one to two gene models for GH6 (CBHII) enzymes are present in the white rot Basidiomycota (Table 11.3). For GH7 exoglucanases (CBHI), a different pattern is seen with variant gene numbers, and in the white rot model fungus *P. chrysosporium*, an expansion of GH7-encoding genes (seven to eight genes) was observed. The genomes of the soil-inhabiting, coprophilic Basidiomycota Agaricomycetes species *Coprinopsis cinerea* and the straw-colonizing species *Volvariella volvacea* both contain an increased number of genes for the CBHs (GH6 and GH7), which indicates substantial cellulolytic activity and decomposition ability of litter lignocellulose polysaccharides. On the contrary, loss of both CBH gene families (GH6 and GH7) is observed in the Agaricomycetes brown rot species *Rhodonia* (*Postia*) *placenta*, *Fomitopsis pinicola*, *Gloeophyllum trabeum*, *Wolfiporia cocos* and *Dacryopinax* sp., and in the ectomycorrhizal species *Laccaria bicolor* (Table 11.3; Floudas et al., 2012; Martin et al., 2008). The brown rot species of the order Boletales (*Serpula lacrymans* and *Coniophora puteana*) have, however, retained their CBH gene models (Table 11.3; Eastwood et al., 2011; Kajisa, Igarashi, & Samejima, 2009).

In addition to hydrolytic enzymes, lignocellulose-degrading Basidiomycota secrete extracellular oxidative enzymes, which apparently attack on crystalline cellulose microfibril regions. These include the lytic polysaccharide monooxygenases (LPMOs, former family GH61), now reclassified to auxiliary activity CAZy family AA9, and cellobiose dehydrogenases (CDH) of family AA3 (Henriksson, Johansson, & Pettersson, 2000; Medie et al., 2012; Vaaje-Kolstad et al., 2010). AA9 LPMO gene models show significant expansions in the Basidiomycota white rot species as well as in the litter-decomposing species, and LPMO enzymes have been identified in fungal cultures on wood (Vanden Wymelenberg et al., 2011).

Combining the genomic information with gene expression studies, it may be concluded that the common ancestor of Basidiomycota Agaricomycetes was able to attack crystalline cellulose by production of the GH6 and GH7 CBHs and AA9 LPMOs, similar to the contemporary white rot species and litter-decomposing soil saprotrophs. Except for *C. puteana*, the brown rot Agaricomycetes genomes include AA9 gene models in lower numbers (Table 11.3), which is in line with their alternative, nonenzymatic Fenton chemistry-based oxidative destruction of cellulose.

Cellobiose dehydrogenase (CDH, CAZy family AA3) has versatile oxidoreductive activities, such as reduction of quinones and dioxygen while using, for example, cellobiose as electron donator. CDH has the combination of FAD and heme as prosthetic ligands (Henriksson et al., 2000), which is unique among extracellular enzymes. As proteins, CDHs are grouped to the glucose–methanol–choline GMC oxidoreductase superfamily of FAD-binding oxidases and dehydrogenases (Pfam family PF00732). White rot species have one CDH gene model (two in *H. irregulare*), whereas in some brown rot species, the gene is absent (Table 11.2).

3.3. Hemicellulose breakdown

Hydrolysis and degradation of plant hemicelluloses are more complicated due to the diverse chemical composition and presence of branched sugar units, acetyl groups and covalent cross-linkages in hemicelluloses, the latter in particular in grass plant biomass (see earlier). Due to this heterogeneity, a diverse group of backbone cleaving and debranching enzymes are needed for decomposition. According to the number of gene models, the enzymatic machinery for hemicellulose degradation is divergent between white rot and brown rot fungi. CAZy families GH10 and GH11 both contain endoxylanase activities, and GH10 enzymes are able to cleave substituted xylan backbones due to lower substrate specificity. The preferred substrates for GH11 enzymes are unsubstituted xylan chains (van den Brink & de Vries, 2011). Several genes (two to six genes, only one in *S. lacrymans*) for GH10 enzyme models are present in the wood-decay and litter-decomposing species (Table 11.3). GH11 genes are much scarcer in the wood-decay genomes with only a single gene present or being completely absent. However, in the litter-decomposing genomes, two to six genes for GH11 enzymes are depicted. A larger array of GH11 endoxylanases indicate adaptation to more efficient degradation of backbone linkages of xylan (β-1,4-D-xylopyranose chains), which is the prominent hemicellulose in grass plants and grass litter (Bidlack et al., 1992).

Xylooligosaccharides are further degraded to monosaccharides by β-xylosidases of which the majority of fungal gene models represent CAZy family GH3 (www.cazy.org; de Vries and Visser, 2001; Shallom & Shoham, 2003). Family GH3 β-xylosidases are common in Ascomycota species, whereas no such enzymes have yet been characterized from Basidiomycota. CAZy family GH43 comprises diverse enzyme activities able to cleave glycosidic linkages of hemicelluloses and pectins. Putative GH43 β-xylosidases are identified more frequently in white rot genomes (2–26 gene models) than in brown rot species (one to seven genes). A striking expansion of GH43 gene models is observed for *Auricularia delicata*, which is in correlation with its white rot decay lifestyle on wood (Table 11.3).

Xyloglucanases, which belong to CAZy families GH 5, 12 and 74 (Grishutin et al., 2004), are, for example, needed for cleavage of coniferous softwood xyloglucan hemicelluloses. White rot and litter-decomposing

Basidiomycota possess putative GH74 enzymes (one to four genes) (Table 11.3). Most brown rot species lack genes encoding GH74, except for *G. trabeum* and *S. lacrymans*, which both possess a single gene for GH74. In contrast, gene models for GH12 enzymes are generally found in the wood-decay and litter-decomposing genomes.

In addition, other hemicellulose-degrading enzymes such as α-arabinofuranosidases (GH51 and GH54), α-glucuronidases (GH67 and GH115), acetyl xylan esterases (CE1), feruloyl esterases (CE1) and glucuronoyl esterases (CE15) are needed for cleaving of the backbone chains and branching side groups (de Vries & Visser, 2001; Saha, 2003; Shallom & Shoham, 2003). These enzyme models are found in variable numbers in the Basidiomycota genomes (Table 11.3). For instance, CE1 feruloyl esterase gene models are more general, although in low numbers (one to four genes) in the white rot and litter-decomposing species, whereas in the brown rot genomes, these genes are absent except for *G. trabeum*.

The backbone of coniferous softwood galactoglucomannan hemicellulose is cleaved by endomannanases of CAZy families GH5 and GH26, and the oligosaccharides are further cleaved to monosaccharides by β-mannosidases (family GH2) and β-glucosidases (several CAZy families). Most wood-decay fungal endomannanases are grouped in the family GH5 (Table 11.3), whereas the genes for GH26 enzymes are rare in these genomes (Floudas et al., 2012).

3.4. Pectin breakdown

CAZy glycoside hydrolases (GH families) and polysaccharide lyases (PL families) are needed for the degradation of plant biomass pectin chains. Endo- and exo-polygalacturonases of the family GH28 cleave the backbone of the smooth regions of pectin chains, whereas endo- and exo-rhamnogalacturonase (GH28), xylogalacturonase (GH28), α-rhamnosidases (GH78), unsaturated glucuronyl hydrolase (GH88) and unsaturated rhamnogalacturonan hydrolase (GH105) enzyme activities are involved in cleaving the intricate branching and hairy regions of pectins (Martens-Uzunova & Schaap, 2009). Gene models for GH28 pectinases are common in high numbers (3–17 genes) in the plant biomass-degrading Basidiomycota genomes (Table 11.3), thus indicating efficiency and flexibility in responding to different types of plant substrates containing chemically variable pectin polysaccharides. Phylogeny of GH28 protein sequences in Ascomycota and Basidiomycota species indicates enzyme grouping according to divergent biochemical functions, with endo-polygalacturonase and endo-rhamnogalacturonase activities forming distinct, apparently ancient clades, while the exo-polygalacturonases are more widely distributed among fungi (Saha, 2003; van den Brink & de Vries, 2011).

4. LIGNIN BREAKDOWN AND LIGNIN-MODIFYING ENZYMES

Most lignin-modifying enzymes are metal-containing oxidoreductases secreted by the wood- and litter-decaying, saprobic Basidiomycota species (Lundell et al., 2010). Oxidation and conversion of lignin even up to carbon–carbon bond cleavage and partial mineralization occur by the biochemical action of the white rot fungi (Hammel & Cullen, 2008; Hatakka & Hammel, 2010; Kirk & Farrell, 1987). Classically, laccases and class II heme-containing peroxidases were considered as lignin-degrading enzymes, but these activities may be expanded to the assisting enzymes that produce H_2O_2, oxidize aromatic alcohols or reduce or oxidize fungal-generated and lignin-derived quinones and phenols (Lundell et al., 2010; Martínez et al., 2009).

The lignin-modifying oxidoreductases are listed in the CAZy database as auxiliary activities (AA) (Levasseur et al., 2013). In addition, the oxalate-degrading enzymes may be included in the pool of enzymes regulating fungal decomposition of lignin and lignocelluloses (Mäkelä et al., 2010). Oxidation and covalent bond cleavage in lignin and its subunits are promoted by lipid peroxidation, which generates peroxyl radicals and enhances the oxidative reactions of lignin-modifying manganese peroxidases (MnPs) (Hofrichter, Lundell, & Hatakka, 2001; Kapich, Korneichik, Hatakka, & Hammel, 2010; Lundell et al., 2010).

4.1. Multicopper oxidases

Laccases of CAZy family AA1 are phenol oxidases belonging to the multicopper oxidase (MCO) superfamily (Baldrian, 2006; Hoegger, Kilaru, James, Thacker, & Kües, 2006) and are generally expressed by lignocellulose and plant biomass-degrading saprobic Basidiomycota and Ascomycota species, as well as plant pathogenic species. Four-copper-containing fungal laccases (blue laccases) are identified as gene models in the white rot and brown rot genomes, except for the white rot model species *P. chrysosporium* (Table 11.2). Expansion to over 10 laccase genes has occurred in the genomes of many white rot species, as well as in the litter-decomposing Basidiomycota, whereas brown rot species harness a lower number (0–7) of laccase-encoding genes. Besides laccases, the AA1 class includes other MCO enzymes such as intra- and extracellular ferro-oxidoreductases, which are important, for example, in iron homeostasis (Hoegger et al., 2006; Larrondo, Salas, Melo, Vicuña, & Cullen, 2003).

Laccases have several functional roles for fungi, not only being involved in the oxidation of extracellular phenols and toxic compounds but also participating in hyphal fusions, basidiocarp and spore formation and pigment synthesis (Baldrian, 2006; Thurston, 1994). Due to the universality of laccases and MCO enzymes in organisms—laccases are not only found in fungi but also in plants, insects, bacteria, and archaea—and their absence in the model white rot lignin-degrading genome of *P. chrysosporium*, the role of laccases in actual oxidation and degradation of lignin is not evident. Laccases are unable to directly oxidize nonphenolic, high-redox potential lignin model compounds and lignins. However, in combination with low-molecular-weight phenolic and nitrogen-substituted compounds, the laccase-mediator-system-promoted oxidation of lignin-like and larger aromatic compounds is possible (Lundell et al., 2010; Martínez et al., 2009; Morozova, Shumakovich, Shleev, & Yaropolov, 2007), and thereby, laccases may be included in the pool of lignin-modifying fungal enzymes.

4.2. Lignin-modifying class II peroxidases

The lignin-modifying fungal peroxidases (PODs) are classified in the CAZy database to the oxidoreductase family AA2. Functionally, PODs are divided to manganese-oxidizing peroxidases (MnPs), lignin peroxidases (LiPs), and versatile peroxidases (VPs), which are categorized as high-redox potential peroxidases due to their ability to abstract electrons from high-redox lignin-like aromatic substrate molecules (Hofrichter, Ullrich, Pecyna, Liers, & Lundell, 2010; Ruiz-Dueñas & Martínez, 2009). Fungal generic peroxidases (GPs) are as well heme-including enzymes but lacking the manganese binding site or protein surface redox centre, thus apparently being unable to oxidize lignin units, and are thereby referred as low-redox potential peroxidases. Fungal-secreted PODs belong to the class II heme-containing "plant peroxidases" or "nonanimal peroxidase superfamily" (Hofrichter et al., 2010; Welinder, 1992; Zamocky, Furtmüller, & Obinger, 2009), which together with the "animal peroxidases" are all grouped in the large Pfam peroxidase

PF00141 protein family. Novel nomenclature and peroxidase enzyme grouping are emphasized in the PeroxiBase database (http://peroxibase.toulouse.inra.fr/; Oliva et al., 2009).

According to genomic data, it is evident that wood white rot lifestyle is correlated with the presence of class II lignin-modifying POD peroxidases (Floudas et al., 2012). The lignin-degrading Basidiomycota genomes possess multiple MnP-, VP- or LiP-encoding genes, whereas the brown rot genomes include only low-redox potential, nonligninolytic general peroxidases (0–2 GP genes) (Table 11.2). Several class II fungal peroxidases are structurally characterized (long-MnPs, VPs, short-MnPs, LiPs and CiP/ARP) (Hofrichter et al., 2010; Ruiz-Dueñas et al., 2013; Ruiz-Dueñas & Martínez, 2009).

Mn binding site in MnPs and VPs consists of three acidic amino acid residues (two Glu, E35 and E39, and one Asp, D179, residues in mature *P. chrysosporium* MnP1), which are crucial for hexacoordination of the Mn^{2+} ion, thereby facilitating rapid electron transfer to the heme porphyrin and oxidized ferryl iron (Sundaramoorthy, Gold, & Poulos, 2010). The oxalate-chelated Mn^{3+} ions act as diffusible charge transfer mediators and attack phenols and larger biopolymers, such as synthetic lignin, milled wood, brown coal, dyes and xenobiotic compounds and wood lignin in biopulping (Hatakka & Hammel, 2010; Hofrichter, 2002; Husain, 2010; Lundell et al., 2010).

Among lignocellulose and wood-decay white rot Basidiomycota, MnPs are the most commonly discovered class II lignin-modifying peroxidases (Table 11.2). The first MnP-encoding gene developed in the early Agaricomycetes lineage (Floudas et al., 2012). Later, the ancient gene diverged by gene duplications to several subfamilies of long- and short-MnP enzymes and, via emergence of a protein surface-exposed radical centre amino acid (tryptophan), to VPs and then to LiPs by modification of the Mn binding site (Ruiz-Dueñas et al., 2013). Gene expansions and functional modifications are seen in the white rot Polyporales species, and atypical short-MnP models with altered Mn binding sites are as well recognized, for example, as five gene models in *Stereum hirsutum* (Ruiz-Dueñas et al., 2013).

Number and character of MnP-encoding genes vary greatly in the white rot genomes (5–16) (Table 11.2). It is noteworthy that short-MnP enzyme-encoding genes are identified in the genomes of the litter-decomposing species *Agaricus bisporus* (two genes) and straw-cultivated mushroom *Volvariella volvacea* (four genes), whereas the saprobic *C. cinerea* and the ectomycorrhizal species *L. bicolor* have no MnP gene models. Instead of MnP-encoding genes, the two latter genomes possess one gene encoding a general peroxidase (GP, low redox potential). The *C. cinerea* GP enzyme (CiP) has been cloned, heterologously expressed, and structurally well characterized (Hofrichter et al., 2010; Kunishima et al., 1994).

LiP is catalytically the most powerful class II fungal peroxidase with the ability to directly oxidize dimeric lignin-like model compounds at the protein surface-exposed tryptophan radical centre (Choinowski, Blodig, Winterhalter, & Piontek, 1999). LiP is expressed and secreted as sets of multiple isozymes by, for example,. *P. chrysosporium*. LiP and VP gene models are so far identified only in a few wood-decay Basidiomycota genomes. The efficient white rot lignin-degrading Polyporales species, such as *P. chrysosporium*, *Gelatoporia (Ceriporiopsis) subvermispora*, *Dichomitus squalens*, *Trametes versicolor*, *Phlebia brevispora* and *Bjerkandera adusta*, as well as the wood-colonizing, white rot Agaricales species *Pleurotus ostreatus* (Ruiz-Dueñas et al., 2011), all contain either LiP or VP genes (Table 11.2). LiP-encoding gene families are significantly expanded in *B. adusta* (12 genes) (Ruiz-Dueñas et al., 2013), *P. chrysosporium* (10 genes) and *T. versicolor* (10 genes) (Table 11.2).

4.3. Other lignin-modifying, fungal-secreted peroxidases

Together with the high-redox potential class II PODs, the heme-thiolate peroxidases (HTPs) and dye-decolorizing peroxidases (DyP) may be included in the pool of fungal-secreted, heme-containing and lignin-modifying peroxidases (Hofrichter et al., 2010), since oxidation of lignin model compounds has been demonstrated for isolated HTPs and DyPs from Basidiomycota species (Ullrich, Nüske, Scheibner, Spantzel, & Hofrichter, 2004). HTP enzymes are characterized as either chloroperoxidases (CPOs) or aromatic universal peroxygenases (APOs and UPOs). In the Basidiomycota genomes, HTP genes are identified in highly variable numbers (3–25), whereas DyP genes are either absent (in many brown rot genomes) or present in low gene numbers (1–5) (Table 11.2). Exceptional expansion of HTP gene family is observed in the litter-decomposing Basidiomycota species *A. bisporus* (25 genes; Morin et al., 2012), and in the wood-decay white rot species *A. delicata*, expansion for both DyP and HTP gene families has occurred (Table 11.2).

A. bisporus HTP-encoding genes are transcribed when the fungal hyphae are colonizing humic-rich straw compost substrate, which indicates a particular role of these enzymes for, for example, successful growth in soil environments, which are extensive with humic substances (Morin et al., 2012). The second function for the HTP enzymes may be modification of fungal-produced secondary metabolites and bioactive compounds, such as is evident in the case of production of hypochlorite and caldariomycin by chloroperoxidase of *Caldariomyces fumago* (Osborne, Raner, Hager, & Dawson, 2006), thus protecting the fungus against other microbes in the highly competed soil environments.

4.4. GMC superfamily oxidases and oxidoreductases

Extracellular generation of H_2O_2 is important for the catalysis of the class II lignin-modifying heme peroxidases and for the generation of highly reactive hydroxyl radicals via Fenton reaction (Hammel & Cullen, 2008). Copper radical oxidases (CROs, CAZy family AA5) and FAD-dependent oxidases such as the GMC (glucose–methanol–choline) superfamily oxidoreductases (CAZy family AA3) have been suggested to play a role in the production of extracellular H_2O_2 for wood-decay fungi (Martinez et al., 2004, 2009). GMC oxidoreductases are catalytically distinct from CROs including various alcohol and sugar oxidases.

Aryl alcohol oxidase (AAO; family AA3_2) is an extracellular FAD-containing GMC oxidoreductase, which is suggested to be involved in generation of H_2O_2 in the redox cycling of aromatic fungal metabolites (Gutíerrez et al., 1994). The defined role of AAO is uncertain, but it has been proposed that AAO enzymes cooperate with aryl alcohol dehydrogenases for continuous H_2O_2 supply at least in some white rot fungi such as *Pleurotus*, *Bjerkandera* and *Trametes* species (Gutíerrez et al., 1994; Martínez et al., 2009). Except for *A. delicata*, AAO gene models are identified in variant numbers (2–14 genes) in the white rot and litter-decomposing Agaricomycetes genomes (Table 11.2), whereas in the brown rot genomes, the gene may be absent. The highest gene number is annotated for the litter-decomposing *C. cinerea*.

FAD-dependent alcohol oxidases (AOX; AA3_3) catalyse the oxidation of primary alcohols to corresponding aldehydes, and pyranose oxidases (POX; AA3_4) oxidize D-glucose, both enzymes generating H_2O_2 as side product (Kroutil, Mang, Edegger, & Faber, 2004; Martínez et al., 2009). In wood-decay Basidiomycota, AOX gene models (1–4) are present, whereas POX genes are only identified in a few white rot genomes (Table 11.2). Expansion of AOX gene family is observed for *S. hirsutum* (seven genes) and *H. irregulare* (16 genes).

Glyoxal oxidase (GLOX; AA5_1) belongs to the enzyme family of copper radical oxidases (CROs) including enzymes like galactose oxidase (GAOX). GLOX reduces O_2 to H_2O_2 and it has broad substrate specificity for the oxidation of simple aldehydes, such as glyoxal and methylglyoxal, to the corresponding carboxylic acids (Whittaker et al., 1996). GLOX and GAOX are copper metalloenzymes containing an unusual free radical-coupled Cu in the active site. Predicted WSC (cell wall integrity and stress-response component) domains in GLOX proteins were suggested to be involved in carbohydrate binding, and they have been detected in three of the CRO-encoding genes of *P. chrysosporium* (Vanden Wymelenberg et al., 2006). In white and brown rot wood-decay and litter-decomposing Agaricomycetes genomes, many GLOX models (two to nine genes) are depicted (Table 11.2). Interestingly, expansion to 11 GLOX-encoding model genes has occurred in the ectomycorrhizal *L. bicolor* genome.

5. *PHANEROCHAETE CHRYSOSPORIUM* AND WOOD DECAY

Due to the limited number of wood- and litter-decaying fungal genome sequences, only recently, transcriptome and proteome studies have elucidated the enzyme profiles employed for plant biomass degradation by Basidiomycota species. The first published basidiomycete genome was from *Phanerochaete chrysosporium*, which is the most intensively studied model fungus for white rot wood decay and lignin degradation (Martinez et al., 2004). *P. chrysosporium* was also the first wood-decay species subjected to thorough transcriptome and secretome studies. The initial studies were performed in carbon- and nitrogen-limited synthetic media and on cellulose (Vanden Wymelenberg et al., 2005) after which the focus has been on complex plant biomass substrates. This ground-breaking data enabled comparative transcriptome and secretome studies with other plant biomass-degrading species of Basidiomycota, such as *P. placenta*, *C. subvermispora*, *A. bisporus* and *Phanerochaete carnosa* (Fernandez-Fueyo et al., 2012; Morin et al., 2012; Suzuki et al., 2012; Vanden Wymelenberg et al., 2011). These functional studies have corroborated enzymatic and biochemical evidence of the individual and dissimilar mechanisms of lignocellulose degradation between species and their lifestyles. In addition, studies on diverse lignocellulosic substrates have revealed quick fungal physiological adaptation to changes in carbon and nutrient source, which is observed as altering transcript and secretome profiles.

5.1. *Phanerochaete* transcriptome and secretome on lignocellulose

Seven cellulose-degrading cellobiohydrolase enzymes (CBHs from CAZy families GH6 and GH7) of *P. chrysosporium* are expressed and produced differentially on various carbon sources, demonstrating the complexity and diversity of cellulolytic strategy of the species (Broda, Birch, Brooks, & Sims, 1995; Ravalason et al., 2008; Suzuki, Igarashi, & Samejima, 2010; Vallim, Janse, Gaskell, Pizzirani-Kleiner, & Cullen, 1998; Vanden Wymelenberg et al., 2009). Cellulose induced concurrent expression of an array of cellulose-attacking genes encoding putative endoglucanases (GH5), CBHs (GH6 and GH7), xylanases (GH10 and GH11) and lytic polysaccharide monooxygenase (LPMOs, CAZy family AA9) (Vanden Wymelenberg et al., 2009). In comparison, xylan hemicellulose as carbon source stimulated production of one putative endoxylanase (GH10) and glucuronoyl esterase (CE15), as well as one CDH and several LPMOs, suggesting that these enzymes are as well involved in the degradation of a wider array of plant polysaccharides than only cellulose (Hori, Igarashi, Katayama, & Samejima, 2011). On the contrary, addition of starch (α-1,4-interlinked polymer of D-glucose) to cellulose cultures of *P. chrysosporium* repressed production of the cellulolytic and xylanolytic enzymes (Hori et al., 2011).

In comparison of *P. chrysosporium* to another species from the genus *Phanerochaete*, *P. carnosa*, production of a similar set of extracellular GH and oxidoreductase enzymes occurred in both species upon growth on spruce wood or cellulose (Mahajan & Master, 2010). However, on spruce softwood, quantitative differences in the gene expression profiles were detected between the species. CAZyme-encoding genes were expressed most abundantly by *P. chrysosporium* (Sato, Feltus, Iyer, & Tien, 2009; Vanden Wymelenberg et al., 2010), whereas in *P. carnosa*, the CAZy AA2 lignin-modifying peroxidase transcripts and H_2O_2-producing enzyme-encoding genes were upregulated (MacDonald et al., 2011). Same abundance of lignin-modifying gene transcripts over polysaccharide-related gene expression was also observed in *P. carnosa* cultures on different hardwood and softwood species (lodgepole pine, white spruce, balsam fir and sugar maple), with the highest relative transcript levels for a single gene encoding a putative LPMO (AA9) (MacDonald et al., 2011). Among the lignin-modifying enzymes, notable upregulation of MnP-, LiP-, GLOX- and P450 monooxygenase-encoding genes was observed. Transcripts encoding CBHs (GH6 and GH7), endoglucanase (GH5) and xylanases (GH10) were among the most abundant of the pool of cellulose and hemicellulose-degrading enzyme-expressed genes.

These data indicate that *P. carnosa* produces similar sets of carbohydrate- and lignin-attacking enzymes on angiosperm hardwood and gymnosperm softwood, but with variable levels of gene expression (MacDonald et al., 2011). While *P. carnosa* secretomes from semisolid white spruce (*Picea glauca*) and microcrystalline cellulose cultures likewise performed similar protein profiles, it was proposed that minor changes in regulation of gene expression enable the growth of *P. carnosa* on various lignocelluloses (Mahajan & Master, 2010). Among the secreted enzymes, putative mannanase (GH5), CBH (GH6 and GH7), xylanase (GH10), endoglucanase (GH16), arabinofuranosidase (GH43) and glucuronoyl esterase (CE15) proteins were detected, as well as putative LPMO enzymes supporting the combination of both hydrolytic and oxidoreductive enzyme attack for degradation of cellulose by white rot fungi. In addition, lignin-modifying MnP, LiP and H_2O_2-producing GLOX and GMC oxidoreductases were identified.

5.2. White Rot fungal secretome and selective degradation of lignin

Studies on transcriptomes and secretomes on lignocelluloses support the view of white rot species degrading cellulose by both hydrolytic and oxidoreductive extracellular reactions, which has led to a novel concept of oxidative cellulose biodegradation. For instance, *P. chrysosporium* produces the cellobiose-oxidizing CDH (AA3_1, AA8) enzyme during growth on red oakwood, wheat straw and cellulose (Salvachúa et al., 2013; Sato et al., 2009; Vanden Wymelenberg et al., 2005). In addition, peptides of CDH and LPMOs have been secreted by the white rot species *B. adusta*, *D. squalens*, *Ganoderma* sp., *P. brevispora* and *T. versicolor* in aspen cultures (Hori et al., 2013). On wheat straw, AA9 LPMO enzymes were also present in the secretomes of *P. chrysosporium* and *Irpex lacteus* (Salvachúa et al., 2013).

The sequenced genome of *Ceriporiopsis* (*Gelatoporia*) *subvermispora* and its transcriptome on aspen hardwood have given insights into the mechanism of selective lignin degradation (Fernandez-Fueyo et al., 2012). In comparison to *P. chrysosporium*, which simultaneously depolymerizes all the plant cell wall polymers, *C. subvermispora* harbours an increased number of AA2 lignin-modifying MnP-encoding genes (13; Table 11.2) of variant protein structures and lipid metabolism-related desaturase-encoding genes. The gene expression data emphasize the importance of the lignin-modifying oxidoreductases operating in connection to lipid peroxidation (Hofrichter et al., 2001; Kapich et al., 2010) in order to generate efficient attack against lignin moieties, together with diminished capacity for cellulose degradation, to facilitate selective white rot fungal decomposition of lignin (Fernandez-Fueyo et al., 2012).

The number and expression patterns of putative cellulase-encoding genes vary between the white rot species *P. chrysosporium* and *C. subvermispora*. For example, *P. chrysosporium* harbours seven protein models of GH7 CBHs (Table 11.3), whereas three GH7 protein models are identified for *C. subvermispora* (Fernandez-Fueyo et al., 2012). Four of the *P. chrysosporium* GH7-encoding genes were significantly upregulated in aspen-containing medium, while only one GH7-encoding gene was induced in *C. subvermispora* under similar conditions. For endoglucanases, significant increase in the amount of transcripts corresponding to two GH5 β-1-4-endoglucanases and two GH12 endoglucanases was detected in cultures of *P. chrysosporium*, while a single GH5 β-1-4 endoglucanase and GH12 endoglucanase gene were induced in the cultures of *C. subvermispora*. *P. chrysosporium* also expressed hemicellulases at a higher level when cultivated on aspen hardwood (Fernandez-Fueyo et al., 2012). Of the putative hemicellulases of *C. subvermispora*, only two to three GH5 and GH10 enzyme-encoding genes, respectively, and one gene for CE1 feruloyl esterase were upregulated. In contrast, the transcripts of genes encoding two GH5, three GH10, GH43, GH53, CE15, four GH74, one CE1 feruloyl esterase and one GH95-encoding gene transcript presented significant accumulation in the cultures of *P. chrysosporium*. The aspen culture secretomes of the two species demonstrated a higher number of CAZy GH glycoside hydrolases for *P. chrysosporium* (18) than produced by *C. subvermispora* (3), thus confirming the transcriptome data. In contrast, both species showed significant expression of multiple putative LPMO-encoding genes on aspen lignocellulose.

In respect to transcriptome features upon fungal degradation of plant biomass, upregulation of numerous HTP heme-thiolate peroxidase (over 20) and β-etherase-encoding genes was observed during mycelial growth of *A. bisporus* on litter compost (Morin et al., 2012). Gene expression pattern distinguishes this litter-decomposing species from the wood-decay white rot and brown rot species. Most possibly, expression of HTP peroxidases is a response to chemical composition of the fungal growth environment abundant with soil humic acids. One of the two *A. bisporus* short-MnP-encoding AA2 genes was also highly expressed on compost (Morin et al., 2012). It should be noted that over 50% of the CAZyme-encoding gene repertoire of *A. bisporus* (Table 11.3) was upregulated in the course of mycelial growth in the lignocellulose-compost cultures.

5.3. Brown rot versus white rot decay of wood

Comparative transcriptomics study of *P. chrysosporium* and the brown rot species *Postia* (*Rhodonia*) *placenta* (Vanden Wymelenberg et al., 2011) supported the biochemical evidence and divergent fungal lifestyles for wood-decay adopted for white rot and brown rot degradation of plant biomass. *P. placenta* produces oxidoreductive oxidase enzymes that putatively generate extracellular Fe^{2+} and H_2O_2, which are needed for the non-enzymatic cellulose depolymerization by Fenton chemistry (see Section 2.2). The role of fungal cellulose-cleaving GH glycoside hydrolases is therefore less significant in brown rot decay of wood compared to the so far better-characterized white rot fungal decay of lignocellulose (Martinez et al., 2009; Vanden Wymelenberg et al., 2010).

When the transcriptomes of *P. chrysosporium* and *P. placenta* were studied in semisolid hardwood aspen (*Populus grandidentata*) and softwood pine (*Pinus strobus*) cultures, a considerable difference in gene expression patterns was detected depending on the wood species (Vanden Wymelenberg et al., 2011). In addition, the transcriptome and secretome data from various lignocellulosic substrates have emphasized the importance of the AA2 oxidoreductase class II peroxidases in white rot fungal degradation of plant biomass (Salvachúa et al., 2013). Interestingly, both transcriptome and proteome studies have highlighted the importance of enzymatic, extracellular H_2O_2 production for both white and brown rot decay mechanisms (Martinez et al., 2009; Ravalason et al., 2008; Salvachúa et al., 2013; Sato et al., 2009; Vanden Wymelenberg et al., 2009, 2010).

Secretome analysis of five white rot (*B. adusta*, *D. squalens*, *Ganoderma* sp., *P. brevispora* and *T. versicolor*) and two brown rot Agaricomycetes (*F. pinicola* and *W. cocos*) cultivated in liquid medium with milled aspen wood as the sole carbon source further corroborates the distinct strategies used by wood-decay fungi for degradation of plant polysaccharides (Hori et al., 2013). CAZy AA6 oxidoreductase family quinone reductases were solely present in the brown rot secretomes of *F. pinicola* and *W. cocos*, which is in accordance to results obtained for *P. placenta* on lignocellulose

(Martinez et al., 2009; Vanden Wymelenberg et al., 2010) and supports the role of quinone reductases in brown rot Fenton chemistry (Jensen, Ryan, Vanden Wymelenberg, Cullen, & Hammel, 2002; Suzuki, Hunt, Houtman, Dalebroux, & Hammel, 2006). However, high expression of a single endoglucanase of CAZy family GH5 was observed for *P. placenta* on aspen and pine cultures, thus suggesting involvement of this particular enzyme in cellulose degradation by the fungus (Vanden Wymelenberg et al., 2011). In addition, putative GH28 pectinases (polygalacturonase and rhamnogalacturonase) were highly expressed in pinewood supplemented cultures of *P. placenta* (Vanden Wymelenberg et al., 2011).

Transcriptome and proteome analyses have revealed dissimilarities in wood degradation mechanisms used by the two brown rot Basidiomycota species *P. placenta* and *S. lacrymans* (Eastwood et al., 2011). *S. lacrymans* apparently employs CAZy GH glycoside hydrolases to depolymerize cellulose, whereas *P. placenta*, as is discussed earlier, degrades cellulose chains mainly nonenzymatically. When compared to glucose cultures, 21% of gene transcripts showing significant upregulation on pinewood cultures of *S. lacrymans* consisted of GH glycoside hydrolases and AA oxidoreductases. A notable increase in the transcript levels was observed for the genes encoding GH3, GH5, GH28 and GH74 enzymes and, interestingly, also for genes that encode putative AA9 LPMOs (Eastwood et al., 2011).

Several studies with *P. placenta* support the ability of brown rot Basidiomycota to modify lignin by demethylation of the lignin aromatic ring ether side groups, which is demonstrated by upregulation of specific AA3 GMC oxidoreductase superfamily methanol oxidase-encoding genes. Methanol oxidases produce H_2O_2 by oxidizing the lignin-derived methanol (Martinez et al., 2009; Vanden Wymelenberg et al., 2010). In contrast, the role of LPMOs in brown rot decay of wood remains yet unclear. Although a few AA9 LPMO-encoding gene models are present in brown rot fungal genomes (Table 11.2), the corresponding protein has so far been identified only from milled aspen wood secretome of *G. trabeum* (Floudas et al., 2012).

6. CONCLUSIONS AND OUTLOOK

As described in the previous parts of this chapter, decay of wood and leaf and forest litter is a major part of the global carbon cycle (Section 1), and fungi are the dominant players in the decomposition processes in terrestrial environments. In particular, Basidiomycota species have been shown to have a strong ability for wood and litter decay, although the physiological variations among them, for example, with respect to white and brown rot decay mechanism, are significant (Section 2). Studies into wood decay and litter decomposition have been performed for many decades, but the recent increase in the availability of plant biomass-degrading Basidiomycota genome sequences (Eastwood et al., 2011; Fernandez-Fueyo et al., 2012; Floudas et al., 2012; Morin et al., 2012; Olson et al., 2012; Suzuki et al., 2012) has enabled a much deeper understanding of the factors involved in the degradation processes and conversion of organic polymers and compounds. Availability of the genome sequences has not only revealed the complexity of the fungal enzyme systems (Fig. 11.2; Sections 3 and 4), but the following post-genomic approaches (transcriptomics and proteomics) have started to reveal the complex regulatory mechanisms that underlie the biological conversion processes (Section 5).

However, as most of the plant biomass-related enzymes that have been discovered in fungal genomes (Fig. 11.2) have not yet been studied biochemically or genetically, their exact roles remain speculative. While assigning the predicted genes to particular enzyme families provides a likely hypothesis about general enzymatic activity, their substrate specificity and enzyme kinetic properties remain largely unknown, which complicates envisaging their exact role in the decomposition of plant biomass. In addition, genomes of Basidiomycota species contain a significant portion of genes (20–40%) without characterized homologues, thus remaining functionally unknown, some of which appear to be specifically expressed during growth on wood or tree litter. For instance, proteomics of the wood white rot decaying species *Ganoderma lucidum*, a model polypore fungus traditionally used in the Far East for medicinal purposes, on sugar cane bagasse revealed the production of 71 proteins, comprising mainly of lignin-modifying oxidoreductase enzymes and glycoside hydrolases (Manavalan, Adav, & Sze, 2011). However, almost one-fifth of the secreted proteins were hypothetical, for which a role in plant biomass degradation is not clear. This indicates adaptations in growth strategies and biochemical mechanisms at the species level that result in this variation in saprobic lifestyles on plant biomass. To obtain a deeper understanding of theses variations, parallel genomic and post-genomic studies on taxonomically diverse Basidiomycota species are required, such as the comparative genomic saprobic Agaricomycetes project (Floudas et al., 2012). Also, elucidating the role of the yet unknown and novel enzymes and assisting proteins will likely further enhance our understanding of the decay mechanisms. The presence of both brown rot and white rot species in naturally decaying wood (Ottosson, 2013;

Ovaskainen et al., 2013) suggests that these two mechanisms in nature may act synergistically, even if they are mediated by different species.

The focus in wood decay and forest litter decomposition has so far been mainly on Basidiomycota species, largely due to their ability to also degrade the recalcitrant heteropolymer lignin, which appears to be absent in most Ascomycota species. However, ecological studies into the occurrence of fungi in wood and litter revealed that Ascomycota species are commonly found on these substrates, such as identification of several new *Penicillium* species isolated from leaf litter (Houbraken et al., 2011).

With such a common occurrence of Ascomycota naturally on plant litter, it is fair to assume that Ascomycota species participate in the overall plant biomass decomposition, although their functional role is largely unknown. Some studies addressing this have been performed in which individual fungal species have been tested for their ability to decay plant biomass substrates. Decomposition of fir needle and birch leaf litter revealed a higher ability for Basidiomycota than Ascomycota, in particular with respect to the lignin fraction (Osono & Takeda, 2006). A similar result was obtained in a study using larch needle litter (Osono, Fukasawa, & Takeda, 2003).

Based on these results, it could be hypothesized that Ascomycota species rely on Basidiomycota to degrade the plant biomass lignin fractions during decay of wood and litter, while Ascomycota mainly contribute to degradation of the plant biomass polysaccharides and other carbohydrates. However, their roles may be more complex than this. Some Ascomycota show good growth on lignin-rich substrates (www.fung-growth.org), although this may be due to residual amounts of carbohydrates or the presence of aromatic monomers and oligomers in the substrates. Also, it has been suggested that organic acids produced by fungi, such as oxalic acid, stimulate the degradation of plant biomass carbohydrates and lignin (Mäkelä et al., 2002). Many Ascomycota are good producers of organic acids (e.g. Schrickx, Raedts, Stouthamer, & van Verseveld, 1995), which could be an additional role they play in the overall decay process.

The complexity of wood and litter decay is underlined by the variety of fungal and bacterial species present in plant biomass during degradation processes. While we are starting to get a better picture of the composition of the microbial communities, and some of the key players in the degradation processes have become more evident due to the increased availability of genome sequences, we still have a long way to go in understanding the biochemical reactions as well as fungal metabolism and genetic regulation. This involves not only detailed analyses of individual organisms, such as fungal species, but also their interactions and ecological functions in microbial communities and in various environments, using, for example, metagenomic approaches. In nature, the combined abilities of the microbial communities create specific and overall degradation processes of the plant biomass polymers, which are more efficient than any man-made approach that has so far been designed. Unravelling these interplays and environmental conversion networks is a major challenge for future research.

REFERENCES

Aneja, M. K., Sharma, S., Fleischmann, F., Stich, S., Heller, W., Bahnweg, G., et al. (2006). Microbial colonization of beech and spruce litter—Influence of decomposition site and plant litter species on the diversity of microbial community. *Microbial Ecology, 52*, 127–135.

Baldrian, P. (2006). Fungal laccases—Occurrence and properties. *FEMS Microbiology Reviews, 30*, 215–242.

Baldrian, P., & Valaskova, V. (2008). Degradation of cellulose by basidiomycetous fungi. *FEMS Microbiology Reviews, 32*, 501–521.

Ballard, R. G., Walsh, M. A., & Cole, W. E. (1984). The penetration and growth of blue-stain fungi in the sapwood of lodgepole pine attacked by mountain pine beetle. *Canadian Journal of Botany, 62*, 1724–1729.

Bao, D., Gong, M., Zheng, H., Chen, M., Zhang, L., Wang, H., et al. (2013). Sequencing and comparative analysis of the straw mushroom (*Volvariella volvacea*) genome. *PLoS One, 8*(3), e58294.

Bidlack, J., Malone, M., & Benson, R. (1992). Molecular structure and component integration of secondary cell walls in plants. *Proceedings of the Oklaholma Academy of Science, 72*, 51–56.

Broda, P., Birch, P. R. J., Brooks, P. R., & Sims, P. F. G. (1995). PCR-mediated analysis of lignocellulolytic gene transcription by *Phanerochaete chrysosporium*: Substrate-dependent differential expression within gene-families. *Applied and Environmental Microbiology, 61*, 2358–2364.

Chen, B., Gui, F., Xie, B., Deng, Y., Sun, X., Lin, M., et al. (2013). Composition and expression of genes encoding carbohydrate-active enzymes in the straw-degrading mushroom *Volvariella volvacea*. *PLoS One, 8*(3), e58780.

Choinowski, T., Blodig, W., Winterhalter, K. H., & Piontek, K. (1999). The crystal structure of lignin peroxidase at 1.70 Å resolution reveals a hydroxy group on the Cβ of tryptophan 171: A novel radical site formed during the redox cycle. *Journal of Molecular Biology, 286*, 809–827.

Clivot, H., Cornut, J., Chauvet, E., Elger, A., Poupin, P., Guerold, F., et al. (2013). Leaf-associated fungal diversity in acidified streams: Insights from combining traditional and molecular approaches. *Environmental Microbiology*. http://dx.doi.org/10.1111/1462-2920.12245.

de Vries, R. P., & Visser, J. (2001). *Aspergillus* enzymes involved in degradation of plant cell wall polysaccharides. *Microbiology and Molecular Biology Reviews, 65*, 497–522.

DiGuistini, S., Wang, Y., Liao, N. Y., Taylor, G., Tanguay, P., Feau, N., et al. (2011). Genome and transcriptome analyses of the mountain pine beetle-fungal symbiont *Grosmannia clavigera*, a lodgepole pine pathogen. *Proceedings of the National Academy of Sciences of the United States of America, 108*, 2504–2509.

Eastwood, D. C., Floudas, D., Binder, M., Majcherczyk, A., Schneider, P., Aerts, A., et al. (2011). The plant cell wall-decomposing machinery underlies the functional diversity of forest fungi. *Science, 333*, 762–765.

Eriksson, K.-E. L., Blanchette, R. A., & Ander, P. (1990). *Microbial and enzymatic degradation of wood components. Springer Series in Wood Science.* Berlin: Springer Verlag.

Falkowski, P., Scholes, R. J., Boyle, E., Canadell, J., Canfield, D., Elser, J., et al. (2000). The global carbon cycle: A test of our knowledge of Earth as a system. *Science, 290*, 291–296.

FAO (2010). *Global Forest Resources Assesment 2010. Main report: FAO Forestry Paper 163.* Rome: Food and Agriculture Organisation of the United Nations. Available at: http://www.fao.org/forestry/fra/fra2010/en/.

Fenton, H. J. H. (1894). Oxidation of tartaric acid in presence of iron. *Journal of the Chemical Society, Transactions, 65*, 899–910.

Fernandez-Fueyo, E., Ruiz-Dueñas, F. J., Ferreira, P., Floudas, D., Hibbett, D. S., Canessa, P., et al. (2012). Comparative genomics of *Ceriporiopsis subvermispora* and *Phanerochaete chrysosporium* provide insight into selective ligninolysis. *Proceedings of the National Academy of Sciences of the United States of America, 109*, 5458–5463.

Floudas, D., Binder, M., Riley, R., Barry, K., Blanchette, R. A., Henrissat, B., et al. (2012). The paleozoic origin of enzymatic lignin decomposition reconstructed from 31 fungal genomes. *Science, 336*, 1715–1719.

Grishutin, S. G., Gusakov, A. V., Markov, A. V., Ustinov, B. B., Semenova, M. V., & Sinitsyn, A. P. (2004). Specific xyloglucanases as a new class of polysaccharide-degrading enzymes. *Biochimica et Biophysica Acta, 1674*, 268–281.

Gutíerrez, A., Caramelo, S., Prieto, A., Martínez, M. J., & Martínez, A. T. (1994). Anisaldehyde production and aryl-alcohol oxidase and dehydrogenase-activities in ligninolytic fungi of the genus Pleurotus. *Applied and Environmental Microbiology, 60*, 1783–1788.

Hammel, K. E., & Cullen, D. (2008). Role of fungal peroxidases in biological ligninolysis. *Current Opinion in Plant Biology, 11*, 349–355.

Haridas, S., Wang, Y., Lim, L., Massoumi Alamouti, S., Jackman, S., Docking, R., et al. (2013). The genome and transcriptome of the pine saprophyte *Ophiostoma piceae*, and a comparison with the bark beetle-associated pine pathogen *Grosmannia clavigera*. *BMC Genomics, 14*, 373. http://dx.doi.org/10.1186/1471-2164-14-373.

Hatakka, A., & Hammel, K. E. (2010). Fungal biodegradation of lignocelluloses. In K. Esser & M. Hofrichter (Eds.), *The Mycota: A comprehensive treatise on fungi as experimental systems for basic and applied research. Industrial applications* (pp. 319–340). Berlin: Springer.

Henriksson, G., Johansson, G., & Pettersson, G. (2000). A critical review of cellobiose dehydrogenases. *Journal of Biotechnology, 78*, 93–113.

Hoegger, P. J., Kilaru, S., James, T. Y., Thacker, J. R., & Kües, U. (2006). Phylogenetic comparison and classification of laccase and related multicopper oxidase protein sequences. *FEBS Journal, 273*, 2308–2326.

Hofrichter, M. (2002). Review: Lignin conversion by manganese peroxidase (MnP). *Enzyme and Microbial Technology, 30*, 454–466.

Hofrichter, M., Lundell, T., & Hatakka, A. (2001). Conversion of milled pine wood by manganese peroxidase from *Phlebia radiata*. *Applied and Environmental Microbiology, 67*, 4588–4593.

Hofrichter, M., Ullrich, R., Pecyna, M. J., Liers, C., & Lundell, T. (2010). New and classic families of secreted fungal heme peroxidases. *Applied Microbiology and Biotechnology, 87*, 871–897.

Hori, C., Gaskell, J., Igarashi, K., Samejima, M., Hibbett, D., Henrissat, B., et al. (2013). Genomewide analysis of polysaccharides degrading enzymes in 11 white- and brown-rot Polyporales provides insight into mechanisms of wood decay. *Mycologia, 105*, 1412–1427.

Hori, C., Igarashi, K., Katayama, A., & Samejima, M. (2011). Effects of xylan and starch on secretome of the basidiomycete *Phanerochaete chrysosporium* grown on cellulose. *FEMS Microbiology Letters*, *321*, 14–23.

Horwarth, W. (2007). Carbon cycling and formation of soil organic matter. In E. A. Paul (Ed.), *Soil microbiology, ecology, and biochemistry* (pp. 303–339) (3rd ed.). Oxford: Academic Press.

Houbraken, J., Lopez-Quintero, C. A., Frisvad, J. C., Boekhout, T., Theelen, B., Franco-Molano, A. E., et al. (2011). *Penicillium araracuarense* sp. nov., *Penicillium elleniae* sp. nov., *Penicillium penarojense* sp. nov., *Penicillium vanderhammenii* sp. nov. and *Penicillium wotroi* sp. nov., isolated from leaf litter. *International Journal of Systematic and Evolutionary Microbiology*, *61*, 1462–1475.

Husain, Q. (2010). Peroxidase mediated decolorization and remediation of wastewater containing industrial dyes: A review. *Reviews in Environmental Science and Biotechnology*, *9*, 117–140.

Igarashi, K., Ishida, T., Hori, C., & Samejima, M. (2008). Characterization of an endoglucanase belonging to a new subfamily of glycoside hydrolase family 45 of the basidiomycete *Phanerochaete chrysosporium*. *Applied and Environmental Microbiology*, *74*, 5628–5634.

Jensen, K. A. J., Houtman, C. J., Ryan, Z. C., & Hammel, K. E. (2001). Pathways for extracellular Fenton chemistry in the brown rot basidiomycete *Gloeophyllum trabeum*. *Applied and Environmental Microbiology*, *67*, 2705–2711.

Jensen, K. A., Jr., Ryan, Z. C., Vanden Wymelenberg, A., Cullen, D., & Hammel, K. E. (2002). An NADH:quinone oxidoreductase active during biodegradation by the brown rot fungus Gloeophyllum trabeum. *Applied and Environmental Microbiology*, *68*, 2699–2703.

Kajisa, T., Igarashi, K., & Samejima, M. (2009). The genes encoding glycoside hydrolase family 6 and 7 cellulases from the brown-rot fungus *Coniophora puteana*. *Journal of Wood Science*, *55*, 376–380.

Kapich, A. N., Korneichik, T. V., Hatakka, A., & Hammel, K. E. (2010). Oxidizability of unsaturated fatty acids and of a non-phenolic lignin structure in the manganese peroxidase-dependent lipid peroxidation system. *Enzyme and Microbial Technology*, *46*, 136–140.

Kirk, T. K., & Farrell, R. L. (1987). Enzymatic "combustion": The microbial degradation of lignin. *Annual Reviews in Microbiology*, *41*, 465–505.

Kluczek-Turpeinen, B., Maijala, P., Hofrichter, M., & Hatakka, A. (2007). Degradation and enzymatic activities of three *Paecilomyces inflatus* strains grown on diverse lignocellulosic substrates. *International Biodeterioration & Biodegradation*, *59*, 283–291.

Korkama-Rajala, T., Müller, M. M., & Pennanen, T. (2008). Decomposition and fungi of needle litter from slow- and fast-growing Norway spruce (*Picea abies*) clones. *Microbial Ecology*, *56*, 76–89.

Kroutil, W., Mang, H., Edegger, K., & Faber, K. (2004). Biocatalytic oxidation of primary and secondary alcohols. *Advanced Synthesis & Catalysis*, *346*, 125–142.

Kubartova, A., Ottosson, E., Dahlberg, A., & Stenlid, J. (2012). Patterns of fungal communities among and within decaying logs, revealed by 454 sequencing. *Molecular Ecology*, *21*, 4514–4532.

Kunishima, N., Fukuyama, K., Matsubara, H., Hatanaka, H., Shibano, Y., & Amachi, T. (1994). Crystal structure of the fungal peroxidase from *Arthromyces ramosus* at 1.9 Å resolution: Structural comparisons with the lignin and cytochrome *c* peroxidases. *Journal of Molecular Biology*, *235*, 331–344.

Larrondo, L. F., Salas, L., Melo, F., Vicuña, R., & Cullen, D. (2003). A novel extracellular multicopper oxidase from *Phanerochaete chrysosporium* with ferroxidase activity. *Applied and Environmental Microbiology*, *69*, 6257–6263.

Levasseur, A., Drula, E., Lombard, V., Coutinho, P. M., & Henrissat, B. (2013). Expansion of the enzymatic repertoire of the CAZy database to integrate auxiliary redox enzymes. *Biotechnology for Biofuels, 6*, 41. http://dx.doi.org/10.1186/1754-6834-6-41.

Liers, C., Ullrich, R., Steffen, K. T., Hatakka, A., & Hofrichter, M. (2006). Mineralization of ^{14}C-labelled synthetic lignin and extracellular enzyme activities of the wood-colonizing ascomycetes *Xylaria hypoxylon* and *Xylaria polymorpha*. *Applied Microbiology and Biotechnology, 69*, 573–579.

Lombard, V., Golaconda Ramulu, H., Drula, E., Coutinho, P. M., & Henrissat, B. (2013). The carbohydrate-active enzymes database (CAZy) in 2013. *Nucleic Acids Research, 2013*, 1–6. http://dx.doi.org/10.1093/nar/gkt1178.

Lundell, T., & Mäkelä, M. (2013). 8.1 Puunlahottajat. In S. Timonen & J. Valkonen (Eds.), *Sienten biologia* (pp. 259–279). Gaudeamus Helsinki University Press (in Finnish).

Lundell, T. K., Mäkelä, M. R., & Hildén, K. (2010). Lignin-modifying enzymes in filamentous basidiomycetes: Ecological, functional and phylogenetic review. *Journal of Basic Microbiology, 50*, 1–16.

MacDonald, J., Doering, M., Canam, T., Gong, Y. C., Guttman, D. S., Campbell, M. M., et al. (2011). Transcriptomic responses of the softwood-degrading white-rot fungus *Phanerochaete carnosa* during growth on coniferous and deciduous wood. *Applied and Environmental Microbiology*, 77, 3211–3218.

Mahajan, S., & Master, E. R. (2010). Lignocellulose-degrading enzymes secreted by *Phanerochaeate carnosa* grown on spruce and microcrystalline cellulose. *Applied Microbiology and Biotechnology, 86*, 1903–1914.

Maijala, P., Fagerstedt, K., & Raudaskoski, M. (1991). Detection of extracellular cellulolytic and proteolytic activity in ectomycorrhizal fungi and *Heterobasidion annosum* (Fr.) Bref. *New Phytologist, 117*, 643–648.

Mäkelä, M., Galkin, S., Hatakka, A., & Lundell, T. (2002). Production of organic acids and oxalate decarboxylase in lignin-degrading white rot fungi. *Enzyme and Microbial Technology, 30*, 542–549.

Mäkelä, M. R., Hildén, K., & de Vries, R. P. (2014). Degradation and modification of plant biomass by fungi. In M. Nowrousian (Ed.), *The Mycota: A comprehensive treatise on fungi as experimental systems for basic and applied research. Fungal genomics*. Berlin: Springer (in press).

Mäkelä, M. R., Hildén, K., & Lundell, T. K. (2010). Oxalate decarboxylase: Biotechnological update and prevalence of the enzyme in filamentous fungi. *Applied Microbiology and Biotechnology, 87*, 801–814.

Manavalan, A., Adav, S. S., & Sze, S. K. (2011). ITRAQ-based quantitative secretome analysis of *Phanerochaete chrysosporium*. *Journal of Proteomics, 75*, 642–654.

Martens-Uzunova, E. S., & Schaap, P. J. (2009). Assessment of the pectin degrading enzyme network of *Aspergillus niger* by functional genomics. *Fungal Genetics and Biology, 46* (Suppl. 1), S170–S179.

Martin, F., Aerts, A., Ahrén, D., Brun, A., Danchin, E. G., Duchaussoy, F., et al. (2008). The genome of *Laccaria bicolor* provides insights into mycorrhizal symbiosis. *Nature, 452*, 88–92.

Martinez, D., Challacombe, J., Morgenstern, I., Hibbett, D. S., Schmoll, M., Kubicek, C. P., et al. (2009). Genome, transcriptome, and secretome analysis of wood decay fungus *Postia placenta* supports unique mechanisms of lignocellulose conversion. *Proceedings of the National Academy of Sciences of the United States of America, 106*, 1954–1959.

Martinez, D., Larrondo, L. F., Putnam, N., Sollewijn Gelpke, M. D., Huang, K., Chapman, J., et al. (2004). Genome sequence of the lignocellulose degrading fungus *Phanerochaete chrysosporium* strain RP78. *Nature Biotechnology, 22*, 695–700.

Martínez, A. T., Ruiz-Dueñas, F. J., Martínez, M. J., del Río, J., & Gutiérrez, A. (2009). Enzymatic delignification of plant cell wall: From nature to mill. *Current Opinion in Biotechnology, 20*(3), 348–357.

Medie, F. M., Davies, G. J., Drancourt, M., & Henrissat, B. (2012). Genome analyses highlight the different biological roles of cellulases. *Nature Reviews. Microbiology, 10*, 227–234.

Mellerowicz, E. J., Immerzeel, P., & Hayashi, T. (2008). Xyloglucan: The molecular muscle of trees. *Annals of Botany, 102*, 659–665.

Morin, E., Kohler, A., Baker, A. R., Foulongne-Oriol, M., Lombard, V., Nagy, L. G., et al. (2012). Genome sequence of the button mushroom *Agaricus bisporus* reveals mechanisms governing adaptation to a humic-rich ecological niche. *Proceedings of the National Academy of Sciences of the United States of America, 109*, 17501–17506.

Morozova, O. V., Shumakovich, G. P., Shleev, S. V., & Yaropolov, A. I. (2007). Laccase-mediator systems and their applications: A review. *Applied Biochemistry and Microbiology, 43*, 523–535.

O'Brien, H. E., Parrent, J. L., Jackson, J. A., Moncalvo, J.-M., & Vilgalys, R. (2005). Fungal community analysis by large-scale sequencing of environmental samples. *Applied and Environmental Microbiology, 71*, 5544–5550.

Oliva, M., Theiler, G., Zamocky, M., Koua, D., Margis-Pinheiro, M., Passardi, F., et al. (2009). PeroxiBase: A powerful tool to collect and analyse peroxidase sequences from Viridiplantae. *Journal of Experimental Botany, 60*, 453–459.

Olson, Å., Aerts, A., Asiegbu, F., Belbahri, L., Bouzid, O., Broberg, A., et al. (2012). Insight into trade-off between wood decay and parasitism from the genome of a fungal forest pathogen. *New Phytologist, 194*, 1001–1013.

Osborne, R. L., Raner, G. M., Hager, L. P., & Dawson, J. H. (2006). *C. fumago* chloroperoxidase is a dehaloperoxidase: Oxidative dehalogenation of halophenols. *Journal of the American Chemical Society, 128*, 1036–1037.

Osono, T., Fukasawa, Y., & Takeda, H. (2003). Roles of diverse fungi in larch needle-litter decomposition. *Mycologia, 95*, 820–826.

Osono, T., & Takeda, H. (2006). Fungal decomposition of *Abies* needle and *Betula* leaf litter. *Mycologia, 98*, 172–179.

Ottosson, E. (2013). Succession of wood-inhabiting fungal communities. PhD thesis. Faculty of Natural Resources and Agricultural Sciences, Department of Forest Mycology and Plant Pathology, Swedish University of Agricultural Sciences, Uppsala, Sweden.

Ovaskainen, O., Schigel, D., Ali-Kovero, H., Auvinen, P., Paulin, L., Nordén, B., et al. (2013). Combining high-throughput sequencing with fruit body surveys reveals contrasting life-history strategies in fungi. *The ISME Journal, 7*, 1696–1709.

Prentice, I. C., Farquhar, G. D., Fasham, M. L., Goulden, M., Heimann, M., Jaramillo, V. J., et al. (2001). The carbon cycle and atmospheric carbon dioxide. In J. T. Houghton, Y. Ding, D. J. Griggs, M. Noguer, P. J. van der Linden, & X. Dai, et al. *Climate change 2001: The scientific basis* (pp. 183–237). *Contribution of Working Group I to the third assessment report of the intergovernmental panel on climate change 2001*, Cambridge University Press.

Prescott, C. E. (2010). Litter decomposition: What controls it and how can we alter it to sequester more carbon in forest soils. *Biogeochemistry, 101*, 133–149.

Ravalason, H., Gwénaël, J., Mollé, D., Pasco, M., Coutinho, P. M., Lapierre, C., et al. (2008). Secretome analysis of *Phanerochaete chrysosporium* strain CIRM-BRFM41 grown on softwood. *Applied Microbiology and Biotechnology, 80*, 719–733.

Ruiz-Dueñas, F. J., Fernandez, E., Martínez, M. J., & Martínez, A. T. (2011). Pleurotus ostreatus heme peroxidases: An in silico analysis from the genome sequence to the enzyme molecular structure. *Comptes Rendus Biologies, 334*, 795–805.

Ruiz-Dueñas, F. J., Lundell, T., Floudas, D., Nagy, L. G., Barrasa, J. M., Hibbett, D. S., et al. (2013). Lignin-degrading peroxidases in Polyporales: An evolutionary survey based on 10 sequenced genomes. *Mycologia, 105*, 1428–1444.

Ruiz-Dueñas, F. J., & Martínez, A. T. (2009). Microbial degradation of lignin: How a bulky recalcitrant polymer is efficiently recycled in nature and how we can take advantage of this. *Microbial Biotechnology, 2*, 164–177.

Saha, B. (2003). Hemicellulose bioconversion. *Journal of Industrial Microbiology and Biotechnology, 30*, 279–291.

Salvachúa, D., Martínez, A. T., Tien, M., López-Lucendo, M. F., García, F., de los Ríos, V., et al. (2013). Differential proteomic analysis of the secretome of *Irpex lacteus* and other white-rot fungi during wheat straw pretreatment. *Biotechnology for Biofuels, 6*, 115. http://dx.doi.org/10.1186/1754-6834-6-115.

Sato, S., Feltus, F. A., Iyer, P., & Tien, M. (2009). The first genome-level transcriptome of the wood-degrading fungus *Phanerochaete chrysosporium* grown on red oak. *Current Genetics, 55*, 273–286.

Schneider, T., Keiblinger, K. M., Schmid, E., Sterflinger-Gleixner, K., Ellersdorfer, G., Roschitzki, B., et al. (2012). Who is who in litter decomposition? Metaproteomics reveals major microbial players and their biogeochemical functions. *The ISME Journal, 6*, 1749–1762.

Schrickx, J. M., Raedts, M. J., Stouthamer, A. H., & van Verseveld, H. W. (1995). Organic acid production by *Aspergillus niger* in recycling culture analyzed by capillary electrophoresis. *Analytical Biochemistry, 231*, 175–181.

Shallom, D., & Shoham, Y. I. (2003). Microbial hemicellulases. *Current Opinion in Microbiology, 6*, 219–228.

Sinsabaugh, R. L., Lauber, C. L., Weintraub, M. N., Ahmed, B., Allison, S. D., Crenshaw, C., et al. (2008). Stoichiometry of soil enzyme activity at global scale. *Ecology Letters, 11*, 1252–1264.

Sjöström, E. (1981). *Wood chemistry. Fundamentals and applications*. New York, NY: Academic Press, Inc.

Stursova, M., Zifcakova, L., Leigh, M. B., Burgess, R., & Baldrian, P. (2012). Cellulose utilization in forest litter and soil: Identification of bacterial and fungal decomposers. *FEMS Microbiology Ecology, 80*, 735–746.

Sundaramoorthy, M., Gold, M. H., & Poulos, T. L. (2010). Ultrahigh (0.93 Å) resolution structure of manganese peroxidase from Phanerochaete chrysosporium: Implications for the catalytic mechanism. *Journal of Inorganic Biochemistry, 104*, 683–690.

Suzuki, M. R., Hunt, C. G., Houtman, C. J., Dalebroux, Z. D., & Hammel, K. E. (2006). Fungal hydroquinones contribute to brown rot of wood. *Environmental Microbiology, 8*, 2214–2223.

Suzuki, H., Igarashi, K., & Samejima, M. (2010). Cellotriose and cellotetraose as inducers of the genes encoding cellobiohydrolases in the basidiomycete *Phanerochaete chrysosporium*. *Applied and Environmental Microbiology, 76*, 6164–6170.

Suzuki, H., Macdonald, J., Syed, K., Salamov, A., Hori, C., Aerts, A., et al. (2012). Comparative genomics of the white-rot fungi, *Phanerochaete carnosa* and *P. chrysosporium*, to elucidate the genetic basis of the distinct wood types they colonize. *BMC Genomics, 13*, 444. http://dx.doi.org/10.1186/1471-2164-13-444.

Thurston, C. (1994). The structure and function of fungal laccases. *Microbiology, 140*, 19–26.

Ullrich, R., Nüske, J., Scheibner, K., Spantzel, J., & Hofrichter, M. (2004). Novel haloperoxidase from the agaric basidiomycete *Agrocybe aegerita* oxidizes aryl alcohols and aldehydes. *Applied and Environmental Microbiology, 70*, 4575–4581.

Vaaje-Kolstad, G., Westereng, B., Hong, S. J., Liu, Z., Zhai, H., Sørlie, M., et al. (2010). An oxidative enzyme boosting the enzymatic conversion of recalcitrant polysaccharides. *Science, 330*, 219–222.

Vallim, M. A., Janse, B. J. H., Gaskell, J., Pizzirani-Kleiner, A. A., & Cullen, D. (1998). *Phanerochaete chrysosporium* cellobiohydrolase and cellobiose dehydrogenase transcripts in wood. *Applied and Environmental Microbiology, 64*, 1924–1928.

van den Brink, J., & de Vries, R. P. (2011). Fungal enzyme sets for plant polysaccharide degradation. *Applied Microbiology and Biotechnology, 91*, 1477–1492.

Vanden Wymelenberg, A., Gaskell, J., Mozuch, M., BonDurant, S. S., Sabat, G., Ralph, J., et al. (2011). Significant alteration of gene expression in wood decay fungi *Postia placenta* and *Phanerochaete chrysosporium* by plant species. *Applied and Environmental Microbiology*, 77, 4499–4507.

Vanden Wymelenberg, A., Gaskell, J., Mozuch, M., Kersten, P., Sabat, G., Martinez, D., et al. (2009). Transcriptome and secretome analyses of *Phanerochaete chrysosporium* reveal complex patterns of gene expression. *Applied and Environmental Microbiology, 75*, 4058–4068.

Vanden Wymelenberg, A., Gaskell, J., Mozuch, M., Sabat, G., Ralph, J., Skyba, O., et al. (2010). Comparative transcriptome and secretome analysis of wood decay fungi *Postia placenta* and *Phanerochaete chrysosporium*. *Applied and Environmental Microbiology, 76*, 3599–3610.

Vanden Wymelenberg, A., Minges, P., Sabat, G., Martinez, D., Aerts, A., Salamov, A., et al. (2006). Computational analysis of the *Phanerochaete chrysosporium* v2.0 genome database and mass spectrometry identification of peptides in ligninolytic cultures reveal complex mixtures of secreted proteins. *Fungal Genetics and Biology, 43*, 343–356.

Vanden Wymelenberg, A. V., Sabat, G., Martinez, D., Rajangam, A. S., Teeri, T. T., Gaskell, J., et al. (2005). The *Phanerochaete chrysosporium* secretome: Database predictions and initial mass spectrometry peptide identifications in cellulose-grown medium. *Journal of Biotechnology, 118*, 17–34.

Vanholme, R., Demedts, B., Morreel, K., Ralph, J., & Boerjan, W. (2010). Lignin biosynthesis and structure. *Plant Physiology, 153*, 895–905.

Welinder, K. (1992). Plant peroxidases: Structure-function relationships. In C. Penel, T. Gaspar, & H. Greppin (Eds.), *Plant peroxidases, topics and detailed literature on molecular, biochemical and physiological aspects* (pp. 1–24). Genève: Université de Genève.

Whittaker, M. M., Kersten, P. J., Nakamura, N., Sanders-Loehr, J., Schweizer, E. S., & Whittaker, J. W. (1996). Glyoxal oxidase from *Phanerochaete chrysosporium* is a new radical-copper oxidase. *Journal of Biological Chemistry, 271*, 681–687.

Yakovlev, I. A., Hietala, A. M., Courty, P.-E., Lundell, T., Solheim, H., & Fossdal, C. G. (2013). Genes associated with lignin degradation in the polyphagous white-rot pathogen *Heterobasidion irregulare* show substrate-specific regulation. *Fungal Genetics and Biology, 56*, 17–24.

Zamocky, M., Furtmüller, P. G., & Obinger, C. (2009). Two distinct groups of fungal catalase/peroxidases. *Biochemical Society Transactions, 37*, 772–777.

Zhang, D., Hui, D., Luo, Y., & Zhou, G. (2008). Rates of litter decomposition in terrestrial ecosystems: Global patterns and controlling factors. *Journal of Plant Ecology, 1*, 85–93.

CHAPTER TWELVE

Heterobasidion annosum s.l. Genomics

Mårten Lind, Jan Stenlid, Åke Olson[1]
Department of Forest Mycology and Plant Pathology, Swedish University of Agricultural Sciences, Uppsala, Sweden
[1]Corresponding author: e-mail address: ake.olson@slu.se

Contents

Abstract

The economically important forest pathogen *H. irregulare* represents one of the most complete sequenced genomes among Basidiomycota/Agaricomycotina, with few gaps and a low nucleotide error rate (1 per 100,000 bp). Ninety percent of the predicted genes have support from microarray analysis or cDNA sequencing although individual gene models would benefit from a curation supported by a high-coverage RNA-seq data. The sequencing of the *H. irregulare* genome has paved the way for many other genetic and genomic studies within the species complex and will be an invaluable resource for the research community for many years.

Advances in Botanical Research, Volume 70
ISSN 0065-2296
http://dx.doi.org/10.1016/B978-0-12-397940-7.00012-4

1. INTRODUCTION

1.1. Relevance and impact

The *Heterobasidion* species complex is economically the most significant pathogen in conifer forests of the northern hemisphere. Losses due to reduction in tree growth and devaluation of timber in Europe exceed €790 million yearly (Woodward, Stenlid, Karjalainen, & Hüttermann, 1998). Hence, a lot of attention has been directed during the past decade to the molecular and genomic study of the interaction between the fungus and its host. As a result, some of the species in this complex are now among the basidiomycetes with the most thoroughly studied genomes, and valuable insights can be obtained from the investigated genomes themselves and from the scientific process that led to their characterisation.

1.2. Overview of species complex

Heterobasidion annosum (Fr.) Bref. *sensu lato* (s.l.) is an aggressive pathogen that causes root and butt rot in primarily conifers of the northern hemisphere (Garbelotto & Gonthier, 2013; Korhonen, Capretti, Karjalainen, & Stenlid, 1998). The species complex is mainly associated with spruce, fir and pine and consists of three European and two North American species (Capretti, Korhonen, Mugnai, & Romagnioli, 1990; Chase & Ullrich, 1988; Korhonen, 1978). In Europe, *H. annosum sensu stricto* (s.s.) is mostly found in pine forests but is able to attack other conifers and broad-leaved trees (Korhonen, 1978). *H. parviporum* Niemelä & Korhonen not only lives almost exclusively on *Picea abies* (L.) Karsten but also attacks *Abies sibirica* Ledeb. in northeastern Europe (Korhonen, Fedorov, La Porta, & Kovbasa, 1997), while *H. abietinum* Niemelä & Korhonen is a pathogen or saprophyte on *Abies* species in southern and central Europe (Korhonen et al., 1998). The North American species were more recently given their taxonomic names *H. irregulare* Otrosina & Garbelotto and *H. occidentale* Otrosina & Garbelotto (Otrosina & Garbelotto, 2010). *H. irregulare* usually attacks pines, junipers and incense cedar, whereas *H. occidentale* can attack many tree genera, although it shows a preference for *Abies*, *Pseudotsuga* and *Tsuga* (Korhonen et al., 1998). *H. irregulare* occurs throughout the North American continent except for the northernmost regions, whereas *H. occidentale* is only present in the western parts.

1.3. Evolutionary history

By comparing sequence differences in 6 different genes among 10–30 individuals of each species of the *Heterobasidion* complex, Dalman, Olson, & Stenlid (2010) estimated the most recent common *Heterobasidion* ancestor to have existed 160 million years ago (mya), on the supercontinent of Laurasia, after its separation from Gondwana. The ancestor evolved into two main clades about 75–85 mya: the *H. irregulare/H. annosum* s.s. clade, primarily associated with ancestral *Pinus* species, and the *H. occidentale/H. parviporum/H. abietinum* clade, associated with *Picea* and *Abies* hosts. This split is not linked to the divergence between *Pinus* and *Picea*, which has been estimated to occur between 128 and 220 mya. However, the major split in the *Heterobasidion* complex does coincide roughly with the split of Laurasia into Euroamerica (later split into Europe and eastern North America) and Asiamerica (later Asia and western North America), divided west of the Ural Mountains. Since *H. annosum* s.s. does not exist in nowadays eastern Asia and *H. irregulare* occur in both eastern and western North America, it is likely that the *Heterobasidion* population left on Asiamerica, east of the divider, and evolved into the *H. occidentale/H. parviporum/H. abietinum* clade (Korhonen, Bobko, Hanso, Piri, & Vasiliauskas, 1992; Otrosina, Chase, Cobb, & Korhonen, 1993), and the population on Euroamerica evolved into the *H. irregulare/H. annosum* s.s. Nucleotide substitution rate data (Dalman et al., 2010) suggest that the *H. annosum* s.s. and *H. irregulare* ancestors split into separate species around 30 mya, which is supported by the fact that the land bridge between Europe and North America disappeared 35–50 mya. Different molecular markers create conflicting phylogeny data for the evolution of the *H. occidentale/H. parviporum/H. abietinum* clade but their most recent common ancestor existed about 30–40 mya. However, *H. parviporum* and *H. occidentale* evolved in allopatry in Asia and western North America earlier than the disappearance of the Bering land bridge.

The dating of distinct evolutionary occasions without calibration with fossil records has been shown to be problematic and often results in controversies (Wheat & Wahlberg, 2013). This is also the case in *Heterobasidion*. The allele age of the homeodomain (HD) genes of the mating-type (*MAT*) A locus was estimated to be 6 × the age of neutral genes (van Diepen, Olson, Ihrmark, Stenlid, & James, 2013). Using a mutation rate of 0.9×10^{-9} substitutions per site and year similar to Dalman et al., (2010), the dating of the oldest divergences gives the pre-*Heterobasidion* population the unrealistic age of more than 600 mya, that is, around the split of Ascomycota and Basidiomycota. This dating might indicate that the *H. annosum* s.l. complex is not as old as indicated by Dalman et al. (2010).

The dating of the oldest divergences goes too far back in time, which illustrates that relying solely on a molecular clock approach introduces large variation depending on mutation rate used and DNA loci of study.

2. INTRASPECIFIC INTERACTIONS

2.1. Mating and sexual compatibility

The bipolar mating system of *Heterobasidion* has arisen through the pheromone/receptor loss of function in mating compatibility. Mating is controlled by the *MAT-A* locus that encodes two pairs of divergently transcribed HD transcription factors with presumably functional redundancy (Olson et al., 2012). In HD2, the DNA-binding HD appeared to be abolished or highly divergent. This finding is surprising because previous studies with the model basidiomycete species *Coprinopsis cinerea* and *Ustilago maydis* have found the HD2 DNA-binding capacities to be more critical than the HD1 motif for function of the heterodimer (Asante-Owusu, Banham, Bohnert, Mellor, & Casselton, 1996; Kües et al., 1994; Schlesinger, Kahmann, & Kamper, 1997).

Analyses of the protein sequences show that both of the HD2-like proteins, Ha-a2 and Ha-b2, have a serine-rich region as observed in other *MAT* HD proteins and assumed to function in transactivation (Spit, Hyland, Mellor, & Casselton, 1998; Tymon, Kues, Richardson, & Casselton, 1992), and both are predicted to be localised to the nucleus. Alleles of both Ha-a2 and Ha-b2 proteins appear to contain coiled-coil motifs in the N-terminus, but no evidence for DNA-binding motifs has been found. Ha-a2 and Ha-b2 are thus predicted proteins that localise to the nucleus, bind with other proteins and may act as TFs but lack the canonical HD DNA-binding domain of typical HD proteins. This observation of an atypical HD2 protein suggests *Heterobasidion MAT* genes may have diverged in function compared with most other mushroom HD loci that usually have DNA-binding domains in both the HD1 and HD2 proteins of the heterodimer (Brown & Casselton, 2001). However, the recently sequenced *Stereum hirsutum* also have divergent HD2 genes that show some similarities with *Heterobasidion* HD2 genes (James et al., 2013). This raises the questions of whether they are of a common origin and when this distinction in the *MAT* locus arose. van Diepen et al. (2013) had shown that the HD genes in *H. annosum* s.l. are under strong negative frequency selection and that trans-specific polymorphism among the alleles has been kept for a significant period of time. The polymorphism in the *MAT-A* locus is on average 6 × as high as in the introns of neutral gene throughout the *H. annosum* s.l. species complex (van Diepen et al., 2013).

2.2. Intersterility and interfertility

Intersterility (IS)/interfertility between the closely related *Heterobasidion* species is controlled by a genetic system consisting of at least five genes. The IS system determines the limits of an interbreeding population and provides the genetic base to define biological species among fungi. It controls the reproduction barriers by preventing mating between individuals of distinct breeding populations. In *H. annosum* s.l., the system has been postulated to consist of five genes, called S, P, V_1, V_2 and V_3, each with a + or a − allele (Chase & Ullrich, 1990). For compatible mating, two individual mycelia must have a common + allele at least at one of these IS loci. All isolates associated with spruce trees belong to the S group (*H. occidentale*/*H. parviporum*) and are thought to possess a + allele at the S-IS locus. In addition, most isolates in the F group (*H. abietinum*) also possess a + allele at the S-IS locus, while all isolates belonging to the P group and associated with pine (*H. irregulare*/*H. annosum* s.s.) are thought to possess a + allele at the P-IS locus, ensuring compatible mating within the groups (Chase & Ullrich, 1990; Stenlid & Karlsson, 1991). Interfertility between the two groups is thought to be mediated by the other three genes, V_1, V_2 and V_3. The V_2 gene is believed to largely define interfertility in western North American populations of *H. irregulare* (Chase & Ullrich, 1990).

It is argued that *Heterobasidion* populations in sympatry have a reason to maintain a more strict IS system than those in allopatry, based on the envisaged risk to dilute favourable adaptive characters. Indeed, it has been shown that the *H. parviporum* population of Scandinavia and northern Europe (A in Fig. 12.1), allopatric to all known *H. abietinum* populations, is considerably more interfertile towards the Pyrenean and Alpine (B and C in Fig. 12.1) *H. abietinum* populations of southern Europe that are more interfertile (61.6–64.2%) than the Alpine *H. parviporum* population in sympatry with *H. abietinum* (19.6–20.3%) (B in Fig. 12.1) (Oliva, Gonthier, & Stenlid, 2011). Curiously enough, the Alpine (C) *H. parviporum* population is equally interfertile towards the Alpine (C) *H. abietinum* population as towards the Pyrenean (B) *H. abietinum* population, despite the fact that the latter is not in sympatry with any *H. parviporum* populations (Oliva et al., 2011). It is hypothesised that this is due to the previous presence of *H. parviporum* and *P. abies* in the Pyrenees, about 0.78–0.13 mya. Apparently, this time frame is not enough for a *Heterobasidion* population to lose an IS gene no longer under selective pressure. The IS genes may also have pleiotropic effects, for example, in host specialisation that can result in selection pressure to maintain certain variants in the population. Furthermore,

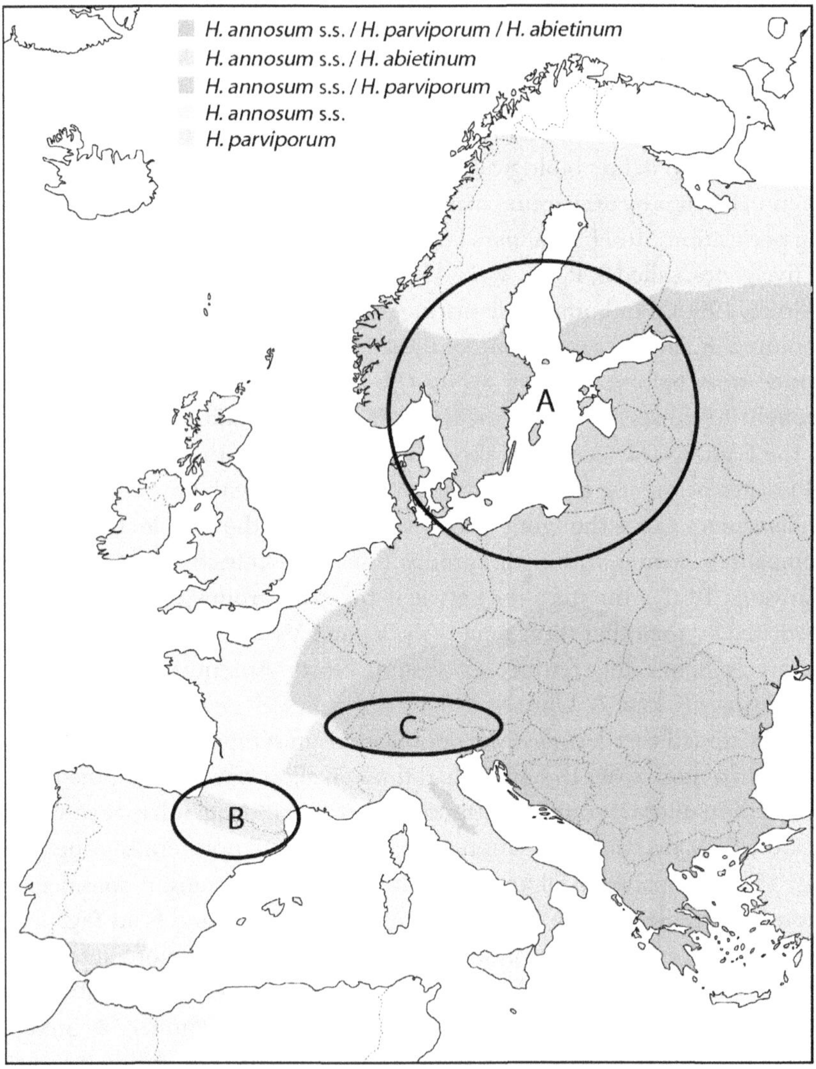

Figure 12.1 The distribution of the European *Heterobasidion* species and the origin of the populations for interfertility tests. (A) represents the origin of the Scandinavian *H. parviporum* population, and (B) the Pyrenean *H. abietinum* and population. (C) Shows the origin of the Alpine populations of *H. parviporum* and *H. abietinum*. (See the colour plate.)

being in allopatry does not guarantee that two populations devolve the IS barrier if the evolutionary distance between them is big. The pine-specific *Heterobasidion* species *H. irregulare* and *H. annosum* s.s. separated from the *picea/abies*-specific species *H. occidentale/H. parviporum/H. abietinum* about 75–85 mya. The level of interfertility between *H. parviporum* and *H. irregulare*, despite their complete allopatry, is only at 12%, roughly similar to the 16% between *H. annosum* s.s. and *H. occidentale*. For sympatric populations of this distant species, like the Scandinavian *H. parviporum* and *H. annosum* s.s., the interfertility is at similar levels of 10% (Stenlid & Karlsson, 1991). The fact that allopatric *H. parviporum* and *H. abietinum* populations display such high interfertility levels as 60% can be contributed to them being evolutionary separated at a much later stage than the other species of the complex. On the other hand, all allopatric *H. irregulare* and *H. occidentale* isolates mated successfully in a test reported by Chase and Ullrich (1990), which they attributed to a high frequency of + alleles of the V_1, V_2 and V_3 genes in these populations.

3. HYBRIDISATION

The IS system can be regarded as a prezygotic-level mechanism to maintain species integrity. Reproduction barriers may also occur at the postzygotic level, by limitations in fitness of the successfully germinated hybrid individuals. Successful fungal hybrids have been observed in the wild for a variety of closely related species (Olson & Stenlid, 2002). However, a common feature between them is that the hybrid has been recovered from a different substrate than that on which its parents are usually observed. This suggests that fungal hybrids need to shift ecological niche in order to avoid competition with their parents.

3.1. Natural hybridisation in north America and Italy

The possibility of introducing exotic pathogen species into naive environments is a very real danger with the *Heterobasidion* species complex. Even though the different species have been separated for up to 75 million years (Dalman et al., 2010), they have retained a high degree of interfertility and a wide range of host species. This means potentially dire consequences if a species should be successfully introduced in a foreign environment. In 2004, Gonthier et al. reported findings of *H. irregulare* fruiting bodies in *Pinus pinea* stands in the Castelporziano woodlands, close to Rome, Italy (Gonthier, Warner, Nicolotti, Mazzaglia, & Garbelotto, 2004). Presumably, the fungus had been introduced through infested wooden material brought by the U.S. Army during the Second World War. According to the initial study, the fungus was not widespread outside Castelporziano in 2004, 60 years after the assumed point of introduction. However, in 2006, *H. irregulare* could be found 24 km NW and 79 km SE from Castelporziano (Gonthier, Nicolotti, Linzer, Guglielmo, & Garbelotto, 2007), indicating that the fungus was widespread already in 2004 or, less likely, that the spread between 2004 and 2006 was very rapid. Evidence suggests that it is not spreading through competition with the native *H. annosum* s.s. population in this area, but rather by being able to sporulate and disperse in a xeric region, traditionally not well suited for colonisation by the European *Heterobasidion* species (Gonthier, Lione, Giordano, & Garbelotto, 2012). This might have ecological reasons, while both deposition rates of spores and infection frequency decline significantly for *H. annosum* s.s. during the dry summer months, the invasive *H. irregulare* seems to maintain the same propagule pressure throughout the year (Garbelotto, Linzer, Nicolotti, & Gonthier, 2010). Also, the native species is strongly associated with certain vegetation types, namely, *P. pinea* stands, and equally strongly disfavoured by deciduous oak forests. *H. irregulare*, on the other hand, is not restricted to any Italian vegetative type and thrives equally well in oak forests and pine stands (Gonthier et al., 2012).

Interfertility between these species is high and molecular evidence of ongoing mating is plentiful, scored per use of markers for nuclear and mitochondrial species-specific alleles (Gonthier & Garbelotto, 2011). Since the initial introduction of *H. irregulare* to Europe in the 1940s, a massive gene introgression has occurred, by far more often from the native *H. annosum* s.s. into the invasive species than the other way around (Gonthier & Garbelotto, 2011). Testing admixed genotypes at different sites in the invaded area with AFLP markers specific for each species discovered that 44 of 48 had a majority of *H. irregulare* AFLP markers and almost exclusively an *H. irregulare* mitochondrion. None of the genes seem to be under positive selection but have rather been maintained in the invading species due to stochastic reasons in the expanding population (Gonthier & Garbelotto, 2011).

A recent population genetic analysis of 10 sequence and 12 SSR loci provides insights on the introduction pathway and spreads patterns of *H. irregulare* in Italy. The analyses show a high level of gene flow among sites and high frequency of admixing of genotypes (Garbelotto, Guglielmo, Mascheretti, Croucher, & Gonthier, 2013). The genetic diversity progressively decreases with increasing distance from the proposed introduction site at Castelporziano. The results suggest that three to four original heterokaryotic isolates were introduced at Castelporziano and that multiple genotypes have since expanded to new forest sites (Garbelotto et al., 2013).

3.2. Artificial hybrids

Although hybridisation within the P IS group is frequent in Italy, hybridisation between the distinct IS groups appears to be extremely rare and has only been documented once (Garbelotto, Ratcliff, Bruns, Cobb, & Otrosina, 1996). At a pine mortality centre in California, United States, an S/P hybrid isolate was collected, and by taxon-specific competitive-priming PCR and isozyme analysis, it was shown that this indeed was a true hybrid containing ribosomal DNA internal transcribed spacer (ITS) regions originating from both S and P isolates (Garbelotto et al., 1996). The isolates contained S-type mitochondria, indicating that the hybrid was generated by an S homokaryon heterokaryotised by a P homokaryon or heterokaryon (Garbelotto et al., 1996). The hybrid isolate appears to be either a heterokaryon undergoing a primary homothallic phase or a diploid with limited ability to exchange nuclei with homokaryons (Garbelotto, Gonthier, Linzer, Nicolotti, & Otrosina, 2004). The thallus of the hybrid comprises only hyphae that are heterozygotic for parental markers. The hybrid seems stable and long-lived, suggesting that hybridisation could play an important role in shaping the species complex (Garbelotto et al., 2004). Based on three housekeeping gene phylogenies, Sedlák and Tomšovský (2014) reported on hybridisation between *H. parviporum* and both *H. annosum* and *H. abietinum* in the Czech Republic. Obviously, the question of hybridisation in the field between *Heterobasidion* species has potentially large practical implications and should be further considered.

When the natural hybrid was inoculated on host where the parental isolates varied in virulence, it showed reduced virulence (Garbelotto, Gonthier, Linzer, & Nicolotti, 2007). However, artificially made S/P hybrids of *Heterobasidion* and their haploid progeny have shown both high and low virulence on *Pinus sylvestris* seedlings analysed *in vitro* (Olson, Lind, & Stenlid, 2005; Olson & Stenlid, 2001). On the same host, the virulence of S/P hybrid isolates has been shown to be linked to their mitochondrial heritage, since virulent S/P hybrid isolates gained their mitochondria from a high-virulent parental isolate and vice versa (Olson & Stenlid, 2001). On the other hand, on nondifferencing host like Sitka spruce, there is no difference in virulence neither between the parents nor their hybrid (Garbelotto et al., 2007). Possible explanations include direct influence of the mitochondria on the pathogenicity or that intracellular conflict/communication among the nuclei and mitochondria is important for the phenotype of hybrid isolates (Olson & Stenlid, 2001).

4. GENOMICS

An organism's genome hides many secrets and interesting features, such as size, structure and content, which is crucial knowledge to better understand the ecology and impact of species. Genetic studies of *Heterobasidion* species have over the last two decades revealed substantial amount of information on this pathogen.

4.1. Genome sequencing and linkage mapping

An AFLP marker-based linkage map of *Heterobasidion* species was originally published by Lind, Olson, and Stenlid (2005). This map was based on a mapping population of 102 single-spore isolates originating from a compatible mating (Olson et al., 2005) between a *H. irregulare* isolate (TC-32-1), later selected for genome sequencing by the U.S. Department of Energy Joint Genome Institute (JGI), and a *H. occidentale* isolate (TC-122-12). The map has been used to locate several traits of interest, such as growth rate (Olson, 2006), virulence (Lind, Dalman, Stenlid, Karlsson, & Olson, 2007; Lind, Stenlid, & Olson, 2007), resistance to overgrowth from the biocontrol agent *Phlebiopsis gigantea* (Samils, Olson, & Stenlid, 2008) and various intraspecific interaction traits (Lind, Dalman, et al., 2007; Lind, Stenlid, et al., 2007). The interfertility between the two isolates was mediated through a common + allele for the V_3 IS locus, while the S, P and V_2 loci were all polymorphic between the parents. This allowed for the mapping not only of these three IS loci but also of the MAT locus, as each progeny isolate will carry the MAT-A (Chase & Ullrich, 1990) allele from either one of the parents. The first efforts to identify these loci with the early linkage map provided a strongly supported position for the S locus, a weaker one for P and no position at all for V_2 and MAT-A (Lind et al., 2005).

In 2007, the JGI agreed to sequence and analyse the genome of one *Heterobasidion* species in collaboration with an international effort. The well-studied and characterised *H. irregulare* isolate TC-32-1 was chosen since it was one of the parental strains of the population used in the construction of the genetic linkage map (Lind et al., 2005). The genome was sequenced with shotgun DNA sequencing of two clone libraries with different insert sizes and traditional Sanger sequencing methods. The first draft assembly was then improved using bacterial artificial chromosome (BAC)-end sequencing and gap closing with PCR amplicon sequencing. This resulted in a 33.6 MB large genome sequence (Fig. 12.2) (Olson et al., 2012). This size is neither

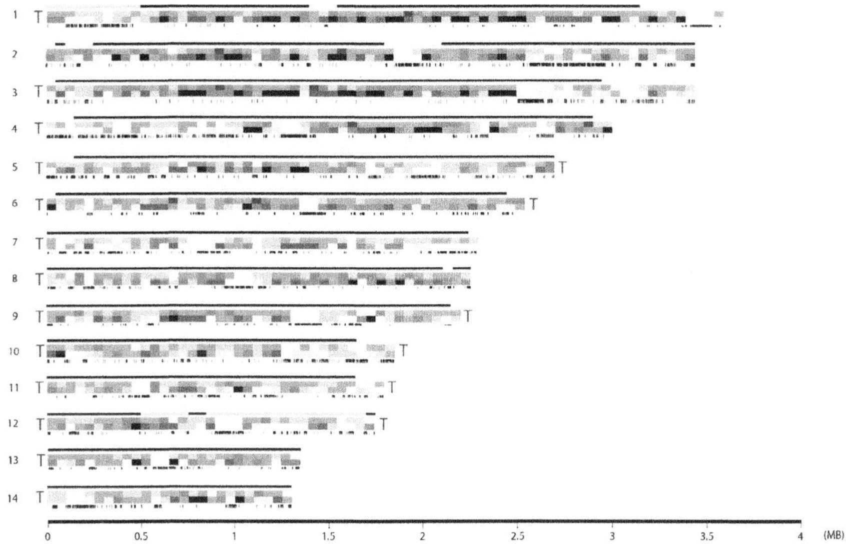

Figure 12.2 The 14 postulated chromosomes of *Heterobasidion irregulare*. The upper black bar of each chromosome denotes linkage map coverage, with pathogenicity QTLs marked in green. The wide yellow-to-brown bar describes gene density (upper half) and gene model quality (lower half) for every 50 kB segment of the sequence. Gene density is calculated in a number of gene models, ranging from over 27 (brown) to under 10 (white). Gene model quality is calculated based on microarray experiments using five probes for each gene model. The colour indicates the percentage of models where all five probes hybridised, ranging from 100% (brown) to below 10% (white). The lowest bar of each chromosome indicates transposon regions, per masked by RepeatMasker. Blue T's stand for an identified telomere region in the corresponding chromosome end. (See the colour plate.)

large nor small within the range of basidiomycetes (Floudas et al., 2012). At the moment, the *H. irregulare* is one of the best sequenced and assembled genomes published, with a low amount of gaps, 90% of the genome covered and six of the scaffolds representing complete chromosomes with sequence spanning from telomere to telomere. Seven other scaffolds have an identified telomere only at one end while one scaffold is still missing both telomere sequences. Repetitive elements make up about 16% of the genome, which is a low proportion compared with many Agaricomycotina species. The high-quality genome assembly was probably facilitated by this. The genetic linkage map also proved to be an invaluable tool to support and validate the assembling of this genome (Olson et al., 2012).

In parallel with the genome sequencing of the *H. irregulare* isolate, a refined genetic linkage map was generated. The second-generation genetic linkage map is based on the alignment of the previously published first-generation map to the parental genomes. 216 of the original 308 AFLP markers were anchored to their respective restriction sites in the genome sequence using an *in silico* approach (Lind, van der Nest, Olson, Brandström-Durling, & Stenlid, 2012). The map resolution was improved by adding 146 sequence-tagged microsatellite markers and 39 sequenced gene markers. The new markers confirmed the positions of the anchored AFLP markers, fused the original 39 linkage groups together into 17 and fully expanded 12 of these to single groups covering entire chromosomes. Map coverage of the genome increased from 55.3% to 92.8%, with 96.3% of 430 markers collinearly aligned with the genome sequence. Just as the *H. irregulare* genome sequencing made it possible to vastly improve the quality of the linkage map, the anchored map in turn also assisted in the assembly process of the genome. It identified several errors in scaffold arrangements and provided linkage information that allowed for combining six major scaffolds previously thought to constitute separate chromosomes into three larger ones. The combined improvement efforts of parallel gap closing, BAC-end sequencing and refining of the genetic linkage map resulted in many resolved assembly problems and that the 39 scaffolds of the first draft assembly were fused into 14 larger (>1 MB) and one very small (~5 KB), which coincides nicely with the proposed 14 *Heterobasidion* chromosomes (Lind et al., 2012; Olson et al., 2012).

This denser, more comprehensive map allowed a more precise sequence-based mapping of three IS loci and the MAT locus (Lind et al., 2012), immediately transferable into possible candidate genes. This demonstrates the benefits of converting anonymous markers into sequence-tagged ones and the power of a sequence-anchored linkage map and its usefulness in the assembly of a whole genome sequence.

4.2. Mapping of quantitative trait loci

The genetic basis for variation in complex traits within a species can be assayed through quantitative trait locus (QTL) analysis and allows linking of phenotypes to specific genomic regions. The genetic linkage map of *Heterobasidion* has been an essential tool not only to disentangle several complex traits such as virulence, somatic incompatibility, growth rate and interaction with *P. gigantea,* the biological control agent of *Heterobasidion,* but also to position single-gene traits like the MAT loci and the IS genes in the genome. QTL mapping of virulence data from Lind, Dalman et al., (2007) on the second-generation map (Lind et al., 2012) resulted in the positioning of three major QTL regions important for pathogenic interaction with Norway spruce and Scots pine (Olson et al., 2012). With the benefit of having the linkage groups anchored to the genome sequence of *H. irregulare,* gene content of these QTL regions could immediately be identified (Olson et al., 2012). The three regions included 178, 142 and 299 predicted gene models, respectively. Another 18 smaller QTL regions were also identified, general for virulence against both hosts and unique for either one. These findings still await further investigations (Mårten Lind, Åke Olson, & Jan Stenlid, Unpublished data) but one 100 kB large area in particular seems interesting for studies on host specificity, as this has proved significant for three virulence traits on Norway spruce in two independent assays, but never in assays on Scots pine.

Such large QTLs require cross verification to identify plausible candidates, such as obtained from expression studies during host infection. Transcript profiling of essentially every gene of an organism can be achieved by microarray or RNA-seq technology (Kohler & Tisserant, 2014, this book). In *Heterobasidion,* a microarray for about 12,000 predicted gene models has been used to quantify the gene expression level in fruit body, during wood degradation, in growth on cellulose or on lignin and during mycelial growth in the cambium of Scots pine. Expression profiles in fruit body and liquid media are distinct from the expression seen on the different plant substrates (Olson et al., 2012). Comparing profiling data with QTL content revealed that a number of these were significantly ($P < 0.05$) regulated during growth in the cambial zone. Among these were a 70-fold upregulated high-affinity sugar transporter and a 9.4-fold upregulated flavin-containing Baeyer–Villiger monooxygenase (Olson et al., 2012). The latter protein is needed for the synthesis of fomannosin, a known *Heterobasidion* phytotoxin (Bassett, Sherwood, Kepler, & Hamilton, 1967).

4.3. Intraspecific genome variation and association mapping in *H. annosum* s.s.

Including several genomes in a genetic analysis increases the possibilities for answering interesting questions and conducting advanced analysis of genome content and gene activity. A population of 23 haploid *H. annosum* s.s. isolates was sequenced to on average about 80–94% genome coverage (Dalman et al., 2013). From the ca. 4.4 million positions with coverage in all individuals, 33,000 high-quality single-nucleotide polymorphic (SNP) sites were extracted that were not singletons and had no gaps in any of the sequences in the alignments. This corresponds to on average 1 SNP per 1000 bp, which gives a high probability of having at least 1 SNP in all genes. This approach shows that analysing small nonrepetitive genomes can generate, in a cost-efficient way, an impressive number of informative SNP by next-generation sequencing (NGS) technology.

Fungal virulence and host resistance in the *Heterobasidion* conifer pathosystem have been estimated by measures of lesion produced in the cambial zone or fungal growth during inoculation experiments according to a method by Stenlid and Swedjemark (1988). There were significant differences in virulence between isolates when analysing the 23 *H. annosum* s.s. isolates on pine and on spruce (Dalman et al., 2013). Twelve SNPs significantly associated with fungal virulence were identified in a genome-wide association study (GWAS) (Dalman et al., 2013). The study identified four and eight SNPs that were associated with fungal growth in spruce and pine, respectively, but no SNPs were associated with lesion length. One of the contigs harboured six of the markers, whereas the remaining six were distributed on separate contigs. The likelihood that the significantly associated SNP is positioned in the gene causal for the analysed phenotype is dependent on the length of the linkage disequilibrium (LD) in the genome. In the six analysed contigs, the LD block size varied between 300 and 30,000 bp. Since the gene density in *H. irregulare* is about 1 per 3000 bp (Olson et al., 2012), the chances are relatively good that a significant marker actually is situated in a gene responsible for the virulence trait. However, there might be genes within the LD blocks that could be causal that have been missed due to stringent criteria used for SNP identification. Interestingly, the low virulent isolates show a deviating allele for the two traits of fungal growth in spruce and in pine (Dalman et al., 2013). The deviating low virulent genotypes originate from all over Europe, for example, Sweden, Italy, England, Belarus and Switzerland, which fit with the absence of population structure in *H. annosum* s.s. (Stenlid, Karlsson, & Högberg, 1994). The GWAS identified genes both with and without homology to known virulence genes from other pathosystems. One highly interesting such pathogenicity candidate gene was calcineurin (CN), an important component in the calcium signalling pathway that is highly conserved in fungi and higher eukaryotes. The calcium signalling pathway controls fundamental aspects of fungal growth, development and virulence. Calcium signalling and CN in particular have been established as an important regulatory component in the human pathogens *Candida albicans*, *C. dubliniensis*, *Mucor circinelloides* and *Aspergillus fumigatus* and more recently in the plant pathogens *A. oryzae*, *Magnaporthe oryzae*, *Ustilago maydis*, *U. horidei* and *Botrytis cinerea* (Harren, Schumacher, & Tudzynski, 2012, and reference within). Transient increase of intracellular Ca^{2+} concentrations is sensed by calmodulin (CaM), a small EF-hand-containing protein that binds Ca^{2+} ions. The Ca^{2+}/CaM complex then activates various target proteins including protein kinases and phosphatases. Functional CN consists of two subunits, a catalytic (CNA) and a regulatory (CNB). *H. irregulare* have one copy of the regulatory subunit (protein ID 151005) and two copies of the catalytic subunit (protein ID 66200 and 148882) (Olson et al., 2012). Activation occurs when Ca^{2+}/CaM binds the CNA adjacent to the CNB binding and displaces an autoinhibitory domain that leads to the activation of the phosphatase and dephosphorylation of the calcineurin substrate. In *Saccharomyces cerevisiae*, one known substrate is the Crz1 transcription factor that can be translocated to the nucleus in order to regulate gene expression of target genes. However, the substrate for the *H. annosum* s.s. calcineurin is not known.

5. GENOME SECRETS

5.1. Gene environment

The size of the gene space of *H. irregulare* does not deviate substantially from other Agaricomycotina species. The 11,464 predicted gene models in *H. irregulare*, out of which about 50% are shared across Basidiomycotina (Olson et al., 2012), are unevenly distributed over the 14 chromosomes (Olson et al., 2012). Regions rich in transposable element (TE) are poor in their gene content (Olson et al., 2012), and similarity to the genes from the National Center for Biotechnology Information (NCBI) was lower for the gene models in such regions (59%) compared with the overall genome (70%). However, in 95% of the gene models without homology (orphans), the presence of a transcribed region was verified by expressed sequenced tags (ESTs) and microarray expression data.

About 16% of the gene models code for predicted secreted proteins, the secretome. Half of these show homology with other proteins, while the remaining 870 are encoded by orphan genes only identified in *H. irregulare* so far (Olson et al., 2012). Of these, 58 are more similar to proteins from other plant pathogenic species than to phylogenetically closer Agaricomycotina species, which is in contrast to the majority of proteins. These genes can be hypothesised to be involved in pathogenicity, but their origin is unclear. They could either have arisen through convergent evolution or have been acquired through horizontal gene transfer (HGT). They could also be of selective advantage for plant pathogens and kept since the last common ancestor, but lost in nonpathogenic fungal species.

A broad spectrum of carbohydrate-active enzymes (CAZymes) could be found within the genome of *H. irregulare* comparable with other white rot Agaricomycotina species (Floudas et al., 2012). *In vitro* growth experiments with *H. irregulare* on simple and complex carbon sources resulted in relatively low growth rates on many substrates compared to other white and brown rot fungi, but higher on starch and guar gum media (Floudas et al., 2012). Being a white rot fungi *H. irregulare* is able to remove lignin during the degradation of wood (Fig. 12.3). This action is considered to be performed in a

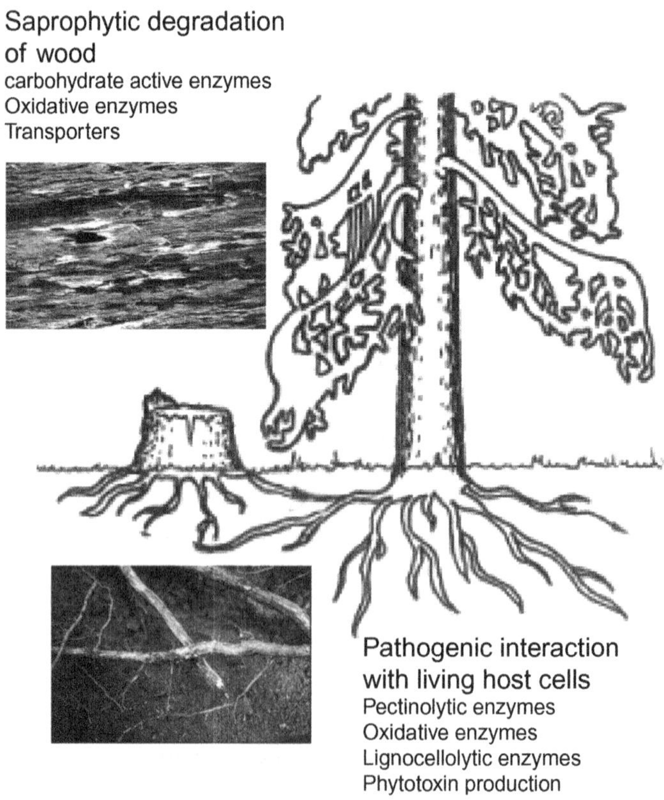

Figure 12.3 *Heterobasidion* stump colonisation and infection of living trees through root contacts, proposed molecular and enzymatic activities important for the specific trophic state. The upper photography illustrates the saprophytic degradation of wood in a dead stump, whereas the lower photography illustrates the pathogenic interaction while *Heterobasidion* infects a living tree from a colonised stump through root contacts.

synergistic action of oxidoreductive enzymes (Lundell, Mäkelä, de Vries, & Hildén, 2014, this book; Lundell, Mäkelä, & Hildén, 2010). Out of 28 analysed gene families encoding oxidoreductases and carbohydrate-active enzymes (CAZymes) that have been implicated in wood degradation, *H. irregulare* has a reduced number of members in 19 and only an increase in three, compared to the closely related white rot fungi *S. hirsutum* (Floudas et al., 2012). This reduction could be interpreted as part of the trade-off for the pathogenic capacity. Many CAZymes active on cellulose are upregulated during early wood degradation. This subset is dominated by the members of the groups GH1, GH5, GH7, GH10, GH12, GH45 and GH61 (Olson et al., 2012). The members of the GH 61 family have a copper-dependent oxidative mechanism rather than hydrolytic enzymatic activity and been predicted to be responsible for the initial step in disrupting crystalline cellulose (Lundell et al., 2014, this book). Out of ten GH61 of *H. irregulare*, six are upregulated on lignocellulosic substrates, including the three proteins that harbour a carbohydrate-binding module (CBM) 1 domain (Yakovlev et al., 2012). Other GH family members that are upregulated in the degradation process are GH5, GH7 and GH15 (Yakovlev et al., 2012).

The first *Heterobasidion* protein that has had its structure determined with X-ray crystallography was a glycoside hydrolase family 7 cellobiohydrolase (*Hirr*Cel7A) (Momeni et al., 2013). The *Hirr*Cel7 A was identified as the protein produced in the highest amount in cultures with wood (spruce heart wood or aspen sawdust) as carbon substrate. This is the only GH7 family gene in *H. irregulare*, whereas the numbers for other analysed white rot fungi range between two and eight and brown rot fungi usually lack GH7 members altogether (Floudas et al., 2012).

Conifer trees active defence in bark and in secondary xylem defence tissue is characterised by the accumulation of polyphenols in the tracheid and ray cell lumens (Hietala et al., 2009). The pathogenicity of *Heterobasidion* is manifested by the ability to overcome this and continue to colonise and degrade plant tissue. The ligninolytic repertoire includes six short hybrid-type Mn peroxidases and one generic peroxidase and one pseudogene (Yakovlev et al., 2013). *H. irregulare* also encodes 18 multicopper oxidases, 13 of which are laccases *sensu stricto* (Floudas et al., 2012). This is the highest amount reported from any white rot fungi and similar to *C. cinerea* (Yakovlev et al., 2013). Twelve of the oxidases are upregulated when *H. irregulare* is growing on reaction zone wood rich in phenolic compounds. This hints about a role in metabolism or detoxification of phenolics such as lignans in host defence tissue (Yakovlev et al., 2013).

5.2. Signal pathways

Mitogen-activated protein kinases (MAPKs) are key signal transduction components in fungi and animals and plants. During the interaction between phytopathogenic fungi and plants, fungal MAPKs help to promote mechanical or enzymatic penetration, stress response differentiation and invasive growth and sporulation (Xu, 2000). MAPK cascades function as a transducer of a phosphorylation/dephosphorylation cycle to channel information. The pathways consist of three interlinked protein kinases that are sequentially activated (Widman, Gibson, Jarpe, & Johnson, 1999). Activated MAPK kinase kinases (MAP3Ks) first phosphorylate two Ser/Thr residence of a MAPK kinase (MAP2K), which in turn activates a MAPK through dual phosphorylation. Active MAPKs can then phosphorylate downstream substrates leading to specific output responses. There are essentially five MAPK pathways, regulating mating, invasive growth, cell wall integrity, hyperosmolarity responses and ascospore formation, of which the latter is only present in ascomycete yeasts and not so well characterised (Hamel, Nicole, Duplessis, & Ellis, 2012). Fungal MAPK families are resolved into four phylogenetically separated clades with 3, 2, 1 and 1 *Heterobasidion* representatives in the clades Kss1/Fus3, Slt2, Hog1 and Ime2 types, respectively (Hamel et al., 2012). MAP2Ks can be divided into three clades with most fungi having one representative in each group, but *H. irregulare* have a species-specific duplication of the MKK1 type (Hamel et al., 2012). Phylogenetic analysis of MAP3K also shows that three well-resolved clades were *H. irregulare* in accordance with most other fungi that have one representative in each group (Hamel et al., 2012).

The importance of Kss/Fus3 cascade as a general multifunctional pathogenicity factor in taxonomically diverse phytopathogenic fungi has been summarised by Hamel et al. (2012). This indicates an ancient function in pathogenicity and suggests that the pathway could be important in *Heterobasidion* as well. Similarly, the Slt2-type kinase seems also broadly conserved in pathogenicity while the Hog1-type kinases are required for virulence in some but not all plant pathogens (Hamel et al., 2012). The core components of the high-osmolarity glycerol (HOG) pathway are all present in *H. irregulare*, but no gene model homologous to the typical transmembrane receptors is present, which indicates that the osmostress signal is received by another yet uncharacterised protein/mechanism (Raffaelo, Keriö, & Asiegbu, 2012).

5.3. Secondary metabolite clusters

Heterobasidion species interact with other microorganisms or its host in all parts of its life cycle. From the moment when a basidiospore lands in a fresh wound or on a stump surface of a newly cut tree, the fungus has to compete for space and nutrients. Natural products such as toxins or hormones are important substances produced by *Heterobasidion* that might change the outcome of the interactions in its own favour. *Heterobasidion* can produce at least 10 different secondary metabolites like the fomajorins (Donnelly, O´Reilly, Polonsky, & Sheridan, 1987) and fomannosin (Kepler et al., 1967), which all belong to the terpenoids, and fomannoxin (Hirotani, O´Reilly, & Donnelly, 1977). These compounds are produced both in axenic cultures and during interaction with plants and other fungi. Hansson et al. (2012a) had identified six fractions from a mycelium-free supernatant of a *H. occidentale* culture with biological activity against four different fungal species, one soil bacteria and the moss *Physcomitrella patens*. One fraction contains fomannosin that was the first toxin that was isolated from *H. annosum* s.l.. Fomannosin was hypothesised to be involved in pathogenesis by Bassett et al. in 1967, although they were not able to detect the compound from infected *Pinus taeda* stems or roots, nor from *H. annosum* s.l. cultures on pine sapwood shavings (Bassett et al., 1967). The toxin was preferentially produced on sugar-rich media, associated with the declining growth phase of the fungus. However, the application of fomannosin to stem wounds in pine seedlings causes needle browning and death and *de novo* synthesis of pinosylvin (Bassett et al., 1967). In addition, six compounds related to fomannosin have been identified and structurally characterised and are proposed as intermediates in a suggested biosynthesis pathway (Hansson, Menkis, Himmelstrand, et al., 2012a). The pathway starts with a cyclisation of trans, trans-farnesyl pyrophosphate to (−)-D6-protoilludene, presumably via a terpene cyclase, and includes several oxidation and dehydrogenation steps. The *H. irregulare* genome harbours three putative terpene cyclases (TCs) in which enzymatic activity is required to synthesise intermediates en route to fomajorins and fomannosin. From gene expression data obtained from a genome-wide microarray study, one out of the three putative TC was significantly downregulated during growth on wood ($P=0.019$) compared with liquid cultures (Olson et al., 2012). This finding fits well with previous failure to detect fomannosin from infected *P. taeda* stems or roots or from *H. annosum* s.l. cultures on pine sapwood shavings (Bassett et al., 1967). Instead, the fungus production of fomannosin during the declining growth phase on sugar-rich media suggests that the primary function of fomannosin can be related to fungal development or microbial interactions, rather than as a phytotoxin. One of the proposed steps in fomannosin biosynthesis requires an oxidation of a ketone to lactone, which is a reaction usually catalysed by Baeyer–Villiger monooxygenase. This type of enzyme has also been implicated in the QTL analysis of virulence in a mapping population of *Heterobasidion* (Olson et al., 2012).

Another toxin produced by *H. annosum* s.l. is fomannoxin, which has a 100-fold greater toxicity to plant cells than fomannosin (Hirotani et al., 1977). This toxin has been isolated from *H. annosum* s.l. infected Sitka spruce stem wood (Heslin, Stuart, Murchú, & Donnelly, 1983). The uptake of fomannoxin by Sitka spruce seedlings resulted in rapid browning of the roots accompanied by chlorosis and progressive browning of needles (Heslin et al., 1983). This, and the production of fomannoxin by actively growing hyphae, suggests a role for fomannoxin during pathogenesis (Heslin et al., 1983). The biosynthetic pathway for fomannoxin has been thoroughly investigated using [1-^{13}C]glucose. The labelling pattern indicates that the benzaldehyde moiety has been synthesised via the shikimic acid pathway and the isoprene motif via the mevalonate pathway (Hansson et al., 2012a). A critical step in the synthesis of fomannoxin and its derivatives is the alkylation of the aromatic ring with an isoprene unit, which gives rise to an early intermediate. Such reactions are usually catalysed by prenyltransferases. The *H. irregulare* genome harbours one gene that encodes a putative aromatic prenyltransferase (JGI Protein ID 108351) (Olson et al., 2012). A phylogenetic analysis of the protein sequence revealed that it belongs to a distinct group of prenyltransferases different from typical dimethylallyltransferase synthases. This gene is located in a cluster consisting of 14 genes with potential function in secondary metabolite production, including two genes with homology to flavin-containing monooxygenases, one dehydrogenase and a major facility transporter protein (Hansson, Menkis, Olson, et al., 2012b).

A substantial number of putative natural product genes were found in the *H. irregulare* genome, suggestive of a biosynthetic capacity greater than evident from the previously mentioned chemical investigations (Olson et al., 2012). Two polyketide synthase (PKS) genes (JGI Protein ID 174227 and 174228) form a small cluster together with a halogenase gene (Fig. 12.4). Protein ID 174227 has an unusual tridomain structure including adenylation, carrier-protein and ketosynthase domains. Further downstream, genes in this cluster indicative for secondary metabolism encode a major facilitator superfamily protein (JGI Protein id. 52305) and a

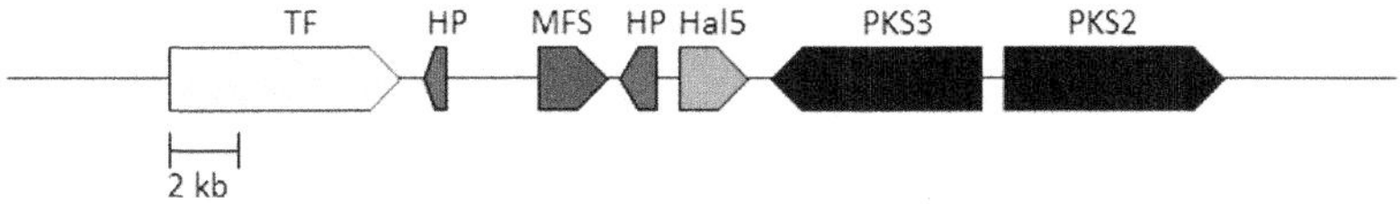

Figure 12.4 Structure and organisation of the secondary metabolite cluster on scaffold 10. The light arrow represents a transcription factor (TF) (JGI Protein ID 118599) containing a Cys_2His_2 zinc finger motif and the dark gray, a major facilitator superfamily (MFS) protein (JGI Protein ID 52305) while the light gray arrow illustrates a halogenase (Hal5) (JGI Protein ID 429470). Black arrows represent the two polyketide synthase (PKS2 and PKS3) genes (JGI Protein ID 174227 and 174228) and the dark gray arrows denote genes encoding hypothetical proteins (HPs).

transcription factor (JGI Protein id. 118599) containing a Cys_2His_2 zinc finger motif. A third PKS gene (JGI Protein ID 50938) for a canonical, non-reducing wA-like synthase is located next to a gene for a dual Cys_2His_2 zinc finger/$Zn(II)_2Cys_6$ binuclear cluster transcriptional regulator (JGI Protein ID 51550), found in other species in the context of natural product genes (Misiek & Hoffmeister, 2008; Zhang, Monahan, Tkacz, & Scott, 2004). No polyketide metabolites have so far been isolated from *H. annosum* s.l., which suggests that the PKS genes are tightly regulated and the conditions where the compounds are produced have not been examined so far.

No genes for multimodular nonribosomal peptide synthetases (NRPSs) were detected. However, 13 genes for tridomain enzymes identically composed of typical NRPS domains (adenylation, thiolation and reduction) are present in the genome and showed high homology to LYS2 from yeast (Ehmann, Gehring, & Walsh, 1999), which functions as α-aminoadipate reductase during L-lysine synthesis in primary metabolism, or to an ochratoxin biosynthesis NRPS-like gene, *npsPN* of *Penicillium nordicum* (Karolewiez & Geisen, 2005). One of the NRPS genes corresponding to JGI Protein ID 153301 was induced during the infection of pine bark ($P=0.033$) and during growth on wood ($P=0.030$), compared with liquid cultures (Olson et al., 2012).

In addition, several genes encoding the so-called tailoring enzymes were identified, that is, enzymes for post-backbone assembly modification, and among those five genes for flavin-dependent halogenases (JGI Protein ID 174229, 181184, 181189, 181191, and 181192). For this enzyme category, involvement in fungal natural product assembly has been experimentally shown (Steffan, Unsöld, & Li, 2007; Wang et al., 2008). While a DMATS may catalyse the prenyl transfer during fomannoxin assembly, halogenated natural products have not been described for the secondary metabolome of *H. annosum* s.l. yet. However, the presence of putative PKS, NRPS and halogenase genes predicts that the biosynthetic capacity of *H. irregulare* is greater than evident from the earlier-mentioned chemical investigations, as no compounds have so far been identified that can be classified as polyketides, nonribosomal peptides or halogenated.

6. FUTURE PROSPECTS

The genome sequencing of *H. irregulare* has paved the way for further genome sequencing effort on other species within the *Heterobasidion* complex. Full genome sequence of several closely related species with different host preferences is a valuable base for answering questions on topics: (1) What are the underlying mechanisms for host specificity in woody plant pathogens? (2) What factors differentiate pathogenic species from saprotrophic ones? (3) What genes are general and specific virulence factors in *Heterobasidion*? Sequencing several genomes within a single species in a population opens many possibilities to analyse the evolutionary forces that shaped the genome of the species. The conducted genome sequencing projects and analyses have resulted in several candidate genes that could potentially be very important for *Heterobasidion* virulence and host preference. However, one remaining issue is how to test hypotheses on their function when no transformation system is available and knock-out/knock-down approaches are missing. Today, new approaches with candidate gene testing in heterologous systems are evaluated and will most likely fill part of the lacking methodology.

ACKNOWLEDGEMENTS

The *Heterobasidion* community including all scientists who contributed to the sequencing, annotation and analysis of the *Heterobasidion irregulare* genome are greatly acknowledged. A special thanks to the staff at Joint Genome Institute for their support. The work of the authors has been financially supported by the Swedish Foundation for Strategic Research.

REFERENCES

Asante-Owusu, R. N., Banham, A. H., Bohnert, H. U., Mellor, E. J. C., & Casselton, L. A. (1996). Heterodimerization between two classes of homeodomain proteins in the mushroom *Coprinus cinereus* brings together potential DNA-binding and activation domains. *Gene*, *172*, 25–31.

Bassett, C., Sherwood, R. T., Kepler, J. A., & Hamilton, P. B. (1967). Production and biological activity of fomannosin a toxic sesquiterpene metabolite of *Fomes annosus*. *Phytopathology*, *57*, 1046–1052.

Brown, A. J., & Casselton, L. A. (2001). Mating in mushrooms: Increasing the chances but prolonging the affair. *Trends in Genetics*, *17*, 393–400.

Capretti, P., Korhonen, K., Mugnai, L., & Romagnioli, C. (1990). An intersterility group of *Heterobasidion annosum*, specialized to *Abies alba*. *European Journal of Forest Pathology*, *19*, 257–262.

Chase, T. E., & Ullrich, R. C. (1988). *Heterobasidion annosum*, root- and butt rot of trees. *Advances in Plant Pathology*, *6*, 501–510.

Chase, T. E., & Ullrich, R. C. (1990). Five genes determining intersterility in *Heterobasidion annosum*. *Mycologia*, *82*, 73–81.

Dalman, K., Himmelstrand, K., Olson, Å., Lind, M., Brandström-Durling, M., & Stenlid, J. (2013). A genome-wide association study identifies genomic regions for virulence in the non-model organism *Heterobasidion annosum* s.s. *PloS One*, *8*, e53525.

Dalman, K., Olson, Å., & Stenlid, J. (2010). Evolutionary history of the conifer root rot fungus *Heterobasidion annosum sensu lato*. *Molecular Ecology*, *19*, 4979–4993.

Donnelly, D. M. X., O'Reilly, J., Polonsky, J., & Sheridan, M. H. (1987). *In vitro* production and biosynthesis of fomajorin D and S by *Fomes annosus* (Fr.) Cooke. *Journal of Chemical Society, Perkin Transactions 1*, *8*, 1869–1872.

Ehmann, D. E., Gehring, A. M., & Walsh, C. T. (1999). Lysine biosynthesis in *Saccharomyces cerevisiae*: Mechanism of α-aminoadipate reductase (Lys2) involves posttranslational phosphopantetheinylation by Lys5. *Biochemistry*, *38*, 6171–6177.

Floudas, D., Binder, M., Riley, R., Barry, K., Blanchette, R. A., et al. (2012). The paleozoic origin of enzymatic lignin decomposition reconstructed from 31 fungal genomes. *Science*, *336*, 1715–1719.

Garbelotto, M., & Gonthier, P. (2013). Biology, epidemiology, and control of heterobasidion species woldwide. *Annual Review of Phytopathology*, *51*, 39–59.

Garbelotto, M., Gonthier, P., Linzer, R., & Nicolotti, G. (2007). Ecological constrains limit the fitness of fungal hybrids in the *Heterobasidion annosum* species complex. *Applied and Environmental Microbiology*, *73*, 6106–6111.

Garbelotto, M., Gonthier, P., Linzer, R., Nicolotti, G., & Otrosina, W. J. (2004). A shift in nuclear state as the result of natural interspecific hybridization between two North American taxa of the basidiomycete complex *Heterobasidion*. *Fungal Genetics and Biology*, *41*, 1046–1051.

Garbelotto, M., Guglielmo, F., Mascheretti, S., Croucher, P. J. P., & Gonthier, P. (2013). Population genetic analyses provide insights on the introduction pathway and spread patterns of the North American forest pathogen *Heterobasidion irregulare* in Italy. *Molecular Ecology*, *22*, 4855–4869.

Garbelotto, M., Linzer, R., Nicolotti, G., & Gonthier, P. (2010). Comparing the influences of ecological and evolutionary factors on the successful invasion of a fungal forest pathogen. *Biological Invasions*, *12*, 943–957.

Garbelotto, M., Ratcliff, A., Bruns, T. D., Cobb, F. W., & Otrosina, W. J. (1996). Use of taxon-specific competitive-priming PCR to study host specificity, hybridization, and intergroup gene flow in intersterility groups of *Heterobasidion annosum*. *Phytopathology*, *86*, 543–551.

Gonthier, P., & Garbelotto, M. M. (2011). Amplified fragment length polymorphism and sequence analyses reveal massive gene introgression from the European fungal pathogen *Heterobasidion annosum* into its introduced congener *H. irregulare*. *Molecular Ecology*, *20*, 2756–2770.

Gonthier, P., Lione, G., Giordano, L., & Garbelotto, M. M. (2012). The American forest pathogen *Heterobasidion irregulare* colonizes unexpected habitats after its introduction in Italy. *Ecological Applications*, *22*(8), 2135–2143.

Gonthier, P., Nicolotti, G., Linzer, R., Guglielmo, F., & Garbelotto, M. M. (2007). Invasion of European pine stands by a North American forest pathogen and its hybridization with a native interfertile taxon. *Molecular Ecology*, *16*, 1389–1400.

Gonthier, P., Warner, R., Nicolotti, G., Mazzaglia, A., & Garbelotto, M. M. (2004). Pathogen introduction as a collateral effect of military activity. *Mycological Research, 108*, 468–470.

Hamel, L.-P., Nicole, M. C., Duplessis, S., & Ellis, B. E. (2012). Mitogen-activated protein kinase signalling in plant-interacting fungi: Distinct messages from conserved messengers. *Plant Cell, 24*, 1327–1351.

Hansson, D., Menkis, A., Himmelstrand, K., Thelander, M., Olson, Å., Stenlid, J., et al. (2012a). Sesquiterpenes from the conifer root rot pathogen *Heterobasidion occidentale*. *Phytochemistry, 82*, 158–165.

Hansson, D., Menkis, A., Olson, Å., Stenlid, J., Broberg, A., & Karlsson, M. (2012b). Biosynthesis of fomannoxin in the root rotting pathogen *Heterobasidion occidentale*. *Phytochemistry, 84*, 31–39.

Harren, K., Schumacher, J., & Tudzynski, B. (2012). The Ca2+/Calcineurin-dependent signalling pathway in the gray mold *Botrytis cinerea*: The role of calcipressin in modulating the calcineurin activity. *PLoS One*, 7, e41761.

Heslin, M. C., Stuart, M. R., Murchú, P. O., & Donnelly, D. M. X. (1983). Fomannoxin, a phytotoxic metabolite of *Fomes annosus*—In vitro production, host toxicity and isolation from naturally infected Sitka spruce heartwood. *European Journal of Forest Pathology, 13*, 11–23.

Hietala, A. M., Nagy, N. E., Steffenrem, A., Kvaalen, H., Fossdal, C. G., & Solheim, H. (2009). Spatial patterns in hyphal growth and substrate exploitation within Norway spruce stems colonization by the pathogenic white-rot fungus *Heterobasidion parviporum*. *Applied and Environmental Microbiology, 75*, 4069–4078.

Hirotani, M., O'Reilly, J., & Donnelly, D. M. X. (1977). Fomannoxin—Toxic metabolite of *Fomes annosus*. *Tetrahedron Letter, 18*, 651–652.

James, T. Y., Sun, S., Li, W., Heitman, J., Kuo, H. C., Lee, Y.-H., et al. (2013). Polyporales genomes reveal the genetic architecture underlying tetrapolar and bipolar mating systems. *Mycologia, 105*, 1374–1390.

Karolewiez, A., & Geisen, R. (2005). Cloning a part of the ochratoxin A biosynthetic gene cluster of *Penicillium nordicum* and characterization of the ochratoxin polyketide synthase gene. *Systematics and Applied Microbiology, 28*, 588–595.

Kepler, J. A., Wall, M. E., Mason, J. E., Bassett, C., McPhail, A. T., & Sim, G. A. (1967). Structure of fomannosin a novel sesquiterpene metabolite of fungus *Fomes annosus*. *Journal of American Chemical Society, 89*, 1260–1261.

Kohler, A., & Tisserant, E. (2014). Deep sequencing of fungal transcriptomes. In F. Martin (Ed.), *Advances in Botanical Research: Genomics of plant-related fungi, 70*, 53–78.

Korhonen, K. (1978). Intersterility groups of *Heterobasidion annosum*. *Communicationes Instituti Forestalis Fenniae, 94*, 1–25.

Korhonen, K., Bobko, I., Hanso, S., Piri, T., & Vasiliauskas, A. (1992). Intersterility groups of *Heterobasidion annosum* in some spruce and pine stands in Byelorussia, Lithuania and Estonia. *European Journal of Forest Pathology, 22*, 384–391.

Korhonen, K., Capretti, P., Karjalainen, R., & Stenlid, J. (1998). Distribution of *Heterobasidion annosum* intersterility groups in Europe. In S. Woodward, J. Stenlid, R. Karjalainen, & A. Hüttermann (Eds.), Heterobasidion annosum. *Biology, ecology, impact and control* (pp. 93–104). Wallingford, Oxon: CAB International.

Korhonen, K., Fedorov, N. I., La Porta, N., & Kovbasa, N. P. (1997). *Abies sibirica* in the Ural region is attacked by the S type of *Heterobasidion annosum*. *European Journal of Forest Pathology, 27*, 273–281.

Kües, U., Göttgens, B., Stratmann, R., Richardson, W. V. J., O'Shea, S. F., & Casselton, L. A. (1994). A chimeric homeodomain protein causes self-compatibility and constitutive sexual development in the mushroom *Coprinus cinereus*. *EMBO Journal, 13*, 4054–4059.

Lind, M., Dalman, K., Stenlid, J., Karlsson, B., & Olson, Å. (2007). Identification of quantitative trait loci affecting virulence in the basidiomycete *Heterobasidion annosum* s. l. *Current Genetics, 52*, 35–44.

Lind, M., Olson, Å., & Stenlid, J. (2005). An AFLP-marker based genetic linkage map of *Heterobasidion annosum* locating intersterility genes. *Fungal Genetics and Biology, 42*, 519–527.

Lind, M., Stenlid, J., & Olson, Å. (2007). Genetics and QTL mapping of somatic incompatibility and intraspecific interactions in the basidiomycete *Heterobasidion annosum* s.l. *Fungal Genetics and Biology, 44*, 1242–1251.

Lind, M., van der Nest, M., Olson, Å., Brandström-Durling, M., & Stenlid, J. (2012). A 2nd generation linkage map of *Heterobasidion annosum* s.l. based on *in silico* anchoring of AFLP markers. *PLoS One, 7*, e18347.

Lundell, T. K., Mäkelä, M. R., de Vries, R. P., & Hildén, K. S. (2014). Genomics, life-styles and future prospects of wood-decaying and litter-decomposing Basidiomycota. In F. Martin (Ed.), *Advances in Botanical Research: Genomics of plant-related fungi, 70*, 329–370.

Lundell, T. K., Mäkelä, M. R., & Hildén, K. (2010). Lignin-modifying enzymes in filamentous basidiomycetes—Ecological, functional and phylogenetic review. *Journal of Basic Microbiology, 50*, 5–20.

Misiek, M., & Hoffmeister, D. (2008). Processing sites involved in intron splicing of *Armillaria* natural product genes. *Mycological Research, 112*, 216–224.

Momeni, M. H., Payne, C. M., Hansson, H., Mikkelsen, N. E., Svedberg, J., Engström, Å., et al. (2013). Structural, Biochemical, and Computational characterization of the glycoside hydrolase family 7 cellobiohydrolase of the tree-killing fungus *Heterobasidion irregulare*. *Journal of Biological Chemistry, 288*, 5861–5872.

Oliva, J., Gonthier, P., & Stenlid, J. (2011). Gene flow and inter-sterility between allopatric and sympatric populations of *Heterobasidion abietinum* and *H. parviporum* in Europe. *Forest Pathology, 41*, 243–252.

Olson, Å. (2006). Genetic linkage between growth rate and the intersterility genes S and P in the basidiomycete *Heterobasidion annosum sensu lato*. *Mycological Research, 110*, 979–984.

Olson, Å., Aerts, A., Asiegbu, F. O., Belbahri, L., Bouzid, O., et al. (2012). Insight into trade-off between wood decay and parasitism from the genome of a fungal forest pathogen. *New Phytologist, 194*, 1001–1013.

Olson, Å., Lind, M., & Stenlid, J. (2005). *In vitro* test of virulence in the progeny of a *Heterobasidion* interspecific cross. *Forest Pathology, 35*, 321–331.

Olson, Å., & Stenlid, J. (2001). Mitochondrial control of fungal hybrid virulence. *Nature, 411*, 438.

Olson, Å., & Stenlid, J. (2002). Pathogenic fungal species hybrids infecting plants. *Microbes and Infection, 4*, 1353–1359.

Otrosina, W. J., Chase, T. E., Cobb, F. W., & Korhonen, K. (1993). Population structure of *Heterobasidion annosum* from North America and Europe. *Canadian Journal of Botany, 71*, 1064–1071.

Otrosina, W. J., & Garbelotto, M. (2010). *Heterobasidion occidentale* sp. nov. and *Heterobasidion irregulare* nom. nov.: A disposition of North American *Heterobasidion* biological species. *Fungal Biology, 114*, 16–25.

Raffaelo, T., Keriö, S., & Asiegbu, F. O. (2012). Role of the HaHOG1 MAP kinase in response of the conifer root and but rot pathogen (*Heterobasidion annosum*) to osmotic and oxidative stress. *PLoS One, 7*, e31186.

Samils, N., Olson, Å., & Stenlid, J. (2008). The capacity in *Heterobasidion annosum* s.l. to resist overgrowth by the biocontrol agent *Phlebiopsis gigantea* is a heritable trait. *Biological Control, 45*, 419–426.

Schlesinger, R., Kahmann, R., & Kamper, J. (1997). The homeodomains of the heterodimeric bE and bW proteins of *Ustilago maydis* are both critical for function. *Molecular and General Genetics, 254*, 514–519.

Sedlák, P., & Tomšovský, M. (2014). Species distribution, host affinity and genetic variability of *Heterobasidion annosum sensu lato* in the Czech Republic. *Forest Pathology*, doi: 10.1111/efp.12102.

Spit, A., Hyland, R. H., Mellor, E. J. C., & Casselton, L. A. (1998). A role for heterodimerization in nuclear localization of a homeodomain protein. *Proceedings of the National Academy Sciences of the United States of America, 95*, 6228–6233.

Steffan, N., Unsöld, I. A., & Li, S.-M. (2007). Chemoenzymatic synthesis of prenylated indole derivatives by using a 4-dimethylallyltryptophan synthase from *Aspergillus fumigatus*. *Chemistry and Biochemistry, 8*, 1298–1307.

Stenlid, J., & Karlsson, J. O. (1991). Partial intersterility in *Heterobasidion annosum*. *Mycological Research, 95*, 1153–1159.

Stenlid, J., Karlsson, J. O., & Högberg, N. (1994). Interspecific genetic variation in *Heterobasidion annosum* revealed by amplification of minisatellite DNA. *Mycological Research, 98*, 57–63.

Stenlid, J., & Swedjemark, G. (1988). Differential growth of S- and P-isolates of *Heterobasidion annosum* in *Picea abies* and *Pinus sylvestris*. *Transactions of the British Mycological Society, 90*, 209–213.

Tymon, A. M., Kues, U., Richardson, W. V. J., & Casselton, L. A. (1992). A fungal mating type protein that regulates sexual and asexual development contains a Pou-related domain. *EMBO Journal, 11*, 1805–1813.

van Diepen, L. T. A., Olson, Å., Ihrmark, K., Stenlid, J., & James, T. Y. (2013). Extensive trans-specific polymorphism at the mating type locus of the root decay fungus *Heterobasidion*. *Molecular Biology and Evolution, 30*, 2286–2301.

Wang, S., Xu, Y., Maine, E. A., Wijeratne, E. M. K., Espinosa-Artiles, P., Gunatilaka, A. A. L., et al. (2008). Functional characterization of the biosynthesis of radicicol, an Hsp90 inhibitor resorcylic acid lactone from *Chaetomium chiversii*. *Chemical Biology, 15*, 1328–1338.

Wheat, C. W., & Wahlberg, N. (2013). Critiquing blind dating: The dangers of overconfident date estimates in comparative genomics. *Trends in Ecology and Evolution, 28*, 636–642.

Widman, C., Gibson, S., Jarpe, M. B., & Johnson, G. L. (1999). Mitogen-activated protein kinase: Conservation of the three-kinase module from yeast to human. *Physiological Review, 79*, 143–180.

Woodward, S., Stenlid, J., Karjalainen, R., & Hüttermann, A. (1998). *Heterobasidion annosum: Biology, ecology, impact and control*. Wallingford, Oxon: CAB International.

Xu, J. R. (2000). Map kinases in fungal pathogens. *Fungal Genetics and Biology, 31*, 137–152.

Yakovlev, I., Hietala, A. M., Courty, P.-E., Lundell, T., Solheim, H., & Fossdal, C. G. (2013). Genes associated with lignin degradation in the polyphagous white-rot pathogen *Heterobasidion irregulare* show substrate-specific regulation. *Fungal Genetics and Biology, 56*, 17–24.

Yakovlev, I., Vaaje-Kolstad, G., Hietala, A. M., Stefanczyk, E., Solheim, H., & Fossdal, C. G. (2012). Substrate-specific transcription of the enigmatic GH61 family of the pathogenic white-rot fungus *Heterobasidion irregulare* during growth on lignocellulose. *Applied Microbiology and Biotechnology, 95*, 979–990.

Zhang, S., Monahan, B. J., Tkacz, J. S., & Scott, B. (2004). Indole-diterpene gene cluster from *Aspergillus flavus*. *Applied and Environmental Microbiology, 70*, 6875–6883.

CHAPTER THIRTEEN

Speciation Genomics of Fungal Plant Pathogens

Eva H. Stukenbrock[1]
MPRG Fungal Biodiversity, Max-Planck-Institute for Terrestrial Microbiology, Marburg, Germany
[1]Corresponding author: e-mail address: eva.stukenbrock@mpi-marburg.mpg.de

Contents

Abstract

Fungi have relatively small genomes, and many species are amendable to experimental and molecular manipulation making them excellent models to unravel the genetic, molecular and cellular aspects of speciation. Pioneering work has been conducted in the two model species *Saccharomyces cerevisiae* and *Neurospora crassa*

Advances in Botanical Research, Volume 70
ISSN 0065-2296
http://dx.doi.org/10.1016/B978-0-12-397940-7.00013-6

where integrated studies using classical genetics, genomic resources and experimental approaches have shed light on mechanisms conferring reproductive isolation between diverged lineages. Experimental evolution, genetic mapping and molecular manipulation can be considerably more challenging in species of fungal plant pathogens. Yet, insight into speciation genetics is fundamental to understand the emergence of new pathogen species. Now, several fundamental questions relating to speciation genetics can be addressed using the growing resource of genome data. Population genomics and comparative genomics approaches offer the possibility of not only recapturing the evolutionary history of species but also identifying signatures of selection and genomic rearrangements associated with speciation events. Thereby, both the history of species and the underlying genetics associated with the establishment of reproductive isolation can be recovered.

1. FUNGAL SPECIATION

Speciation of fungi has been studied using both experimental and retrospective approaches. Experimental studies have focused on mechanisms of speciation and have directly addressed the underlying genetics of reproductive isolation and species formation. Retrospective studies have used DNA sequence analyses including phylogenetics and coalescence-based approaches to reconstruct the evolutionary history of species including the demography of speciation and speciation times. The genetic and evolutionary studies of speciation provide different insights into speciation, and together, they allow us to understand both how and when new species emerged.

Experimental studies have mainly been carried out in the two model organisms *Saccharomyces cerevisiae* and *Neurospora crassa*. Experimental evolution, crossing experiments and genetic mapping in these two ascomycetes have been used to pin down genetic traits directly involved in the formation of reproductive barriers and thereby speciation (Anderson et al., 2010; Dettman, Anderson, & Kohn, 2008; Dettman, Sirjusingh, Kohn, & Anderson, 2007; Turner, Jacobson, & Taylor, 2011). In model systems of fungal plant pathogens, experimental approaches have been far less applied in studies of speciation. The *Armillaria* root rot pathogen *Armillaria mellea* has however served as an experimental model to study incompatibility patterns between sympatric and allopatric lineages (Anderson, Korhonen, & Ullrich, 1980; Anderson & Ullrich, 1979). Simple crossing experiments have here indicated significant differences in the evolution of reproductive barriers between sympatric and allopatric lineages of *A. mellea*.

Speciation studies of fungal plant pathogens have mainly used retrospective approaches to infer species histories in natural and managed ecosystems (Couch et al., 2005; Gladieux et al., 2011; Stukenbrock, Banke, Javan-Nikkhah, & Mcdonald, 2007; Zaffarano, Mcdonald, & Linde, 2008). Central questions in these studies have been how and when a particular crop pathogen emerged as a disease-causing agent. By analysing DNA sequence data from closely related species, it has ben possible to infer speciation times and reconstruct the ancestral demography of several important plant pathogens including the rice blast pathogen *Magnaporthe oryzae* (Couch et al., 2005), the leaf blotch pathogen *Zymoseptoria tritici* (synonym *Mycosphaerella graminicola*) (Stukenbrock et al., 2007) and the barley scald pathogen *Rhynchosporium secalis* (Zaffarano et al., 2008). From these studies, it has become clear that the dynamic of pathogen species over time is high and that new reproductively isolated lineages can emerge instantaneously. Several examples have documented host-driven speciation of pathogenic species. However, hybridization also appears to be widespread and constitutes a distinct route of species formation.

While tools and model systems are available to decipher species histories, we still know very little about the underlying genetics of speciation for fungal plant pathogens. Important questions to address are the following: Which genetic traits allow the rapid isolation of diverged populations? Which evolutionary forces drive rapid speciation in plant pathogenic fungi? How does the environment affect speciation? Comparative genomics of closely related species and population genomics offer new avenues to address these questions that relate to both species histories and mechanisms of speciation in fungal pathogens. Genome sequencing of several fungal plant pathogens has brought new insights into species-specific or lifestyle-specific traits and set the stage for directed functional studies of such species-specific characters for individual species.

In this chapter, I will consider speciation theory in the light of genomics and discuss recent advances and possibilities in the fields of experimental evolution and comparative and population genomics of fungal plant pathogens.

2. THE GENETICS OF SPECIATION: A POPULATION GENETICS PERSPECTIVE

2.1. Isolation with migration

Allopatric populations are geographically separated from each other, while parapatric or sympatric populations coexist in the same habitat. Parapatric populations occupy distinct niches in their environment and are thus separated by ecological factors, while sympatric species share the same ecological niche. In natural habitats, heterogeneity in the landscape can generate spatial variation between populations, and this spatial fragmentation can provide the starting point for genetic divergence and parapatric speciation. Over time, spatially fragmented metapopulations can adapt to local environmental conditions and thereby independently accumulate distinct mutations. In the case of fungal plant pathogens, metapopulations may be the product of host jumping or adaptation to heterogeneously distributed host species (see review by Stukenbrock and Mcdonald (2008)). Under both scenarios, the diverging populations can still coexist in the same environment but be separated according to their host niches. Gene flow will be the genetic constraint counteracting local adaptation and divergence by repeatedly breaking down coevolved gene combinations in the distinct populations. Most fungal species produce enormous numbers of propagules that are dispersed by air, soil or water. Consequently, levels of gene flow locally and even more distantly can be significant. Still, divergence of fungal plant pathogens can occur in the presence of gene flow as documented by the finding of many closely related species inhabiting the same geographic location but occupying distinct niches. How reproductive barriers are established in the presence of gene flow is poorly understood but may result from strong divergent selection on metapopulations.

However, natural selection can directly promote the formation of reproductive barriers between sympatric or parapatric species, a phenomenon called reinforcement. This implies that particular genes involved in mating or progeny formation and fitness are affected by natural selection to work only in certain genetic combinations. Reinforcement of reproductive isolation can thereby allow the divergence of coexisting species even in the presence of gene flow between diverged populations or species. As mentioned in the preceding text, such reinforcement of reproductive barriers has been clearly demonstrated in sympatric lineages of *A. mellea*; however, the loci responsible for reproductive isolation have so far not been identified (Anderson et al., 1980). A recent study of *N. crassa* and its sister species *N. intermedia* allowed the identification of 11 QTL loci responsible for hybrid abortion in crosses of sympatric *N. crassa* and *N. intermedia* isolates (Turner et al., 2011). The genomic footprints of reinforcement in *Neurospora* are signatures of positive selection at loci that are strongly associated with mating and abortive hybrid fruiting body development.

Genetic divergence between allopatric populations accumulates as a consequence of prolonged independent evolution. Between such physically isolated species, gene flow is prevented or strongly reduced by geographic barriers allowing neutral and nonneutral variations to accumulate. Because allopatric species never or rarely encounter each other, reproductive isolation is not a trait selected for by natural selection. Genomic divergence between allopatric species will reflect the adaptation to local environmental conditions, but will not bear traces of reinforcement selection.

Coalescence theory and coalescence models have been widely applied to study the divergence of species. In its simple form, coalescence theory assumes divergence of populations in the absence of gene flow, thus resembling scenarios of allopatric speciation (Wakeley, 2009). However, more recent models allow for the presence of gene flow during isolation of species (Gavrilets, Li, & Vose, 2000; Nielsen & Wakeley, 2001). One of these is the "isolation with migration" (IM) model, which models the divergence of an ancestral population into two or more descendant populations (Hey, 2010; Nielsen & Wakeley, 2001). Migration, modelled as rates of gene flow, can occur between these descendant species in spite of their divergence as isolated lineages. Thereby, IM models allow for more biologically relevant scenarios of species divergence. Depending on the particular IM model, several parameters can be inferred, such as ancestral and present-day population sizes, rates and times of gene flow and speciation times. Together, these parameters can provide insight into the species dynamic at early stages of population divergence or speciation. Previous studies applying IM models have used relatively short sequences sampled from population data. For example, speciation histories in model systems of fungal plant pathogens were inferred using IM models and population genetic data sampling (Gladieux et al., 2011; Stukenbrock et al., 2007). Recently, the IM model has been modified for whole-genome data, providing a very powerful approach to infer divergence times and migration rates across whole-genome sequences (Mailund et al., 2012).

2.2. Genetic divergence

Genetic divergence between populations comprises both neutral and nonneutral variations (Fig. 13.1). Neutral genetic variation between spatially separated populations is generated by the stochastic accumulation of mutations (genetic drift) and the hitchhiking of neighbouring neutral variation at loci under selection (genetic draft) (Barton & Charlesworth, 1984; Gillespie,

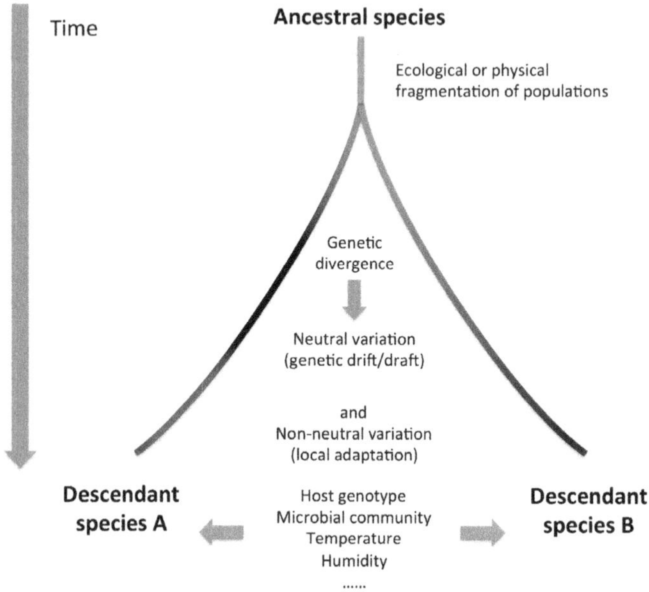

Figure 13.1 Illustration of species divergence during allopatric and ecological speciation. The descendant species A and B diverge from a common ancestral species. After the populations are isolated, genetic divergence accumulates as both neutral genetic variation and nonneutral variation. Local adaptation involves not only specialization to the host genotype but also a range of biotic and abiotic factors.

2000). Gene variants can disappear completely as a consequence of these stochastic processes. As a result, the composition of alleles in spatially separated populations can be drastically changed after a period of isolation if the effect of genetic drift or draft is strong. The effect of genetic drift will be strongest in small and asexual populations, for example, founder populations or populations that have undergone a genetic bottleneck. This is important since it implies that demography and reproductive mode can play determining roles in speciation. Whether genetic drift and genetic draft can be directly instrumental in speciation of sexual species is otherwise still controversial (Barton & Charlesworth, 1984).

The ecological environment of fungal plant pathogens is composed of a range of biotic and abiotic factors operating inside and outside the host plant. In the field, fungal pathogens exist in communities of other microorganisms that are associated with the host plant. Some may be antagonistic and others beneficial or neutral to the growth and reproduction of fungal plant pathogens. There may also be significant differences in local abiotic conditions such as temperature, humidity and soil pH that can greatly impact survival and propagation of fungal pathogens. Variation in each of these ecological conditions can drive divergent adaptation between populations of plant pathogens. Recent studies based on population genetic sampling and analyses have documented adaptive divergence between populations of plant pathogens according to temperature gradients (Laine, 2008; Stefansson, Mcdonald, & Willi, 2013; Zhan & Mcdonald, 2011). To what extent climatic conditions can mediate extensive genetic divergence and speciation is not known. This question is however relevant given the perspective of global climate change and the anthropogenic global spread of fungal pathogens.

A particular environmental factor, which furthermore strongly can promote divergent evolution in pathogens, is the difference between agroecosystems and natural ecosystems. These two types of ecosystems impose very different types of selection pressure on pathogen populations: Aside from the genetic and spatial homogeneity of agroecosystems, the physiology of crop species is often markedly changed compared with uncultivated species. Furthermore, nutrient composition and supplies in agricultural fields are different from those of natural ecosystems. Finally, the supplies of chemicals including pesticides constitute a significant ecological difference between ecosystems, which can drive extensive genetic differentiation between pathogen populations of cultivated and uncultivated hosts (Stukenbrock & Mcdonald, 2008).

One of the most important ecological factors in the evolution of a plant pathogen is the host plant, and genetic divergence and speciation can be tightly linked to host specialization. Some determinants of host specificity are effector genes that encode small secreted proteins that impact or inhibit host defences (De Jonge, Bolton, & Thomma, 2011). A large number of effector genes have been identified in a range of pathogenic species including biotrophs, hemibiotrophs and necrotrophs and in both specialist and generalist pathogens (see reviews by De Jonge et al., 2011; Friesen, Faris, Solomon, & Oliver, 2008; Stergiopoulos & De Wit, 2009). Pathogens can produce a wide range of effectors during their interaction with the host from the initial stage of infection to colonization of different host tissues. When populations of plant pathogens diverge on distinct hosts, their repertoire of effectors will diverge correspondingly as a consequence of local adaptation to distinct host genotypes. In this way, mutational changes in single genes may be sufficient to confer isolation between populations separated by their ability to infect distinct hosts (Giraud, Gladieux, & Gavrilets, 2010). Effectors are therefore not themselves pathogen "speciation genes", but can act to enhance reproductive isolation between populations. There is amble evidence for host-driven evolution of effector genes; however, the direct involvement of effector-related mutations in speciation has only been documented in the rice blast pathogen *Magnaporthe oryzae* (see succeeding text) (Couch et al., 2005).

While much attention is given to effectors for their determining role during the initial establishment and development of pathogens, little is known about host specialization at other stages of infection. Host specialization involves adaptation not only to the host immune system but also to the host as a nutrient resource and a substrate for reproduction. The correlation between host specialization and spore production has been addressed primarily in basidiomycete pathogens. These studies have illustrated that host specialization involves not only virulence-related traits but also traits responsible for reproduction and dispersal (Biere & Honders, 1996; López-Villavicencio et al., 2005). The underlying genetic traits involved in spore production, how the host affects these and how they vary remain to be described. At the genome level, ecological specialization including host specialization can be detected as signatures of strong positive selection in genes related to growth and reproduction in the particular environment or host species.

2.3. Genetic incompatibility and reproductive isolation

Reinforcement of reproductive barriers can involve both premating (prezygotic) isolating mechanisms and postmating (postzygotic) isolating mechanisms (Coyne & Orr, 1998). Prezygotic isolating mechanisms can evolve by changes in genes involved in mating, for example, traits affecting gametic compatibility or local adaptation that promotes assortative mating patterns between populations of fungal plant pathogens. In plant pathogens, assortative mating can, for example, be a product of divergent selection of "host-specialization genes" that prevent mating between isolates infecting distinct hosts, for example, different alleles of effector genes (Gavrilets, 2004). The overall effect of "host-specialization genes" on speciation is not known, but should be considered in the context of (1) how strongly the trait affects a component of premating isolation and (2) how much total reproductive isolation already exists when a "host-specialization gene" diverges between populations.

Postzygotic isolation occurs when hybrids are less fit than their parents. Fitness problems in hybrids can result from different chromosomal arrangements or ploidy levels in the parental species or from fundamental incompatibilities between allelic combinations (Burke & Arnold, 2001). The latter mechanism is also called Dobzhansky–Muller incompatibilities. The Dobzhansky–Muller speciation model describes how postzygotic isolation and hybrid sterility or inviability result from the presence of incompatibilities or negative epistatic interactions between two or more allelic variants (Bateson, 1909; Dobzhansky & Dobzhansky, 1937; Muller, 1942). According to the Dobzhansky–Muller model, allele combinations from diverged parents were never "tested" by natural selection and are likely to not function properly in hybrid offspring. In fungi, the role of Dobzhansky–Muller incompatibilities during incipient speciation has been elegantly demonstrated in *S. cerevisiae* and *Neurospora crassa* (Anderson et al., 2010; Dettman et al., 2007, 2008). These studies utilized experimental evolution, divergent adaptation and crossing experiments to show that reproductive barriers in these two ascomycetes can evolve as a result of both negative epistasis in hybrids and divergent adaptive evolution (Anderson et al., 2010; Turner et al., 2011). The importance of negative epistatic interactions in fungal plant pathogens is so far not known. However, it seems very likely that specialization and divergence on distinct host species also may result in mutations that confer negative epistasis in hybrid progenies. Genetic incompatibilities may also evolve as a by-product of prolonged asexual reproduction. This is the case when asexually diverging lineages accumulate so much genetic differentiation that mating or the generation of viable offsprings is no longer possible (Kohn, 2005).

2.4. Hybridization

During allopatric or ecological speciation, divergent selection and gradual evolution of genic incompatibilities occur over many generations. However, the formation of new species can also occur instantaneously as is the case of hybridization. Hybridization is the successful mating between individuals of otherwise genetically isolated lineages. If the parental species are similar in genome content and sequence, the hybridization can occur through recombination between homologous chromosomes resulting in progenies with the same chromosome number as their parents (homoploid hybridization). Alternatively, hybridization can also result in polyploid or aneuploid progenies when chromosome numbers are distinct as documented in many hybrid plant species (Rieseberg & Willis, 2007).

Usually, the genetic forces that have evolved to maintain the genetic integrity of distinct species counteract the generation of hybrids. These are, as described in the preceding text, prezygotic and/or postzygotic reproductive barriers that either prevent interspecific mating or result in inviable or sterile hybrid progenies. The existence and occasional strong success of hybrid species however illustrate that reproductive barriers can be permissive and that interspecific crossing in some cases can result in the formation of viable and fertile progenies. Hybrid speciation requires successful mating between distinct species and self-fertile hybrids that can evolve reproductive barriers to become sufficiently isolated from the parental species. Evidently, hybrids must also be fit and able to explore and occupy a niche in their environment even in competition with their parents.

How can reproductive isolation between parental species be permissive? As mentioned in the succeeding text, reproductive barriers evolve under different conditions in allopatric and parapatric/sympatric species. In allopatric species, a genetic basis for reproductive isolation is not directly required because populations already are physically isolated. Reproductive barriers therefore evolve as a by-product of genetic divergence rather than as reinforcement of reproductive isolation. Consequently, these barriers may be less strong, and when allopatric species are brought into contact, they may be able to interbreed as shown in species such as *Armillaria mellea* and *Neurospora crassa* (Anderson et al., 1980; Dettman, Jacobson, Turner, Pringle, & Taylor, 2003). Consistently, these studies have revealed that reproductive barriers are stronger between parapatric and sympatric species than between allopatric species. Hybridization therefore likely occurs more readily between allopatric species. This idea should receive special consideration given the fact that the human-mediated dispersal of fungal pathogens greatly has enhanced that potential for allopatric species to encounter each other (Fisher et al., 2012).

Hybrids may be self-fertile and propagate through mating within the hybrid population, or propagation may occur by backcrossing to parental individuals. However, extensive backcrossing will over time dilute the hybrid genome with parental alleles, and ultimately, the hybridization event may only be visible as traces of past introgression. But if hybrids are able to occupy a distinct niche or even outcompete the parental species, they may be able to persist and establish genetic integrity as new species.

One paradoxical issue to consider in the study of hybrid formation is how genetic incompatibilities are overcome. A hybrid genome consists of genes, which have not coevolved in the same genetic background and therefore were not optimized in parallel by natural selection. Indeed, many examples of hybrid sterility and inferior fitness observed in plants, animals and fungi support the hypothesis that incompatibilities between individual genes with major effects, or larger number of genes with a cumulative effect, are responsible for the deleterious effects in hybrids (Orr, 1995).

In fungi, a high tolerance for structural changes and genomic plasticity may enable genic incompatibilities to be overcome more easily (Croll & Mcdonald, 2012; Raffaele & Kamoun, 2012). Genome comparisons have confirmed previous karyotyping studies demonstrating considerable amounts of within-species variability in chromosome structure and chromosome numbers (see review by Zolan, 1995). To what extent chromosome rearrangements and genome plasticity affect homologous recombination between individuals is not known.

Another paradoxical issue of hybrids is how they can become established in an environment where the parental species also exist. The parental species are expected to be considerably better adapted in their environment relative to hybrids of intermediate genotypes and phenotypes. Hybrids may however contain new gene combinations that contribute new phenotypic characteristics allowing them to exploit other environmental niches not occupied by the parents. In yeast, experimental evolution studies revealed that hybrid progenies under certain environmental conditions can exhibit an even higher fitness than their parents (Greig, Louis, Borts, & Travisano, 2002). In particular, under intermediate or fluctuating conditions, hybrids may have a fitness advantage relative to their parents.

3. THE GENOMICS OF SPECIATION

We are in the position to acquire new insight in speciation mechanisms and species histories of fungal pathogens using genome data. Whole-genome sequencing can be applied to formulate and test hypotheses about species formation and speciation mechanisms. The genome sequences of most known agricultural fungal plant pathogens have been sequenced and characterized. Comparative genomics in several systems have allowed the identification of species-specific characters and provided insight into the diversity of evolutionary mechanisms of fungal pathogens (e.g. Klosterman et al., 2011; Ma et al., 2010; Schirawski et al., 2010)

Table 13.1 Summary of speciation scenarios and their genomic signatures

	Genomic footprints between genomes
Ecological speciation	*Allopatric species:* Increased accumulation of non-synonymous mutations in genes involved in local adaptation. Selective sweeps may be visible as regions depleted of variation.
	Sympatric species: Local divergent adaptation can be reflected as an accumulation of non-synonymous mutations in a particular lineage or as a selective sweep in distinct genomic regions. Reinforcement selection may be identified in genes involved in pre- or post-zygotic mating.
Hybridization	Comparative analyses between hybrid species and their progenitors will reveal mosaics of nucleotide divergence in the hybrid. Divergence patterns will alternate between low- and high-divergence levels between hybrid and parental genomes. Block sizes reflect the recombination history of the hybrid and will be larger the younger the hybrid is. Genomic incompatibilities will create strong linkage between allelic combinations.
Clonal speciation	The population of clonal species will include individuals of only one mating type. Population genomic analyses will reveal a genome wide loss of genetic diversity when compared to related sexual species.
Chromosomal speciation	Chromosomal speciation results from large chromosomal rearrangements such as inversions, translocations and deletions. Recombination will be suppressed in these regions resulting in an increased level of divergence.

(Table 13.1). Many fungal plant pathogens have particular genomic regions, which are enriched in pathogenicity related genes. These regions include accessory chromosomes, repetitive islands and AT-isochors that evolve more rapidly and undergo more structural rearrangements. Thereby, new genetic diversity can evolve at a faster rate (see reviews by Croll & Mcdonald, 2012; Raffaele & Kamoun, 2012). Genome rearrangements can be instrumental in speciation as has been demonstrated in species of *Drosophila* (Kulathinal, Stevison, & Noor, 2009; Mcgaugh & Noor, 2012). In the genome of closely related *Drosophila* species, large chromosomal inversions are responsible for the suppression of recombination and thereby an increased divergence rate in the inverted chromosomal regions. It is possible that chromosomal speciation also plays a significant role in the divergence of new species. Detailed descriptions of these rearrangements and their effect on recombination and sequence evolution can be used to test whether genomic rearrangements can drive species formation.

Sequencing of multiple individuals of the same species, population genomics, has furthermore been used to characterize genome-wide footprints of natural selection (Stukenbrock & Bataillon, 2012; Stukenbrock et al., 2011). With this type of data, "speciation genes" should be recognizable through their accumulation of divergent adaptive mutations in sister species. A genome scan was applied to genomic data of *Zymoseptoria* spp. to search for positively selected "host-specialization" or "speciation genes" (Stukenbrock et al., 2010, 2011). Indeed, positively selected genes in *Zymoseptoria* appear to have functional relevance in spore production and host infection (Stephan Poppe, Petra Happel and Eva H. Stukenbrock, in preparation) supporting the use of evolutionary predictions to identify speciation-related genes.

The evolutionary modes of populations and species have furthermore been addressed using whole-genome coalescence analyses (Stukenbrock et al., 2011). As described in the succeeding text, this type of analyses allows not only inference of speciation times but also estimates of demographic changes and can be applied to identify particular genomic regions affected by introgression, incomplete lineages sorting or natural selection (Stukenbrock et al., 2011).

3.1. Comparative genomics and the inference of species histories

A number of whole-genome coalescence approaches have been developed to reconstruct species histories and evolution. These methods, so far mainly applied in primate genomics (Dutheil et al., 2009; Hobolth, Christensen, Mailund, & Schierup, 2007; Hobolth, Dutheil, Hawks, Schierup, & Mailund, 2011; Locke et al., 2011), use whole-genome sequence data to make inferences about the coalescence times of diverged lineages and relate genome-wide properties such as estimates of effective population sizes, incomplete lineage sorting and recombination rates. The power of whole-genome approaches comes from the fact that different genomic regions have different evolutionary histories, particularly in sexually reproducing species where recombination scrambles genetic variation between individuals. Different parts of the genome bring information about the traits that underwent selection or modifications during speciation.

Variation across genome sequences arises from the combined and variable effects of natural selection, genetic drift and recombination. Some regions may be prone to strong purifying selection, that is, genes containing housekeeping genes; others may be more affected by the stochastic processes of genetic drift, while other loci are linked to strongly selected genes and thereby undergo genetic hitchhiking. Of particular interest in studies of speciation genomics are those genomic regions or individual genes, which have been affected by natural selection in a particular species. If the genealogy and coalescence time of such regions coincide with the canonical species tree, they may indeed represent "speciation traits" (Dutheil & Hobolth, 2012). They will reflect the functional and phenotypic traits that have been selected for during the divergence of the new species.

3.2. Natural selection and genomic footprints

Several methods relaying on divergence estimates, SNP data or estimates of nonsynonymous and synonymous divergence are developed to recognize outlier loci in the genome. Measures that can reflect adaptive evolution include Fst, Tajima's D, statistics based on the site frequency spectrum and, for coding sequences, estimates of nonsynonymous and synonymous divergences or polymorphisms. More elaborated evolutionary models such as branch models and codon models detect natural selection, which has acted in particular branches or only on specific codons (Yang, 2007; Yang & Nielsen, 2002). At the population level, strong effects of natural selection may also be visible as a depletion of variation within and in the surroundings of a selected allele. This is the case when an advantageous allele has been swept to fixation in a population. Comparative population genomics approaches have the power to contrast within and between species variations and thereby uncover the effect of natural selection in current-day populations and the past effect of natural selection during species divergence. Ultimately, such studies can deliver inferences of speciation scenarios and provide candidate genes involved in ecological speciation and reproductive isolation.

3.3. Chromosomal rearrangements and their role in speciation

Structural variation between genomes has been proposed as a mechanism of reproductive isolation between diverged lineages and has been termed "chromosomal speciation". The best-studied example of chromosomal speciation is in *Drosophila*. Chromosomal inversions are present in the closely related parapatric species *Drosophila pseudoobscura* and *D. persimilis* (Kulathinal et al., 2009). Hybrids of *D. pseudoobscura* and *D. persimilis* are rarely found in nature, and male hybrids are sterile. Kulathinal and colleagues have investigated the genomic consequences of three large paracentric chromosomal inversions and their role in speciation (Kulathinal et al., 2009). Inverted chromosomal regions are characterized by a significantly higher divergence and a strong suppression of recombination, which suggests that gene flow, during the initial species divergence, has affected the inverted regions to a much lesser extent (Kulathinal et al., 2009; Mcgaugh & Noor, 2012). Recombination is suppressed in these regions due to fatal consequences of meiotic crossing over between heterokaryons, as the products of recombination will be either acentric or dicentric. Consequently, genes located inside the inversions are "protected" from gene flow with other populations and will diverge to a higher extent. In *Drosophila*, genes located in the inverted regions include those involved in male sterility, male mating success, female species preferences and hybrid inviability. Higher divergence in these genes may indeed promote reproductive isolation between the two *Drosophila* species.

To what extent chromosomal rearrangements can lead to hybrid inviability of fungal pathogens is a crucial question. Unexpectedly high levels of structural variation within species observed in several sexual species show that individuals can mate despite variation in their chromosomal composition. Recombination may however be suppressed in affected genomic locations, as found in *Drosophila*, leaving these regions more prone to accelerated evolution.

4. SPECIATION IN PLANT PATHOGENIC FUNGI: EXAMPLES FROM GENOME STUDIES

4.1. Speciation and adaptive evolution of the rice blast genome: Repetitive elements and clonal divergence of *Magnaporthe oryzae*

The rice blast pathogen *Magnaporthe oryzae* exemplifies the successful emergence and spread of a clonal pathogen. *M. oryzae* originated from a species complex of grass-infecting pathogens called *M. grisea* (Couch et al., 2005). The emergence of the rice-infecting *M. oryzae* lineage was associated with (1) loss of the avirulence gene *AVR-Co39* enabling a host jump and the invasion of cultivated rice *Oryzae sativa* and (2) strong clonal propagation on the new host (Couch et al., 2005). In this case, speciation appears to have happened following the occurrence of only one or few mutational events, namely, the deletion of a virulence-related gene. *M. oryzae* has propagated asexually on rice and avoided the reintrogression of *AVR-Co39* by recombination with the ancestral species. While the asexual behaviour of *M. oryzae* has allowed and promoted the establishment and divergence on rice, it has however also entailed a strong loss of genetic variation in the new rice-infecting lineage (Couch et al., 2005).

Previous studies have addressed the correlation between host specialization and reproductive isolation by generating experimental crosses of *Magnaporthe* isolates from rice and wild hosts: goose grass, weeping love grass and finger millet and cultivated rice varieties (Leung, Borromeo, Bernardo, & Notteghem, 1988; Valent, Farrall, & Chumley, 1991). Indeed, these studies have documented a severe reduction in both virulence and fertility in hybrid progenies, suggesting that speciation is determined by the presence or absence of particular virulence-related genes. Rather, negative epistatic interactions in the hybrid genomes play a significant role in the establishment of reproductive barriers. Such genomic incompatibilities may also have been conducive for the divergence of *M. oryzae* on rice. The genetic origin of hybrid inferiority and thereby reproductive isolation between species in the *M. grisea* species complex still need to be identified.

Comparative genomics and recent population genomic analyses have shed light on the ability of adaptive evolution and host specialization in clonal populations of *M. oryzae* (Dean et al., 2005; Xue et al., 2012). Xue and co-workers sequenced two Chinese field isolates of *M. oryzae* in order to assess the within-species level of genome variation and address the underlying genetics of host specialization in *M. oryzae* (Xue et al., 2012). Genomes of the two field isolates did not differ significantly in size from the reference isolate; however, in terms of gene content, the three isolates contained hundreds of isolate-specific genes predicted to be involved in infection and virulence behaviour. The authors conducted functional analyses of several isolate-specific genes and could indeed demonstrate a role in virulence for most of these genes. Interestingly, virulence-related genes are enriched in repetitive regions of the *M. oryzae* genome. Thus, the dynamic of these genes may be strongly correlated with the activity of transposons. Only 30% of transposable elements are conserved between the three isolates, suggesting strong present-day activity of these repetitive elements in shaping the *M. oryzae* genome and probably driving the rapid emergence of new host specificities and new specialized lineages.

Genome studies in other asexual fungal pathogens have likewise revealed increased levels of repetitive elements and genome plasticity, for example, in the *Verticillium* wilt pathogen *Verticillium dahliae* (De Jonge et al., 2012) and the white mould fungus *Sclerotinia sclerotiorum* (Amselem et al., 2011). Possibly, the flexibility in genomes of asexual fungi provides an alternative route for adaptive evolution and lineage divergence to occur (Croll & Mcdonald, 2012). Population genomic analyses will in the future be able to shed light on genome evolution in asexuals and elucidate the role of genome plasticity and repetitive elements in speciation.

4.2. Insight from comparative population genomics and coalescence analyses: Ecological speciation of the wheat pathogen *Zymoseptoria tritici* during host domestication

Similar to speciation of *M. oryzae*, emergence of the prominent wheat pathogen *Zymoseptoria tritici* (synonym *Mycosphaerella graminicola*) was associated with host domestication and adaptation to an agricultural environment (Stukenbrock et al., 2007). *Z. tritici* also originated from a species complex of grass pathogens specialized to infect a range of different host species. The speciation of *Z. tritici* did, in contrast to *M. oryzae*, however, not involve loss of sex or loss of genetic variation (Stukenbrock et al., 2011). The *Zymoseptoria* species complex includes a group of coexisting closely related species, which to some extent have overlapping host ranges. The fact that these species have evolved as evolutionarily independent lineages suggests that ecological or geographic isolation at some time in the past has isolated the *Zymoseptoria* species. Sampling of "wild" relatives of *Z. tritici* at its centre of origin has allowed a detailed investigation of the evolutionary history of the wheat-associated lineage and its relatives (Stukenbrock et al., 2007, 2010, 2011).

The species history of *Z. tritici* and the two closest known relatives *Z. pseudotritici* and *Z. ardabiliae* was inferred by hidden Markov model coalescence analyses using whole-genome sequences and the genome of an out-group species *Septoria passerinii* (Dutheil et al., 2009; Stukenbrock et al., 2011). A main goal of this analysis was to assess the distribution of incomplete lineage sorting among the three *Zymoseptoria* species. In any comparison of more than two closely related species, it is possible to compute the proportion of incomplete lineage sorting in a genome alignment. Incomplete lineage sorting refers to ancestral polymorphisms in the descendant species. Such ancestral polymorphisms have persisted in the descendant species as a consequence of stochastic processes and may have a different genealogy than the canonical genealogy that reflects the species tree. Why is incomplete lineage sorting interesting? The amount of incomplete lineage sorting is a function of the ratio of ancestral effective population size over the time between the two speciation events. For a given time span between the two speciation events, a large amount of incomplete lineage sorting reflects a large ancestral effective population size (Dutheil & Hobolth, 2012). The amount of incomplete lineage sorting can be regarded as a proxy of ancestral effective population size. A large amount of incomplete lineage sorting reflects large ancestral effective population sizes relative to the time between two speciation events. In terms of species histories, the amount and distribution of incomplete lineage sorting in a genome alignment therefore provide insight into the demographic and evolutionary processes underlying speciation.

To characterize genome-wide patterns of adaptive evolution and identify footprints of positive selection in "host-specialization" or "speciation" genes, a comparative population genomics study was conducted using whole-genome alignments of 13 fungal individuals. The sequencing of several genomes from each of three species allowed us to identify and map polymorphisms and substitutions across the genome alignment. The high nucleotide similarity and genome colinearity between the *Zymoseptoria* species enabled the comparison of >9000 homologous genes in which the frequency of nonsynonymous and synonymous polymorphisms and substitutions could be assessed for the computation of adaptive evolution. Signatures of positive selection were only found in a small set of genes. The deletion of five of these candidate genes in *Z. tritici* has so far revealed reduced necrosis and impaired pycnidia (asexual fruiting body) formation of three of the mutants in wheat (Poppe, Happel, & Stukenbrock, in preparation). Interestingly, the mutants were able to establish biotrophic growth in wheat but were mainly affected in their ability to reproduce inside the host, suggesting that the candidate genes do not encode effectors. Rather, adaptive mutations in these genes reflect host specialization at later stages of the pathogen development, namely, asexual reproduction. The comparative genome study in *Zymoseptoria* spp. has so far focused on homologous genes but has not considered the component of species-specific genes. It is indeed likely that the gain or loss of particular genes has enabled colonization of new hosts as in *M. oryzae* (Couch et al., 2005). The identification of such species-specific genes is to be conducted.

4.3. The happy rendezvous of old relatives: Hybrid speciation in pathogenic fungi

A recent population genomics study of the *Z. pseudotritici* ascomycete pathogen not only revealed recent hybrid speciation species but also illustrated the temporal dynamic of pathogenic species in natural vegetation (Stukenbrock, Christiansen, Hansen, Dutheil, & Schierup, 2012). The detailed analysis of five genome sequences of *Z. pseudotritici* showed a mosaic of large segments up to 100 kb without a single polymorphism interspersed between segments comprising two distinct haplotypes. Such a pattern is consistent with a recent homoploid hybridization event between two individuals. The nonvariable segments represent sequences inherited by the hybrid swarm from only one of the parental strains, while the variable segments represent the sequences passed on to the hybrid swarm from both parental strains. From the size distribution of variable chromosome blocks untouched by recombination in the five genomes, it was possible to infer that the first recombination event, that is, the hybridization event, occurred approximately 380 sexual generations ago. Both parental haplotypes present in the *Z. pseudotritici* genome have the same genetic distance to the closest known relative *Z. tritici*. This suggests that the common ancestor of the two parental species diverged from *Z. tritici* before the divergence of the two parental lineages. In summary, after the divergence of the *Z. tritici* lineage, the ancestor of *Z. pseudotritici* diverged into two reproductively isolated lineages. Recently, two individuals from these two diverged lineages underwent sexual recombination to form the *Z. pseudotritici* hybrid. The example illustrates a rapid dynamic of species in the *Zymoseptoria* species complex (Fig. 13.2). This may not be unusual, but the fact that the hybridization in this case was associated with such a strong loss of variation and the absence of backcrossing allows us to compute in detail the recombination history of the hybrid. Whether the parental strains were outcompeted by the hybrid species or just occupy different niches, where they were not sampled, is still an open question.

Population genetic and genomics studies of several other fungal pathogens suggest that hybrid species are not rare. Fungal hybrids can, due to their relatively small genome sizes and their experimental properties, serve as excellent model systems to study genome evolution in hybrid species. Many questions remain to be addressed such as: How does natural selection act on hybrid genomes to circumvent severe losses of genetic variation? How are genomic incompatibilities accounted for? To what is extent and how is backcrossing with parental species prevented? Experimental studies in yeast

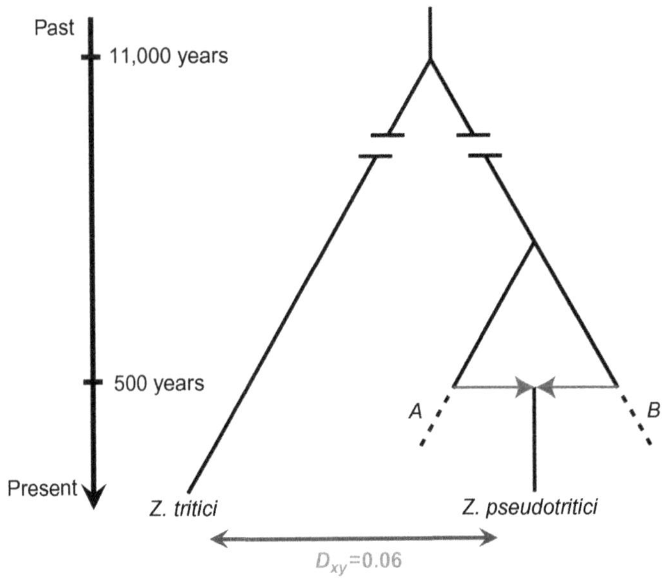

Figure 13.2 A population genomics study of the grass pathogen *Zymoseptoria pseudotritici* illustrates the dynamic of species of fungal plant pathogens. The species history of *Z. pseudotritici* as reconstructed within the last 11,000 years illustrates divergence of parental species (allopatrically or ecologically driven divergence) followed by hybridization and strong population expansion. *Modified from Stukenbrock et al. (2012).*

have aimed to address these questions and have provided solid evidence for the survival and success of hybrids under certain environmental conditions (Greig et al., 2002). Closely related species of fungal plant pathogens may similarly serve as models in experimental evolution and hybridization studies to address how hybrid pathogens acquire new virulence traits, how they perform on old and new hosts and if ecological speciation succeeds hybridization by enhancing the isolation between hybrids and their parental species.

5. EXPERIMENTAL EVOLUTION AND THE IDENTIFICATION OF "DOBZHANSKY–MULLER INCOMPATIBILITIES": INTEGRATING EXPERIMENTAL, GENOMIC AND TRANSCRIPTOMIC APPROACHES IN MODEL FUNGI

Empirical studies to test the role of "Dobzhansky–Muller" incompatibilities have only been conducted in model species such as *S. cerevisiae* and *N. crassa* (Anderson et al., 2010; Dettman et al., 2008; Lee et al., 2008).

In *S. cerevisiae*, "Dobzhansky–Muller" incompatibilities were identified by experimentally evolving *S. cerevisiae* populations in either high-salt (S) or low-glucose medium (M) to simulate ecological divergence (Dettman et al., 2007). After 500 generations, the evolved populations had gained a fitness advantage in their respective environments relative to the progenitor strain. Both mitotic fitness and meiotic efficiency were significantly reduced in S–M hybrids most likely due to a clash of adaptive factors in the hybrid strains. A deficit in sexual formation of hybrids can either be explained by (1) the occurrence of mismatch DNA repair due to sequence divergence between the evolved parents, (2) genomic incompatibilities resulting from chromosomal alterations or (3) the presence of genetic incompatibilities between differentially adapted alleles in the hybrid genome. For the first two possibilities, the overall level of genomic divergence, including chromosomal changes, was minimal and insufficient to explain the reduced ability to undergo meiosis between the evolved strains. This leaves antagonistic interactions between differentially evolved alleles as a cause of the deficit in sexual reproduction. Anderson and colleagues conducted a gene expression analysis to identify misexpressed genes in the hybrid strains and whole-genome resequencing to identify adaptive mutations in the evolved strains (Anderson et al., 2010). The authors were able to identify both differential gene expression in hybrids and a "Dobzhansky–Muller" incompatibility between evolved alleles of a proton efflux pump PMA1 and a global regulator of mRNA encoding mitochondrial proteins MKT1 (Anderson et al., 2010, Parreiras, Kohn, & Anderson, 2011). The interaction of diverged alleles of these two genes in hybrid progenies was responsible for a strong reduction in fitness.

Although experimental evolution may be considerably more laborious in plant pathogenic species that rely on propagation on the host, similar experimental evolution studies would greatly contribute to our understanding of plant pathogen divergence and the evolution of reproductive isolation between ecologically isolated species. Experimental evolution studies will not only allow the recognition of such "speciation traits" in fungal plant pathogens but also reveal their possible contribution to decreased fitness in hybrids.

6. CONCLUSIONS: SPECIES AND SPECIES DYNAMICS OF FUNGAL PLANT PATHOGENS

The study of speciation of fungal plant pathogens has an important applied aspect. Agricultural plant production is continuously challenged by the emergence of new fungal diseases. A fundamental problem is that rapid evolution of fungal plant pathogens allows them to overcome plant resistances and to acquire resistance towards applied fungicides. New species do not fall from the sky but evolve from existing species, for example, the rice blast pathogen emerging from a pathogen population infecting another grass host. Their potential fitness advantages and ability to invade agroecosystems are enabled or promoted by new mutations and the population dynamic of the species. We need to understand how pathogen speciation occurs and what promotes the onset of speciation in agroecosystems to circumvent the emergence of new pathogenic species on crops. From the aforementioned examples, it is clear that homogenous crop populations can promote the evolution and specialization of both clonal and sexual lineages and lead to the formation of new pathogen species. Likewise, new species may emerge from the hybridization between introduced and native pathogen species. Improved quarantine regulations and the development of mixed cultivars could act to slow down the emergence of new disease-causing species.

Aside from the importance to agricultural research, studies of plant pathogen speciation also allow us to address basic questions about parasite evolution and speciation. Fungal plant pathogens can serve as excellent models to address fundamental questions such as the influence of host–pathogen dynamics on speciation, the contribution of minor versus major genetic changes in speciation and the underlying genetic processes affecting species with and without sexual reproduction.

Genome analyses have changed our view on adaptive evolution of fungal plant pathogens. As summarized in the preceding text, a significant and general finding is the presence of extreme levels of genome plasticity in fungi. It has been suggested that genome plasticity enhances the ability of adaptive evolution in pathogenic species, in particular for species without sex (Croll & Mcdonald, 2012; Raffaele & Kamoun, 2012). But genome plasticity may also play a significant role in the speciation of fungal plant pathogens, for example, through genomic incompatibility between populations following major chromosomal rearrangements or, inversely, the formation of hybrids in spite of diverged chromosome structure in the parental species.

An important conclusion from studies of plant pathogen speciation is that speciation can occur rapidly. Distinct host species appear to act as strong reproductive barriers between diverged lineages. However, beside host-driven divergence and ecological speciation, hybridization also contributes to the emergence of new lineages. The study of *Z. pseudotritici* illustrates the dynamic in one plant pathogen lineage including the split of populations, isolation, local adaptation and finally hybridization. The study demonstrates that species are not static entities in nature. Species are dynamic in time and space and we should study them as so.

7. FUTURE PERSPECTIVES

Genome studies will continue to add new insights into speciation processes of fungal pathogens. Comparative genomics and population genomics approaches including coalescence analyses provide powerful tools to reconstruct species histories. An important aspect that is often missing is the sampling of appropriate species representatives and the identification of sister species. The most powerful comparisons are possible by including several species representing a range of divergence times, for example, 5000 years, 50,000 years, 500,000 years and 5 million years. The closest relatives can provide insight into the actual speciation processes, while the more distant relatives provide useful anchoring points to enable accurate comparisons. Finally, it should be mentioned that the interpretation of genome patterns and evolutionary analyses relies on insight into the population genetics of the included species.

The importance of experimental evolution should also be emphasized here as a new direction in studies of plant pathogen speciation. The difficulties in propagating pathogens over several generations on host plants may be circumvented by improved model systems using species with fast generation times and uncomplicated fast inoculation and isolation approaches. Additionally, *in vitro* approaches can be developed to mimic pathogen divergence according to environmental differences.

Lastly, a main focus in plant pathogen research is virulence and the adaptation to host defences. However, in addition to the host immune system, a fungal pathogen explores a number of other ecological niches. We know very little about the adaptation of pathogens to distinct host tissues, to developmental differences between host species and to reproduction and dispersal within or outside their host. Molecular studies are so far mainly focusing on dissecting the underlying genetics of virulence. However, with the available molecular tools and experimental approaches, it will in the future be possible to disentangle other details of the molecular ecology of plant pathogens and to integrate this knowledge into studies of adaptive evolution and speciation of fungal plant pathogens.

ACKNOWLEDGEMENTS

I would like to thank Linda Kohn, Bruce McDonald and Julien Dutheil for their helpful discussions and comments to a previous version of this chapter. Research of EHS is supported by the Max Planck Society, Germany.

REFERENCES

Amselem, J., Cuomo, C. A., Van Kan, J. A. L., Viaud, M., Benito, E. P., Coulox, A., et al. (2011). Genomic Analysis of the necrotrophic fungal pathogens *Sclerotinia sclerotiorum* and *Botrytis cinerea*. *PLoS Genetics*, 7, e1002230.

Anderson, J. B., Funt, J., Thompson, D. A., Prabhu, S., Socha, A., Sirjusingh, C., et al. (2010). Determinants of divergent adaptation and Dobzhansky-Muller interaction in experimental yeast populations. *Current Biology*, *20*, 1383–1388.

Anderson, J. B., Korhonen, K., & Ullrich, R. C. (1980). Relationships between European and North American biological species of *Armillaria mellea*. *Experimental Mycology*, *4*, 78–86.

Anderson, J. B., & Ullrich, R. C. (1979). Biological species of *Armillaria mellea* in North America. *Mycologia*, *71*, 402–414.

Barton, N. H., & Charlesworth, B. (1984). Genetic revolutions, founder effects, and speciation. *Annual Review of Ecology and Systematics*, *15*, 133–164.

Bateson, W. (1909). Heredity and variation in modern lights. In A. C. Seward (Ed.), *Darwin and modern science* (pp. 85–101). Cambridge, UK: Cambridge University Press.

Biere, A., & Honders, S. (1996). Host adaptation in the anther smut fungus *Ustilago violacea* (*Microbotryum violaceum*): Infection success, spore production and alteration of floral traits on two host species and their F1-hybrid. *Oecologia*, *107*, 307–320.

Burke, J. M., & Arnold, M. L. (2001). Genetics and the fitness of hybrids. *Annual Review of Genetics*, *35*, 31–52.

Couch, B. C., Fudal, I., Lebrun, M.-H., Tharreua, D., Valent, B., van Kim, P., et al. (2005). Origins of host-specific populations of the blast pathogen *Magnaporthe oryzae* in crop domestication with subsequent expansion of pandemic clones on rice and weeds of rice. *Genetics*, *170*, 613–630.

Coyne, J. A., & Orr, H. A. (1998). The evolutionary genetics of speciation. *Philosophical Transactions of the Royal Society, B: Biological Sciences*, *353*, 287.

Croll, D., & Mcdonald, B. A. (2012). The accessory genome as a cradle for adaptive evolution in pathogens. *PLoS Pathogens*, *8*, e1002608.

Dean, R. A., Talbot, N. J., Ebbole, D. J., Farman, M. L., Mitchell, T. K., Orbach, M. J., et al. (2005). The genome sequence of the rice blast fungus *Magnaporthe grisea*. *Nature*, *434*, 980–986.

De Jonge, R., Bolton, M. D., & Thomma, B. P. H. J. (2011). How filamentous pathogens co-opt plants: The ins and outs of fungal effectors. *Current Opinion in Plant Biology*, *14*, 400–406.

De Jonge, R., Van Esse, P. H., Karunakaran, M., Bolton, M., Santhanam, P., Saber, M. K., et al. (2012). Tomato immune receptor Ve1 recognizes effector of multiple fungal pathogens uncovered by genome and RNA sequencing. *Proceedings of the National Academy of Sciences*, *109*, 5110–5115.

Dettman, J., Anderson, J., & Kohn, L. (2008). Divergent adaptation promotes reproductive isolation among experimental populations of the filamentous fungus *Neurospora*. *BMC Evolutionary Biology*, *8*, 35.

Dettman, J. R., Jacobson, D. J., Turner, E., Pringle, A., & Taylor, J. W. (2003). Reproductive isolation and phylogenetic divergence in *Neurospora*: Comparing methods of species recognition in a model eukaryote. *Evolution*, *57*, 2721–2741.

Dettman, J. R., Sirjusingh, C., Kohn, L. M., & Anderson, J. B. (2007). Incipient speciation by divergent adaptation and antagonistic epistasis in yeast. *Nature, 447*, 585–588.

Dobzhansky, T. G., & Dobzhansky, T. (1937). *Genetics and the origin of species.* New York, NY: Columbia University Press.

Dutheil, J. Y., Ganapathy, G., Hobolth, A., Mailund, T., Uyenoyama, M. K., & Schierup, M. H. (2009). Ancestral population genomics: The coalescent hidden Markov model approach. *Genetics, 183*, 259–274.

Dutheil, J., & Hobolth, A. (2012). Ancestral population genomics. In M. Anisimova (Ed.), *Methods in Molecular Biology: Vol. 856. Evolutionary genomics* (pp. 293–313). New York, NY: Humana Press Inc.

Fisher, M. C., Henk, D. A., Briggs, C. J., Brownstein, J. S., Madoff, L. C., McCraw, S. L., et al. (2012). Emerging fungal threats to animal, plant and ecosystem health. *Nature, 484*, 186–194.

Friesen, T. L., Faris, J. D., Solomon, P. S., & Oliver, R. P. (2008). Host-specific toxins: Effectors of necrotrophic pathogenicity. *Cellular Microbiology, 10*(7), 1421–1428.

Gavrilets, S. (2004). *Fitness landscapes and the origin of species.* Princeton, NJ: Princeton University Press.

Gavrilets, S., Li, H., & Vose, M. D. (2000). Patterns of parapatric speciation. *Evolution, 54*, 1126–1134.

Gillespie, J. H. (2000). Genetic drift in an infinite population: The pseudohitchhiking Model. *Genetics, 155*, 909–919.

Giraud, T., Gladieux, P., & Gavrilets, S. (2010). Linking the emergence of fungal plant diseases with ecological speciation. *Trends in Ecology and Evolution, 25*, 387–395.

Gladieux, P., Vercken, E., Fontaine, M. C., Hood, M., Jonot, O., Couloux, A., et al. (2011). Maintenance of fungal pathogen species that are specialized to different hosts: Allopatric divergence and introgression through secondary contact. *Molecular Biology and Evolution, 28*, 459–471.

Greig, D., Louis, E. J., Borts, R. H., & Travisano, M. (2002). Hybrid Speciation in Experimental Populations of Yeast. *Science, 298*, 1773–1775.

Hey, J. (2010). Isolation with migration models for more than two populations. *Molecular Biology and Evolution, 27*, 905–920.

Hobolth, A., Christensen, O. F., Mailund, T., & Schierup, M. H. (2007). Genomic relationships and speciation times of human, chimpanzee, and gorilla inferred from a coalescent hidden Markov model. *PLoS Genetics, 3*, e7.

Hobolth, A., Dutheil, J. Y., Hawks, J., Schierup, M. H., & Mailund, T. (2011). Incomplete lineage sorting patterns among human, chimpanzee, and orangutan suggest recent orangutan speciation and widespread selection. *Genome Research, 21*, 349–356.

Klosterman, S. J., Subbarao, K. V., Kang, S., Veronese, P., Gold, S. E., Thomma, B. P. H. J., et al. (2011). Comparative genomics yields insights into niche adaptation of plant vascular wilt pathogens. *PLoS Pathogens*, 7, e1002137.

Kohn, L. M. (2005). Mechanisms of fungal speciation. *Annual Review of Phytopathology, 43*, 279–308.

Kulathinal, R. J., Stevison, L. S., & Noor, MaF (2009). The genomics of speciation in *Drosophila*: Diversity, divergence, and introgression estimated using low-coverage genome sequencing. *PLoS Genetics, 5*, e1000550.

Laine, A. L. (2008). Temperature-mediated patterns of local adaptation in a natural plant-pathogen metapopulation. *Ecology Letters, 11*, 327–337.

Lee, H. Y., Chou, J. Y., Cheong, L., Chang, N. H., Yang, S. Y., & Leu, J. Y. (2008). Incompatibility of nuclear and mitochondrial genomes causes hybrid sterility between two yeast species. *Cell, 135*, 1065–1073.

Leung, H., Borromeo, E. S., Bernardo, M. A., & Notteghem, J. L. (1988). Genetic analysis of virulence in the rice blast fungus *Magnaporthe grisea. Phytopathology, 78*, 1227–1233.

Locke, D. P., Hillier, L. W., Warren, W. C., Worley, K. C., Nazareth, L. V., Muzny, D. M., et al. (2011). Comparative and demographic analysis of orang-utan genomes. *Nature, 469*, 529–533.

López-Villavicencio, M., Enjalbert, J., Hood, M. E., Shykoff, J. A., Raquin, C., & Giraud, T. (2005). The anther smut disease on *Gypsophila repens*: A case of parasite sub-optimal performance following a recent host shift? *Journal of Evolutionary Biology, 18*, 1293–1303.

Ma, L.-J., Van Der Does, H. C., Borkovich, K. A., Coleman, J. J., Daboussi, M.-J., Di Pietro, A., et al. (2010). Comparative genomics reveals mobile pathogenicity chromosomes in *Fusarium*. *Nature, 464*, 367–373.

Mailund, T., Halager, A. E., Westergaard, M., Dutheil, J. Y., Munch, K., Andersen, L. N., et al. (2012). A New isolation with migration model along complete genomes infers very different divergence processes among closely related great ape species. *PLoS Genetics, 8*, e1003125.

Mcgaugh, S. E., & Noor, MaF (2012). Genomic impacts of chromosomal inversions in parapatric *Drosophila* species. *Philosophical Transactions of the Royal Society, B: Biological Sciences, 367*, 422–429.

Muller, H. J. (1942). Isolating mechanisms, evolution and temperature. *Biological Symposia, 6*(811), 71–125.

Nielsen, R., & Wakeley, J. (2001). Distinguishing migration from isolation: A Markov chain Monte Carlo approach. *Genetics, 158*, 885–896.

Orr, H. A. (1995). The population genetics of speciation: The evolution of hybrid incompatibilities. *Genetics, 139*, 1805–1813.

Parreiras, L. S., Kohn, L. M., & Anderson, J. B. (2011). Cellular effects and epistasis among three determinants of adaptation in experimental populations of *Saccharomyces cerevisiae*. *Eukaryotic Cell, 10*, 1348–1356.

Raffaele, S., & Kamoun, S. (2012). Genome evolution in filamentous plant pathogens: Why bigger can be better. *Nature Reviews Microbiology, 10*, 417–430.

Rieseberg, L. H., & Willis, J. H. (2007). Plant speciation. *Science, 317*, 910–914.

Schirawski, J., Mannhaupt, G., Munch, K., Brefort, T., Schipper, K., Doehlemann, G., et al. (2010). Pathogenicity determinants in smut fungi revealed by genome comparison. *Science, 330*, 1546–1548.

Stefansson, T. S., Mcdonald, B. A., & Willi, Y. (2013). Local adaptation and evolutionary potential along a temperature gradient in the fungal pathogen *Rhynchosporium commune*. *Evolutionary Applications, 6*(3), 524–534.

Stergiopoulos, I., & De Wit, P. J. (2009). Fungal effector proteins. *Annual Review of Phytopathology, 47*, 233–263.

Stukenbrock, E. H., Banke, S., Javan-Nikkhah, M., & Mcdonald, B. A. (2007). Origin and domestication of the fungal wheat pathogen *Mycosphaerella graminicola* via sympatric speciation. *Molecular Biology and Evolution, 24*, 398–411.

Stukenbrock, E. H., & Bataillon, T. (2012). A Population genomics perspective on the emergence and adaptation of new plant pathogens in agro-ecosystems. *PLoS Pathogens, 8*, e1002893.

Stukenbrock, E. H., Bataillon, T., Dutheil, J. Y., Hansen, T. T., Li, R., Zala, M., et al. (2011). The making of a new pathogen: Insights from comparative population genomics of the domesticated wheat pathogen *Mycosphaerella graminicola* and its wild sister species. *Genome Research, 21*, 2157–2166.

Stukenbrock, E. H., Christiansen, F. B., Hansen, T. T., Dutheil, J. Y., & Schierup, M. H. (2012). Fusion of two divergent fungal individuals led to the recent emergence of a unique widespread pathogen species. *Proceedings of the National Academy of Sciences, 109*(27), 10954–10959.

Stukenbrock, E. H., Jorgensen, F. G., Zala, M., Hansen, T. T., Mcdonald, B. A., & Schierup, M. H. (2010). Whole-genome and chromosome evolution associated with

host adaptation and speciation of the wheat pathogen *Mycosphaerella graminicola*. *PLoS Genetics*, *6*, e1001189.
Stukenbrock, E. H., & Mcdonald, B. A. (2008). The origins of plant pathogens in agro-ecosystems. *Annual Review of Phytopathology*, *46*, 75–100.
Turner, E., Jacobson, D. J., & Taylor, J. W. (2011). Genetic architecture of a reinforced, postmating, reproductive isolation barrier between *Neurospora* species indicates evolution via natural selection. *PLoS Genetics*, 7, e1002204.
Valent, B., Farrall, L., & Chumley, F. G. (1991). *Magnaporthe grisea* genes for pathogenicity and virulence identified through a series of backcrosses. *Genetics*, *127*, 87–101.
Wakeley, J. (2009). *Coalescent Theory. An Introduction.* Greenwood Village: Roberts and Company Publishers.
Xue, M., Yang, J., Li, Z., Hu, S., Yao, N., Dean, R. A., et al. (2012). Comparative analysis of the genomes of two field isolates of the rice blast fungus *Magnaporthe oryzae*. *PLoS Genetics*, *8*, e1002869.
Yang, Z. (2007). PAML 4: Phylogenetic analysis by maximum likelihood. *Molecular Biology and Evolution*, *24*, 1586–1591.
Yang, Z., & Nielsen, R. (2002). Codon-substitution models for detecting molecular adaptation at individual sites along specific lineages. *Molecular Biology and Evolution*, *19*, 908–917.
Zaffarano, P. L., Mcdonald, B. A., & Linde, C. C. (2008). Rapid speciation following host shifts in the plant pathogenic fungus *Rhyncosporium*. *Evolution*, *62*, 1418–1436.
Zhan, J., & Mcdonald, B. A. (2011). Thermal adaptation in the fungal pathogen *Mycosphaerella graminicola*. *Molecular Ecology*, *20*, 1689–1701.
Zolan, M. E. (1995). Chromosome-length polymorphism in fungi. *Microbiological Reviews*, *59*, 686–698.

AUTHOR INDEX

Note: Page numbers followed by "*f*" indicate figures and "*t*" indicate tables.

B

C

D

G

I

J

K

N

O

Q

R

T

U

V

X

Y

Z

SUBJECT INDEX

Note: Page numbers followed by "*f*" indicate figures and "*t*" indicate tables.

A

B

H

M

N

O

P

S

T

U

V

W

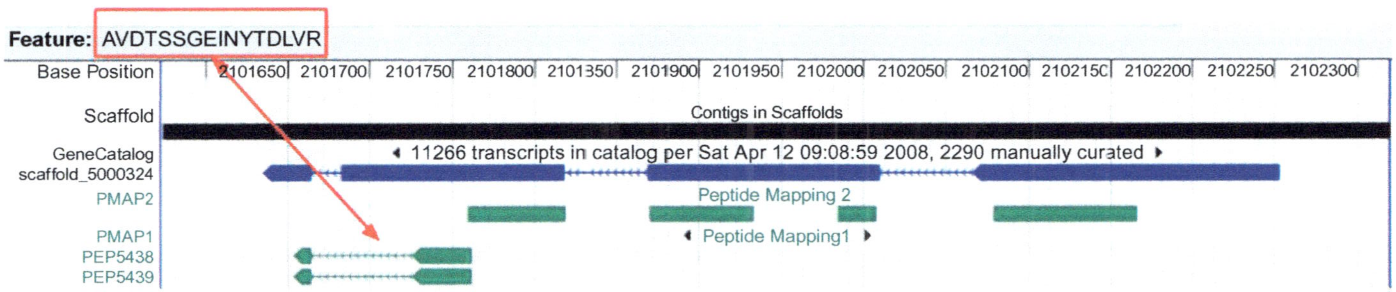

Igor V. Grigoriev *et al.*, Figure 1.4, page 33.

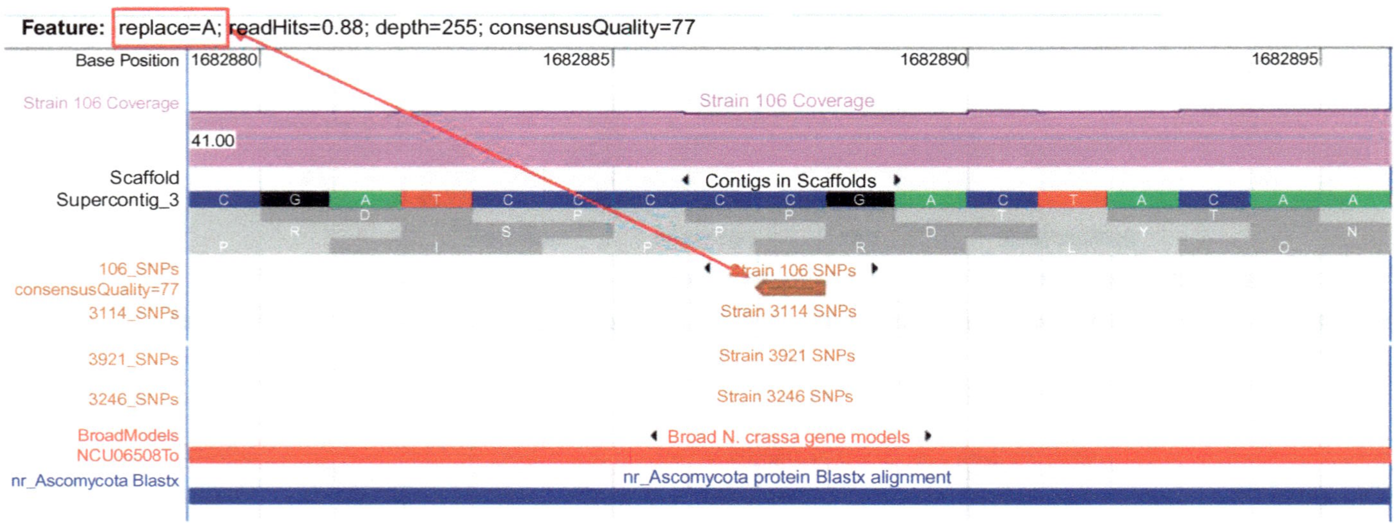

Igor V. Grigoriev *et al.*, Figure 1.5, page 36.

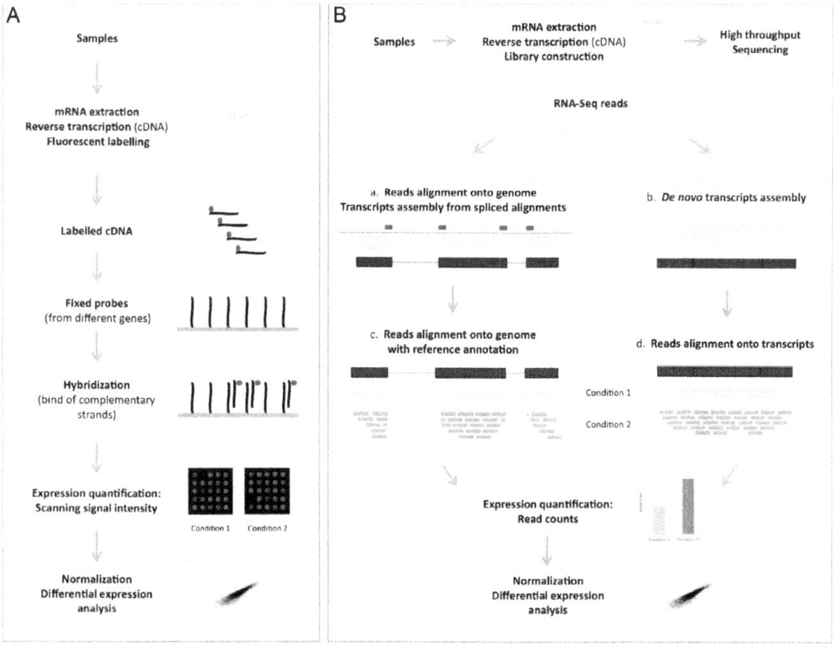

Annegret Kohler and Emilie Tisserant, Figure 2.1, page 55.

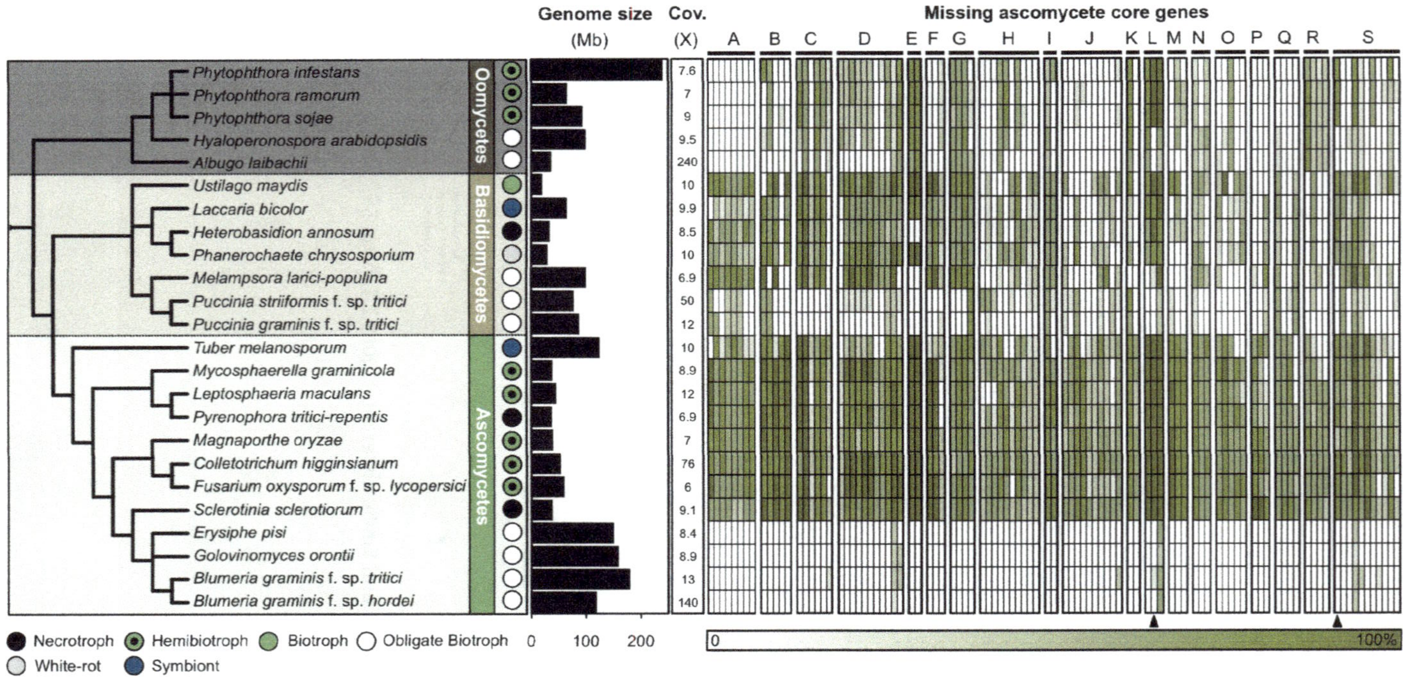

Stéphane Hacquard, Figure 4.1, page 116.

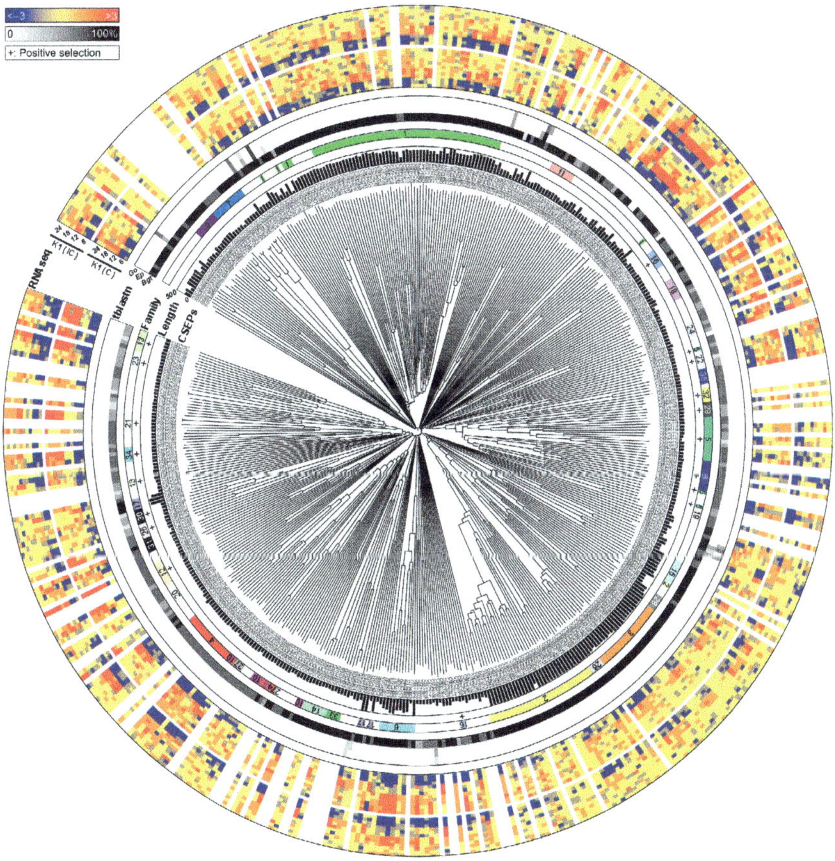

Stéphane Hacquard, Figure 4.2, page 126.

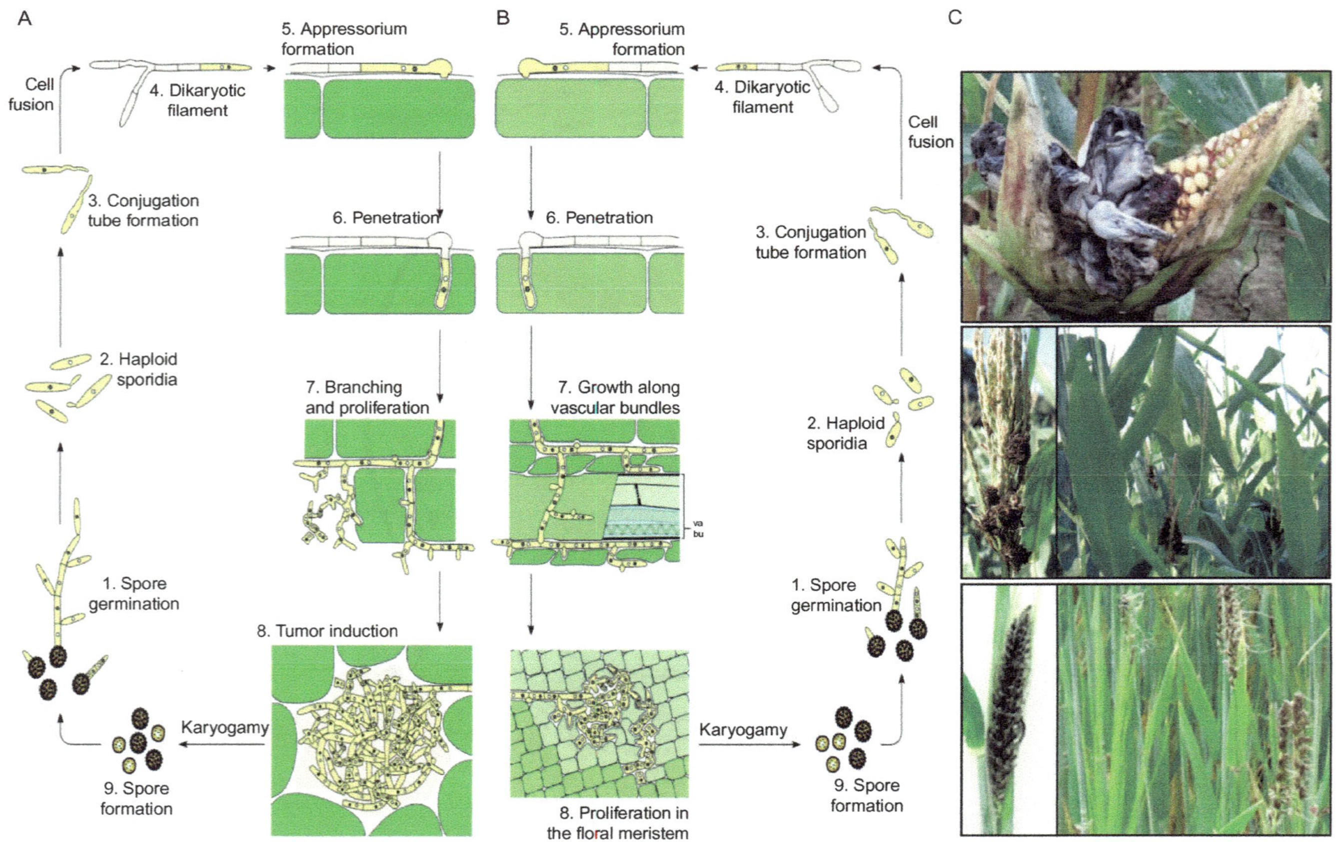

Gunther Doehlemann *et al.*, Figure 5.1, page 145.

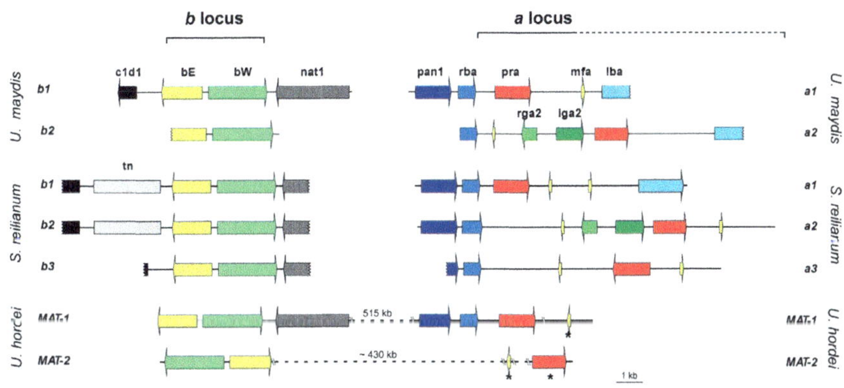

Gunther Doehlemann *et al.*, Figure 5.2, page 150.

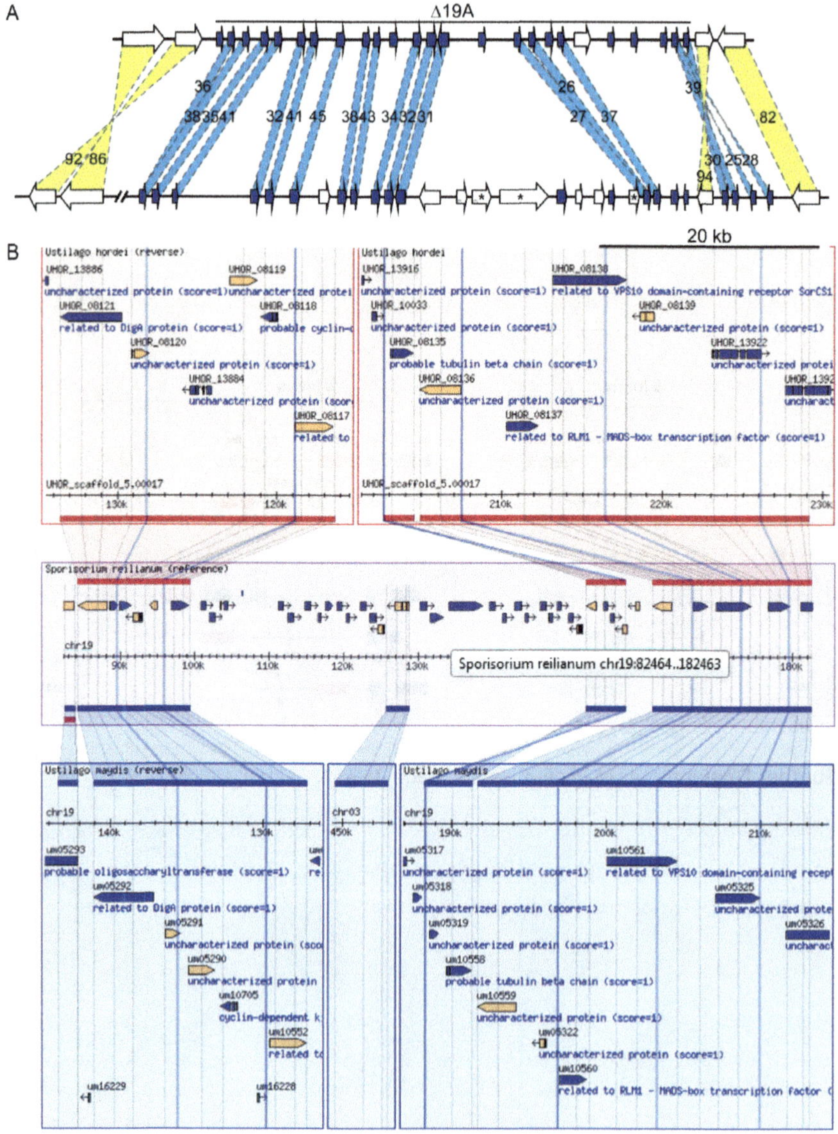

Gunther Doehlemann *et al.*, Figure 5.3, page 153.

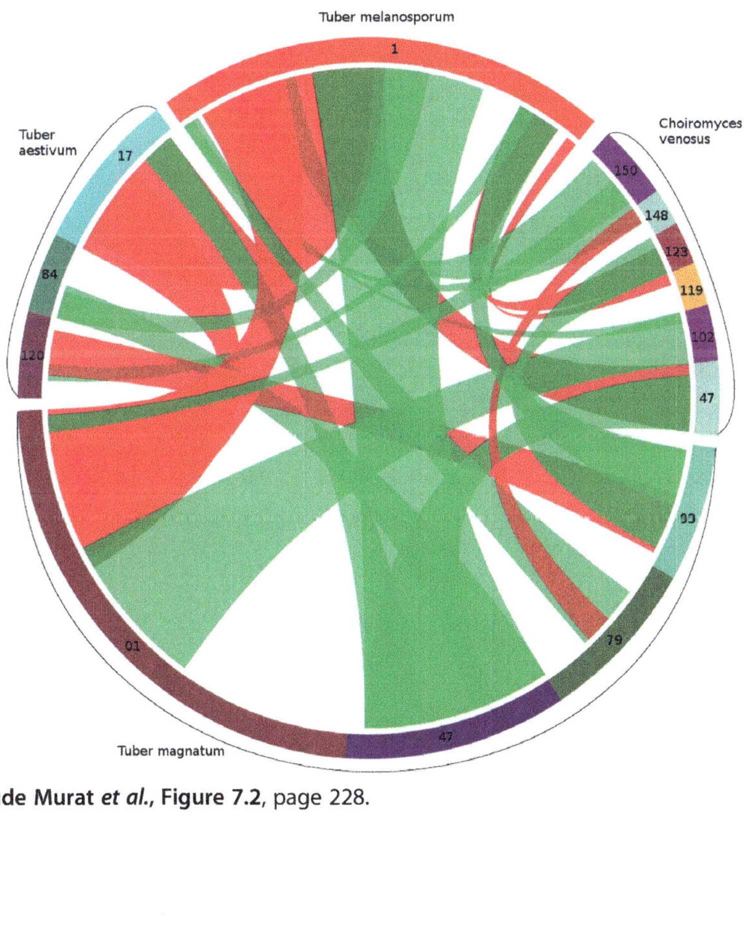

Claude Murat *et al.*, **Figure 7.2**, page 228.

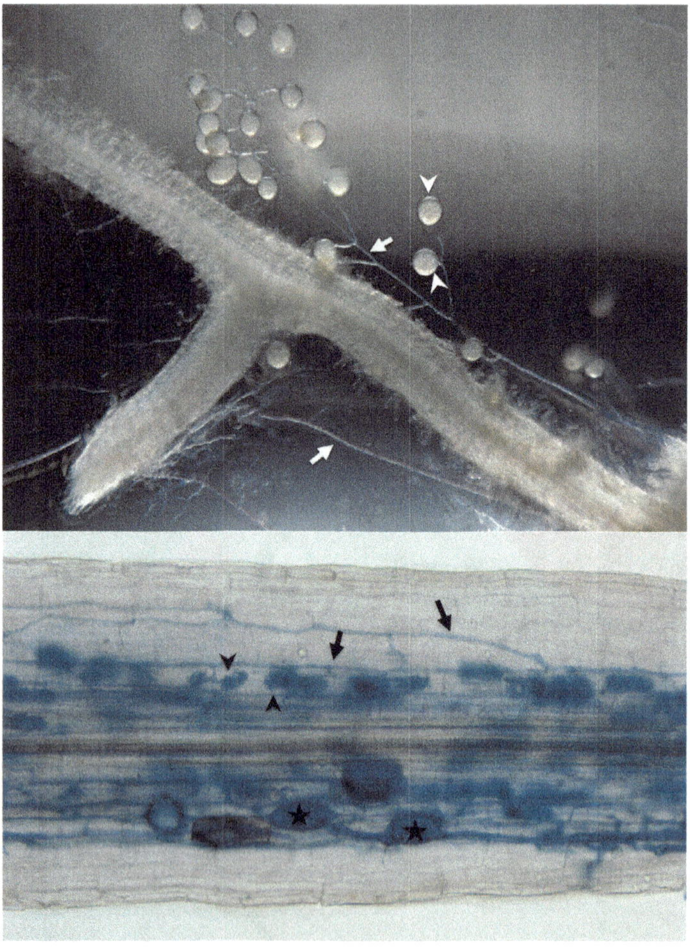

Christophe Roux *et al.*, Figure 9.1, page 264.

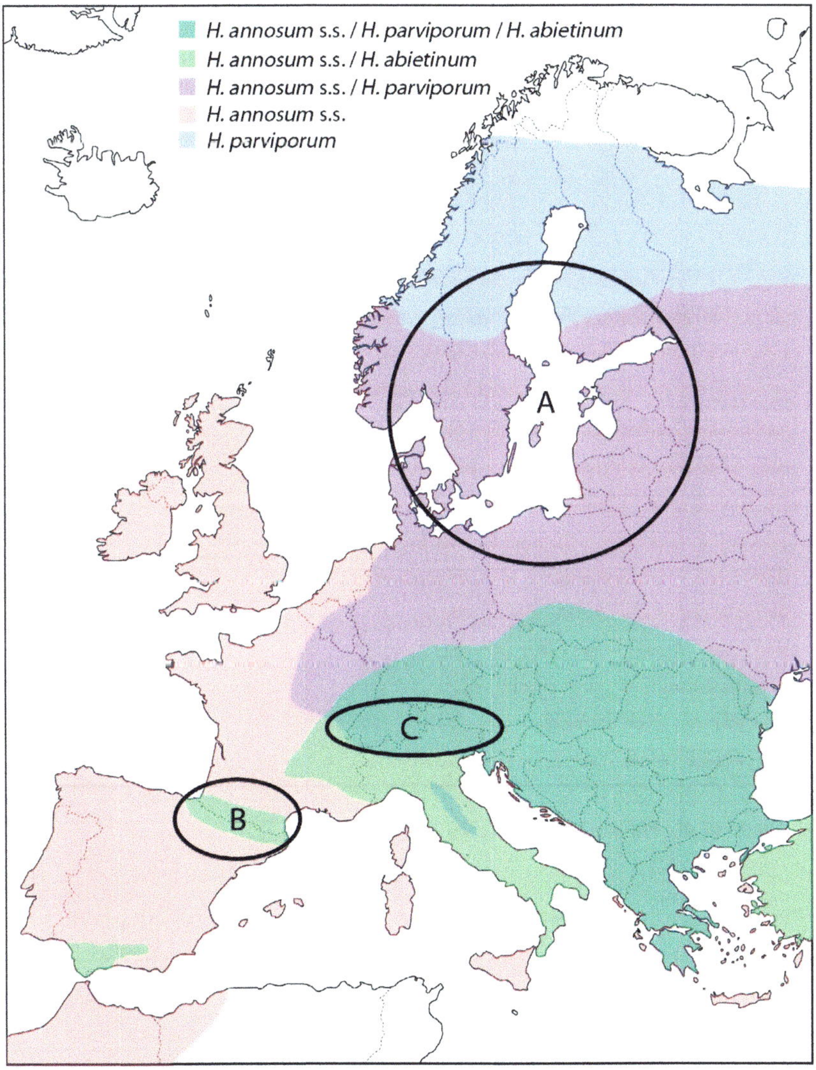

Åke Olson *et al.*, Figure 12.1, page 376.

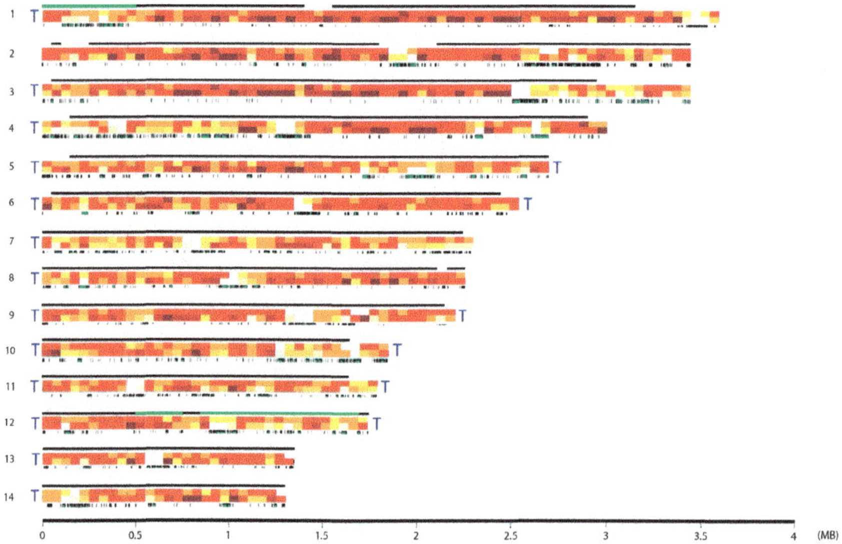

Åke Olson *et al.*, Figure 12.2, page 381.

Edwards Brothers Malloy
Ann Arbor MI. USA
May 27, 2014